D1016815

Birds of the Great Basin

Max C. Fleischmann Series
in Great Basin Natural History

Books in the
Great Basin Natural History Series

Trees of the Great Basin
Ronald M. Lanner

Birds of the Great Basin
Fred A. Ryser, Jr.

Geology of the Great Basin
Bill Fiero

Shrubs of the Great Basin
Hugh N. Mozingo

Fishes of the Great Basin
William & John Sigler

The Sagebrush Ocean
Stephen Trimble

Birds of the Great Basin

A Natural History

Fred A. Ryser, Jr.

Drawings by Jennifer Dewey

University of Nevada Press
Reno & Las Vegas

Great Basin Series Editor: John F. Stetter

Cover photograph: Golden Eagle. *Stephen Trimble.*
Back cover photograph: Tail feathers of female American Kestrel. *Stephen Trimble.*

Copyright © University of Nevada Press 1985
Drawings copyright © Jennifer Dewey 1985
All rights reserved

Composed and printed in the United States of America

Library of Congress Cataloging-in-Publication Data

Ryser, Fred A., 1920–
Birds of the Great Basin.
(Max C. Fleischmann series in Great Basin natural history)
Bibliography: p.
Includes index.
1. Birds—Great Basin. I. Dewey, Jennifer. II. Title. III. Series.
QL683.G65R97 1985 598.2979 84-25763
ISBN 0-87417-079-6
ISBN 0-87417-080-X (pbk.)

2 4 6 8 9 7 5 3

To my wife Janet and the memory of our sons Chip and Robin

Contents

Acknowledgments

AUTHORSHIP is not a solitary endeavor, and many people have contributed, in one way or another, to the genesis of this book. During my thirty years of studying natural history in the Great Basin, my two most constant field companions were Hugh Mozingo, a peripatetic botanist, and Peter Herlan, the late curator of natural history at the Nevada State Museum. Some of my fieldwork was accomplished in the company of either a physician or a physiologist—John M. Davis or Jack Knoll. During the years of 1965, 1966, and 1967, Jack and I systematically retraced, after the lapse of one hundred years, the footsteps of the great nineteenth-century ornithologist, Robert Ridgway, along the lower reaches of the Truckee River below the Big Bend.

I have spent a considerable amount of time either birding or working on bird problems with some of my graduate students—particularly David Salter, Tom Burns, Ron Panik, John Gustafson, Verne Woodbury, and David Worley. In addition, I have exchanged information about Great Basin birds or birded with good people like Ingrid Hanf, Bob and Jessie Alves, John Kartesz, Dick Ankers, Ella Knoll, Al Knorr, John Biewener, Chuck Lawson, Bill Mewaldt, Gary Herron, and others. Over the years I have enjoyed some stimulating teacher-pupil relationships and acquired some new knowledge from students enrolled in my university courses in ornithology. Finally, this book is based not only on my field observations but also on the published observations of legions of keen amateur and professional ornithologists.

I owe a special debt of gratitude to Donald Baepler and Robert Laxalt. These two men—one an ornithologist and then chancellor of the University of Nevada, the other a writer and then director of the University of Nevada Press—selected me to write about birds after the trustees of the Fleischmann Foundation generously granted the Press money to produce a series of books on the natural history of the Great Basin. Thanks are due for their trust and support.

During the four years I spent researching and writing this book, many obstacles were placed in my path. I probably would not have persisted had it not been for Rick. But the Fleischmann Natural History Series editor and now director of the University of Nevada Press, John Fredrick Stetter—better known as Rick—was superb with the material, moral, and motivational support he afforded me. Although his letters to sundry administrators went to naught, in all other respects his support was highly effective. Rick and I share a common love for books and knowledge.

Three scientists reviewed part or all of the manuscript prior to publication. William H. Behle of the University of Utah and George T. Austin of the Nevada State Museum reviewed the entire manuscript. Both of these reviewers have made notable contributions to the science of ornithology, and both possess firsthand knowledge of Great Basin ornithology. William R. Dawson of the University of Michigan reviewed the chapters on the fire of life and the water of life. Some of Dawson's experimental studies in these areas are classics. I sincerely appreciate the help I received from these three scientists. However, I remain solely responsible for any errors of commission or omission in the manuscript.

While working on the manuscript, I had to rely heavily on the resources of the Life Science and Health Library at the University of Nevada at Reno, as well as on my personal library. This branch library remains the most pleasant place for me on campus and is my home away from home. Anne Amaral has been a jewel. Anne and her colleagues, Betty Hulse and Dorothy Good, form an outstanding trio of librarians. These three are extremely helpful in an efficient and gracious manner. They could well have served as prototypes for Marion the librarian in *The Music Man*.

Holly Carver gave a virtuosic performance as the copy editor for this book. Operating with the precision and deftness of a premier neurosurgeon, Holly undangled participles, unsplit infinitives, transplanted misplaced phrases, and excised redundant or superfluous words and phrases. Overall, she tightened and lightened the text, without reshaping my prose style. Holly did not leave a single man standing as she ruthlessly eliminated all instances where I chauvinistically equated men and mankind with *Homo sapiens*. In their places she substituted more proper designations such as people, humans, and civilization. Even such apparently innocuous designations as cattlemen and sheepmen came under her close scrutiny. My personal experience with copy editors is limited, but to me Holly, a talented editor with a love for birds, will always be the best there is.

There are other talented people whose contributions to this book must be acknowledged. Jennifer Dewey created the original pencil sketches. Tony Diebold, Bob Alves, Keith Giezentanner, Stephen Trimble, John

Running, and David Worley took the color photographs. Marianne Martin did the illustration on Mallard reproductive behavior. Genevieve Gray of Tucson, Arizona, constructed the index. Barbara Christy did my typing. Mary Hill of Austin, Texas, did the proofreading. Robert Blesse worked on the design of the book, as did Christine Rasmuss Stetter. In addition, Christine contributed the maps. Financial affairs were in the capable hands of Kenneth R. Robbins. Nicholas Cady gave me the gifts of sage counsel and general editorial assistance. Ian Newton graciously gave permission to use data on the mean body weights of accipiters and their prey from his book *Population Ecology of Raptors*.

When all is said and done, my greatest debt remains the debt of gratitude I owe to my family—particularly to my wife. Not only did Janet give me loving support, understandingly tolerating the endless times my work kept me away from normal family life, but she allowed me to gamble with our retirement. She accepted the financial losses and the added work years I was forced to take in order to write this book. Susan and Mark Byars, my daughter and son-in-law, were also entirely supportive of my activities. However, I feel compelled to apologize to my two-year-old grandson, Kristopher Mark Byars, for the sad fact that I have not had time, as yet, to properly function as a grandfather. Grandfathering takes a lot of time if it is done right—with all that playing, exploring, birding, butterfly collecting, fishing, and reading aloud. Finally, my number 1 son Ronald has enthusiastically supported my book-writing efforts.

FRED A. RYSER, JR.
RENO, NEVADA

1

The Great Basin: Birds and Environments

GREAT BASIN AVIFAUNA

To the uninitiated, the Great Basin does not appear to possess much bird life. It is possible for the casual observer to completely traverse the Basin on Interstate 80 or U.S. 50 and hardly see a bird. The intercity highways were built in the valleys, where the going was the easiest; consequently, the roads interface with desert shrublands and, to a lesser extent, with desert woodlands. These areas appear openly desolate; and one's eyes and mind can soon be overwhelmed by all that bare ground, which is so very wide open to the sun and wind and often completely possessed by light and heat or cold.

Whether seen or not, numerous kinds of birds frequent these desert shrublands and woodlands. This is magnificent country for raptors, and during the winter eagles, hawks, and falcons often concentrate in such numbers as to apparently defy the limitations imposed by a pyramid of biomass. The pyramid of biomass is a generalization in ecology that the food supply must always greatly outweigh the animals feeding on it. So very much prey is needed to support the energy needs of raptors that these predators of necessity must be few and far between in any viable community of organisms. The shrublands are also the natal home of an array of distinctive birds, including the Black-throated Sparrow, Sage Sparrow, Lark Sparrow, House Finch, and Burrowing Owl. In sagebrush-dominated shrublands, the courtship displays of Sage Grouse are performed on ancestral strutting grounds. Some of the finest bird actors in the world—a troupe of corvids featuring the Black-billed Magpie, Common Raven, and Pinyon Jay—are on stage in the shrublands and woodlands. Saline or alkaline desert lakes, sloughs, marshes, alkali or salt flats, sinks, and meadows interrupt the shrublands of the Great Basin and welcome large flights of migrating grebes, waterfowl, shorebirds, and others. These wet places also provide nesting grounds for some highly interesting and troubled species, such as American White Pelicans, White-faced Ibises, Long-billed Curlews, Sandhill Cranes, Snowy Plovers, and American Avocets. Then, every few miles eastward or westward, mountain ranges trending north and south hold aloft woodlands, forests, canyons, and even some arctic-alpine habitat. These mountains afford food, shelter, water, and coolness during the summer for a fascinating deployment of such species as Clark's Nutcrackers, Northern Goshawks, Red Crossbills, American Dippers, Water Pipits, and Mountain Bluebirds.

Although the Great Basin, at its largest extent, spans about six hundred miles of longitude and four hundred miles of latitude and possesses many and varied types of isolated environments, it does not have a single endemic or unique species of bird; that is, not a single species of bird is completely restricted in its geographic occurrence to just the Basin. Consequently, whatever uniqueness the Basin avifauna possesses is due to the presence not of endemic species but of a special assemblage of species. This lack of endemism is not surprising. Birds possess great mobility because of their strong flight. As homeotherms (warm-blooded animals) they have very high metabolic rates and a well-developed regulatory physiology, which confer tolerance and hardness in the face of a wide array of environmental stresses. Because of these attributes, birds are able to migrate and live in summer and in winter in vastly different ecologic worlds. Since their comings and goings are not greatly influenced by topographic or geographic features—such as mountains, deserts, and even large expanses of water—birds need not be greatly circumscribed in their distribution.

GREAT BASIN BOUNDARIES

Great Basin birds do not have a single great basin within which to live. Rather, scores of basins or areas of interior drainage, lying between scores of mountain ranges trending north and south, were collectively called the Great Basin by John C. Frémont, who led expeditions into this region in 1843 and 1845. The name Great Basin has stuck, despite being a complete misnomer.

The Great Basin forms the northern part of the Basin and Range Province. To the west of the Basin lies the Sierra-Cascade Province, to the north the Columbia Plateau Province, and to the east the Colorado Plateau Province. Hence, the Sierra Nevada, the Snake River drainage, and the Wasatch Range form recognizable, partial "rims" for three sides of the Basin.

There is no neat hydrographic or physiographic way of recognizing a southern boundary for the Great Basin, since the Basin is merely the north-

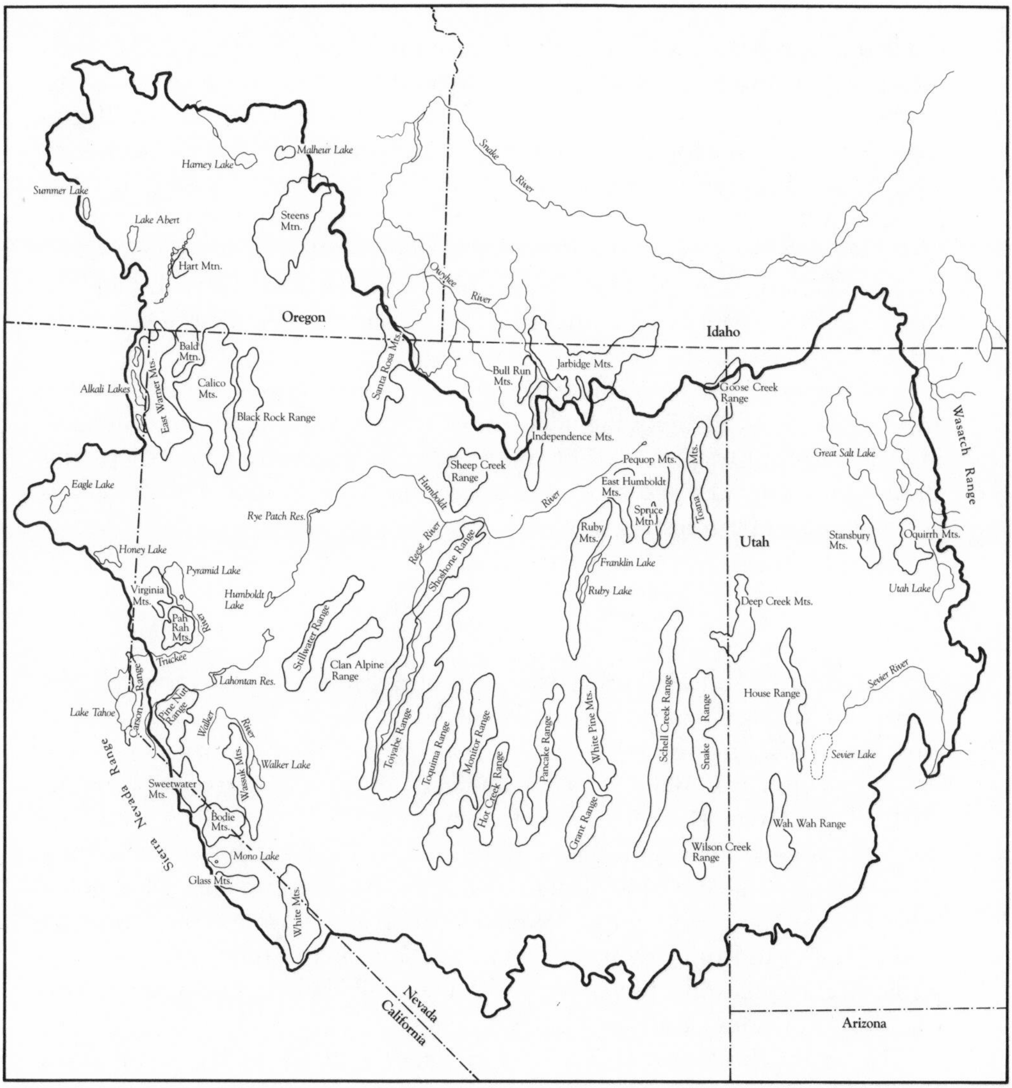

The Great Basin, showing principal mountain ranges, rivers, and lakes.

ern half of the Basin and Range Province—a province which extends down into Mexico. From the standpoint of natural history, however, an actual southern boundary to the Basin can be recognized much more readily on biological than on geological grounds. As a consequence, in this book I intend to set the southern boundary at the level of Owens Valley in California, southern Nye and Lincoln counties in Nevada, and Iron County in Utah. At this level the Great Basin Desert, a cold desert, merges into the Mohave Desert, a hot desert. Going south into the hot desert, by adding many species of birds and further environmental considerations, would greatly overburden the carrying capacity of a single book; the natural history of the birds of the Mohave Desert merits a book by itself. Thus, the area involved in this book includes bits of eastern California, quite a chunk of southeastern Oregon, a bit of southeastern Idaho, most of northern and all of central Nevada, and most of western and central Utah. This area contains a large and interesting avifauna.

GENERAL ENVIRONMENTS

No organism is an island unto itself. No organism lives in a vacuum. The science of ecology had its genesis in this great truism and is dedicated, in part, to elucidating the relationships which exist between organisms and their environments. Before we can understand or even appreciate how the birds of the Great Basin make their livings and lead their daily lives, we must become familiar with their environments. We must know something about the physical and biological nature of these environments, what problems are rampant, and what resources are available for exploitation by birds. Then, and only then, are we in a position to appreciate the morphological (structural), physiological (functional), behavioral, and ecological adaptations of Great Basin birds.

There are three general types of environments for birds in the Great Basin—desert, montane, and aquatic environments.

The Great Basin Desert

About 50 percent of the land area of the Basin is occupied by a cold desert called the Great Basin Desert. By definition, a desert is an arid region; by one definition, it is a region so arid that the potential annual evaporation is at least two times greater than the yearly precipitation. A cold desert is a desert in which the winters are severe, and snow and freezing temperatures regularly occur.

There are several forces which singly or in combination are responsible for the presence of a desert. The Great Basin Desert is primarily the result of a rain shadow cast by mountain ranges and mountain systems. The prevailing air movements into the Great Basin are from the Pacific Ocean to the west. The Coast ranges, the Cascades, and, much more important, the Sierra Nevada system relieve the air of much of its moisture before it reaches the Great Basin. In other directions, the Basin is both far removed from oceanic sources of moisture and separated from these sources by intervening mountain ranges and systems.

Since the Great Basin Desert is both a high-altitude and high-latitude desert, it is a cold desert. Much of this lies four thousand feet or more above sea level, and the entire desert lies north of the thirty-seventh parallel. So the Great Basin Desert is high enough above sea level, far enough removed from the equator, and well enough insulated from the ameliorating effects of the Pacific Ocean to be a cold desert. However, it is important to remember that the designation of cold desert refers only to the winter climate of that desert. The winters are cold and the frost-free period is relatively short in the Great Basin, but the summers are hot—often intensely so.

Great Basin Mountain Ranges

As I have previously noted, the bowl of the Great Basin is not smooth but corrugated by scores of mountain ranges. Up on the sides of the higher mountain ranges, arid desert environments yield to wetter montane ones. Some of the larger mountain canyons and ravines are wet enough to contain permanent streams. The streams and runoff from the mountains flow into the valleys, and what water is not used for domestic, industrial, or agricultural purposes eventually disappears into natural sinks or reservoirs or lakes.

The Great Basin ranges are of various shapes and sizes. They commonly run up to fifty miles or more in a general north-south direction. Their east-west widths vary from a few miles on up to fifteen miles or so. Numerous peaks stand seven thousand to eight thousand feet above sea level. A considerable number of peaks are ten thousand feet in altitude, and several are in excess of thirteen thousand feet.

Unlike the desert, where a maze of roads and tracks covers the valleys, a very limited road network services the mountain ranges in the Basin. Seldom does a mountain road run all the way across a range; usually the road gets only partway up a canyon or on a ridge before ending. Many of the roads require high-clearance, four-wheel-drive vehicles. As a consequence, the serious bird-watcher must do a lot of walking and climbing in order to study montane birds, and the avifaunas of many mountain ranges have not yet been well studied.

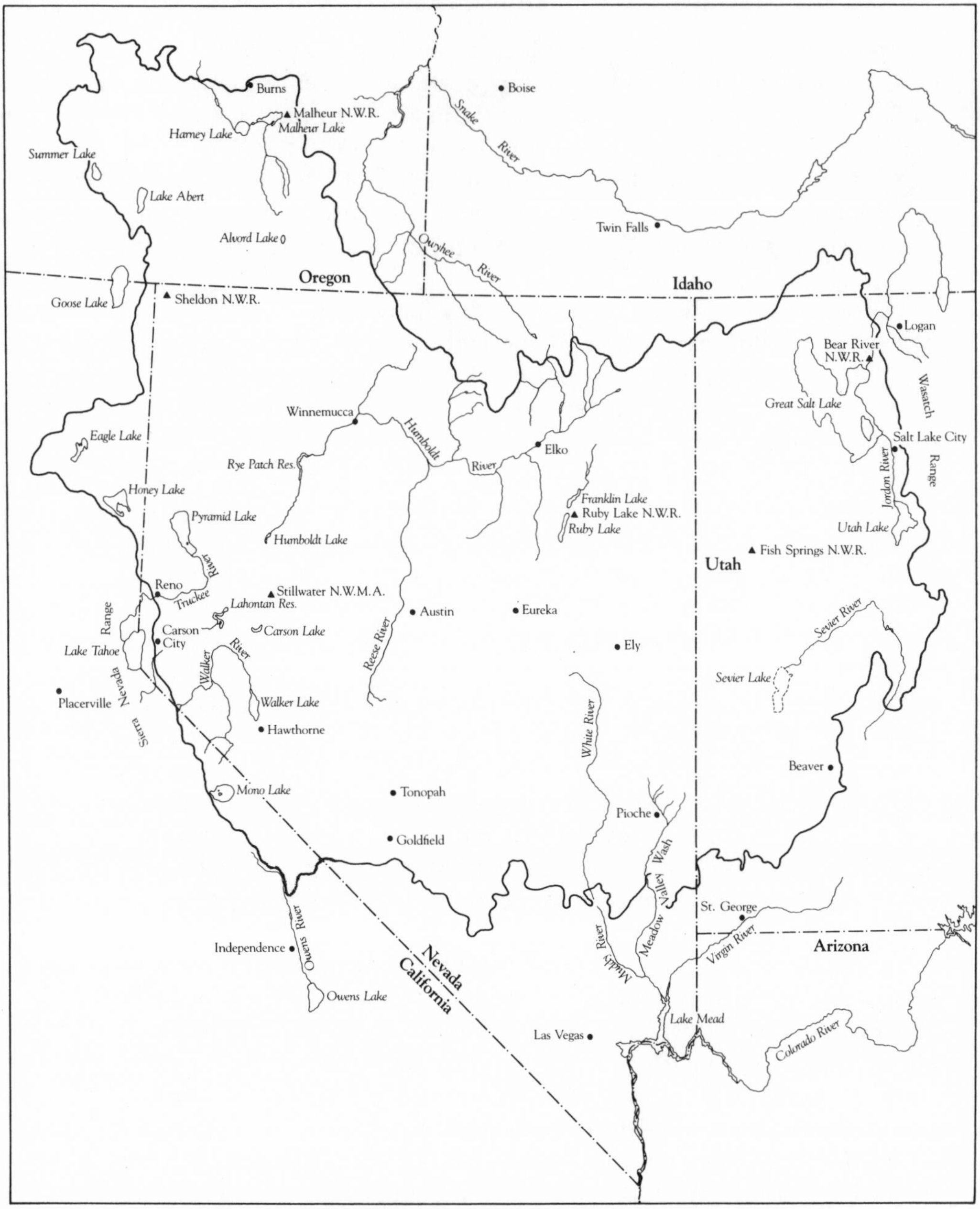

The Great Basin. Abbreviations: N.W.R. = national wildlife refuge; N.W.M.A. = national wildlife management area.

Great Basin Wet Areas

Despite its aridity, the Great Basin has wet areas of major importance to waterfowl, shorebirds, and marshbirds. The American Avocet reaches its peak nesting abundance in the Basin. The breeding colony of American White Pelicans at Pyramid Lake has been the largest in North America. Troubled species such as the White-faced Ibis, Sandhill Crane, Snowy Plover, and Long-billed Curlew have significant nesting populations in the Basin. The marshes and open waters at sites such as the Bear River and Malheur National Wildlife Refuges and the Stillwater Wildlife Management Area function as major way stations for throngs of migrating water birds and as vital nesting grounds for many others. Mono Lake functions as the breeding grounds for approximately 20 percent of the total world population of California Gulls and serves as the major way station in the West for southward-migrating Wilson's Phalaropes.

Although some fresh water does exist, there is relatively little freshwater habitat in the Great Basin. Most of the lakes, marshes, and sinks are areas of internal drainage and lack outlets. Since they lack outlets, they serve as evaporation pans in concentrating minerals, and their waters are alkaline or saline in nature. However, they are still highly desirable habitats for water birds. Lake Tahoe—which sits in the mountainous western rim of the Great Basin and drains, via the Truckee River, into Pyramid Lake—has very fresh water, as do most of the small mountain lakes in the Basin with outward drainage. The water of Malheur Lake is relatively fresh, since Malheur drains into Harney Lake, which then functions as the evaporation pan for the system. The water of Utah Lake is also relatively fresh, since it drains into Great Salt Lake via the Jordan River, and Great Salt Lake functions as the evaporation pan.

Back at the end of the Pleistocene epoch (the Ice Age), two large lakes occupied much of what is now the Great Basin. The eastern end of the Basin was occupied by Lake Bonneville, whose desiccating and desiccated remnants include Great Salt Lake, Utah Lake, Sevier Lake, and the Great Salt Lake Desert. The western end was occupied by Lake Lahontan, whose desiccating and desiccated remnants include Honey Lake, Pyramid Lake, Walker Lake, and Winnemucca Lake. The waters of these remnant lakes are also alkaline or saline.

Periodically, in the past, some of the Great Basin lakes may have freshened up somewhat, either by having temporary outlets into evaporation pans or by drying up. When a lake is evaporated to dryness during drought times, all its mineral matters are precipitated. Then, given a renewed supply of water, the lake begins again as a freshwater lake.

Although we do not know what all the changes have been, it is apparent that water bird habitats have altered extremely in the Basin since the time of Lake Lahontan and Lake Bonneville. Obviously, there has been a great decline in the extent of feeding grounds, the amount of food, and the number of nesting sites. This is very apparent in the case of the American White Pelican; one can find evidence of its prehistoric nesting sites on old beachlines on dry hills which were once islands in Lake Lahontan.

ALTITUDINAL ZONATION OF ENVIRONMENTS

With the exception of oceanic islands, the number of different kinds of birds present in a geographic region is related to the number and complexity of environments in that region. The greater the variety of environments, the greater the number of ecological niches or ways of making a living. Oceanic islands have disharmonic avifaunas; that is, the few species present leave many niches unexploited. Avifaunas in continental regions are usually harmonic; that is, there are enough species present to fill and exploit most of the available niches. The Great Basin avifauna is essentially a harmonic one. However, the extremely successful introduction of the Chukar in recent years—without supplanting any native species—shows that a few niches may still be available for exploitation.

To a considerable extent, the rich diversity of bird species in the Basin is due to the mountain ranges. Mountain ranges add a new dimension to environmental diversity—one which goes beyond the diversity conferred by the latitudinal and longitudinal sprawl of the Basin. A bird living in a closed conifer forest on a Great Basin mountain is actually living in a climatic life zone comparable to the taiga or great circumboreal conifer forests of northern North America and Eurasia. By moving a short distance higher up and crossing the upper tree line into the alpine zone on that mountain, the bird enters a climatic life zone comparable to the tundra of arctic North America and Eurasia.

Each mountain range provides an altitudinal series of climates. In general, as the altitude increases the winters become longer and colder, the summers shorter and cooler, and the precipitation greater. Back in 1920, an entomologist by the name of A. D. Hopkins calculated actual values for altitudinal, latitudinal, and longitudinal changes in climate as they affect life history events—such as germination, flowering, egg laying, hatching, and migration—of plants and animals. Hopkins' bioclimatic law states that, if local conditions such as slope and exposure are comparable, on the average, spring life history events will occur one day later and fall events one day earlier with each increase in altitude of one hundred feet.

During its evolution, the vegetation of the Great Basin exploited the series of climates present on the mountain ranges and in the process became altitudinally zoned. Animals exploiting the emerging vegetation zones also became zoned. The altitudinal zonation of vegetation involved two parameters, both of which promoted environmental and niche diversity and, consequently, bird species diversity: altitudinal changes in the life forms of the dominant vegetation and altitudinal changes in the plant species present. Since these two vegetation parameters so vitally shape the environments and niches of birds, we must briefly consider them.

Altitudinal Changes in Life Forms

In order to possess enough altitudinal span to provide for the entire vegetational sequence of life forms, a Great Basin mountain has to peak at somewhere around ten thousand feet. The complete sequence of life forms of the dominant vegetation in the Basin, progressing from valley to mountaintop, is as follows: shrub or shrub-grass; low tree (widely spaced trees); sometimes, shrub-grass again; tall tree (closely spaced trees); stunted tree (widely spaced trees); and, finally, low herbaceous plants (grasses, sedges, wild flowers). Shrubland birds, grassland birds, woodland birds, forest birds, and alpine-tundra birds are deployed along this life-form sequence.

Over much of the Basin the dominant plant species in both the lower and the upper shrub-grass zones is big sagebrush, *Artemisia tridentata*. The single-leaf pinyon pine, *Pinus monophylla*, and Utah juniper, *Juniperus osteosperma*, are the dominant species in the woodlands of low, widely spaced trees. The northward distribution of these two species comes to an end in northern Nevada. North of the Truckee River in western Nevada and north of the Humboldt River in central and eastern Nevada, the pinyon-juniper woodlands are "absent or poorly developed." In 1951, plant ecologist W. D. Billings noted the absence of pinyon-juniper from the aforementioned part of the Great Basin. He also noted that on certain mountain ranges there, for example, the Santa Rosa, sagebrush-grass ran up the mountainsides to elevations of almost ten thousand feet, uninterrupted except by high-altitude aspen groves. Billings then speculated, "It appears that the pinyon-juniper vegetation is merely superimposed upon a large sagebrush-grass zone which has wide elevational tolerance."[1] Regardless of why these trees are missing, their absence denies habitat to woodland species. For example, the economy of the Pinyon Jay is closely tied to pinyon pine seeds, and the Townsend's Solitaire heavily utilizes juniper berries for winter food. However, having a sagebrush-grass shrubland in place of a pinyon-juniper woodland is not all bad, since the former affords prime habitat for Sage Grouse, Sage Sparrows, Sage Thrashers, Brewer's Sparrows, and others.

Altitudinal Changes in Plant Species

The second altitudinal parameter which promotes environmental and niche diversity for birds in the Great Basin follows from altitudinal changes in the kinds of plants present. Different species of plants have different climatic requirements and thus different altitudinal tolerances. As a consequence, the species of plants change as a function of altitude. Pinyon pine and limber pine may occur on the same mountain, but the pinyon pine will be at lower altitudes than the limber pine.

The richness of altitudinal sequencing of plant species is further enhanced by a longitudinal effect within the Basin. Mountain ranges in the Great Basin possess Basin Range plant zones. In addition, some eastern ranges—such as the Snake and Deep Creek ranges—have some Wasatch plant zones; and some western ranges—such as the Carson and Sweetwater ranges—have some Sierran plant zones. As a consequence of encroachment of Sierran plant species from the west and Wasatch plant species from the east, Billings recognized three montane zonal series of vegetation in the Great Basin: the mountain ranges in the extreme western part have a Sierran Series; the interior ranges have a Basin Range Series; and the eastern ranges have a Wasatch Series of plant zones.[1]

VERTICAL AND HORIZONTAL DIVERSIFICATION

Not only are there altitudinal sequences of environments and ecological niches for birds in the Basin, but at any given altitude there are also horizontal and vertical arrays of niches for birds. There is vertically distributed environmental diversity because the plants of any one locale are of various kinds, sizes, and shapes. There is horizontally distributed environmental diversity because the topography, moisture conditions, and soil often vary greatly from one spot to the next—creating a patchwork of vegetation. Vertical stratification and horizontal patchiness within an area support a variety of life-styles among birds.

In shrublands the ground-level vegetation is available for nesting, feeding, sleeping, loafing, and courting purposes. Above the ground there are usually several layers of shrubs available for these purposes. Finally, the airspace between and above the shrubs can be utilized as a hunting ground for flying insects or other prey or as a courtship area.

There is a greater variety of vertically distributed ecological niches in woodlands and forests than in shrublands. In a forest the ground-level plants are present; there are usually several understory layers of shrubs, saplings, and sometimes vines; and finally the canopy itself is stratified, since it is usually formed by trees of various kinds, shapes, and sizes. Because there is an overall lack of height, complexity, and lushness in the vegetation, Great Basin forests do not possess anywhere near the potential for vertical niche diversity that the deciduous forests of the eastern United States possess.

Vegetation is seldom uniform over an extended area but varies along with local topographic, hydrologic, and soil conditions. The plant species present and their relative abundances change from place to place. Sometimes the changes are quite obvious, as when shadscale, *Atriplex confertifolia*, replaces big sagebrush as the dominant shrub on drier, somewhat salty soils in the Great Basin; or when greasewood, *Sarcobatus vermiculatus*, takes over on clay flats or alkali flats; or when iodine bush, *Allenrolfea occidentalis*, takes over on moist alkali flats.

Environmental changes are most striking where shrublands abut on streams, lakes, or sinks. At these meetings even the life form of the dominant species of plants changes. Within a few feet of the stream banks, the life form of shrubs gives way to that of the trees in the vegetation along the streams. Marshes are often present where shrublands meet lakes and sinks. In these meetings the life form of the dominant plants changes from that of bushy shrubs to that of tall, thin emergent marsh plants.

ENVIRONMENTAL STRESSES

The Great Basin is no avian Garden of Eden. Living conditions are often stressful. Yet a fascinating array of birds inhabits the lands and waters of the Basin. Some birds meet the fiercest challenges of climate and land head-on—and feed and court and even nest or overwinter in the harshest of environments. To know these birds, we must first meet the land and feel its muscles and become aware of the stresses which buffet the birds from all sides. Only then can we appreciate how very well birds have made this their land.

Prolonged hot spells occur in the Great Basin during the summer; air temperatures in the shade daily rise to well above 90° F and often exceed 100° F. Exposure to the sun's rays is heavy. The summer skies are usually clear, and surfaces and bodies exposed to direct sunlight superheat to well above air temperatures. The scanty vegetation in the shrublands poorly insulates the ground from the sun, and ground surfaces become blisteringly hot.

As the temperature of the bare ground and of rocks and other objects on the ground climbs, they reradiate increasing amounts of heat back into the environment and seriously add to the heat burden directly imposed by the sun. And the superheating ground reradiates long-wave heat radiation, which superheats the air directly above it. Objects viewed through intervening columns of superheated air are distorted, and mirages may occur. Distant objects are fuzzy and pulsating in appearance. Objects on the ground on exposed surfaces such as playas or dry lake beds and asphalt roads often appear to be suspended in the air. Also, with so much light-colored ground exposed, much of the sunlight is reflected back into the environment, and glare effects can be strong.

Prolonged cold periods occur in the winter, with minimal nightly temperatures well below the freezing point of water. Due to the strong, often gusty winds, the chill factor is often high. In general, the montane environments are colder than the desert ones.

Although ice may form during the nights, there are relatively few cold spells during the desert winter when ice is thick enough on ponds and sloughs to deny access to swimming waterfowl or support the weight of human ice skaters. The larger bodies of water—such as Great Salt Lake, Pyramid Lake, Lake Tahoe, and Walker Lake—never freeze over. Nor do the hot springs in geothermal areas. However, the small montane lakes, ponds, and streams will be closed by ice and snow all winter long.

When snow falls in the valleys it usually melts, except in deep shade, within a few days. In the mountains, at higher elevations, snow may be present most of the year. On the highest mountains—such as Wheeler Peak in the Snake Range—small snow fields may be present the year around. Deep snow is a problem only at higher elevations in the mountains. An occasional avalanche may occur in the higher mountains, such as the Ruby Mountains.

In the valleys the annual precipitation is very low. Much of the Great Basin Desert receives only five to ten inches of annual precipitation; the Walker Lake depression in Nevada and the salt desert near Wendover, Utah, average less than five inches a year. The eastern end of the Great Basin is wetter, and at certain locales here the annual precipitation may average over fifteen inches.

Fresh water is a limited commodity in the Great Basin. There are minor streams flowing out of the mountain canyons, but there are only two rivers of any real length—the Humboldt River in Nevada and the Sevier River in Utah. Since most of the major lakes and minor lakes lack outlets and function as evaporation pans, they are heavily mineralized. Only a few lakes—Tahoe, Malheur, and Utah—have outlets and therefore contain fresh or relatively fresh water. Many of the springs are geothermal ones, and their waters are heavily charged with minerals.

The low places in the valleys, being undrained basins, have extremely salty or alkaline soils. The playas and alkali and salt flats support little or no vegetation. Salt-tolerant plants may be present around the edges of playas and salt flats, but they will add to the salt burden of animals using them. These places provide little in the way of fresh water, food, shade, or cover for birds. But some birds, such as the tiny Snowy Plover, nest on the harshest of alkali flats.

Windy days are frequent the year around. Wind accelerates the evaporation of water, promoting dehydration and desiccation and bringing an added chill factor to the spring and winter. Wind can interfere with flight and feeding and represents a threat to aboveground nests.

Compared to those in many other deserts, sandstorms are relatively infrequent in the Great Basin Desert. They are mainly a problem in areas around playas and areas where humans have disturbed or removed the vegetation. Sand dunes are not a widespread feature, although sandy areas with low dunes are scattered over the Basin. The larger dune systems include Sand Mountain and the Winnemucca Dunes in Nevada and the Little Sahara Sand Dunes north of Delta, Utah.

On sunny days when the air directly above the ground superheats, it becomes unstable. When the winds or breezes of the desert freshen, some of this unstable air may spill over an embankment or obstacle. Then a narrow but high, rapidly rotating column of air may form—much like a miniature cyclone—and propel itself across the ground, rotating sometimes clockwise, at other times counterclockwise. This is a dust devil. Since some dust devils generate considerable force, they pose a threat to bird nests in their paths.

On many days, the weather can be downright unseasonable. Summery weather can occur in the winter, and wintry weather is common in the late spring. The transition from drought to flash flood requires but a moment's time. The sleet and snowstorms which occur in May, and even early June, wreak havoc upon nesting birds. During the summer, the minimum temperature at night is often 40 to 50° F lower than the daytime maximum—and untended eggs or nestlings may fry by day and freeze by night.

Because of the lack of overall height, the paucity of foliage density, and the wide spacing of plants in the valley and foothill vegetation, adequate cover is seldom available. Good cover is important when protection from predators, wind, rain, or snow is needed. And the scanty vegetation offers little protection from the sun. In the valleys and foothills, it is often difficult to place a nest so that its eggs or young are protected from the sun. Little deep shade is available for times when the environment is loaded with direct solar heat and reradiated heat. At these times, parent birds are often under great heat stress as they attempt to shade their eggs or young—by standing and interposing their bodies between the sun and the nest.

In the valleys, there is a scarcity of elevated nesting and perching sites in the form of tall trees and cliffsides. Utility poles and transmission towers are used as nesting sites by raptors and ravens, as well as hunting perches on cold mornings or during the winter day.

But the Great Basin remains a land of many moods. All is not harshness—many days and nights are mellow and gentle and bountiful. Many of the problems confronting Great Basin birds are problems that must be solved by birds the world over. They must cope with diseases, accidents, predators, and the destruction wreaked by humans upon the land and air and water; they must successfully molt, migrate, and reproduce. But life in the Basin demands more than the ordinary from birds, for there will be times when they will be under tremendous heat loads or experience drastic heat losses or water losses. Their key to survival in the Great Basin lies in successfully remaining in heat balance and water balance. The story of how this is done will be told in the next two chapters.

2

The Fire of Life

HOMEOTHERMISM

Birds and mammals remain at almost constant warm temperatures in the ever changing thermal environments on earth. As air, soil, rock, and other organisms heat and cool by the moment, day, and season in these ever shifting thermal environments, the fire of life burns with relative constancy only within feathers and fur. Among all of life, only birds and mammals are warm-blooded. Only birds and mammals are homeotherms, organisms capable of physiologically regulating their body temperatures; and they maintain relatively high, constant body temperatures. Yet homeotherms encounter problems in remaining in heat balance in environments as thermally stressful as those of the Great Basin, and overheating or overcooling can have serious or fatal consequences. An array of structural, physiological, behavioral and ecological adaptations is involved in maintaining heat balance. Let us look at these mechanisms, in an attempt to understand how it is that the fire burns with constancy within birds in the harsh environments of the Basin.

Homeotherms are not at the mercy of their thermal environments. During cold periods they are warmer than their environments, and in hot periods they are cooler than their environments. Thus, homeotherms are able to exploit most of the environments on earth.

The blessings of homeothermism can best be appreciated by viewing life from the standpoint of cold-blooded animals, poikilotherms, whose ranks include all the invertebrates, fish, amphibians, and reptiles. Poikilotherms are closely controlled by their thermal environments. They cannot efficiently regulate their body temperatures, although a few insects and reptiles do practice behavioral temperature regulation, such as moving in and out of the shade, when under mild thermal stress. Poikilotherms can exist as active adults only during the time of day and year when their environments provide them with the proper amount of heat to sustain their body temperatures at proper operational levels. Because of their almost total reliance on environmental heat for the warmth of life, poikilotherms are sometimes referred to as ectotherms.

In contrast, homeotherms, whose high standard or basal metabolic rates generate much internal heat, are called endotherms. Further, their feathers and fur afford excellent insulation, and their high-capacity circulatory systems allow them to efficiently regulate rates of heat exchange with their environments. However, some entomologists classify certain insects—such as some of the bees, moths, and beetles—as endotherms. These insects can elevate their thoracic temperatures, while in cool environments, by contracting their flight muscles and generating myogenic heat through shivering. And they can physiologically control, to some extent, rates of heat exchange with their environments.

In cold climates poikilotherms, at best, can lead active lives for only a few months or weeks of the year. In hot climates they must often avoid direct sunlight or confine their activities to the cooler nights. The general lack of, say, insects and lizards abroad during wintertime is very noticeable, as is the general lull in poikilotherm activity on cold or rainy days or during the midday heat of the summer. Some poikilotherms are extremely heat-sensitive. Butterflies, creatures of the sunlight, must immediately seek the warmth of the ground when the sun disappears behind clouds, or cold paralysis soon sets in. A few adult poikilotherms survive long periods of thermal stress, such as winter, by inactivity in some sheltered microclimate. Most adult poikilotherms die during these periods, relying on some resistant stage in the species' life history, such as eggs or pupae, to renew the population when the mild seasons of the year return.

Because they can regulate their body temperatures, homeotherms are active the day and year around even in temperate and polar regions. Some may migrate, some may hibernate or aestivate, but there are no massive seasonal die-offs. During hibernation or aestivation a homeotherm seeks a sheltered, mild microclimate where it can enter into a period of inactivity or torpidity. In this mild microclimate the species' body temperature decreases significantly, and all its vital processes, such as circulation and respiration, slow down. The animal is in a type of deep sleep—its energy expenditures are greatly reduced. Hibernation allows a homeotherm to avoid the cold weather and often reduced food supplies of winter. Aestivation allows it to avoid meeting the impact of a hot, dry summer.

BODY TEMPERATURE

We have already noted that there are only two classes of homeotherms, birds and mammals, and that homeotherms have much higher standard metabolic rates—rates of energy utilization and heat production when at rest and not under thermal stress—than do any of the poikilotherms. In small birds, the fire of life burns hotter than in any other group of homeotherms. The standard metabolic rate of a small bird may be as much as 70 percent or so higher than that of a mammal of comparable body weight. Most small birds regulate their body temperatures at 40 to 41° C, which is three to four degrees higher than those of most mammals.

As we shall see, birds do not always maintain their body temperatures at one level. During heat stress or vigorous muscular activity, the deep-core body temperature may rise above the normally regulated level; during cold stress or inactivity, it may drop below this level. Then, too, the lower legs and feet are not well insulated, and the temperatures here are more variable.

HEAT FLOW

Birds are continually exchanging heat energy with their environments. In order to comprehend the problems of heat balance and matters of cold and heat stress confronting Great Basin birds, we must first consider the four physical processes involved in heat flow: conduction, convection, radiation, and the evaporation of water. Through the first three processes heat flows "downhill," that is, from a hotter place to a colder one. The evaporation of water requires heat—the so-called heat of vaporization. This is the only process in which heat can flow "uphill," that is, from a colder place to a hotter one. And this is the process in operation during the desert summer which allows birds and other animals to survive at air temperatures higher than their body temperatures. Therefore, the problems of heat balance and water balance are intricately interwoven, although I shall discuss them in separate chapters in this book.

Conduction

Conduction takes place through solids, gases, and liquids but does not involve the bulk movement of any of these three types of matter. All matter is composed of molecules which are in motion, colliding with neighboring molecules. The molecules in hot matter are moving faster than the molecules in cold matter. So heat flow is merely the energy which faster molecules impart to slower molecules when they collide, and heat (energy) flows from the heat source (faster molecules) to the heat sink (slower molecules).

Conduction takes place through the tissues of the bird's body and even through the feathers; it takes place between the bird's feet and the substratum upon which they are resting, typically the ground or a branch. A boundary layer of air surrounds the bird and is in contact with its skin and feathers. Heat exchanges between the skin and feathers and this boundary layer are by conduction.

Matter varies greatly in its ability to conduct heat. Silver is an excellent conductor of heat; air is a very poor conductor; the down feathers of ducks are almost as poor a conductor as air. Water is a much better conductor; its coefficient of thermal conductivity is about twenty-eight times greater than that of air. If water instead of air comes into contact with the skin, the heat exchange rates escalate. Water birds, such as grebes, who spend a lot of time swimming in cold water are very densely insulated with down feathers on the underside of the body; they may also possess a thick layer of subcutaneous fat.

Convection

Convection takes place in gases and liquids and involves the actual movement of the gases and liquids. In forced convection either air or water flows over a bird's body, carrying along heat energy. Forced convection can result from wind or water currents flowing by the bird, or it can result when the bird flies or swims through still air or water.

Free convection does not require relative movement between the bird's body and air or water. As air or water is heated or cooled, its density changes, and hence its buoyancy; and it is replaced by the surrounding air or water, which is at a different temperature and density.

Finally, a bird has a high-capacity circulatory system, and the mass movement of blood circulating throughout the body transports heat. The circulatory system is responsible for internal convection within the body of a bird.

Radiation

All matter in the universe, including birds, is continually radiating energy. In the solar system, the main source of this electromagnetic energy is the sun. At temperatures above absolute zero, the electrons in matter are in constant motion and are emitting electromagnetic waves. Some of these wavelengths constitute heat energy. The intensity and wavelengths of the emitted energy depend on the absolute temperature of the surface of the radiating body and on its capacity to radiate energy. The hotter the surface, the shorter the wavelengths of emitted radiation.

The wavelengths which constitute thermal radiation include those which we recognize as visible light and infrared radiation. Heat flow via radiation differs from that of both conduction and convection in that no conducting or transporting medium is required, since electromagnetic waves can cross empty space. Further, thermal radiation leaves a radiating body in all directions—regardless of the direction of temperature gradients. However, a colder body will end up with a net thermal gain through its radiational exchanges with a hotter environment, and a hotter body will end up with a net thermal loss through its radiational exchanges with a colder environment.

Evaporation of Water

Since the evaporation of water requires heat of vaporization in order to take place, heat is lost from every evaporating surface. At 0° C the heat of vaporization is 595 calories per gram of evaporated water. In physiological studies on homeotherms, the value of 580 calories per gram of evaporated water is used, representing evaporation at a surface temperature of about 35° C.

As we have seen, evaporative cooling is the only way heat can be moved "uphill" when the environmental temperature exceeds the body temperature. Being able to move heat "uphill" during the hot summer days makes life possible for desert birds. However, evaporative cooling removes water as well as heat from the body; and, while trying to prevent overheating by stepping up their evaporatory losses, birds can get out of water balance and become dehydrated. The respiratory surfaces in the lungs, where the exchanges of molecular oxygen and carbon dioxide between the air and blood take place, must be kept moist in order to function properly. Therefore, every time a bird exhales, the expired air is saturated with water vapor, removing both water and heat from the body. On hot days the respiratory heat losses help prevent overheating, but on colder days these losses can constitute a serious drain on the heat economy of a bird.

BODY SIZE AND HEAT BALANCE

Small birds the size of Bushtits, chickadees, and sparrows have a much more difficult time practicing homeothermism than do birds the size of magpies and ravens in the rapidly fluctuating thermal environments of deserts and mountains. For body size affects heat flow and the heat exchanges between a bird and its environment—all to the detriment of small birds.

The thought that body size could influence a species' thermal relations with its environment has been with us for well over a hundred years, as naturalists have long pondered the so-called Bergmann's rule. This rule, derived from an 1847 paper by Carl Bergmann, notes that, in a widely distributed species of homeotherms, the populations living in colder climates tend to have larger body sizes than populations of the same species living in warmer climates and that, among closely related species, those living in colder climates tend to have larger body sizes than those living in warmer climates. The main theoretical explanation of Bergmann's rule has long been that large-bodied homeotherms have relatively less surface area per unit of metabolically active tissue to lose heat through than do small-bodied homeotherms and are therefore more heat-efficient in cold climates.

In recent years Bergmann's rule has generated considerable controversy. Arguments have sprung up over whether it is a widespread phenomenon among homeotherms; whether it is limited to homeotherms or whether some poikilotherms might possibly conform to it; and whether the value of large versus small body size to a homeotherm in a cold climate is entirely or even mainly due to the reduced ratio of surface area to volume. There is no disagreement over the fact that in bodies of a similar shape, the larger ones have less surface area in proportion to volume, since with increasing size the surface area increases only as the product of two dimensions, while the volume is increasing as the product of three dimensions. And there is no denying that heat flow is proportional to the amount of surface area available.

Regardless of arguments about the intricacies of how body size functions in a cold climate, a large bird does have a much less difficult time remaining in heat balance in a cold environment than a small bird. Not only does a large bird have greater resistance to both cold and heat, but it can be more frugal in its expenditure of water for evaporative cooling in a hot environment and more frugal in its production of additional metabolic heat in a cold environment. As another advantage, a large body will support a thicker and heavier layer of insulation than a small body. With a thicker layer of feathers, heat has a longer trapped-air pathway to traverse between the skin and the outer surface of the plumage, and a lower temperature gradient can be maintained. This reduces the bird's convective heat exchange with its environment. Also, a large body diameter produces a thicker boundary layer, reducing the bird's sensitivity to forced convection.

Having a large body size is not superior in all ways to having a small one. A small body fits into many bits of cover and microclimate which are too restricted to accommodate a large body. Then, too, a small body requires much less food from its environment than a large one. However, I cannot see how having a greater food requirement constitutes much of a handicap to a large bird. Even casual observations show that there is, and of necessity must be, an adequate fit between the size of a bird and the size of its food staples. You are not going to see Golden Eagles preying on ants or flies or shrews; their food staples are animals the size of ground squirrels and rabbits and larger. To remain in energy balance, a bird must obtain more energy from metabolizing its food than it expended in procuring and eating that food.

One of the reasons that birds come in various sizes may be due in part to the fact that food comes in various sizes. This may be particularly true of raptors, who have to capture, subdue, and kill animal prey. For example, in the Great Basin three species of accipiters can be found hunting in the same general areas—the large Northern Goshawk, the medium-size Cooper's Hawk, and the smaller Sharp-shinned Hawk. Further, each of these three species shows sexual size dimorphism, with the females being considerably larger than the males. They all employ traditional hunting techniques. They still-hunt from concealed perches, or they hedge-hop by alternating flapping and gliding flight. Accipiters suddenly pounce on their prey or impetuously dash after it, dodging in and out of trees and shrubs. They prey mainly on birds, although they take some mammals. These three accipiters are certainly much alike, but it is their size differences which limit food competition among them, by spreading their predation out over a wide range of sizes of prey. Small, medium, and large species of birds are all exploited when these three accipiters are on the loose.

PLUMAGE COLORATION AND HEAT BALANCE

Ornithologists have long wondered why birds come in various colors. Many birds are relatively inconspicuous when motionless, well concealed from predators, because of their color patterns. Most species possess some degree of obliterative shading. The plumage is countershaded, with the upperparts being darker than the underparts. Countershading tends to reduce the relief of the bird and flatten its silhouette, since the brightest illumination normally comes from above; and the bird blends into the darker ground or lighted sky when viewed from above or below. Obliterative shading is often reinforced by cryptic coloration. Then the coloration closely matches the predominant color or color pattern of the bird's normal habitat. In addition, dark disruptive markings may be present to break up the outline of the bird by disrupting its surface continuity and obscuring conspicuous anatomical features such as the eyes. Take a close look at Great Basin species such as the Killdeer or the Snowy Plover, and you will see all three features of coloration in one bird: obliterative shading, cryptic coloration, and dark disruptive markings—as complete or partial black breast-rings and dark lines through and between the eyes.

It is obvious that color does not always function to render a bird less conspicuous. To the contrary, color and color patterns often render a bird more conspicuous. Note the snow-white swans and egrets and the coal-black ravens. Note the brightly colored air sacs of male grouse, the white outer tail feathers of juncos, and the red-and-yellow epaulets of male Red-winged Blackbirds. Coloration which promotes the conspicuousness of a bird is called phaneric coloration. This serves a variety of functions; for example, it acts as a warning signal; enhances a movement or posture in a threat, territorial, courtship, or copulatory display; acts as a signaling device to enable flocking birds to maintain visual contact, particularly when flying; and attracts others to the site where a foraging bird has located food.

But does color stop here? Does it merely function in the visual world to demote or promote conspicuousness? No, it does more than meets the eye. Color also plays a role in the heat economy of birds.

Sunlight consists of many different wavelengths of radiant energy, and the plumage color of a bird is of importance in how that bird absorbs and reflects some of the heat-carrying wavelengths. A raven appears coal-black to the human eye because its plumage does not reflect visible light. A swan or egret appears to be snow-white because its plumage reflects, equally well, all the wavelengths of visible light. When perched in still air, a black or dark-colored bird can utilize sunlight for heating purposes more effectively than can a white or light-colored bird. Not only does black plumage absorb heat from the infrared wavelengths of sunlight, as does white plumage, it also does not reflect away heat carried by visible light.

Laboratory experiments have indicated that the ability to heavily absorb visible light and the shorter wavelengths of near infrared radiation makes a significant difference in the heat economy of a small bird. Black male Brown-headed Cowbirds and gray Zebra Finches could reduce their oxygen consumption, and hence their metabolic heat production, by 26 percent at air temperatures of 10° C when provided with simulated sunlight. So could the slightly smaller and brown-colored female cowbirds. However, albino Zebra Finches could reduce their metabolic heat production only by 6 percent, because of their inability to absorb much visible light or near infrared radiation.[2]

Radiation penetrates less readily down into black plumage than into white plumage, and the radiative heating of black plumage is more affected by convective cooling than is white plumage. Therefore, when winds are strong or when a bird is flying or gliding rapidly through the air in sunlight, a black bird may assume less of a radiative heat load than a white bird.[3]

It is important to note that, when we say a bird absorbs heat radiation, this does not mean heat is transmitted through the feather layer and skin, on into the body of that bird. During most of the time in most environments, the air temperature is lower than the body temperature of a bird; and a temperature gradient runs "downhill" from the body of a bird, through its skin and feather layer, to the outside air. When impinging radiant energy is absorbed by the outer surface of the feathers, this surface warms up somewhat, and the temperature gradient from the body to the outer surface of the feathers becomes less steep. This slows down convective heat losses from the body. The 26 percent decrease in heat production shown by the birds in the experiments we just considered was due not to the body gaining heat from the impinging radiation but, rather, to a slowdown in the rate of convective heat losses from the body.

However, if the air temperature is high enough, as on a summer day, the outward-extending temperature gradient is not merely reduced but is reversed. The temperature gradient now extends inward instead of outward, and heat flows "downhill" from the environment into the bird's body. In experiments on Red-winged Blackbirds subjected to heat radiation, the temperature gradient was reversed at an air temperature of 30° C. But this value was for blackbirds at rest. When birds are flying or the wind is blowing there are forced convective heat losses, and a higher air temperature is needed to reverse the temperature gradient, so that the impinging heat radiation flows into the bird's body.[4]

Plumage color is important when a bird is in direct sunlight, but in the shade or at night it has no effect on how a bird absorbs radiant energy from its environment. Nor does color, in or out of the sun, have any effect on the rate at which a bird radiates heat energy to its environment. Because the surface of the sun is so very hot, the laws of physics dictate that most of its heat radiation will be in the form of short wavelengths—in the form of visible light and near infrared radiation. And, as we have seen, color affects the efficiency with which bodies absorb or reflect short wavelengths. However, the surface temperature of a bird or even the temperature of the desert sand, on a sunny day, is very low compared to that of the sun. Therefore, when birds and the objects in their environment radiate or reradiate heat energy, the laws of physics dictate that they do so in the form of infrared radiation of long wavelengths. Experiments have shown that color has no effect on how efficiently a body absorbs or reflects long wavelengths. Long-wave heat radiation is color-blind, and the outer surfaces of birds and other animals function as "black bodies" in the presence of long wavelengths—they are both excellent absorbers and radiators of long wavelengths. So, in the shade or at night, a black bird is not at a disadvantage compared to a white bird. Its net radiant energy exchanges with objects in its environment are neither greater nor lesser because of its blackness.

With the thought in mind that dark coloration can be a real disadvantage to birds in hot, highly sunlit environments, ornithologists have long pondered the puzzling fact that so many of our desert birds are black or dark-colored. During the summer in the desert, not only are the days usually sunny, but the air and ground temperatures are extremely high. There is little vertical height or lushness to the vegetation and virtually no deep shade. In the Great Basin Desert, many of the birds are black or very darkly colored—vultures, eagles, ravens, magpies, starlings, cowbirds, and blackbirds, for example. The vast majority of the remaining species are dark (brown, blue, or gray) on their upperside. Outside of water birds, very few birds in the Great Basin Desert are white or light-colored.

Birds avoid the effects of direct sunlight in the summer by confining most of their activities to the early morning or late afternoon hours. During the heat of the day, they decrease their metabolic heat production by remaining relatively inactive and avoid direct sunlight by utilizing whatever shade is available. Vultures, eagles, hawks, and ravens are often active during the heat of the day, but their soaring and gliding flight does not require much muscular exertion and does not generate much extra metabolic heat, and soaring and gliding step up their heat losses by forced convection, while soaring allows them to reach cooler altitudes. And even in the desert the disadvantage of dark coloration is limited to the short periods of time when the air temperature approaches or exceeds the body temperature.

The avian blackness which fascinates me most is that of the coal-black Common Raven, for I have met it afield during the bitter subarctic winter as well as during the blistering desert summer. The Common Raven thrives in both hot and cold climates. Blackness aids its heat economy most of the time. Because of its largeness and blackness, the raven can be seen from afar. Being conspicuous does it little harm, since it has no real enemies outside of humans. And conspicuousness helps ravens keep visual contact with one another. A raven descending from the sky to carrion on the ground is visible a long way off.

THERMONEUTRALITY

There is a range of ambient (environmental) temperatures over which a homeotherm is neither under heat stress nor under cold stress; this range is called the thermoneutral zone. In the thermoneutral zone a resting bird's rate of metabolic heat production, as measured by its rate of oxygen consumption, is at a minimal level. In this zone, a bird regulates its body temperature by controlling the rate at which it exchanges heat with its environment; it controls the rate of heat loss and gain by employing circulatory, feather depression or erection, and postural responses. In general, the zone of thermoneutrality spans a greater range of ambient temperatures in large birds than in small birds, as it does in birds from cold climates compared to birds from warm climates.

The lower end of the thermoneutral zone is called the lower critical temperature; the upper end is the upper critical temperature. At ambient temperatures below the lower critical temperature, a bird regulates its body temperature by adjusting its heat production; at ambient temperatures above the upper critical temperature, a bird can regulate its body temperature by adjusting its evaporatory heat loss. Even during the Great Basin summer, at night and in the early morning, ambient temperatures are often below the lower critical temperature. However, during the summer from midmorning until evening, the ambient temperatures are often above the upper critical temperature.

HEAT STRESS

The basic fundamentals of heat stress can be appreciated now that we have examined the thermal relations between a bird and its environment. Even before the air temperature reaches the level of the body temperature, a bird may be gaining heat faster than it is losing it. Metabolic heat is continually being produced within the bird's body, mainly within the musculature; and the more active the bird, the greater the amount of heat generated. Furthermore, since bare soil and rocks cannot regulate their "body" temperatures, their surfaces may soon become so hot in sunny weather that, during a bird's radiant energy exchanges with them, the bird ends up with a net gain of heat instead of a net loss. Also, the heated ground and rocks will start to warm up the air by reradiating long-wave thermal radiation. As all these thermal exchanges progress, not only will the temperature gradient through the feather layer of the bird become reduced, but eventually it will be reversed. Heat from the environment will then flow into the bird's body, raising its temperature—unless the bird increases its evaporative heat losses to move heat "uphill," against the temperature gradient.

The figure on page 26 depicts two birds, one in the sunlight and one in the shade. The bird in the sunlight is receiving thermal radiation of various wavelengths from direct, reflected, and scattered sunlight. It is also receiving long-wave infrared thermal radiation from the opaque objects around it and from the atmosphere. Its feet are engaged in conductive heat exchanges with the ground, and it will experience a heat gain if the ground temperature is higher than the temperature of the feet. Internally, it is producing metabolic heat. As far as its heat losses are concerned, it is radiating long-wave infrared radiation from its body surface. As water is evaporated from its res-

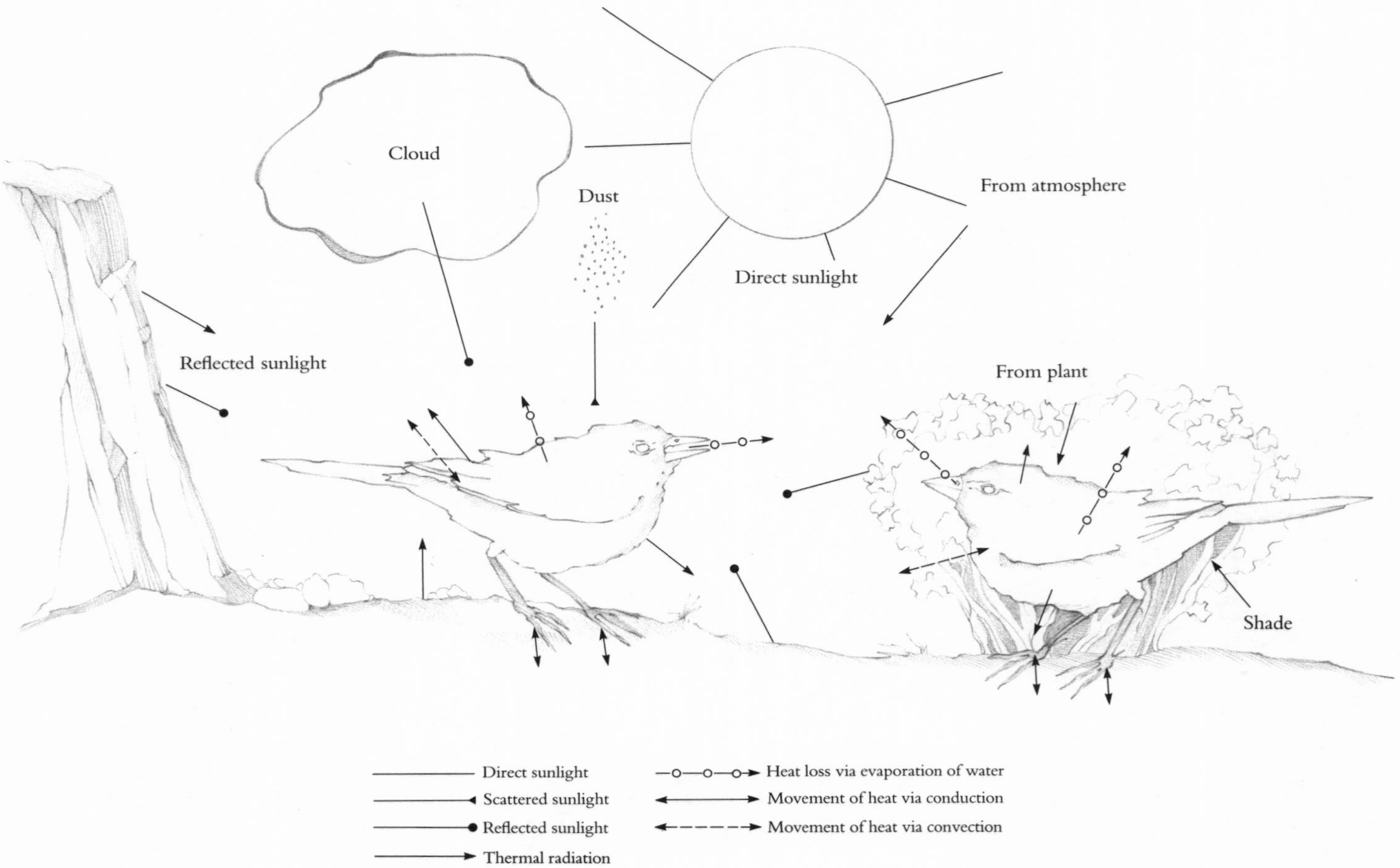
Cloud
Dust
From atmosphere
Direct sunlight
Reflected sunlight
From plant
Shade
Direct sunlight
Scattered sunlight
Reflected sunlight
Thermal radiation
Heat loss via evaporation of water
Movement of heat via conduction
Movement of heat via convection

piratory passages and skin, it is losing heat of vaporization to its environment. As long as the temperature gradient through its feather layer extends outward, it will lose heat by free convection to its environment. If the wind is blowing, forced convection may increase its heat losses.

The heat exchanges of the bird in the shade are somewhat different. It is not receiving direct sunlight; also, the ground underneath and around it is not as hot as the ground exposed to direct sunlight. The air temperature immediately around it probably does not differ appreciably from the air temperature in the sun, since only a limited volume of air is in the shade. Overall, however, the heat burden of the bird in the shade is much less than that of the bird in the sun.

Midday Inactivity in the Shade

There is a very obvious lull in activity during the middle of a summer day in the desert shrublands and woodlands. Not only birds but also diurnal mammals, reptiles, and insects observe this midday pause in activity, which makes possible two means of reducing the heat burden. Muscular inactivity holds metabolic heat production to a minimal level, and it allows birds to remain in the shade.

The most vigorous muscular activity of birds—flight—greatly increases metabolic heat production. Since there are major differences in aerodynamic efficiency between birds with various wing shapes, wing sizes, and wing loadings, the metabolic cost of flight varies considerably. Metabolic measurements are difficult to make on flying birds. However, the limited data available indicate that, during flight, the metabolic rates of various birds increase from fivefold up to twelvefold above the resting rates. Of course, evaporative and forced convection heat losses will also increase during flight, but probably not as steeply as the metabolic heat gains.

Compared to flapping flight, soaring and gliding are metabolically much less demanding, and soaring and gliding birds frequent the desert skies during the middle of the summer days. And this is the time when the sky escalators or thermals—ascending columns of heated air—ridden by soaring birds are the strongest. Turkey Vultures, Golden Eagles, Red-tailed Hawks, Swainson's Hawks, Ferruginous Hawks, and Common Ravens will be seen soaring and gliding in the midday sky of the Great Basin.

Deep shade is difficult to come by in the desert shrublands and woodlands. Small birds are better off than larger birds, since they fit into thinner bits of shade. Also, depending on their method of foraging, it is easier for them to move around somewhat and still remain mostly in the shade. In open areas, during the midday, it is not unusual to see a large bird, such as a raven, resting on the ground in the scant, thin shadow cast by an upright object such as a fence post. Utility poles which are properly aligned with the midday sun are often used for shade perching by Common Ravens, Black-billed Magpies, Brewer's Blackbirds, Red-tailed Hawks, and other birds. As you drive along Interstate 80 or U.S. 50, which trend in an east-west direction, during a summer afternoon, note how practically every perching bird will be on the cross arm of a utility pole in the shadow cast by the top of the pole. As you pass grazing cattle, you may note that ground-foraging birds such as cowbirds, blackbirds, and starlings often stay in the shadows cast by the cattle in the midday heat.

Gaping, Panting, and Gular Fluttering

During summer days in the Great Basin, it seems as if every other bird you see is going around with its bill open and its wings drooping away from the sides of the body, exposing the more thinly feathered underwing areas. Gaping birds frequently engage in bouts of panting, and some birds also gular flutter—by vibrating the floor of the mouth-throat region. Gaping, panting, and gular fluttering are all means by which birds increase evaporative heat losses from their respiratory and digestive tracts. When the heat burden is so onerous that the "downhill" methods of moving heat by conduction, convection, and radiation are not adequate to keep the bird in heat balance, it must either increase its evaporatory heat losses or perish by overheating.

While gaping with their mandibles widely separated, birds expose the moist linings of their mouth and throat cavities. This enhances evaporative heat losses. I have even seen gaping on a mild September morning, when a flock of starlings was feeding in bright sunlight on my lawn in Reno—vividly attesting to the heating powers of sunlight.

Under heat stress, birds pant—much like dogs—while gaping. With its bill wide open, the bird breathes in and out rapidly—at sixteen to twenty-seven times or so its resting breathing rate—greatly increasing the rate of evaporation from the moist linings of the mouth and throat region and respiratory system.

Researchers have detected two general patterns of panting in the species of birds they have studied. Some birds—including quail, cormorants, and the poorwill—gradually increase the rate of panting as their body temperatures increase under heat stress. Other birds—including pigeons, doves, owls, and pelicans—have two distinct levels of rates, the normal breathing rates and the panting rates. In this second pattern, the breathing rate is quite abruptly elevated to the panting level when the bird is under heat stress.

Panting increases the evaporatory water and heat losses by moving a greater volume of air in and out of the respiratory system per unit of time than does normal breathing. In physiology, panting is referred to as hyperventilation. If you have ever voluntarily hyperventilated, by breathing very rapidly for a while, you know that this has severe physiological consequences. Weird things begin to happen to you; you become light-headed and dizzy, and numbness and tingling sensations develop in your arms and legs. Hyperventilation removes carbon dioxide from the lungs faster than normal breathing and drops its level in the lungs and arterial blood—a condition called hypocapnia. With this drop in carbon dioxide levels the blood becomes slightly more alkaline, and hypocapnic alkalosis has developed, accompanied by both cardiovascular and neurological changes. The intriguing question immediately arises, how can birds get away with all that summer panting?

Experiments have shown that some birds develop hypocapnic alkalosis, while other birds don't. Recent research has shown that it is the very severely heat-stressed birds which develop it and that less severely heat-stressed birds avoid it by superimposing panting on normal respiration. When small volumes of air are rapidly moved in and out of the upper end of the respiratory system, the tidal effects of panting do not reach down into the lungs, where the exchanges of carbon dioxide and molecular oxygen occur between the air and blood. Consequently, carbon dioxide is not flushed out by shallow panting, as it is by normal breathing, which involves larger volumes of air moving in and out.

Recent experiments on domestic pigeons by Bernstein and Samaniego have contributed to our knowledge of how panting functions during heat stress. At body temperatures of 41 to 42° C, the experimental pigeons did not pant. Under heat stress, when body temperatures ranged from 42 to 45° C, the pigeons superimposed panting on normal breathing. This compound type of ventilation did not lead to hypocapnic alkalosis. But at body temperatures of 46° C the panting became deeper, normal breathing ceased altogether, and the deeper panting flushed out carbon dioxide so rapidly that hypocapnic alkalosis developed.[5] Of course, at 46° C a bird is up against the lethal limit for body temperature, and the deep panting is strictly an emergency measure.

Some birds supplement panting by fluttering or vibrating the gular region—a thin, membranous area representing the floor of the mouth and throat cavities (on the outer surface of a bird, the gular region includes what we call the chin and throat). The hyoid apparatus apparently powers the movements of the gular region. The hyoid, a skeletal element located in the throat region, helps anchor or support the tongue and upper ends of the digestive and respiratory tracts. Since some muscles insert on the hyoid, it is movable. Gular fluttering enhances evaporation from the moist linings of the upper part of the digestive system, not the respiratory system, and it does not lead to hypocapnic alkalosis. Since only a small amount of muscle and bone mass is involved in gular fluttering, the process costs little in terms of energy expenditures and extra metabolic heat.

Only certain kinds of birds practice gular fluttering. In the Great Basin, these include pelicans, cormorants, herons, owls, doves, quail, poorwills, and nighthawks. It appears as if in some doves, owls, and quail gular fluttering may be synchronous with panting; that is, it may occur at the same frequencies or rates.

One of the most intensely studied species in terms of gular fluttering is the Common Poorwill, a bird found throughout the Great Basin in the summertime. In a classic laboratory study, Lasiewski and Bartholomew discovered that, at ambient temperatures of up to 38° C, the poorwill regulated its body temperature with its mouth closed. At ambient temperatures of 38 to 39° C, it could dissipate enough heat to remain in heat balance by just opening its mouth without any gular fluttering. At ambient temperatures above 39° C, the poorwill employed gular fluttering, intermittently at lower temperatures but almost continuously as the temperature reached 46.5° C. Depending on the individual bird, the rate of vibration was always between 590 and 690 flutters per minute—regardless of the amount of stress imposed by the ambient temperatures. However, the amplitude of the flutters and the amount of gular region actually fluttering increased as the heat stress increased. With a small amount of heat stress, only the mouth area fluttered, but as the stress increased, progressively greater amounts of the throat and esophageal areas became involved in the movements.[6]

For any vibrating system, there is a natural rate or resonant frequency at which that system oscillates with the least amount of energy expenditure and the least generation of metabolic heat. Birds who employ resonant or harmonic frequencies can be recognized, since they pant or flutter over a very narrow range of frequencies. Studies have shown that cormorants, pelicans, poorwills, nighthawks, roadrunners, pigeons, doves, and owls are among the birds who gular flutter at resonant rates. Roadrunners, pigeons, doves, and owls have panting rates which are synchronous with the gular flutter rates, and they both pant and gular flutter at the same resonant rates. By using resonant rates, birds avoid generating much in the way of additional heat when they turn on their panting and fluttering systems.

Frequent Bathing

Bathing is an effective and inexpensive method of cooling off during hot weather. Wherever water can be found, Great Basin birds are frequent bathers during the summer. Irrigation ditches as well as overhead spray systems supplement the natural sources of bathwater. Bathing is a very effective method of increasing evaporative heat losses, since all the surface areas of a bird—body, head, wings, and legs—are simultaneously employed as evaporatory surfaces.

Some of the shorebirds nesting in open, hot, barren areas will engage in frequent belly soaking to increase evaporative cooling when the ambient temperatures are high. Snowy Plovers will stand in pools of saline or alkaline water. Killdeers may belly-soak just before commencing incubation duty. Instead of sitting on the eggs, the Killdeer will crouch over them. It can lower its wet belly feathers onto the eggs to instigate evaporative cooling and prevent the eggs from overheating in the intense heat of the day.

Urohidrosis

Some vultures, storks, and other birds enhance their evaporative heat losses, when under heat stress, by excreting on their legs. The legs are effective evaporatory surfaces since they are unfeathered and well supplied with blood and, hence, with heat to be dissipated. This practice of excreting on the legs has been referred to as urohidrosis.

Among Great Basin birds, the Turkey Vulture practices urohidrosis. Laboratory studies carried out by Hatch showed that heat-stressed Turkey Vultures practiced both urohidrosis and panting, usually getting urohidrosis under way before starting to pant. Although the vultures excreted upon both legs, one at a time, they tended to favor one leg over the other. After three to four excretions, the amount of liquid decreased, but even after excreting thirty to forty times during a four-hour period, the vultures were still able to discharge at least a small drop on almost every attempt.[7]

Heat Storage

Birds cannot maintain constant body temperatures the clock around. During periods of inactivity, especially sleep, the body temperature may be one or several degrees C lower than it is when the bird is active during the day. When a bird is engaged in vigorous muscular activity or under severe heat stress, its body temperature may rise 2 to 4° C above the regulated level of an unstressed, normally active bird. Elevations of body temperature above normally regulated levels are referred to as hyperthermia. This is merely a symptom of increased heat storage within the body of a bird.

In heat-stressed birds, there are advantages to heat storage and the subsequent hyperthermia. Depending on the intensity of the ambient temperature and on whether the temperature gradient through the bird's feather layer extends outward or inward, hyperthermia decreases the steepness of the temperature gradient and thereby increases the rate of convective heat loss from the bird or decreases the rate of heat gain from the environment. In both instances, the heat economy of the bird is improved. Decreasing the steepness of the temperature gradient through the feather layer is particularly significant to small birds, because they possess such high surface area to volume ratios. At an ambient temperature of 46° C, raising the body temperature of a small bird from 40 to 43° C will reduce its heat gain from the environment by 50 percent.[8]

Heat storage is not only of importance in influencing the rates of heat loss and gain; it also affects the water economy of the bird. The stored heat is heat that does not have to be dissipated by evaporative cooling, and it can be dumped back into the environment by convective cooling later on in the day when ambient temperatures are lower. The savings in water due to heat storage are more significant in large birds than in small ones, because large birds have much greater storage capacities for heat.

COLD STRESS

Birds are under cold stress over much of the Great Basin year. The high montane environments are cool or cold the year around, except when under the midday sun of summer. Even during summer nights in the Great Basin Desert, ambient temperatures are often low enough that birds must increase their heat production above resting levels to regulate their body temperatures. Once the environment has cooled below a bird's lower critical temperature, the bird must increase its heat production to balance increases in heat loss; otherwise its body temperature will drop. The fire of life must burn hotter. Since most heat is myogenic heat, derived from muscular activity, the bird begins to shiver.

It is interesting that even in cold climates birds' lower critical temperatures—which can vary with seasonal changes in insulation and metabolism—are not all that low. In many instances the lower critical temperature is in the range of 20 to 30° C; relatively few birds have lower critical temperatures below 10° C. The fifty-gram American Dipper, who frequents the cold montane streams of the Basin and feeds while walking underwater, has a lower critical temperature of 11.5° C.[9] That of the much larger Common Raven is 0° C.[10]

Since ambient temperatures are usually higher or lower than the thermoneutral zone of most birds, it would appear as if, most of the time, birds in the Great Basin are either gaping and panting or shivering. This is not true when birds are active during the daytime, since they are then producing myogenic heat beyond the standard metabolic levels and do not have to shiver as they do when inactive or asleep. However, birds do not always turn up their insulation to its highest setting before beginning to shiver. Some gradually turn up their insulation while simultaneously increasing their shivering.

Shelter

Just as shade is used for protection from the sun, shelter is used for protection from wind, rain, snow, and heat sinks. In good shelter, birds can reduce their forced convective heat losses to the wind and, at night, their radiative losses to the cold sky. Warmer microclimates occur in well-sheltered spots. Cavities, crevices, old nests, and other confined spaces offer sleeping and nesting quarters to help insulate birds from the colder macroclimate.

Most of the trees in the Great Basin are conifers, so the forests and woodlands shelter birds from the wind and sky the year around with their foliage. In a recent study in Michigan, W. A. Buttemer found that American Goldfinches roosting in evergreens during winter nights reduced their energy expenditure to nearly one-third below that of goldfinches roosting in open sites. The mountainsides are scored with systems of canyons, side canyons, and ravines affording shelter from any wind direction. Washes, embankments, depressions, ridges, and patches of dense vegetation interrupt the wind in the valleys. When strong winds are blowing, birds even forage in shelter—within dense vegetation or along the lee side of any windbreak. Wind protection is important for night roosting. In a large Common Raven roost on the western edge of Malheur Lake, where the ravens roost in dense marsh vegetation, a record count of 836 ravens was made in January 1977. When strong winds blow out of the southeast, the ravens shift their roost site close in to the lee side of a low ridge there.[11] Birds are very sensitive to forced convective heat losses and usually seek to alleviate them.

Overhead shelter from the night skies can be of importance to birds. The cold, clear night skies that predominate in the Great Basin function as a vast heat sink, and radiation losses to them can be large. One of the most interesting behavioral adaptations to these stressful night skies is shown by the Calliope Hummingbird, the smallest bird in North America. Weighing only about 3.5 grams, these tiny homeotherms face major difficulties in playing out their roles in the Great Basin. The female usually builds her nest on the horizontal limb of a conifer, where there is another limb immediately overhead, protecting her back from the night sky when she is in the nest trying to keep eggs warm.

Some of the very best shelter for holding down heat losses when nesting or sleeping is provided by the enclosed, well-insulated spaces in natural or woodpecker-built cavities in trees, in rock crevices, and in the abandoned nests of Black-billed Magpies and Cliff Swallows. In these confined spaces the free and forced convective, conductive, and radiative heat losses are all minimized, and, as a bonus, protection from predators is maximized. Woodpeckers, nuthatches, chickadees, creepers, titmice, swallows, kestrels, and others use cavities for nesting and sleeping. The abandoned, domed stick nests of Black-billed Magpies are used by a variety of Great Basin birds for nesting, sleeping, and storm shelter purposes. The gourd-shaped mud nests of Cliff Swallows are used by House Sparrows for nesting and by House Sparrows and Rosy Finches for sleeping during the winter.

The Great Basin is literally riddled with abandoned mine tunnels and shafts. Not only do they provide protection from the wind, rain, snow, and night sky, but the ambient temperatures inside them are milder than the outside temperatures—cooler in the summer and warmer in the winter. Several species of birds, especially Say's Phoebes, nest in the outer ends of mine tunnels. The most interesting tenants of these tunnels and shafts are the Rosy Finches. During the cold part of the year in western Nevada—at sites on Peavine Mountain, in the Washoe Mountains, in the Pine Nut Range, and elsewhere—Rosy Finches sleep in mine tunnels and shafts and rest in them during the day when they are not out foraging. They particularly like the more vertical shafts, where they roost on the wooden beams bracing the shafts. When I am in the field during the late fall to early spring, I investigate the vertical shafts I pass by tossing small rocks into them. If a flock or part of a flock of Rosy Finches is within the shaft, the falling rocks will cause them to come boiling out in a spectacular eruption.

When inactive or sleeping under cold stress, birds improve the insulative value of their plumage by fluffing up their feathers. They reduce the amount of exposed surface area by compacting their bodies—holding their wings in against their sides and retracting their necks. Since lower legs and feet, being typically unfeathered, are major avenues of heat loss, birds minimize this by squatting or hunkering down on their legs. Some shore-

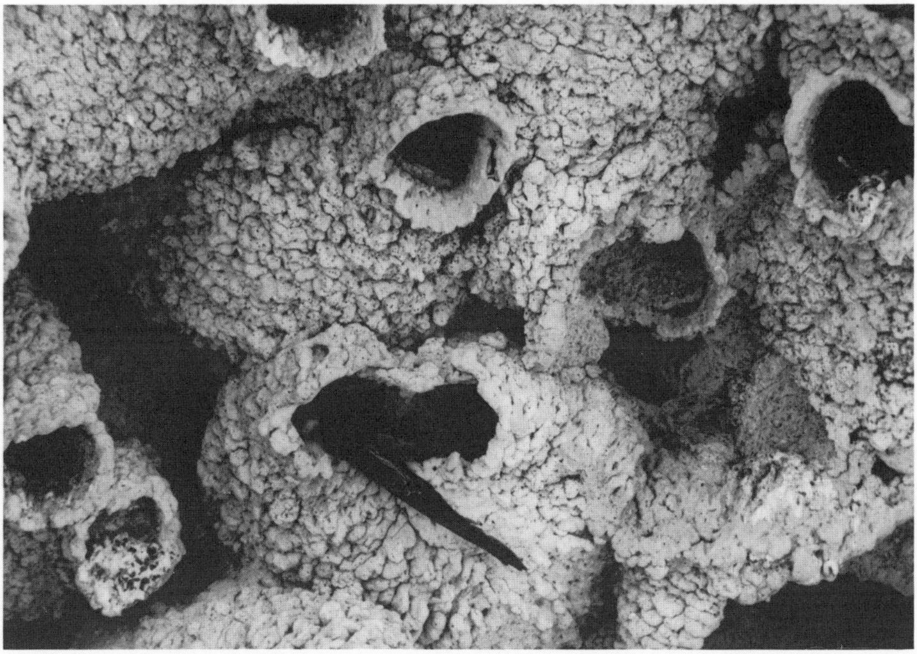

The gourd-shaped clay and mud nests of a Cliff Swallow colony afford shelter to other birds and even bats. *Stephen Trimble*

birds, ducks, and other birds will stand on one leg at a time, with the other leg and foot retracted into the belly feathers. However, infrared radiometer studies have shown that the greatest heat losses are from the head, belly, and back regions.[12] A small bird will cut down on this heat loss by sleeping with its head tucked under a wing. A larger bird, such as a duck, will turn its head back over one shoulder and bury its bill and chin in the scapular or back feathers. With the basal part of its bill tucked under feathers, the bird gains the further advantage of recycling body heat by breathing in air prewarmed by its body.

Huddling could be of considerable value in holding down heat losses while sleeping, since it will reduce the amount of exposed surface area on each of the huddling birds. Unfortunately, we do not have many observations of sleeping birds made under undisturbed, natural conditions. There is fragmentary evidence, based on observations of eleven creepers roosting together inside a hollow beam, that Brown Creepers may huddle. Many roosting congregations of birds appear to be quite closely packed at night, but they may not be in actual physical contact with each other.

Migration

Many species of birds avoid the long period of winter cold in the Great Basin by migrating to warmer climates. There are groups of birds where all or most of the species migrate out of the Basin to overwinter; these include the flycatchers, swallows, vireos, wood-warblers, sparrows, tanagers, orioles, shorebirds, swifts, and hummingbirds. In certain groups there are trade-offs, with some species migrating out of and other species migrating into the Basin to overwinter. Among the hawks, the Swainson's Hawk migrates to South America to overwinter, while the Rough-legged Hawk and Merlin come down from the north to overwinter in the Great Basin. In some species partial migration takes place, with some of the population remaining behind each year to overwinter. Song Sparrows do this. And every winter some American Robins remain in the Basin; large numbers may overwinter during mild winters when food is plentiful. Then, there are weather migrations; geese and ducks will move out of Great Basin locales during extended periods of closed water, only to promptly return in a few days or weeks when the ice melts and frees the water. During some years there is a profound failure of food in the north. Then winter irruptions will occur, and large numbers of Bohemian Waxwings or a sprinkling of Snowy Owls will appear in the Great Basin.

The least understood migrations in the Basin are the altitudinal migrations. Species who can be found nesting in the Great Basin mountains during the summer will show up in the valleys only during the cold part of the year—such as Dark-eyed Juncos, White-crowned Sparrows, Brown Creepers, White-breasted Nuthatches, and Goshawks. Other species present mainly in the mountains during the summer, such as Cooper's and Sharp-shinned hawks, will be present in numbers in the valleys during the winter. There is also a downward shift, away from the higher altitudes, on the part of Mountain Quail, Rosy Finches, and Cassin's Finches. In years of food failure, flocks of Evening Grosbeaks come out of the mountains and inundate the valleys of western Nevada. Of course, it is impossible to always separate the results of altitudinal migration from those of latitudinal migration. Some of the Dark-eyed Juncos, White-crowned Sparrows, Cooper's Hawks, Sharp-shinned Hawks, and others may be coming in from further north, and some of our local birds may be leaving the mountains and moving further south.

Although the winters in the valleys are not as cold as those in the mountains, altitudinal migration may be more the product of deep snow and the general lack or inaccessibility of food in the mountains than of coldness.

Cold Torpidity and Hibernation

It is often difficult or impossible for birds to obtain sufficient food or to meet the high energy demands of practicing homeothermism in the cold. One way to solve this problem is to temporarily abandon homeothermism and go into cold torpidity or even hibernation until conditions improve. During cold torpidity and hibernation, there is profound hypothermia with major drops in the body temperature; most muscular activity ceases; and physiological processes such as breathing and circulation slow down. Cold torpidity is a short-range phenomenon which takes place overnight or over a wet or cold day or two, whereas hibernation is a long-range, seasonal process.

Cold torpidity or hibernation is known to take place in several groups of birds—including some poorwills, nighthawks, swifts, hummingbirds, and swallows. I shall discuss this further in my natural history accounts of these birds.

3

The Water of Life

WATER is the principal component of life. In the arid, highly evaporative environments of the Great Basin, water is often difficult to obtain and, once obtained, difficult to retain. During the Great Basin summer, the problem of remaining in water balance may become critical.

Water is eminently suited to be the principal ingredient of living matter. Water is a common and abundant substance, except where flaws in the hydrologic cycle, which circulates water from ocean to air to ground to ocean, have created deserts. Water is a superb solvent and dissolves many substances or puts them into colloidal suspension. This high capacity of water to separate substances into tiny particles in solution or suspension provides fantastic amounts of particle surface area, so that the chemistry of life can proceed, as the surfaces of the various reactants come in contact with each other.

Water has high specific heat, so that the addition or subtraction of much heat is required to change its temperature. The high specific heat of their water content helps buffer birds from the effects of sudden changes in ambient temperatures. Water has great fluidity and confers fluidity on blood, allowing birds to have a rapid circulatory system to carry oxygen and nutrients around the body. The water content of blood and its high specific heat also allow effective internal convection or the transport of heat to and from the body core and body surface. Finally, water's ability to change physical states, from a liquid to a gas, allows birds to survive the desert summer—for during the process of evaporation, as water changes from a liquid to a gas, heat (heat of vaporization) is subtracted from the body. Thus, the evaporation of water provides birds with their only effective mechanism for dumping heat into a hot environment; and it is the only mechanism by which heat can be moved "uphill" into an environment hotter than the birds' own bodies.

The problem of remaining in water balance during the desert summer is compounded by the general unavailability of drinking water and by water's reluctance to stay put in hot, dry environments. Water is continually evaporating from a bird, day and night, summer and winter; and the hotter and drier the ambient air, the greater its evaporating and water-holding capacities. Although birds have dry skins, lacking sweat and sebaceous glands, there is some movement of water through the skin and, consequently, evaporation from the outer surfaces of birds. But their greatest evaporatory losses are from the respiratory tract.

Respiratory exchanges in the avian lung take place in tiny tubes called air capillaries, which function like the alveoli in the mammalian lung. The walls of the air capillaries must be kept moist, for molecular oxygen has to go into solution in this water, before it can pass through the air capillary and blood capillary walls into the bloodstream. Therefore, every time a bird breathes in dry air, the air passing over the moist respiratory passages strips away water and becomes saturated before it is exhaled. Of course, these evaporatory water losses subtract heat from the bird's body. This is beneficial during the heat of the summer day, since it helps the bird remain in heat balance, but it is detrimental during much of the day and during the night, when ambient temperatures are lower and the high body temperature must be maintained. When under heat stress, the bird purposively increases its heat losses by increasing its evaporatory losses by panting and/or gular fluttering—in order to maintain heat balance, the bird may jeopardize its water balance. But, if the bird goes too far into negative water balance and becomes too dehydrated, its body chemistry is compromised and it will die.

Most of our knowledge of the lethal effects of hot, dry environments and negative water balance is derived from laboratory studies. Laboratory studies are much less physically demanding than field studies, and observations can be made under controlled conditions in designed experiments. In the field one has to be satisfied with what Mother Nature provides, and it is difficult to detect and capture dehydrated birds, before they either recover or become carrion for the scavengers. But one may have nagging doubts about how laboratory responses—evoked under artificial conditions from forcibly restrained wild birds—actually relate to free-living birds. So we are indeed fortunate to have the benefit of field observations by Miller and Stebbins on water balance problems of birds in the desert environments of Joshua Tree National Monument in southern California.

Miller and Stebbins observed that some of the small migrants, particularly warblers and flycatchers, suffered severely when crossing the deserts in the fall. During August and September, shade temperatures often exceeded 100° F during the day for days on end; the air was very dry, with extremely low relative humidities; and surface water was often difficult to find. Although warblers and flycatchers are insectivorous, consequently obtaining much water of succulence from their food, they often were not able to remain in water and heat balance. In writing about the warblers, Miller and Stebbins stated, "We found dead birds, dried and emaciated; we saw others, scarcely able to fly, so anxious for food, water, and shade that in the heat of the day they came into the shelter of our camps almost under our feet; and we took specimens that were in such poor condition that they almost certainly would have succumbed before they could travel the remaining 500 miles southeastward across the Colorado Desert basin."[13] Most of the weakened and dead warblers found belonged to the genera *Vermivora*, *Oporornis*, and *Wilsonia*. Among the small *Empidonax* flycatchers, the Willow Flycatcher, in particular, suffered severely and showed indications of succumbing to the heat and water stresses. Exposure in the dry desert heat seemed to quickly "dry the soft bill, mouth borders and feet of the small flycatchers."

Birds remain in water balance as long as they can compensate for their losses with gains of equal magnitude. Many structural, functional, behavioral, and ecological adaptations come into play in the water exchanges of birds. In order to examine these adaptations, I shall discuss the water economy of Great Basin birds first in terms of water losses, then in terms of water gains.

WATER LOSSES

Evaporation

The major water loss experienced by birds is pulmocutaneous, by evaporation from the respiratory passages and from the skin. Although the respiratory surfaces are recessed into the body, they must be kept moist in order to function properly, and birds must move large amounts of ambient air through their lungs in order to obtain the molecular oxygen to support their high metabolic rates. The cutaneous losses are affected somewhat by the birds' dry skin, which is relatively impervious to water, and by their feathers, which retard air movements that could speed up evaporation.

The rate at which pulmocutaneous evaporation takes place is influenced by a number of variables. The hotter and drier the ambient air, the greater its evaporative power and capacity to hold water vapor. A volume of air heated to 40° C, the core temperature of a bird's body, can hold about twice as much water before becoming saturated than can the same volume of air at a temperature of 25° C. And dry air has greater evaporative power than moist air at the same temperature, since more of its total holding capacity for water vapor is unoccupied. For example, at a relative humidity of 15 percent, 85 percent of the holding capacity for water vapor is unoccupied and can function in evaporation.

It is important to remember that the evaporative power of ambient air usually varies during the course of a day. The relative humidity cycle is dependent on the temperature cycle and is inversely related to it; as the air temperatures change during the course of a day, the moisture-holding capacity of the air changes as well. On most days the high in the temperature cycle occurs in midafternoon, whereas the high in the relative humidity cycle occurs at night.

Pulmonary water losses are not entirely dependent on the temperature of the ambient air. Ambient air at temperatures less than the 40° C of a bird's body can be warmed up somewhat while passing through the respiratory system, and ambient air at temperatures greater than 40° C can be cooled down. However, the situation is not all that simple, since the incoming air evaporates water from the moist nasal passages and respiratory surfaces, and the removal of heat as heat of vaporization cools these surfaces down. Then, as the saturated air is exhaled back over these precooled nasal passages, the air temperature drops. This decreases the air's carrying capacity for water vapor, and the exhaled air parts with some of the water it is carrying before it clears the nasal passages. Because the nasal passages of birds are not very constricted or very extensive in size, the decrease in air temperature and the resulting reclamation of water during exhalation are not as great as they are in mammals, who generally have longer and narrower nasal passages.

There are other variables which affect evaporatory water losses beyond the temperature and relative humidity of the ambient air. One of these is convection—particularly forced convection. If the boundary layer of air in contact with the evaporatory surface of the skin is continually being stripped away and replaced by dry air, the rate of evaporation will increase. Still another important variable is the level of activity of the bird—working through its effect on respiration and forced convection. Finally, there are some quite minor variables, which I shall not consider, which affect the rate of evaporation by influencing the rate at which water vapor diffuses out of the boundary layer of air into the ambient air.

Over the years, avian physiologists merely assumed that most of the evaporative water losses of birds accompanied respiration and that cutaneous water losses were slight. After all, birds have a dry skin which is thickly covered by feathers. Experimentally, it was easier to measure total evaporative water losses than to separately measure pulmonary and cutaneous losses. When these losses were finally measured separately in a few species, researchers found that, in the Ostrich, the largest living bird, cutaneous water losses were less than 2 percent of the total evaporative water losses, when measured in a resting bird at an ambient temperature of 40° C. However, small and medium-size birds, measured at rest at ambient temperatures of 30 to 35° C, had cutaneous water losses which ranged from 45 to 74 percent of the total evaporative water losses.[14]

Activity can have a dramatic effect on evaporative water losses. Not only does running or flying promote forced convection, but it demands greatly increased rates of oxygen consumption by birds. During flight, metabolic rates may increase five- to twelvefold or more over resting rates. This means that breathing has to increase manifold and that pulmonary evaporative water losses are going to skyrocket. The increased ventilation of the lungs—needed to supply the additional molecular oxygen for flight—is also influenced by the circumstance that birds in flight do not remove molecular oxygen as effectively from the air before exhaling it as do birds at rest. The high respiratory water losses during flight may show up in losses of 5 to 8 percent of the body weight per hour at ambient temperatures of 35 to 37° C.[15] Of course, the muscular activity associated with running or flying greatly increases heat production within the bird's body, and the evaporative water losses associated with exercise do remove heat from the body and function in temperature regulation.

Evaporative water losses are correlated with body size, and there is an inverse relationship between body size and evaporative water losses. Smaller birds have relatively higher evaporative water losses than larger birds. During an early, classic study involving twelve species of birds, Bartholomew and Dawson discovered that the body weights associated with relatively high evaporative water losses lie below the 40- to 60-gram range. They confined birds in a chamber at an ambient temperature of 25° C and at relative humidities ranging from 28 to 60 percent and measured their rates of evaporative water losses. They discovered that the 147-gram California Quail's mean rate of loss was about 3.5 percent, the 41-gram Loggerhead Shrike's 5.8 percent, and the 19-gram House Finch's 17.2 percent of its body weight per twenty-four hours. However, the tiny House Wren, weighing slightly less than 10 grams, had a mean evaporative water loss at 25° C of over 35 percent of its body weight per twenty-four hours.[16]

Of course, 25° C is a moderate ambient temperature, and resting birds should be under little or no heat stress. If the ambient temperature does increase above the zone of thermoneutrality or if the birds are very active, they may store heat for a while and experience some hyperthermia. However, given enough of a heat load, birds must either greatly increase their evaporatory water losses by panting or gular fluttering or else die. Under severe heat stress, evaporative water losses during panting and/or gular fluttering can be explosive. Recent interesting studies on this matter have involved the Galah, a cockatoo of some 270 grams inhabiting the arid regions of Australia. During experiments conducted in the summer, "evaporative water losses under standard conditions increased nearly 40-fold between 20.8° and 48.0° C." This great increase accompanied panting and tongue movements which vibrated the gular area.[17]

Urinary and Fecal Losses

Birds also lose water in the urinary and fecal material voided from their bodies. Urine coming from the kidneys, fecal material coming down the gastrointestinal tract, and genital products coming from the gonads are all dumped into the cloaca. From this chamber, which opens to the outside of the body through a vent, the droppings of a bird are voided; each dropping usually consists of darker fecal material and cream-colored urine.

The main excretory product (nitrogen-containing compound) in bird urine is the white, pasty uric acid and urate salts, although significant amounts of ammonia and minor amounts of other nitrogen-containing compounds may be present. Uric acid is highly insoluble in water, although its urate salts are somewhat more soluble. Therefore, the avian kidney cannot reabsorb a lot of water, when urine is being produced in the kidney tubules, without precipitating the uric acid and clogging the tubules. So the uric acid and urates are moved along in colloidal solutions through the kidney tubules, collecting ducts, and ureters to the cloaca. And urine enters the cloaca as a liquid or semiliquid.

If the bird is under water stress and there is need to conserve water, water will now be reabsorbed, and the urine will be voided as a semisolid. Some water reabsorption may occur in the cloaca, but urine from the cloaca may move forward, through the anal opening, out of the cloaca and into the end of the large intestine. The urine may even move as far up the gastrointestinal tract as the junction of the colon and the ileum. Some birds possess large intestinal outpocketings, and some urine may enter these. In the cloaca and intestine there is some breakdown, presumably by bacteria, of uric acid. This breakdown could free sodium and facilitate the reabsorption of water by the cloacal and intestinal walls.[18]

When birds are provided with drinking water or have a high water intake, the water content of their droppings may constitute as much as 80 to 90 percent of the weight or mass of their excrement. When they are under water stress, the cloacal and intestinal water reabsorption may decrease the water content of their excrement to only 50 to 70 percent. In some arid land species, the water reabsorption is even greater. The tiny Brewer's Sparrow, who is so common and abundant in the sagebrush shrublands of the Great Basin, can reduce the water content of its excrement from 93 to 33 percent when under water stress.[19]

In absolute terms, as well as relative terms, the urinary and fecal water losses of a small bird can be impressive. A case in point is that of the 18-gram Sage Sparrow, another common summer resident of the sagebrush shrublands of the Great Basin. When provided with unlimited drinking water, Sage Sparrows voided droppings which were approximately 76 percent water, and their estimated daily cloacal water loss was 4.9 grams, 27 percent of their body weight. When put on a regimen of only 2 milliliters of drinking water per day, their daily cloacal water loss dropped to 1.8 grams—a 63 percent reduction. However, this savings was accomplished to a great extent by reducing the amount of dry fecal material voided, since the water content of the individual droppings only decreased from 76 to 66 percent.[20]

Regurgitation

Another avenue of water loss from a bird during the nesting season is regurgitation. In some species the parents swallow food while foraging and regurgitate this food later on to feed the young. The practice of regurgitation may carry water away from the crop and esophagus of the parent. In many cases the young are fed moist food, such as insects, and regurgitation may then have little influence on the water economy of the parent. As far as I know, no one has yet attempted to assess the influence of regurgitation on the water economy of a parent bird.

In the Family Columbidae, to which the pigeons and doves belong, regurgitation could represent a burden on the water economy of the parents. The parents produce and secrete pigeon milk in their crops and regurgitate it to feed the young. The water content of pigeon milk varies between 65 and 81 percent.[21] For the first few days of life, the young receive nothing but pigeon milk. From about the fourth day on, seeds are mixed in with it. So dependency on pigeon milk decreases as the young grow older.

Columbids normally lay two eggs per clutch. In a study by Nice of Mourning Doves, the body weight of the two nestlings increased from 8.1 to 34.3 grams and from 6.2 to 37.8 grams during the first five days of life—an average increase of 5.8 grams per day.[22] Obviously, each parent Mourning Dove would lose much more than 5.8 grams of water per day in supporting a 5.8-gram weight increase by each nestling. Pigeon milk contains no carbohydrates; it has a protein content of 13.3 to 18.6 percent and a fat content of 6.9 to 12.7 percent.[21] Since energy transformations within the animal body are relatively inefficient, much of the fat and protein in the pigeon milk would be used for energy purposes by the growing young and not show up as body weight gains. Much more than 5.8 grams of pigeon milk would be required to produce a gain of 5.8 grams of body weight per day.

Egg Laying

Another source of water loss from the adult female bird occurs during the production of eggs. The eggshell and its membranes contain only traces of water; but the contents of an egg, the yolk and albumen, are very watery—the albumen may be three times or so more watery than the yolk. The water content of eggs varies between 70 and 81 percent by weight in the various species of birds. It is somewhat related to the ecological conditions under which birds live as well as their metabolic requirements. The greater the water content, the lesser the nutritional and caloric content of the egg. Waterfowl, which produce active chicks, have the least watery eggs (70 percent). Land birds like the quail and pheasant, which produce active chicks, lay eggs with an intermediate water content (74 percent). Land birds like the blackbird and finch, which produce inactive nestlings, lay eggs with the highest average water content (81 percent).[23]

The size of the egg, in addition to its water content, helps determine the impact of egg laying on a female's water economy. Egg water losses are greater in small than in large birds, since there is an inverse relationship between body size and egg size. Birds weighing between 2 and 180 grams lay eggs that weigh about one-ninth of their body weight; 400- to 1,500-gram birds lay eggs that weigh about one-fifteenth of their body weight; 2,600- to 12,000-gram birds lay eggs that weigh about one-twenty-eighth of their body weight.[23]

The number of eggs in a clutch and the frequency of laying also influence water loss. The many birds which lay an egg a day will lose water each day they lay an egg. Some birds like hummingbirds, swifts, and doves lay eggs at intervals of from thirty-eight to forty-eight hours. This spaces out the water losses somewhat. Egg-laying water losses are long continued in female gallinaceous birds like quail, grouse, and pheasants, which have large clutches of twelve to twenty eggs or so.

WATER GAINS

Metabolic Water

Off and on, over the years, the popular press has learnedly informed the general reading public of the existence of certain small desert rodents, called kangaroo rats and gerbils, who have the unique ability to produce water within their bodies. And these desert rodents surely do this, but not uniquely so. All animals, even humans, produce water in their bodies during the oxidation of foodstuffs. This is the so-called metabolic water, which is the inevitable consequence of the complete oxidation of organic compounds.

The quantity of metabolic water formed during the oxidation of a particular foodstuff depends on the amount of hydrogen present in that foodstuff. Oxidation of 1 gram of carbohydrates yields 0.56 gram of metabolic water. Oxidation of 1 gram of fat yields 1.07 grams of metabolic water. The formation of metabolic water from the oxidation of proteins is a more complex story. Nitrogen is present in proteins, and it must be excreted by the animal before it reaches toxic levels; thus the main function of an animal's urinary system is to get rid of nitrogen. The quantity of metabolic water derived from the oxidation of a protein depends on the type of nitrogenous waste product an animal forms. Mammals, excreting urea, derive 0.39 gram of metabolic water per gram of protein. Birds, excreting uric acid, derive 0.50 gram of metabolic water per gram of protein. However, the hydrogen excreted in urea and uric acid cannot be used to form metabolic water. Animals forming urea lose two atoms of hydrogen for each nitrogen atom excreted, whereas animals forming uric acid lose only one atom of hydrogen for each nitrogen atom excreted.

At first inspection, it would appear as if birds under heat and water stresses would be best off feeding on fatty foods. After all, each gram of fat provides 1.07 grams of metabolic water. But upon closer inspection this is not true, since fats also provide much more energy per gram than do carbohydrates and proteins. Since a bird can expend only a limited quantity of energy in its metabolism per unit of time, it is best off oxidizing carbohydrates. The oxidation of carbohydrates provides a little more water per calorie of energy released than do fats and proteins.

Small birds have higher metabolic rates than large ones and therefore produce relatively more metabolic water than large birds. But then, as we have already seen, the evaporative water losses of small birds are relatively higher than those of large birds. No bird has a high enough metabolic rate to rely solely on metabolic water gains to keep it in water balance. Evaporative water losses alone are always higher than metabolic water gains, and the deficit between the two is always much greater in small birds than in large ones. A resting, 19-gram House Finch at 25° C may gain enough metabolic water to equal about 5 percent of its body weight over twenty-four hours, but during this time its evaporative water losses would approximate 17 percent of its body weight. A bird cannot survive on metabolic water alone.

Water of Succulence

Birds derive water from their food, not only indirectly through the oxidation of its carbohydrates, fats, and proteins but also directly from its preformed water—water of succulence. Food items vary greatly in their moisture content. Animal food represents a very wet type of food for insectivorous and carnivorous birds. Insects vary in their moisture content—ranging from about 50 to 90 percent by weight water. So even a diet of drier insects provides birds with food that is at least 50 percent water. Vertebrates also constitute very succulent food. A diet of mice or rats will provide a raptor with food that is about 67 percent by weight water.

Certain types of plant food have a high water content. Fruit may consist of over 90 percent by weight water. Some of the vegetative parts of green plants can be very moist. However, since succulent desert plants can have very high salt contents, they can add to the salt burden of the bird.

Dry seeds constitute a very poor source of water of succulence for birds. In the desert seeds are subjected to intense heat, low relative humidities, and tremendously high evaporation potentials. The moisture content of exposed seeds may drop below 10 or even 5 percent. During the night, when the relative humidity increases, seeds may rehydrate somewhat. Seeds represent a major source of food for birds over much of the world; but in deserts most birds living away from water during the summer feed on succulent food, and relatively few are primarily granivorous in their feeding habits.

Some granivorous desert birds nest and forage at a distance from surface water. The Mourning Dove is a seedeater who often nests in the Great Basin Desert miles away from surface water. This dove, a swift and powerful flier, visits distant water holes to drink during the early morning and late afternoon hours. A capacious drinker, it is able to swallow enough water in a continuous draft or two to completely rehydrate itself. The nestling doves remain in water balance because they are fed watery pigeon milk by their parents. There are several species of small birds living in the Great Basin Desert who, if not under heat stress, can remain in water balance without drinking water despite their granivorous food habits. The Black-throated Sparrow and Brewer's Sparrow can then get by on metabolic water and the limited amount of water of succulence found in seeds. They remain in water balance by minimizing their cloacal water losses through efficient reabsorption of water from urinary and fecal material and by minimizing their evaporative water losses, probably in part by cooling exhaled air and reclaiming water from it in their nasal passages.

Surface Water

The obvious way to make up water losses and remain in water balance is to drink water. But freshwater sources are often few and far between in deserts, since aridity is what a desert is all about. In the Great Basin Desert, there are only a few freshwater lakes and a few rivers of any length. Many of the streams coming out of the mountain ranges do not carry water very far out onto the desert floor, and these streams often dry up along their lower reaches during the summer. There are relatively few freshwater springs and seepage areas. Ranching has had some impact on the water supply since impoundments have been dug, and fresh water is pumped out of the ground and put in stock tanks, flows in irrigation ditches, and comes out of overhead irrigation sprinklers, providing drinking and bathing water for birds.

Some birds, like the Mourning Dove, live away from water but regularly visit water holes to drink during the summer. But many of the birds inhabiting the intermontane valleys which constitute the Great Basin Desert are found only in close proximity to freshwater sources during the summer. Such a bird is the House Finch. A permanent resident in the valleys of the Basin, it is always found close to water, most commonly in areas where there is enough moisture to support trees or thickets. Since many birds found in the Great Basin Desert are restricted to the vicinity of water during the summer, it is difficult to envision them as true desert birds although they live in desert surroundings. Birds like the Brewer's Blackbird, Bullock's Oriole, House Wren, Song Sparrow, Lazuli Bunting, Wilson's Warbler, Barn Swallow, and Killdeer just exploit the moist interruptions of the Great Basin Desert, not the dry areas that make this a desert.

There are many sources of brackish water in the Great Basin Desert, and one wonders whether birds can replenish their water supplies by using some of this brackish water as drinking water. The problem in drinking salt water lies in getting rid of the salt, which is absorbed along with the water in the intestine, and remaining in salt balance. In order to gain water while preventing an overload of salt, the kidneys must be able to excrete the extra salt in the urine at concentrations greater than those at which the salt arrived in the drinking water.

The avian kidney is not as well designed as the mammalian kidney for concentrating electrolytes such as salt in the urine. Salt is concentrated by the loop of Henle in the kidney tubule, but many of the tubules in the avian kidney lack a loop of Henle. The best avian kidney can concentrate electrolytes only up to several times their concentration in the body fluids of birds.

The ability of birds to drink salt water has been explored experimentally in the laboratory. The birds are put under water stress and given various concentrations of sodium chloride to drink. The highest concentration of salt water at which they can maintain their body weight is then determined. Some of the species tested are found in the Great Basin: Mourning Doves and White-crowned Sparrows can successfully handle water about 25 percent as salty as seawater; House Finches, California Quail, and Sage Sparrows can process water 40 percent as salty as seawater; and Black-throated Sparrows can process water about 75 percent as salty as seawater. But it remains to be seen whether birds in nature drink brackish water, which frequently contains other chemical substances in addition to sodium chloride.

Members of about thirteen orders of birds have salt glands. These glands, derived from the nasal glands, are located above each eye. Salt glands, which are most highly developed in marine species of birds, allow them to drink seawater by highly concentrating the absorbed salt and secreting it with little loss of water. The ducts coming from the salt glands open into the nasal cavity, and the highly concentrated salt solution flows out through the external nostrils. Functional salt glands have been found in certain grebes, penguins, cormorants, pelicans, ducks, geese, herons, hawks, eagles, falcons, rails, coots, plovers, gulls, terns, and others. They could be of value to birds living away from water in desert regions, but only a few species—the Ostrich and Greater Roadrunner, for example—possess them.

4

The Diversity of Birds

Birds come in various kinds, and a considerable diversity of bird life occurs in the Great Basin. You may not be familiar with many of these birds. If so, as you read the natural history accounts which follow, or when you are birding, you should have a good field guide at hand. The guides which are of value in identifying Great Basin birds include *A Field Guide to Western Birds*, by Roger Tory Peterson; *Birds of North America*, by Chandler S. Robbins, Bertel Bruun, Herbert S. Zim, and Arthur Singer; *Field Guide to the Birds of North America*, edited by Shirley L. Scott for the National Geographic Society; and *The Audubon Society Field Guide to North American Birds: Western Region*, by Miklos D. F. Udvardy. The illustrations in the first three guides are original art; those in the fourth are photographs.

The living and extinct birds of the world constitute a group of animals of the Class Aves. Aves is the plural of the Latin word *avis*, meaning bird. In using the word *avis* to designate a bird, the Romans were alluding to the common observation that when a bird moves about it usually flies; and hence it travels without (*a*) a path or way (*via*), through the pathless skies. We often use the term avian when referring to birds and the term avifauna when referring collectively to all the birds of a given region—such as the Great Basin avifauna.

There are approximately 8,700 recognized species of birds in the world today—it is impossible to give a precise number. Among ornithologists the "splitters" recognize some populations of birds as being full species, whereas the "lumpers" recognize them as merely subspecies. In recent years the lumpers have been in ascendancy. There is some controversy today between evolutionists and certain religious fundamentalists over whether the species had their origin via organic evolution or came directly from the hands of a divine creator. As far as North American birds are concerned, a third force is also at work—the Committee on Classification and Nomenclature of the American Ornithologists' Union.

In the 1970s, the committee made a number of changes in classification and nomenclature. Some "enlarged" species were created which involved the Great Basin avifauna. Some of these generative acts which affected our avifauna are as follows: the creation of a Northern Oriole by lumping the Bullock's Oriole and Baltimore Oriole; of a Common Flicker by lumping the Yellow-shafted Flicker, Red-shafted Flicker, and Gilded Flicker; of a Dark-eyed Junco by lumping the Slate-colored Junco, White-winged Junco, Oregon Junco, and Guadalupe Junco; and of a Yellow-rumped Warbler by lumping the Audubon's Warbler and Myrtle Warbler. Then, pervasive changes in the classification and nomenclature of North American birds were announced in July 1982, in the thirty-fourth supplement to the AOU checklist—in advance of the new sixth edition of the checklist, which appeared in print in 1983. These changes will be noted in appropriate places throughout this book.

On the basis of patterns of structural and functional (biochemical and physiological) similarities shared by different species of birds, the Class Aves can be divided into a number of major subgroups called orders. Approximately twenty-seven different orders are recognized by ornithologists, the exact number depending on individual splitting or lumping propensities. However, close to 60 percent of all the species of birds in the world today are assigned to just one order—the Order Passeriformes. Of the twenty-seven orders, seventeen are represented by one or more species which regularly occur as residents, migrants, or visitants in the Great Basin. These seventeen orders are as follows:

1. Order Gaviiformes (loons)
2. Order Podicipediformes (grebes)
3. Order Pelecaniformes (pelicans and cormorants)
4. Order Ciconiiformes (herons, egrets, bitterns, and ibises)
5. Order Anseriformes (waterfowl = swans, geese, and ducks)
6. Order Falconiformes (vultures, Ospreys, kites, hawks, eagles, and falcons)
7. Order Galliformes (gallinaceous birds = grouse, quail, and pheasants)
8. Order Gruiformes (cranes, rails, and coots)
9. Order Charadriiformes (shorebirds = plovers, curlews, snipe, sandpipers, avocets, stilts, and phalaropes; gulls and terns; and auks, murres, and puffins)
10. Order Columbiformes (pigeons and doves)
11. Order Cuculiformes (cuckoos and roadrunners)
12. Order Strigiformes (owls)
13. Order Caprimulgiformes (poorwills and nighthawks)

14. Order Apodiformes (swifts and hummingbirds)
15. Order Coraciiformes (kingfishers)
16. Order Piciformes (woodpeckers)
17. Order Passeriformes (passerines or perching birds = flycatchers, larks, swallows, jays, ravens, magpies, chickadees, nuthatches, creepers, dippers, wrens, thrashers, thrushes, kinglets, waxwings, vireos, wood-warblers, blackbirds, tanagers, finches, and sparrows)

As you have probably noted, all order names have the latinized ending of *iformes*. An anglicized order name can readily be made by dropping the terminal *es* from the scientific name of the order and replacing it with *s*. The Order Gaviiformes then becomes gaviiforms, and the Order Passeriformes becomes passeriforms.

Based on patterns of shared structural and functional similarities, each order can be divided into smaller subgroups called families. Each family has a latinized scientific name ending in *idae*. For example, the jays, crows, ravens, nutcrackers, and magpies belong to the Family Corvidae. Family names can be anglicized by replacing the terminal *ae* with *s*, and the Family Corvidae then becomes corvids.

Sometimes subpatterns of shared structural and functional similarities allow a family to be divided into subfamilies. Each subfamily has a latinized scientific name ending in *inae*. For example, the Family Anatidae is separated into the Subfamily Anserinae (whistling-ducks, swans, and geese) and the Subfamily Anatinae (ducks). Subfamily names can be anglicized by replacing the terminal *ae* with *es*: Anatinae then becomes anatines.

Families of birds are always broken down into genera. Most living families of birds possess enough structural and functional diversity that they are divided into several to many genera. However, some families are so small and possess so little diversity that but a single genus can be recognized within each family. Such a family is the Family Gaviidae, the loons. There are only four living species of loons in the world today, and they are so closely alike in structure and function that all four are put in the same genus, *Gavia*.

The final assignment of a hierarchical rank to birds on the basis of subpatterns of structural and functional similarities is that of species. Some genera contain so little internal diversity that they are monotypic, and all the birds present in such a genus belong to one species. But subpatterns of diversity do usually exist within living genera of birds, and each such genus is broken into as many species as there are subpatterns.

The question often arises, what is a species? Classifiers of organisms, taxonomists or biosystematists, have never been able to come up with a clean, clear definition of what a species is. The reason they have been unable to do so is because the diversity of life arises from a process called organic evolution. Organic evolution does not proceed in steps of uniform magnitude, nor at the same tempo, for each evolving gene pool of organisms.

A species is essentially a population or a group of populations of organisms which possess recognizable, unique structural and/or functional characteristics and are reproductively isolated from all other closely related populations. To qualify as a full species, the population or populations of organisms must be reproductively isolated from all closely related populations; how much they differ from closely related populations is less important. Genetically fixed isolating mechanisms must be present to prevent interbreeding or hybridization from regularly occurring when closely related species come together in nature. In birds, these isolating mechanisms can be structural, ecological, or behavioral mechanisms which prevent the physical act of copulation; or they can be physiological mechanisms which result in reduced hybrid viability or in hybrid sterility if copulation does occur and offspring are produced.

The Committee on Classification and Nomenclature of the American Ornithologists' Union, in its recent operation of enlarging species, was acting on evidence that some formerly recognized species were successfully hybridizing when they came together in nature. The committee has even lumped or merged some genera and some families. The rationale for lumping or merging genera or higher categories has nothing to do with reproductive isolation. It is based on the magnitude of the structural and functional differences separating groups of species, on whether these differences are great enough to warrant recognition at the generic or higher levels.

Species of North American birds possess unique names. Each species has a common or English name and a scientific or binomial name. We are indeed fortunate that the American Ornithologists' Union is a strong, vital group and that its Committee on Classification and Nomenclature can successfully dictate what the scientific and common names are to be. Thus, we have a uniform system of names and can communicate without confusion. The value of having a uniform set of names and a single classification scheme greatly outweighs one's personal gripes about some of the committee's recent decisions.

The official common and scientific names of birds are listed in the *Check-list of North American Birds.*[24] The sixth edition, prepared by the Committee on Classification and Nomenclature and published in 1983, contains many major changes involving the classification of birds and their scientific and common names. It will require a major effort on the part of some of us older birders to become accustomed to many of the sixth edition's changes. It will require a major effort on the part of neophyte birders to learn the old as well as the new classification and names so that they can utilize the ornithological literature published prior to 1983.

Major changes in classification involved reducing some families to subfamilies, lumping the wood-warblers, tanagers, grosbeaks, sparrows, blackbirds, and orioles into a single family—Emberizidae—and dismembering the Family Fringillidae. The committee's objective with regard to common names has been to replace meaningless modifiers and adopt English names from "a global viewpoint." Nevertheless, the new checklist contains a plethora of meaningless modifiers—such as all those in the Common Goldeneye, Common Moorhen, Common Snipe, Common Tern, Common Barn-Owl, Common Poorwill, Common Grackle, Common Redpoll, and others. However, the committee had second thoughts about the Common Flicker it created in 1973 and has once again changed its English name—this time to Northern Flicker. Some of the changes in English names certainly have global implications, for example, changing the name European Wigeon to Eurasian Wigeon. But changing the English names of wide-ranging species to conform with names used in Great Britain constitutes more of an anglophilic viewpoint than a global one—such as changing the name Common Gallinule to Common Moorhen and the name Marsh Hawk to Northern Harrier. There are also numerous changes in the hyphenation of compound English names. The Barn Owl now becomes the Common Barn-Owl, the Poor-will becomes the Common Poorwill, and so on. The committee continues to use the possessive case in situations where bird names commemorate human names—such as Clark's Nutcracker, Lincoln's Sparrow, and so on.

The ornithological literature published prior to 1983 constitutes a priceless heritage and an irreplaceable resource for students of natural history, but the recent changes in classification and nomenclature cannot be ignored. In this book I have incorporated the new changes, but, in each natural history account, where change has occurred I have explained what has happened. Thus, you will be able to more readily utilize pre-1983 literature and guides.

The scientific name of a species consists of two latinized words or latinized compound words. The first, the name of the genus to which the species belongs, is always capitalized. The second word identifies the species within the genus and is never capitalized—regardless of whether it is derived from a proper noun or not. For example, Lewis' Woodpecker, named after the Lewis of Lewis and Clark fame, is *Melanerpes lewis*. To be complete, the scientific name of a species should also include the name of the taxonomist who first named and described that species. For example, the complete name of Lewis' Woodpecker is *Melanerpes lewis* (Gray). The original author's name is enclosed in parentheses only if a change in genus name has occurred since the original naming. In the above case, Gray originally named the Lewis' Woodpecker as *Picus lewis* Gray. For brevity we seldom include the author's name as part of a bird's scientific name. Subspecies are designated by a trinomial instead of a binomial name. The Yellow-rumped Warbler's species name is *Dendroica coronata*, and the subspecies of the Yellow-rumped Warbler which we call the Audubon's Warbler has the scientific name of *Dendroica coronata auduboni*. The AOU Committee on Classification and Nomenclature did not list subspecies in the sixth edition of the checklist, since it did not have time to consider fully the more than two thousand species in the geographic area covered. To fill this void, it recommended the continued use of the fifth edition plus the thirty-second and thirty-third supplements to the checklist for subspecies taxonomy.

When storing bird specimens in museum cabinets or when putting together a bird guide, we usually organize the species in a phylogenetic scheme or progression. In this chapter my list of the seventeen orders of birds found in the Great Basin is organized along phylogenetic lines. The listing progresses from what is considered to be the most primitive order of the seventeen (Gaviiformes) to the least primitive order (Passeriformes). The closer together any two orders occur in the listing, the more closely akin they are considered to be, since phylogenetic listings are supposedly based on evolutionary relationships.

Providing there is a general acceptance of what the actual evolutionary relationships are between birds, a phylogenetically organized system in museum cabinets or bird guides greatly facilitates the retrieval of stored specimens or information. You know exactly where to go to find what you are looking for. The Order Gaviiformes will not only be number 1 on my list, but specimens of loons will be described and pictured on the initial pages of any bird guide, and specimens of loons will be stored at the beginning of museum collections.

Once again, meaningful uniformity can serve a purpose. But in the natural history accounts which follow this chapter, I shall take a few liberties with the AOU-approved phylogenetic scheme of organization for North American birds. This is not done to satisfy any stray whim on my part: it makes for a more logical and integrated development of topics. The quick retrieval of specimens or bits of information is of less consequence here. Greater insight can be gained by rearranging birds according to their general ways of life, if possible, and better relating features of ecology and behavior.

The general scheme of organization which I shall follow is given below.

To clearly show how I have departed from phylogenetic order, I have enclosed the phylogenetic sequence number for each order in parentheses after the order name. You will note that my first liberty is to separate the orders of birds which in general frequent water or other wet environments from those which frequent land environments. Among the water birds I have taken the further liberty of grouping gaviiforms and podicipediforms as divers, pelecaniforms and coraciiforms as specialized fishermen, and ciconiiforms and gruiforms as marshbirds. Among the land birds I have taken the liberty of putting the two great orders of raptors, the falconiforms and strigiforms, next to each other; of grouping the galliforms and columbiforms, the upland game birds; and of combining the caprimulgiforms and apodiforms, orders with the unique ability to practice cold torpidity and hibernation.

WATER BIRDS

Divers:	Gaviiformes (1) and Podicipediformes (2)
Specialized fishermen:	Pelecaniformes (3) and Coraciiformes (15)
Marshbirds:	Ciconiiformes (4) and Gruiformes (8)
Waterfowl:	Anseriformes (5)
Shoreline birds:	Charadriiformes (9)

LAND BIRDS

Diurnal birds of prey:	Falconiformes (6)
Nocturnal birds of prey:	Strigiformes (12)
Upland game birds:	Galliformes (7) and Columbiformes (10)
Cuckoos:	Cuculiformes (11)
Hibernators:	Caprimulgiformes (13) and Apodiformes (14)
Woodpeckers:	Piciformes (16)
Perching birds:	Passeriformes (17)

In many of the natural history accounts in this book, I refer to observations on Great Basin birds made by Robert Ridgway in the 1860s. In some instances Ridgway is quoted extensively. At the age of seventeen, he was appointed zoologist to an expedition exploring a railroad route along the fortieth parallel of latitude (between thirty-nine degrees and forty-two degrees north latitude). Clarence King was the geologist-in-charge of this United States Geological Exploration of the Fortieth Parallel. On about July 12, 1867, the survey team entered Nevada from California. Fieldwork—

initiated from various camps in west central Nevada and from winter quarters in Carson City—continued until June 21, 1868. The survey team then worked its way across the Great Basin, arriving at Salt Lake City in early October of 1868. Fieldwork continued in the Salt Lake Valley area and at points eastward in the Wasatch Mountains until August 16, 1869.

Thus, during a period of twenty-five months, Robert Ridgway collected and observed birds in the Great Basin. His fieldwork was extensive enough that considerable insight on the past status of birds in the Basin can be gained from his observations, which appeared in print in 1877 as *Part 3: Ornithology* in volume 4 of the official report on the geological exploration of the fortieth parallel.[25] Since this government publication is difficult to come by, I have quoted it at length in the natural history accounts.

Upon returning to Washington from the Great Basin, Ridgway became a staff member of the Smithsonian Institution. He went on to become one of the leading ornithologists of his time. During the first two decades of the twentieth century, eight volumes of his great taxonomic work, *The Birds of North and Middle America*, were published. After Ridgway's death, Herbert Friedmann continued the series by writing three more volumes.

5

Divers

THERE ARE two orders of birds in the Great Basin, highly adapted for underwater diving and swimming, who mainly make their living underwater. These are the Order Gaviiformes, the loons, and the Order Podicipediformes, the grebes. The legs of loons and grebes are located so far back on their body that they are virtually helpless on land and are seldom seen there. The webbed feet of the loons and the lobate-webbed toes of the grebes aid them in diving and swimming.

FAMILY GAVIIDAE

There is but one family of loons in the Order Gaviiformes—the Family Gaviidae. Gaviidae is certainly not an important Great Basin family. The desert lakes of the Basin do not afford suitable breeding habitat for loons. The mountain lakes are either too tiny, too disturbed by humans, or otherwise inadequate to meet the nesting and feeding requirements of loons. In earlier days the Common Loon nested at Eagle Lake, California, on the western edge of the Basin. It also nested at several small mountain lakes to the west of Eagle Lake, just outside the Basin.[26]

Today the Common Loon is found only as a visitant at Great Basin lakes during spring and fall migrations or as a straggler in winter or summer. Its breeding range lies mainly in Alaska and Canada and along our northern border states. There are three other species of North American loons, all of which have been occasionally recorded from the Basin.

Common Loon
Gavia immer

The Common Loon may not be as common in the eastern end of the Great Basin as in the western end. Utah authorities report it as an "uncommon winter visitant and transient throughout the state, occasionally occurring even in the summer but as a nonbreeder."[27] In contrast, along the western side of the Basin it is a common transient and, during spring and fall migrations, can often be found in numbers on Pyramid Lake, Lake Tahoe, Topaz Lake, and Walker Lake. Even Virginia Lake, located within the city of Reno, is regularly visited by migratory loons.

It is interesting that the great waterfowl refuges in the Basin—Malheur, Stillwater, Ruby Lake, and Bear River—all list the Common Loon as a rare, or even occasional, transient on their official checklists. Yet Pyramid and Walker lakes, both desert lakes, are quite attractive to migratory loons. Of course, they are both deep-water lakes, compared to the shallow bodies of marsh water at the refuges named.

A loon is certainly not designed for land life. Its legs cannot hold it off the ground, since they are positioned at the end of, instead of underneath, the body. On land a loon moves by pushing itself along on its belly or by employing its wings as crutches. In recognition of their helplessness on land, loons build their nests immediately adjacent to water.

In water the strong rear-action legs and large webbed feet power the loon in its swift dives and rapid pursuit of swimming fish. The Common Loon has been known to reach depths of two-hundred feet while diving. The next time you are loon watching, note the dives. In plunging from the surface of the water, the loon enters its dive by lowering its head as it pushes its body forward. On occasion, a loon will gradually sink down in the water by decreasing its buoyancy. This it does by expelling air from its breathing system.

Birds have the most complex breathing system of all animals. Air is circulated not only to the lungs but also to air sacs which lie beyond the lungs, among the body organs and inside some of the long bones. Common Loons are not as pneumatic as most birds; anatomical studies have shown that none of the long bones is penetrated by air sacs. Of course, if you are making your living as a diver you do not want to be too inflated. If its buoyancy regulation goes haywire it is difficult, even impossible, for the loon to dive. Years ago, such a loon was encountered on Pyramid Lake on one of our field trips. Despite being chased by a boat and repeatedly attempting to dive, the loon was just too buoyant to successfully dive.

Another type of behavior a loon watcher in the Great Basin will witness is the so-called rolling preen. Since the loon is seldom ashore, it must perform its body care functions afloat. In preening, it first rolls over on its side—maintaining one leg up in the air, the other leg in the water. Paddling with the down foot, it slowly circles in the water, while preening its feathers with its bill.

Other Gaviids

All three remaining species of North American loons have been recorded from Great Basin sites. These species breed far north of the Basin in arctic regions and are only of accidental occurrence here. The Red-throated Loon, *Gavia stellata*, has been reported from Malheur and Bear River National Wildlife Refuges and from Walker Lake, Nevada. The Arctic Loon, *Gavia arctica*, has been reported from Churchill, Douglas, Elko, and Washoe counties in Nevada and from Beaver County in Utah. There is a Yellow-billed Loon, *Gavia adamsii*, record from Lake Tahoe.

Many bird-watchers live for the day when they can report seeing a bird of "accidental" or "occasional" occurrence. This is great sport, but over-eagerness to obtain a noteworthy record can prompt misidentifications. Loons can best be identified as to species when in their nuptial (alternate) plumages. When they are in winter (basic) plumages, species separation under field conditions is more difficult, since greater reliance must be placed on characteristics more difficult to see—such as the size, shape, and color of the bill. Whenever possible, close-up photographs should be taken for documentation.

FAMILY PODICIPEDIDAE

The grebe family is a small but important Great Basin family. Although divers and swimmers like the loons, grebes find the desert lakes, ponds, and marshes of the Basin more to their liking than do the loons. Five of the six North American species of grebes have been found in the Basin.

Pied-billed Grebe
Podilymbus podiceps

The Pied-billed Grebe is widely distributed over the three Americas—in North, Central, and South America. It is a common summer resident on suitable streams and ponds in the Great Basin. Pied-bills can be found the year around over most of the Great Basin, but in greatly reduced numbers during the winter in the northern, central, and eastern parts. In the Reno area, in the western part of the Basin, it appears to be as abundant in the winter as it is in the summer, if not more so.

The Pied-billed Grebe is not highly gregarious, as are our other grebes, and in the field you will seldom encounter more than one or several at a time. I have yet to see a tight flock of these grebes in the Great Basin. Occasionally, tight flocks of migratory Pied-billed Grebes have been seen elsewhere, and there have been reports of up to twenty thousand present on the Salton Sea in California in November. The Pied-bill is a solitary nester. Usually a small pond will have only one nesting pair on it. On more extensive breeding habitat, territorial behavior will keep the pairs spread out.

Courtship displays of Pied-billed Grebes are mutual displays, but they are much less rigidly fixed or stereotyped than some of those of the more social Western and Eared grebes. Wetmore witnessed pursuits or chases: one member of a pair of grebes, presumably the male, would dive after the other member; upon surfacing again, he would chase the female, with flapping wings, over the surface for up to one hundred feet or so. Another display witnessed by Wetmore was the circle display. During this the members of a pair would closely approach each other in the water but would stop short and swing around until they were tail to tail. Both would then turn their heads from side to side for a second or two, before swinging around to face each other again. This display would be repeated a number of times, as if the birds were on pivots. Sometimes their movements were not synchronized with each other.[28]

Most of the courtship movements of the Pied-billed Grebe are less structured than the circle display. Grebes, apparently males, will fidget with nesting material—carrying it in their bills and sometimes diving underwater with it. At times, two grebes have been seen billing, approaching each other and touching bills for a few seconds. Billing may be used as a nest-relief ceremony by Pied-billed Grebes, since the male and female touch bills as they pass each other when exchanging places at the nest.

The nest of the Pied-bill is usually a floating structure anchored to emergent vegetation, but in shallow water it may be anchored to the bottom. Both members of the pair work at nest building, and they may work on several structures before completing one for nesting. Pied-billed Grebes vigorously defend their territories from trespass by other Pied-billed Grebes, Eared Grebes, American Coots, and Ruddy Ducks. Usually the male will defend the territory, but he is sometimes aided by the female. Making underwater attacks, the territorial grebe will peck at the feet of the intruder being chased.

Eared Grebe
Podiceps nigricollis

The Eared Grebe inhabits western North America and parts of Eurasia and Africa; it is a breeding bird at many of the lakes and impoundments in the Great Basin. Eared Grebes can be found the year around in the western part of the Basin, where, in the valleys, the winters are relatively mild. The larger lakes such as Pyramid and Walker never freeze over, and the smaller lakes are closed only for a few days at the most during the coldest weather. Eared Grebes are reported to be rare during the winter in the colder eastern part of the Great Basin.[27]

Large numbers of Eared Grebes pass through the Great Basin during spring and fall migrations. Then, rafts of hundreds or even thousands of grebes can be seen on lakes and impoundments. The largest rafts of these grebes in the Basin form on Great Salt Lake and on Mono Lake. At times these concentrations contain hundreds of thousands of grebes. The causeway leading to Antelope Island in Great Salt Lake affords good vantage points. I have seen spring and fall congregations of ten thousand to twenty thousand grebes from there.

Eared Grebe watching is most rewarding in the spring when courtship is under way. Some early glimpses of courtship may even be had during migration. The courtship displays of grebes are of a type called mutual displays. During these, the male and female each perform the same movements and make the same vocalizations. Mutual displays are often associated with sexual monomorphism—where the male and female look alike, as in grebes. The following account of courtship in Eared Grebes is based on a study carried out by McAllister.[29] It should give you some appreciation of how intricate and stereotyped bird displays may be.

When seeking a mate, an unpaired grebe engages in advertising. It swims up to other grebes, with its body feathers fluffed out, crest and neck feathers raised, neck erect, bill straight forward and closed, and calls *poo-eee-chk*.

Habit preening is the most common display. Grebes float along in pairs, preening quickly and vigorously. The two grebes simultaneously preen identical feather areas on their bodies.

Head shaking occurs by itself and as part of the penguin dance. A pair of grebes swim rapidly along, high in the water, one grebe ahead and slightly to one side of the other. The body feathers and crest are raised and the neck stretched high, as each turns its head from side to side six to twelve times. Then they usually dive underwater.

During the penguin dance a pair of grebes will stand upright in the water, while facing each other close together and treading water. While upright they utter a shrill chittering call and vigorously shake or turn their heads from side to side. Dropping down into a swimming position, they slowly head shake eight to twenty times while facing each other. Rarely, they will turn while standing upright and run a short distance on the water. Finally, this display may involve habit preening either before or during the head shakes.

The cat attitude is a rarer courtship display. Here, the head is held low with the neck and crest feathers elevated. The wings are raised off the back, with the elbows bent down and almost touching the water. The cat attitude is used by an advertising grebe when approached by another grebe. The approaching grebe dives and swims underwater toward the advertising grebe, who assumes the cat attitude, while the approaching grebe rises above the water into the penguin attitude. Then the two engage in the penguin dance.

Finally, there is a pivoting display. Here, two grebes starting from a bill-to-bill position pivot until they are tail to tail. Thus, they perform a number of half circles together and often engage in head shaking while tail to tail.

The Eared Grebe is a colonial nester. The colonies, often containing hundreds of nesting grebes, are located on lakes and impoundments with shallow margins of emergent and floating vegetation. The floating nests are built in water one to three feet deep, among emergent vegetation, and are separated from each other by only several feet of water. Eared Grebes nest at many locales within the Great Basin; they can readily be found at federal bird refuges and management areas—including Bear River, Ruby Lake, Stillwater, and Malheur.

The earliest visible sign of nesting involves a display called soliciting. While soliciting, a grebe of either sex will assume a low position in the water, with head and bill extending straight ahead on the surface. An eerie, trill-like call accompanies this behavior. Eventually, grebes will solicit from their nest platforms. Nest-building activity slowly awakens within the colony—gradually involving more birds and more of their time. The colony will often build nest platforms and then abandon the site and the platforms. Every few days, for as many as six times, a colony has been known to abandon nest platforms and move on to another patch of emergent vegetation to start building all over again. Some egg laying is under way during this time, and the eggs are abandoned along with the nest platforms.

The female apparently does all the actual work of nest building, although she is accompanied by her mate while working. The final nest is built on a platform or foundation which may be formed by bending over several stalks of emergent vegetation, such as reeds. Debris, particularly algae, is brought up and piled on the platform to make a nest. The nest cup or bowl is a very shallow depression. Often the depression lies below the water level, so that the eggs sit in some water. At best, the floating nest of the Eared Grebe is but a very flimsy structure, and the entire structure may tilt conspicuously as the grebe comes aboard.

The normal clutch of an Eared Grebe numbers about three or four eggs. The parents take turns incubating the eggs. Since they commence in-

Eared Grebe

cubating with the first egg laid, the young hatch asynchronously; the incubation period for an individual egg is about three weeks. Except in sudden emergencies, the incubating parent always covers the eggs with wet vegetation before leaving the nest. The practice of covering unguarded eggs with wet vegetation is an old family trait in grebes—as is the practice of building floating nests.

Grebe chicks are precocial. But, although they can soon swim and dive, they are often carried around on the back of a swimming adult. This, too, is a family trait.

As we have noted, migratory flocks of Eared Grebes are often huge—numbering way up in the thousands. Such a movement was in progress during the night of December 12, 1928. By 2 A.M. on the morning of December 13, a vast number of Eared Grebes were passing over the south central edge of the Great Basin. At this time an extremely heavy snowfall was in progress, and the good burghers of the sleepy little town of Caliente, Nevada, were awakened by heavy thumpings on their roofs. December 13 was a little early for Santa and his reindeer!

Well, it turned out to be snowing Eared Grebes.[30] When daylight arrived, thousands of grebes were found on the ground and on the flat roofs of buildings in Caliente. Thousands of grebes who had been buried by the heavy snow were working their way out of it. The main impact had been centered on Caliente, but grounded grebes were scattered out for twenty miles or so in all directions. Many had been crippled in landing and were dead. Thousands died. According to E. C. D. Marriage, editor of the *Caliente Herald*, thousands of grebes were saved by being put in the creek. Many thousands remained in the streams and ponds in the Caliente area and in nearby Meadow Valley Wash.

The Eared Grebes had been migrating southward on a broad front that night, since reports from Enterprise, Utah, about forty-five miles east of Caliente, related that several hundred grebes had dropped out of the sky onto the streets there the same night. So the migratory front of the grebes was at least sixty-five miles wide in an east-west direction. Further, several hundred more grebes dropped into the snow at Uvada and Modena, Utah, about ten to fifteen miles further north than Enterprise, that same night.[30]

Later, it was surmised that the grebes had been forced to the ground by the very dense snowfall. Of course, this doesn't explain why so many of them landed precisely in four towns in a very sparsely settled region. It is difficult for me to visualize a snowfall too dense for the birds to fly through being solely responsible for grounding so many thousands of grebes in a little town like Caliente. Rather, I think it more likely that the grebes were attracted by the town lights. According to editor Marriage, the railroad yard at Caliente was brightly lit that night.

Migratory movements require orientation. At night some birds employ celestial navigation; others may use ground landmarks. When they cannot see the stars or the ground on stormy or overcast nights, they become disoriented. In trying to descend and land, they are often attracted to ground lights. The Caliente grebes may well have had some difficulty flying through the dense snowfall; but, no doubt, they were attracted to Caliente in large numbers by the railroad yard lights. A similar happening of great magnitude occurred in Reno some years ago on an overcast spring night, and I shall discuss it in my narrative on the Red-necked Phalarope.

The legs of a grebe are placed far back on its body. This is ideal for diving and swimming but virtually useless for locomotion on land—thus grebes must take off and land on water. Water dissipates the momentum of landing, and it allows a grebe to taxi or skitter along rapidly enough to develop sufficient lift on the wings to take off. The amount of lift generated by a wing is proportional to the square of the speed of the air flowing past the wing. Hence, lift can be greatly increased by taxiing into the wind. The great danger of a grebe landing on the ground is that it may injure itself when it hits; grebes are not built for landing at low speeds. Even if a grebe escapes injury while landing on the ground, it is virtually impossible for it to move fast enough on land to get airborne again. However, there is a report in the literature of a Horned Grebe seen taking off from a hard beach by lumbering into a brisk wind. Sometimes grebes, deceived by the appearance of surfaces at night, land on highways and parking lots instead of water.

Western Grebe
Aechmophorus occidentalis

The Western Grebe is found only in western North America. In the Great Basin, it is a common colonial breeder on the larger lakes and impoundments. During migration scores to hundreds of widely scattered Western Grebes may be seen, feeding during the day, on a single body of water. Westerns are occasionally seen during mild winters in the eastern part of the Great Basin, but overwintering is more common in the western part.

The Western Grebe is a large, long-necked grebe. One of its vernacular names has been that of Swan Grebe. Its courtship displays have been closely observed and described by Nero during field studies in southern Saskatchewan. Some of its display movements are graceful and oddly beautiful, others are dashing and spectacular. In describing mutual presentation, Nero wrote:

> One bird would dive, emerge with a mass of water weeds in its beak and swim about holding this sodden load aloft as if with great pride. This often led to a mutual presentation of nest-materials (for of such is the nest usually made), both members of a pair holding a beakful of weeds while rapidly approaching each other, then silently, while facing each other, rising higher and higher out of the water; finally, in a strikingly impressive ceremony, they touch breasts and bring their beaks together, for several seconds resting against each other, while literally standing on top of the water. Then they gradually sink down and resume swimming or loafing together. This unusual behavior is apparently the "true courtship" of the western grebe, leading to actual nest-building, mating and egg-laying.[31]

There are other stimulating displays which involve paired birds. Frequently, the male will bring up a fish and either proffer it in his bill to his mate or lay it on the water for her; this is courtship feeding. When a pair swim together, the male may take the lead in the high-arch position, with tail up, head held high, and bill directed downward toward the water. Pairs may engage in mutual bob preening or habit preening—involving stereotyped, often intense movements that mimic actual feather preening. Starting from the high-arch position, they will reach back to pluck at back or wing feathers, and then both return to the high-arch position. We believe displays such as these—which are highly stereotyped, often exaggerated, often intense, rhythmically repeated, and mimic ordinary movements such as feather preening—have come about through an evolutionary process called ritualization. I shall have more to say about ritualization as my natural history accounts of Great Basin birds unfold.

The most spectacular display of the Western Grebe, that of the race, run, or water dance, is one familiar to much of the general public, since they have seen it in nature films in the theater and on television. The race is not considered to be a courtship display, since it occurs throughout the breeding season, is always initiated by ritualized aggressive displays, and often involves combinations of birds other than a paired male and female. Nero believes it is a social display functioning to release nervous energy.

The race is initiated by two or sometimes more than two grebes, swiftly swimming toward each other, alternating two types of aggressive displays: in mutual threat-pointing, with heads extended forward and bills pointing at each other, the birds utter a clicking or buzzing sound; in mutual dip-shaking, each bird rapidly dips its bill into the water, then raises its head and shakes its bill from side to side. Shortly before their beaks meet, each bird makes a right-angled turn; then off they go, side by side, in an upright fashion. While racing their wings are partly outspread, and their feet, pounding the surface of the water, produce a "roar like a motor." After racing fifty to one hundred feet, both suddenly dive underwater. Surfacing again, they converge on each other, treading water, each holding its head high and stiff, while progressively elevating themselves higher out of the water. On close approach, each begins head-turning in a stiff fashion, toward and away from each other, while uttering a whistle "like that of a boiling teakettle." Then, the display terminates as the birds move apart, or, if paired, they swim away together.[31]

The Western Grebe is a colonial nester; often hundreds or even thousands of birds nest in the colonies on our better lakes. The nest is usually a floating structure, either anchored to emergent vegetation such as tules or çattails or attached to the underwater roots of aquatic plants. Both sexes participate in nest building and in incubating the eggs. The newly hatched chicks are highly precocial and can swim shortly after leaving the shell. Chicks are often carried on the backs of their parents, between the wings or even under the wings. Sometimes they are out of sight, or just their heads will stick out above the feathers between the parent's wings.

Small fish comprise the principal item in the diet of the Western Grebe, with a lesser reliance on insects. The grebes dive and pursue fish underwater. Apparently, they spear the fish with their rapierlike bills, for recently captured fish found in the stomach of a Western all had a small hole passing completely through the center of their bodies.[32] Hence, the generic part of the scientific name of this grebe is a fitting one: *Aechmophorus* is derived from Greek words meaning spear bearing.

While on the subject of stomach contents, it is appropriate to consider another characteristic of the grebes: their stomachs almost always contain wads of feathers. An examination by Wetmore of nearly four hundred stomachs of various species of grebes revealed that masses of feathers were almost always present: one mass in the gizzard or muscular stomach proper, another mass in an outpocketing of the glandular stomach or ventriculus.[33] Grebes purposely swallow feathers, and feather eating in the Western begins early in life. In reference to the Western Grebe, Chapman observed, "I am at loss to understand why chicks not more than three days old should have their stomachs tightly stuffed with a ball of their parents feathers. In the stomach of one I found a compact wad of 238 feathers, and in another there were no less than 331. All were the smaller body feathers of the adult Grebe."[34] An interesting development in feather eating has taken place in the Western Grebe. It has become incorporated into the ceremony of courtship feeding; on occasions, a male Western will present feathers to his mate to be eaten.

We are still not absolutely sure of the function of feather eating in grebes. Since prey is swallowed whole, an obvious thought is that feather masses would help protect the stomach from puncture by large, sharp fishbones. Pellet casting by Horned and Eared grebes has been observed in the field, but these grebes feed more on insects and small invertebrates and less on fish. Further, Lawrence, following field studies at Clear Lake, California, maintained that "the Western Grebe does not regurgitate undigested bones, scales and chitinous parts. These sharp objects have been found in many states of erosion by digestive action" in the stomachs of grebes.[32] Hence, Wetmore may have been on the right track when he suggested that the function of the feather masses is to delay the passage of larger fishbones and scales through the stomach. The hydrochloric acid there would have longer to partially dissolve and soften the larger bones, before they continued down into the alkaline environment of the small intestine where such action would come to an end.

Grebes have long suffered under the heavy hand of human society. In western North America, plume hunters raided the larger breeding colonies, each year killing thousands upon thousands of Western and Eared grebes for the millinery and garment trades. During the early 1900s, many bitter propaganda and political battles were fought by the Audubon societies and their supporters against the millinery and garment industries and their lackeys—the plume hunters. This slaughter of birds provided reason and impetus for the creation of the National Audubon Society. A major victory for the grebes occurred in 1908, when the Audubon societies prevailed upon President Theodore Roosevelt to set aside the Klamath Lake Reservation and the Malheur Lake Reservation, protecting the grebes' once vast breeding grounds at these places.

According to Finley, the Western Grebe was "the greatest sufferer at the hands of the market hunter. This diver, of the glistening-white breast and the silvery-gray back was sought not without reason. The grebe hunters call the skin of this bird fur rather than feathers, because it is so tough it can be scraped and handled like a hide, and because of the thick warm plumage that seems much more like the fur of a mammal than the skin of a bird. These skins when prepared and placed on the market in the form of coats and capes, brought the prices of the most expensive furs."[35]

Only the breast of the grebe was used; that of a Western Grebe commanded anywhere from fifteen to fifty cents on the New York feather market. Chapman wrote about his encounter with an old plume hunter at Lower Klamath Lake in the summer of 1906 as follows:

> Living in a house-boat hidden somewhere in tules, this degenerate representative of the pioneer trapper seemed far from the world of millinery adornment, but no stockbroker kept his eye on the "tape" more keenly than he did on the quotation of the New York feather market, with which the dealers regularly supplied him, and the moment the figures promised a profit, he took to the field.
>
> It appeared that for several preceding seasons Grebes' breasts had brought only fifteen cents each, and at this price the birds were not worth killing. Hence their abundance during the visit of Finley and Bohlman. In the meantime, the demands of fashion had advanced the price to fifty cents per breast, a sum sufficient to tempt the hunter, and in a few weeks he had wiped out the increase of years.[34]

Over the years, grebes have suffered from the degradation and destruction of their breeding and feeding habitats due to human activities. Since World War II, in many parts of the world, birds have been affected by chlorinated hydrocarbon contamination of their food webs. A striking example of this involved the Western Grebes at Clear Lake, California. Here, from 1949 to 1957, the lake was sprayed three times with a very, very low level of DDD, a close relative of DDT. DDT and its metabolites, such as DDD and DDE, are fat-soluble. Once they enter a plant or animal body, they are retained in its fats and oils and are scarcely metabolized or excreted.

When microscopic plants are exposed to chlorinated hydrocarbons, they continue to concentrate them in their fats and oils. The microscopic plants are fed on by microscopic animals, who during their lifetimes remagnify the chlorinated hydrocarbon levels in their food. Small animals are fed upon by larger animals, who in turn remagnify still further the previously magnified levels. Finally, when the Western Grebe comes along, high up in the food webs, to feed on medium-size fish, it is feeding on highly contaminated food. In fact, some of the body fat of dead Westerns collected at Clear Lake contained eighty thousand times the total level of DDD sprayed there. As a consequence of this contamination, the Western Grebe population at Clear Lake dropped from over one thousand pairs to twenty-five pairs by 1959—with no production of young taking place. DDT and its close relatives interfere with reproduction primarily by promoting eggshell thinning, and thus the breakage of eggs soon after laying, or by causing the death of the embryo before hatching.[36] In recent years our federal government has greatly restricted and almost eliminated the use of chlorinated hydrocarbons as pesticides. So our environments are slowly recovering, and birds such as the Western Grebes at Clear Lake are coming back.

Like other divers, Western Grebes must take off from water. When they alight on the ground, they cannot move fast enough or taxi to get airborne again. Nor can they take off from snow or ice. Robert Ridgway, the zoologist on the United States Geological Exploration of the Fortieth Parallel, collected a Western Grebe in the Great Basin in 1868 of which he wrote, "The specimen in the collection was found 'snow-bound' in the sage-brush near Carson City, being discovered by its tracks in the deep snow, where it had scrambled along for a hundred yards or more. It was headed toward the Carson River, and had evidently come from Washoe Lake, almost five miles distant, and becoming exhausted by the long flight had fallen to the ground."[25]

Other Podicipedids

There are several sight records of the Red-necked Grebe, *Podiceps grisegena*, from the east side of Great Salt Lake and from Utah Lake; a prehistoric specimen was found in Lovelock Cave in northern Nevada; and a specimen was collected at Summer Lake, Oregon.

The Horned Grebe, *Podiceps auritus*, is a regular but rare migrant and winter visitant in the Great Basin. It is listed as an occasional breeding bird on the official checklist from Malheur National Wildlife Refuge. Evidently, the Horned Grebe formerly nested in northern Utah.[24]

6

Specialized Fishermen: Pouch Users and Plungers

AN ORDER of pouched fishermen occurs in the Great Basin. This is the Order Pelecaniformes. Two families are represented by species here—Pelecanidae, containing the pelicans, and Phalacrocoracidae, containing the cormorants. Pelecaniforms are water birds who mainly live on fish. They all possess a gular or throat pouch, which reaches its maximum size in pelicans. They all possess totipalmate feet; all four toes on each foot are directed forward and connected by a web.

Coraciiformes, another order of specialized fishermen, is present in the Great Basin. Worldwide, this order possesses some seven to ten families, but only a single species of kingfisher from the Family Alcedinidae occurs here. Although I have characterized kingfishers as plunging fishermen, it is true that a number of orders of birds contain species which plunge from the air after fish—including terns in the Order Charadriiformes and the Osprey in the Order Falconiformes.

FAMILY PELECANIDAE

The pelicans of the world are divided into six to nine species, depending on whether the avian taxonomist classifying them is a lumper or a splitter. The Brown Pelican is restricted to coastal waters in North America, Central America, and South America. All the remaining species of pelicans are whitish in color, have black wing tips, are inland birds, and are found in various regions of the world. Although their breeding habitat is inland lakes, some of these white pelicans may overwinter in coastal waters.

There are two species of pelicans in North America—the American White Pelican and the Brown Pelican. In the Great Basin the American

White Pelican is a regular breeding species and migrant, and the Brown Pelican is of accidental or occasional occurrence.

American White Pelican
Pelecanus erythrorhynchos

The American White Pelican is a breeding species at a few localities in western and north central United States and in Canada, and it sporadically nests on the Gulf Coast. Over the rest of North America, it may appear as a migrant, winter visitant, or straggler.

In the western end of the Great Basin, the American White Pelican is a breeding species at Pyramid Lake in Nevada, Honey Lake in California, and Pelican Lake in Oregon. At times in the past, the breeding colony on Anaho Island in Pyramid Lake may well have been the largest in North America. In the eastern end of the Great Basin, the American White Pelican is a breeding species on Gunnison Island in Great Salt Lake.[27]

During the nesting season, parties of white pelicans from Anaho Island and Gunnison Island forage far afield for food; the birds have been seen flying to and from bodies of water 100 or so airline miles away. From Pyramid Lake fishing parties visit places on Washoe Lake, the Humboldt River Sink, and Lahontan Reservoir approximately 50 miles away; the Stillwater marshes 60 miles away; and Walker Lake 100 miles away. From Gunnison Island fishing parties visit the Bear River National Wildlife Refuge some 35 miles away, and "undoubtedly, some pelicans fly as far as 90 miles to Utah Lake, a former nesting site, to obtain food."[37] Thus, some white pelicans make round trips of between 100 to 200 airline miles from their colonies to fish. Over the rest of the Basin, American White Pelicans can be seen during migration or during the postbreeding dispersal movements which carry pelicans, particularly juvenile ones, in various directions from the breeding colony.[38]

The American White Pelican has had serious difficulties contending with the spread of civilization over North America. Many breeding colony sites have been abandoned by the pelicans, and at other sites breeding does not occur every year. Each year large numbers of nonbreeding adults are seen at and away from the breeding colonies—sometimes far away.

To prosper, the American White Pelican needs inaccessible islands on which to nest and productive, shallow-water fishing grounds. It is a colonial nester and a ground nester. On islands pelican nests are safe from coyotes and other marauding ground predators. Further, inaccessible islands or protected island refuges are needed to shield the pelicans from disastrous disturbances by humans.

Shallow-water fishing grounds provide most of the food required by

American White Pelicans

the hundreds or thousands of adult and young pelicans living on a breeding island. Since the American White Pelican captures its fish from the surface by thrusting its head and neck underwater, its prey consists of fish that frequent shallow water or come within about a yard of the surface in deeper water. As we shall see, when obliged to fish in deeper water, white pelicans may attempt cooperatively to drive fish into shallow water where they can reach them. Pelicans from the Anaho colony feed principally on tui chub and carp; the tui chub is a deep-water fish found at Pyramid Lake, and the carp occurs in shallow water away from the lake. As a consequence, pelicans concentrate on fishing for carp away from Pyramid Lake, except in June when the tui chub enter shallow water to spawn.[39]

During historic times, both nature and civilization have delivered hard blows to the American White Pelican. Reclamation and irrigation projects as well as periodic droughts and desiccation have driven the pelicans away from some of their nesting sites. Nesting islands have become peninsulas connected to the mainland. Lakes have been converted to dry beds or farmlands. Fish have disappeared along with the waters. Humans have destroyed many birds and their eggs and young. This happened on Hat Island in Great Salt Lake. According to Behle, during the nesting season of 1918 "representatives of the state department of fish and game journeyed to Hat Island and, according to Charles G. Plummer who visited the island less than a week later, they shot and clubbed to death literally hundreds of young and adult pelicans and nearly the entire population of herons."[40] Human fishermen often begrudge these birds their daily fish. Even when intruders are not bent on mischief when they put ashore on breeding islands, the pelicans suffer greatly. They flush wildly from their nests with the near approach of a person or boat, and the eggs and young nestlings are devoured by the ever present gulls or overheat when exposed to the sun.

Thousands of years ago, the Great Basin was probably a haven for the American White Pelican. When the two great Pleistocene lakes, Bonneville and Lahontan, were in place, innumerable islands and fishing waters were present for pelicans. Many of today's hills and foothills were then islands. Desiccation followed the Pleistocene, and as the waters receded still other islands appeared. If you patrol the old beachlines on these arid hills of today, you may discover traces of ancient pelican nestings. Laura Mills took E. Raymond Hall to such an ancient nesting site on Rattlesnake Hill at Fallon, Nevada. Hall reported on this nesting ground as follows:

> With the aid of Messrs. J. R. Alcorn and Vernon Mills, and their wives, we recovered about a gallon measure of bones. Roughly three out of four were those of the White Pelican (*Pelecanus erythrorhynchos*), more

> than half of them from immature birds. The remaining bones were mostly of adult Double-crested Cormorants (*Phalacrocorax auritus*), although one bone of a Canada Goose (*Branta canadensis canadensis*) was found, and there were many bones of fishes of several sizes and of more than one species. The bones are fragile, and some seem to be partly fossilized; minerals from the enveloping sediments appear to have replaced some of the original substance of the bone. About a third of the bones are more or less water worn. The egg shells are in fragments whose curvature and texture suggests that they were laid by pelicans and not cormorants.[41]

Although we do not know much about those long-ago days and how many nesting colonies were among the waters of Lake Lahontan and Lake Bonneville, I do know that Rattlesnake Hill was not alone, for on a spring day in the 1960s the late Ira La Rivers and I found another ancient nesting colony. We had four-wheel-drive problems that day on wet, alkaline flats, but we did find an ancient pelican nesting colony on an old beachline on a ridge outside of Hazen, Nevada, about twenty airline miles or so from the Rattlesnake Hill colony.

During spring migration, white pelicans begin to arrive at Pyramid Lake and Great Salt Lake from mid to late March on. Knopf studied the courtship behavior of this species at Gunnison Island in Great Salt Lake. Unmated pelicans form flocks and soar over the island. Sometimes a few pelicans drop away from the soaring flock on partially folded wings and pull out of their stoops with a whooshing sound when about fifty to twenty-five yards above the ground. Off and on, pelicans break away from a soaring flock and join the flocks on the ground below. When a new arrival approaches a flock on the ground, bill-jabbing exchanges occur between the new arrival and the flock members. A female responds to a jabbing attack by bringing her bill in against her breast in a bow. A male may court a bowing female by moving his head back and forth over her head, displaying the sides of his expanded gular pouch while grunting. Sometimes a female moves away, followed by a male, in a strutting walk. Other males may join a strutting walk, forming a courting party; the males within a courting party often exchange jabs. Sometimes pursuit flights follow strutting walks, and unmated males join the courting male in pursuit of the female.[42]

On an island such as Anaho or Gunnison, the pelicans form a number of nesting colonies. The number may vary from year to year, and the localities of some of the colonies may change. The flocks which form the various breeding colonies arrive in the spring at different times. Therefore, although the individuals within a single colony will be synchronous in their breeding, some colonies will be egg laying and getting breeding under way sooner than others.

The nests within a pelican colony are built close together, but still each is out of reach of pelicans sitting in adjacent nests. Nests vary architecturally from slight depressions on the ground to a mound of dirt and debris. The usual clutch size is two eggs. White pelican eggs are peculiar in possessing a rough shell—roughened by calcareous deposits which crack or flake off irregularly. The eggs are frequently stained. Ridgway wrote after surveying pelican eggs at Anaho Island, "These eggs were, with scarcely an exception, conspicuously blood-stained, caused in part by their large size, but chiefly by the roughness of their calcareous coating; the haemorrhage being in some instances so copious that half the surface was discolored."[25] The parents take turns incubating the eggs. Head-up and bow nest-relief displays occur when the parents exchange places at the nest. "The incubation period of the White Pelican at Anaho Island was found to be approximately one month, which seems to be about average for other localities. During the incubation period at least one parent is always in attendance, either squatting on the eggs or standing over and shading them."[38]

Hatching is asynchronous, with the first egg hatching one or a few days before the second one. The young are naked and relatively helpless at hatching. The parents feed the nestlings regurgitated fish from the gular pouches; by positioning its open bill sideways, a nestling can reach in and help itself. The nestlings must be brooded to protect them from the cold and shaded to protect them from the sun. As the nestlings become older and more mobile, they can actively seek the shade cast by the body of a parent or by a shrub or boulder. When under heat stress the nestlings, as well as adult pelicans, increase their evaporatory heat losses by panting and fluttering their gular pouches. The well-vascularized gular pouch affords a very large surface area from which to promote evaporation.

Sibling aggression is usually intense in the American White Pelican. The older and larger sibling pecks and bullies its younger nest mate and grabs more than its rightful share of the food. The battered and abused younger nestling may eventually succumb to injuries or starvation. As the young pelicans become bigger and stronger, feeding becomes rougher on both parents and young. The young shove their heads, necks, and upper bodies into the mouths and throats of their parents to feed on regurgitated fish. They are insatiable and sometimes keep their heads down the throats of their parents for over two minutes at a time. Seldom do they back out willingly, and sometimes they are so tenacious that a parent is forced to violently terminate the feeding by vigorously whipping or snapping its head and neck before it can shake or toss a youngster free. A young pelican appears exhausted, even dazed, after the termination of a feeding bout and may sprawl out on the ground without moving. Without a doubt, it has experienced some oxygen deprivation while buried in the throat of its parent. Sometimes a tantrum will ensue, and the young bird will grovel on the ground and peck or bite at its own wings or body.

After they are three or four weeks old, the young leave the nests during the day to form small social groups or pods, but they return to their nests to spend the night. Soon they abandon their nests altogether and remain in their social groups. Woodbury visited a pod during the night on Anaho Island and wrote:

> On one occasion, a small pod of about a dozen young from four to five weeks old was observed at night. There were no adult birds attending them and their behavior was most unusual. When first observed with the use of lanterns, some of the young were lying prostrate as if dead, while others were sleeping in the normal squatting fashion. When approached closely and awakened, the young birds scrambled noisily toward us and huddled around our legs. This behavior is quite unlike that displayed during the daytime when they make a hasty but exceedingly clumsy retreat, stumbling over anything in their paths and even climbing over one another, as an intruder approaches.[38]

Pelicans subsist mainly on fish. The American White Pelican fishes by swimming on the surface of the water and plunging its head and neck underwater to capture fish that are within a yard or so of the surface. It does not chase fish by diving underwater from the air, as does the Brown Pelican, or by diving from the surface, as does the cormorant. It usually uses its lower bill and gular pouch as a dip net, although it can capture a fish between its two mandibles. Fish are swallowed headfirst; a bird never carries fish around in its pouch while swimming or flying. If the pouch contains water, it is emptied before the fish is swallowed. In deep water white pelicans feed as individuals and quietly cruise around, bill on the alert close to the water, waiting for a fish to wander up to the surface. Closer to shore, in water of moderate depth, a party of pelicans may fish cooperatively. Facing shoreward, they form a semicircle around a school of fish. Then, holding this alignment, they swim toward shore while thrashing the water with their wings and feet. The moving semicircle herds the fish into water so shallow that they cannot avoid being netted by the pelicans. On occasions pelicans have been seen starting with a circle and driving the fish inward as they tighten the circle.

Pelicans take off and land on water. They run or kick with their feet on the water to gain speed and hence sufficient lift for the takeoff. They fly with their necks folded and their heads resting back between their shoulders. White pelicans often fly in formation, either in a line or in a reverse V. They alternate flapping and gliding, in unison with the leader of the formation. At times the leader will suddenly rise or drop a few feet in the air, and each following bird will repeat this maneuver when it reaches the location where the leader did it. Pelicans often soar at great heights. At times, I have been aware of their presence in the summer sky above Pyramid Lake, although they were too high to be distinctly visible to the naked eye, for as the flight circled, tiny mirrors would suddenly flash in the sky, as their white bodies turned into the sun at the proper angle to reflect light down to my eyes.

Woodbury has an interesting account of aerial maneuvers by the pelicans of Anaho Island, carried out prior to feeding their young:

> When the first adult foraging parties begin returning to the island, usually in late morning or early afternoon, they frequently perform a most remarkable display of aerial maneuvers. They begin soaring to increasingly greater heights over the lake, flashing their white wings in the sunlight as they glide and turn in unison. This often continues until the birds are so high as to be barely visible to the human eye. Then, suddenly, the formation is broken, and they begin descending in a rapid twisting motion toward the island. As they approach the island the sound of air rushing over the partly folded wings can be clearly heard. They then circle their nesting site and begin dropping in among the anxiously awaiting hungry young. Once on the ground, their grace and agility disappear as they waddle about in search of their offspring. The advanced young birds, that have assembled in pods, scramble about and virtually assault the returning fish-laden adults in attempts to be fed. The ravenous young birds seem to try to obtain food from any of the returning adults. The parent birds, however, are more particular and seem to seek out and feed only their own offspring. The experiments of Hall (1925), using marked birds, showed that adults feed only their own young.[38]

The American White Pelican is not weather-wise and may fly during thunderstorms. There are several reports in the ornithological literature of lightning striking flocks of flying pelicans. A flock of seventy-five white pelicans was flying over farmlands in Nebraska during an April storm, when a bolt of lightning knocked thirty-three pelicans out of the flock. When examined, some of the dead pelicans had singed feathers.[43] Another tragedy struck a flock of pelicans near Granger, Utah, during an August thunderstorm. The entire flock of twenty-seven pelicans was knocked out of the air, following a loud clap of thunder, when flying at five hundred feet. An observer found sixteen of the dead birds scattered over an area of less than ten acres in extent, but other dead pelicans had been carried away before the count was made.[44]

A bleak future confronts the American White Pelican. As old colony sites are abandoned, relatively few new ones are established. Within established colonies, more breeding populations are decreasing than increasing in size. During historic times in the Great Basin, major breeding colonies have disappeared from Oregon and Utah. A tremendous breeding colony existed at Malheur Lake in Oregon before the lake dried up in the early 1930s. The official Malheur National Wildlife Refuge checklist now classifies the white pelican as common in the summer but not a breeding species. In 1982 I saw white pelicans nesting at Pelican Lake, Oregon—southeast of Malheur. In Utah, the breeding colony site on Rock Island in Utah Lake has been abandoned. At Great Salt Lake, pelicans once nested on Egg Island and Hat Island but are now restricted to Gunnison Island.[45] I do not know of any breeding sites in Nevada abandoned during historic times, since there is no documentation that white pelicans actually bred at Walker and Washoe lakes—as some people have claimed.

A new Great Basin breeding colony was discovered in 1976 at Hartson Reservoir, adjacent to Honey Lake in Lassen County, California. The colony was nesting on a thinly vegetated peninsula on the eastern shoreline of the reservoir and on an adjacent island. On July 6, 705 young pelicans were counted here, but this was only a partial count, since the less advanced nestlings were not included.[46]

Both the postbreeding and migratory movements of white pelicans from the Great Salt Lake and Pyramid Lake regions have been studied. Color-marked birds, as well as banded birds, have been used in these studies. Banding returns show that many pelicans from Great Salt Lake engage in a postbreeding movement into southern and central Idaho prior to migrating southward. The returns show that most of the Utah birds overwinter in various places in Mexico, with fewer overwintering in California.[45]

The postbreeding movements of young pelicans from Pyramid Lake have been studied by using dyes. Color marking was not a success in 1962, when an unstable pink dye was used in marking the pelicans. However, a durable green dye was used in 1963 and again in 1964. Observations of these green-marked pelicans showed that there were some postbreeding movements northeast and northwest from Pyramid Lake into Utah, Idaho, and Oregon. However, most of the young pelicans moved westward into central California. Following the postbreeding movements, migration occurs. Band returns on Pyramid Lake pelicans show that most of them overwinter in Mexico, particularly in western coastal regions, with fewer overwintering in California.[38] This is the same overwintering pattern shown by Utah pelicans.

Brown Pelican
Pelecanus occidentalis

The Brown Pelican, a bird of coastal waters, is of accidental occurrence on inland waters—such as in the Great Basin. There are a few sight records for Brown Pelicans from both sides of the Basin: from the mouth of the Jordan River, Farmington Bay, and Utah Lake in Utah and from the Stillwater region in Nevada. Since the Brown Pelican by nature is a bird of marine waters, it will not occur often on inland, freshwater bodies. When you are watching white pelicans in the Great Basin, especially during migration, keep your eyes open for Brown Pelicans. They differ strikingly from white pelicans—in color and in their spectacular dives from the air after fish.

FAMILY PHALACROCORACIDAE

The Family Phalacrocoracidae contains the cormorants. There are six species of North American cormorants, only one of which occurs in the Great Basin.

Double-crested Cormorant
Phalacrocorax auritus

The Double-crested Cormorant is a nesting species over much of North America and Cuba. It is a nesting species in the western and northeastern part of the Great Basin, but in the rest of the Basin it occurs only as a transient visitant. A few cormorants overwinter in the Basin, particularly in the western part during milder winters. Often a few cormorants can be seen at Virginia Lake in Reno during the winter.

Along with the white pelicans, the cormorants in the Great Basin had more extensive nesting and fishing habitat available for their use prior to shrinkage of the two great Pleistocene lakes—Bonneville and Lahontan. In recent times, the cormorants have had to contend not only with continuing desiccation within the Basin but also with ever increasing human activity. Therefore, cormorants now exist in fewer numbers at fewer colonies than in the immediate past. In 1975, Mitchell stated that the "population of Double-crested Cormorants in Utah has been steadily decreasing for the past fifty years. . . . of the thirteen colonies that have existed at one time or another within the state, only five are still in use."[47]

The courtship displays of the Doublecrest center around pursuit flights, with the male pursuing the female by flying, swimming, and diving. If more than one male joins in the pursuit, the female often flies away. The male will perform a variety of swimming and diving maneuvers around the female when he overtakes her and she is receptive.[48] During this performance the female may join in at times, as in a mutual display. The male's performance is more vigorous than graceful. There is much splashing as he beats the water with his wings while standing erect or while partially out of the water as he plunges forward in a series of jumps. He may swim on a zigzag course with his neck and head underwater. He often dives and brings up vegetation, which he presents to the female by dropping it in the water near her or tosses in the air. Females often join in the diving, particularly when being pursued. Either or both of the birds may spin in the water.

Eventually the mature males, three years and older in age, will each establish a territory on the colony's breeding ground. The individual territories are often small, with barely sufficient room for a nest and a perching place. From his territory the male displays in an attempt to attract an unpaired adult female, and when he is successful pair bonding occurs there. While displaying the male assumes a strained crouching position; jerking his wings with each note, he utters a loud, rapid series of *oak, oak, oak, oak* notes. When approached by a receptive, acceptable female, the male ends his song. He may straighten up, open his bill, and with head movements utter a clicking sound with each forward movement. Sometimes he maintains the crouch but, twisting his head slightly, lunges toward the female while uttering a series of nasal gurglings. If the female is ready to accept the pair bond, she approaches within touching distance, and "the two birds then lock beaks or caress each other on the head, neck or back."[48]

The Double-crested Cormorant nests in colonies; only rarely is it a solitary nester. Some colonies are ground-nesting ones, and others are tree-nesting ones. The birds favor islands and islets for ground nesting and trees standing in water for tree nesting. In recent years, as more and more colonies have been forced out of their traditional nesting localities because of human disturbances, questions have arisen about how rigidly fixed the nesting tradition is within a colony. Can alternations between ground nesting and tree nesting occur? Is a cormorant conditioned at birth to perpetuate the mode of nesting practiced by its natal colony? I am not aware of any controlled experiments which would answer these questions, but the history of the cormorants at Pyramid Lake indicates that nesting traditions can change rapidly.

Ridgway visited Pyramid Lake in 1867 and again in 1868 to study and collect birds. He went out to Anaho Island both years and observed American White Pelicans, Great Blue Herons, and California Gulls nesting there—but no Double-crested Cormorants. Concerning the cormorants at Pyramid Lake he noted, "Small congregations were frequently to be seen during the summer-time, perched upon the snags far out in the lake, the latter being nearly submerged cotton-wood trees which marked, at that time, the former course of the river when the lake occupied more restricted limits. On these tree-tops many of their nests were found, these being composed of sticks, and containing one to three eggs each."[25] In 1924 Hall visited Pyramid Lake. The cormorants were no longer nesting in trees. These trees were now over two miles in from the shore, due to desiccation of the lake. Hall discovered that the cormorants were now nesting off the north shore of the lake—on ledges on the towering Pinnacles and on rocky islets rising out of the water there. No cormorant nesting was reported from Anaho Island.[49] This was still the situation in 1927, when Gromme visited the lake and reported that the "nesting colonies of this species were all situated among Pinnacles at the north end of the lake."[50] Then in 1940, for the first time, Bond reported a colony of about fifty cormorant nests on the southwest shore of Anaho Island, mixed in with nests of California Gulls.[51]

In the years following 1940, although some nesting continued at the Pinnacles, the main nesting of the cormorants at Pyramid Lake occurred on Anaho Island, a rather large island about one by one and a half miles in extent. By 1957 the number of cormorant nests on Anaho had reached a high of 1,580. Over the years, the number of cormorant colonies on the island has varied from one to six. The locations of the colonies have varied from the southwest shore, to the southeast tip, to the east shore, to the "high precipitous cliffs on the southwest edge of the island," back to the east shore.[38] Thus, the cormorants of Pyramid Lake have shown remarkable adaptability in utilizing a wide array of nesting sites. They have nested colonially on flooded trees, on ledges on high pinnacles rising out of the lake, on the tops of rocky islets, on the ground on various shores, and on steep cliffsides on a large island. Colony traditions have been not only flexible but downright changeable.

Both sexes participate in nest building. A new nest may be built, or an old nest may be renovated by adding a new top and lining. The nests within a colony are usually not uniformly dispersed but are clumped in groups of about four to thirty-three, with some unoccupied spaces separating the groups. Before starting the construction of a new nest, the pair accumulate a pile of building material—such as aquatic plants, weeds, and sticks—at the building site. From then on the nesting material, and soon the nest, must be guarded by the pair. Cormorants not only pull apart abandoned nests for nesting material but will pilfer it from active nests which are left unguarded. They may continue to add material to the nest all during the nesting season. Reused nests tend to get larger with each year of use, and ground nests become moundlike. Not only do nests become larger with time, they also become filthier—cormorants do not overindulge in nest sanitation practices. Soon befouled with excrement, decomposing fish, and often decomposing young cormorants who have died in the nest, the nests may also be alive with flies, fleas, and lice.

Incubation usually starts after the third egg of the clutch has been laid, but sometimes it begins sooner. The parents take turns incubating, and the incubation period varies from three and one half to four weeks. Frequently, the parent not on duty will stand on the edge of the nest or beside the nest while its mate incubates. Nest-relief ceremonies accompany the changeover at the nest. The relief may bring nesting material to the nest. Often the two birds caress and pet each other, or they both stand on the nest, croaking in low tones, before the changeover is accomplished.[48] The eggs as well as the young nestlings are very vulnerable to predation by gulls and other nest robbers. Eggs and nestlings are usually protected from predators by the constant presence of at least one of the parents at the nest. But when the adults in the colony are scared away by some disturbance, such as human intrusion, the ever present circling gulls, who are not as timid as the cormorants, drop out of the sky onto the unprotected eggs and nestlings. There is a large breeding colony of California Gulls on Anaho Island, and gull predation on Anaho's cormorants has been severe.

The young cormorants are fed regurgitated, semiliquid food by their parents. When feeding newly hatched young, the parent regurgitates the food into the nestlings' mouths. After its second day of nest life, a young bird must help itself to the regurgitated food in the parent's mouth. At first the parent will guide the food transfer by carefully placing its open mouth over the heads of the young birds. When the young become more capable of movement, the parent merely has to open its mouth and pump up the food. The young are so greedy and insatiable that it is often difficult for an openmouthed parent to avoid receiving more than one young head at a time! Nestlings not only beg and push and shove for food, but they strike at their parents with their bills if they do not receive prompt service. The parents tolerate this unruly behavior; as a consequence, the younger and weaker nestlings often do not obtain enough food to survive. The presence of younger and weaker nestlings in a brood is often the product of asynchronous hatching—which occurs when incubation commences before the full clutch of eggs is laid.

Young cormorants are altricial; that is, when newly hatched they are helpless, naked, and blind. Altricial young are called nestlings, in distinction from precocial young, who are referred to as chicks. The eyes of a nestling cormorant open within four or five days of life, and down starts to appear by the end of the first week. A black down plumage is completed by the age of two weeks. Nestling cormorants cannot regulate their body temperature when subjected to heat or cold stress and must be brooded by their parents. Heat death is a major threat confronting young cormorants in the Great Basin, where insolation can be intense and reradiation from the ground and rocks extremely high. This problem is further magnified by the open, unsheltered positions of the nests and by the blackness of the young birds' naked skin and their down plumage. Parent cormorants can often be seen standing at the nest with partially spread wings, shading the nestlings from the sun. When the parents are flushed from their nests, the young can quickly overheat and die. As the nestlings grow older, their temperature regulation will improve. They will acquire better plumage insulation, and they will start to increase their evaporatory heat losses by panting and gular fluttering.

In the Great Basin the Double-crested Cormorant feeds mainly on fish such as carp, perch, chub, suckers, and minnows—getting more trash fish than game fish. It fishes by diving from the surface of the water or by slowly submerging and then swimming after fish. The cormorant is a foot-propelled swimmer; its wings are not normally used underwater. All four toes on each foot are joined in a web. These totipalmate feet are great for swimming but can be awkward when used on land and in trees, especially when they are associated with legs placed far back on the body, as on cormorants. When perched upright, a cormorant can brace itself with its short, stiff tail. Cormorants may have trouble landing in treetops—they sort of crash in—and getting settled. It is difficult for them to perch among small, scattered limbs where the stiff tail is of no use. On land they employ a waddling gait. In getting airborne, they drop from elevated perches, leap into the air from the ground or rocks, or taxi for a few feet with beating wings on the water.

Often a perched cormorant will assume a spread-wing posture, with the wings extended out at right angles to the body. There is considerable controversy over the function, if any, of wing spreading. The simplest and most common explanation is that the bird does this to better air out and dry its wings. The cormorant possesses a preen gland and frequently waterproofs its plumage with oil from this gland. The plumage of the cormorant appears to be as waterproof as that of other water birds which do not utilize wing spreading but merely flap or shake their wings to dry them.

Like pelicans, cormorants possess a gular or throat pouch, but theirs is smaller and less elastic. The gular pouch of the Double-crested Cormorant is half-feathered and the bare part is orange in color. In flight cormorants appear quite gooselike; they fly with powerful wingbeats and extended necks and are often arrayed in oblique lines or in reverse V formations. When over water they often fly low. Like their relatives the pelicans, they, too, may court disaster by flying in stormy weather. Sprunt saw a bolt of lightning strike four Double-crested Cormorants out of a flock in South Carolina.[52]

FAMILY ALCEDINIDAE

Although a few species of alcedinids inhabit the temperate zones, kingfishers are best represented in the tropical and subtropical parts of the world. These are noisy, often brilliantly colored, large-billed birds. They are cavity or hole nesters. They possess syndactyl feet—the front three toes are united over part of their length.

Belted Kingfisher
Ceryle alcyon

The breeding range of the Belted Kingfisher extends over North America from central Alaska and central Canada to Mexico. This is an uncommon breeding species near streams and lakes in the Great Basin. Kingfishers overwinter in the Basin in areas where there is open water for fishing. Prior to 1982 the scientific name of this kingfisher was *Megaceryle alcyon*.

The name of kingfisher well fits this master fisherman—its headlong aerial dives into the water after fish are dashingly and skillfully executed. The Belted Kingfisher has two techniques for initiating its plunge-diving: along its defended territory it has regular fishing perches, where it can sit and scan the water below and await the arrival of prey; and it patrols from the air, hovering here and there, thirty to forty feet above the surface, scanning the water below. Since the kingfisher detects fish by eye and catches them by plunge-diving, it can hunt only over clear, relatively calm water of a depth sufficient to check its dive. The kingfisher enters the water headfirst, catches the fish in its bill, resurfaces, breaks clear of the water, and flies off to a perch to feed. Back at the perch the fish is beaten senseless, then swallowed headfirst. The undigested scales and bones are regurgitated as pellets.

The plunge-diving of the Belted Kingfisher is not only employed to capture fish but is also used as an escape tactic when the bird is under attack by a raptor. Raptors, particularly accipiters like the Sharp-shinned and Cooper's hawks, are prone to attack kingfishers. When pursued by an accipiter, the kingfisher will plunge-dive underwater at the last moment to avoid capture. It may have to dive repeatedly if the hawk remains nearby to launch subsequent attacks.

The Belted Kingfisher is a solitary bird, except during the mating season. It nests in burrows excavated in dirt banks. Both sexes participate, with bills and feet, in excavating the nest burrow—once a burrow is constructed it may be used for a number of years in a row. Kingfishers prefer to burrow in a bank close to their fishing territory. But, if a bank of proper height and soil condition is not present near the fishing waters, they may excavate a burrow as far as a mile away from the fishing territory.

The kingfisher does not attempt to keep its preserve a secret—if disturbed or alarmed, the bird sounds its loud, harsh, rattling note. You can gain some knowledge of the boundaries of a kingfisher's feeding territory by pursuing the bird. The kingfisher will fly from perch to perch; then, at the edge of its territory, it will swing out in a wide detour around your position and fly back near its starting point.

7

Marshbirds

MEMBERS of two orders of marshbirds are found in the Great Basin. Our herons, egrets, bitterns, ibises, and storks are members of the Order Ciconiiformes, and the cranes, rails, and coots belong to the Order Gruiformes. Although not a member of the Great Basin avifauna, a stray Greater Flamingo of the Order Phoenicopteriformes occasionally shows up here. These marshbirds have toe, foot, and leg adaptations for wading or for walking over muddy ground.

FAMILY ARDEIDAE

The heron family is represented by eight species of regular occurrence and two species of irregular or accidental occurrence in the Great Basin. These birds hunt for aquatic prey while wading or standing in water. They possess powder-down feathers—feathers which are never molted but continue to grow at the base and fray into powder at the tip. The powder is spread over the plumage to soak up dirt and grease and then removed from the plumage with the comblike claw of the bird's middle toe.

American Bittern
Botaurus lentiginosus

The American Bittern occurs as a breeding species over much of North America—outside of the Far North. It is a common breeding species at marshes throughout the Great Basin and a rare to common overwintering species at these marshes. Apparently, the status of the bittern has not changed much since the 1860s, for Ridgway wrote, "The common Bittern was constantly found in all marshy situations in the Interior, where it appeared to be resident all the year."[25]

The American Bittern is a much more conspicuous bird than its diminutive relative, the Least Bittern. It does not lead such an overwater existence but frequents the marsh edges and meadows, where it is more visible. The spring song of the male is quite unique. Brewster has written of this:

> Standing in an open part of the meadow, usually half concealed by the surrounding grasses, he first makes a succession of low clicking or gulping sounds accompanied by quick opening and shutting of the bill and then, with abrupt contortions of the head and neck unpleasantly suggestive of those of a person afflicted by nausea, belches forth in deep, guttural tones, and with tremendous emphasis, a *pump-er-lunk* repeated from two or three to six or seven times in quick succession and suggesting the sound of an old-fashioned wooden pump.[53]

The pumping sounds have also been compared to the sound of a wooden mallet driving a stake; this has led some to call this bittern the stake driver.

The pumping or stake-driving song of the male is believed to function as a dual-purpose advertising song—announcing in a threatening vein the whereabouts of a territorial male to other males and announcing in an attractive vein the whereabouts of an available territorial male to unpaired females. When defending his territory or advertising, the male may expose two white tufts of feathers which are "attached to the skin on each side of the breast near where the humerus enters the body and beneath the shoulder of the folded wing by which they are ordinarily concealed." These feathers are erected during display; "as they rise above the shoulders these ruffs spread toward each other at right angles to the long axis of the bird's body until, at their bases, they nearly meet in the centre of the back."[54]

There is circumstantial evidence that polygamy of the type called polygyny, where the male has more than one mate, may be practiced by some American Bitterns. Although this species is usually a solitary nester, small groupings of nests have been found with only one male nearby—suggesting polygyny. Then, too, the female builds the nest, incubates the eggs, and cares for the young all by herself.

The American Bittern is a solitary hunter, feeding on typical heron fare such as small fish and frogs. Its hunting techniques are simple and based on stealth—it wades or walks slowly and stands and waits.[55] It is well known for its concealing posture. Frozen in place, with neck and bill pointed skyward and brownish plumage compressed, it is well camouflaged. It may even sway in rhythm with the windblown vegetation.

Least Bittern
Ixobrychus exilis

The Least Bittern is the smallest of all the herons. It is such a quiet, retiring, inconspicuous marshbird that its comings and goings and local and geographic distributions are poorly understood. As a breeding species it extends from southeastern Oregon and southeastern Canada into South America.

The Least Bittern is a rare migrant and breeding species in the Great Basin. In earlier days, it nested in the eastern part of the Basin in marshes around Great Salt Lake. A set of eggs was collected in 1884 at Hot Springs Lake north of Salt Lake City.[56] In recent years it has been a rare nesting species in the northwestern part of the Basin—at Honey Lake, California, and Malheur National Wildlife Refuge. I am not aware of any breeding records for the Nevada part of the Great Basin.

The paucity of occurrence records would seem to indicate that the Least Bittern has always been a rare species in the Great Basin. Ridgway encountered but a single Least Bittern—in some willows along the lower end of the Truckee River—during his years here. During the passage of well over one hundred years, only a handful of records have accumulated from Great Basin localities in Utah, Nevada, California, and Oregon. However, when a species is judged to be rare in occurrence, the question always arises, is this judgment the product of the actual rareness of the species or, rather, of the rareness of potential observers? Given our present knowledge, I believe that this bittern is indeed a rare bird in the Great Basin, but it may not be quite as rare as the handful of records, slowly accumulated over the years, would indicate. Further, I suspect that it may be nesting at other places in the Basin besides the northwest corner. My suspicions are fueled by a combination of thoughts: the inconspicuousness of this bittern; its penchant for nesting over water in the remotest parts of marshes; the vastness of the Great Basin and the limited accessibility of its marshes; the very few competent, off-road field observers working in the Basin; the lack of any real interest in this species; and the lack of any comprehensive search program for this bittern. So, when you are afield in marshes in the Great Basin during the nesting season, keep a conscious eye out for Least Bitterns.

The Least Bittern is seldom seen except by those who seek it out. It lives an overwater life, hidden from casual view, in the dense emergent vegetation of marshes. To enter its world, you must wade or use a boat. When you approach it, a bittern prefers to run away instead of taking to the air. When it does fly, it is weak and fluttering, moving low over the top of the emergent vegetation. With extended neck and dangling legs, it flies but a short distance before dropping back down into the vegetation. When startled it may utter a loud cackle. When walking or running with its straddling gait, above the water, through the emergent vegetation, the bittern uses plant stalks as stepping-stones. Its legs are spread apart, each foot grasping one or several stalks, as it steps along. When threatened by danger, especially on its nest, it may freeze in place, with its bill pointed straight up toward the sky. In this pose, with its brownish coloration, it is extremely well camouflaged. A bittern may even sway from side to side, as if it is vegetation stirred by the wind.

Little in the way of courtship display has been observed in the Least Bittern. The male has been heard uttering a guttural, dovelike *uh-uh-uh-oo-oo-oo-oo-ooah* spring song, to which the female may respond with several short notes of *uk-uk-uk*.[57] The nest is a frail platform built above the water—often on bent-down, dead stalks of emergent vegetation. The male apparently does much of the nest building. Both sexes incubate the eggs and feed the young. A nest-relief ceremony has been seen. The bird on the nest erects its crown feathers and calls *gra-a-a*, and the relief may also erect its crown and body feathers. When the relief comes on the nest, both birds open their mandibles and shake their bills from side to side, producing a rattling sound. "This rattling seems very important in nest relief but also occurs when a lone bird of either sex returns to the nest after drinking."[58]

The Least Bittern feeds on typical heron food—fish, amphibians, insects, crustaceans, and small mammals. It may feed on the eggs and nestlings of other marshbirds. During a study of a nesting colony of Yellow-headed Blackbirds, Roberts reported that a pair of Least Bitterns built their nest in the midst of the blackbird colony. Not only were the adult blackbirds greatly disturbed by the presence of the bitterns, but day after day the eggs and new nestlings disappeared from the blackbird nests until all were gone.[59]

The hunting techniques of the Least Bittern are few. It feeds in small pools in openings among the emergent vegetation, using its straddle technique to walk slowly on the stalks of vegetation at the edges of the pools. It may stand and wait, with its legs spread apart, its head and neck lowered out over the pool, and its bill almost touching the water. Wing flicking may startle motionless prey into moving. After making a single capture, a Least Bittern may retreat back into the vegetation and move on to another pool—not tarrying long at any one pool.[55]

Great Blue Heron
Ardea herodias

The breeding range of the Great Blue Heron spans the width of North America, extending from southern Canada to northern South America. This heron is a common breeding species at many of the lakes, streams, and marshes in the Great Basin. Lesser numbers overwinter. In the eastern part of the Basin in Utah, this heron has been categorized as a "casual winter resident."[27] Some Audubon Christmas Bird Counts do produce good Great Blue Heron counts in late December at Utah sites. The December 1980 tally was: Bear River Refuge, sixty-one; Fish Springs Refuge, twenty-five; Provo, forty-eight; and Salt Lake City, nine.[60] The Great Blue Heron is a common winter figure in the western part of the Basin; and this long-legged, ungainly fisherman can be seen hunting, even away from wet areas and irrigation ditches, in dry pastures and fields. At Malheur National Wildlife Refuge in the northwest corner of the Great Basin, the Great Blue Heron is reported to be uncommon during the winter.[61] But, during the 1980 Christmas Bird Count, twenty-six Great Blue Herons were counted at the refuge.[60]

Great Blue Herons usually build their nests in colonies. Sometime after returning to the heronry in the spring, the male seizes a small territory, which usually incorporates an old nest. Pair bonding occurs when a female is attracted to a male's territory. The new pair carry out courtship here, and both defend the territory from trespass by other Great Blues. Territorial defense consists of threat displays, clapping the mandibles, and a vocal bark.[62] The courtship displays witnessed in a Michigan study involved "Erecting plumes and crest, Shaking head from side to side, Walking around each other in the nest, Clapping mandibles together (loud hollow sound), Grasping each other's mandibles and seesawing back and forth, Howling, Shaking stick in nest, Preening (individual and mutual), and one bird stroking the other with bill on throat, nape, and back."[62]

The Great Blue Heron prefers to nest in the crowns of tall trees, but if trees are not available the bird will build in shrubs or even on the ground. Sometimes a flimsy new nest is built. Often an old nest is repaired, and it becomes more massive and substantial with the passing years. As many as several dozen nests may be built in a single large tree. Sometimes Great Blue Herons, as part of a mixed-species nesting colony, share the trees with other species of herons and water birds and even with hawks, owls, and vultures. During nest building the male brings sticks to the nest and presents them to the female, who then incorporates them into the structure. Stick presentation at the nest is ceremonially accomplished: as the male approaches the nest with a stick in his bill, the waiting female bows to him and greets him with a loud howl, before taking the stick from him.

Some Great Blue Heron colonies in the Basin may be located in tall trees—such as the heronry in a grove of cottonwoods near Stillwater, Nevada. But often our heronries are located in shrubs or on the ground on lake islands. At least since the time of Ridgway, Great Blues have nested on Anaho Island in Pyramid Lake along with the pelicans, cormorants, gulls, and terns. In May of 1868, Ridgway found them nesting on the tops of large greasewood bushes, about 5 feet above the ground, on Anaho Island. When he climbed the large formation called the Pyramid—which gives the lake its name and which was then an island close to the eastern shore—he collected four Great Blue Heron eggs from a "nest on the 'Pyramid,' among the rocks, about 150 feet above the surface of the lake."[25] Great Blue Herons have also nested on the ground on Anaho Island among bushes or on dense weed beds. And heronries are located on some of the islands in Great Salt Lake. "There have been at one time or another six separate locations on the islands used by the herons as nesting sites, namely Gunnison Island, the south end of Fremont Island, Egg Island, White Rock, Hat Island and Badger Island."[45] On some of these islands, as on Anaho Island, large greasewood bushes have been used as nest sites, and ground nesting—between rock masses and in tall grass—has occurred.

The sexes take turns incubating the eggs. At the changeover at the nest, a relief ceremony may occur, with both birds erecting their head, lower neck, and scapular plumes and sounding an *arre-arre-ar-ar-ar-ar* call.[62] The incubation period varies from twenty-five to twenty-nine days. There is an interval of two days, sometimes three, between the laying of consecutive eggs in a clutch.[63] The eggs hatch asynchronously, and the young, although possessing down, are not precocial. Sibling aggression can be severe, with the older and stronger young viciously pecking the smaller young and taking more than their share of food. This "frequently results in the death of the younger one or two birds."[45] The parents feed the nestlings regurgitated food. When the nestlings are very young, the parent places the regurgitated food in their open bills. Later on, the young peck at the parent or grasp its bill crosswise. The parent then lowers its bill, the young bird grasps its tip, and the parent regurgitates food into the mouth of the young or onto the floor of the nest.

Great Blue Herons are fishermen but are not above taking amphibians, reptiles, mammals, birds, crustaceans, and even insects. Their usual hunting technique is easy to observe—they stand quietly in shallow water, waiting for game to move within striking distance of their long necks and bills. Sometimes they stalk their prey by wading slowly through the shallow water. Occasionally, they will hunt ashore; in the Great Basin they can be seen hunting in the pastures and meadows in cattle country. Since prey is swallowed whole, part of it does not get digested. Pellets of undigested material are regurgitated.

Although the Great Blue Heron is a colonial breeder and has a close-knit family life during the breeding season, over most of the year it is a solitary figure. It fishes by itself and stays by itself. However, small, loose flocks have been seen during migration.

Sadly enough, this consummate fisherman, who fishes but to live, is seldom given a friendly greeting by members of the human fishing fraternity. In fact, it is often persecuted to death. It is true that this heron can cause some damage at a fish hatchery rearing pond, but in nature it doesn't get much in the way of game or trophy fish in the shallow waters in which it casts its bill. It has often been held responsible for our mistakes in overfishing and in practicing poor fishing management. Back in 1893, the state of Utah appropriated the grand sum of three hundred dollars so that a bounty of ten cents a head could be paid on Great Blue Herons. This motivated a party of bounty hunters to set out in late June for the south end of Utah Lake. There, according to Behle, "early in the morning they drove young, still unable to fly, from the nests and killed them with oars. Some 3,000 young were thus killed. They collected from the State Treasury $300.00 which wiped out that particular fund. That did not stop the slaughter of herons."[45] Today, however, the Great Blue Heron populations in the Basin appear to be in good health.

Great Egret
Casmerodius albus

The Great Egret is a cosmopolitan species occurring on all the continental land masses in the world, except for the Far North, and on many islands. In the eastern part of the Great Basin, it is present only as a "rare transient" with "no satisfactory evidence of breeding in Utah."[27] However, it is a breeding species in the central and western part of the Basin.

During the late 1800s and early 1900s, Great Egret heronries over much of the world were decimated or wiped out, when hunters killed the adult birds during the nesting season for their plumes. With the death of the parent birds in a heronry, the eggs were destined to rot and the nestlings to starve. The soft, pure white, flowing nuptial plumes of the Great Egret and other egrets, called aigrettes, were much in demand in the millinery trade as ornaments for hats; since they lack the locking mechanisms of ordinary contour feathers, aigrettes are plumelike instead of stiff-vaned. Eventually, worldwide indignation, followed by the passage of legislation, resulted in better protection for the birds. Great Egrets staged a recovery which peaked in the mid 1930s.[57] Since then they have once again been on the decline in many regions of the world. Ever-increasing human populations and activity are causing the destruction and degradation of the breeding habitats and aquatic feeding grounds of this species.

The Great Egret usually nests in colonies, although solitary nesting may occur. It often nests in heronries with other species of herons and water birds. It builds its nest in emergent vegetation over water or in shrubs or trees. In the Great Basin, tule beds are often used by nesting colonies.

Breeding commences with the males establishing territories. Aggressive displays, aerial chases, and some fighting occur between territorial males. Each male builds a nest platform or repairs an old nest on his territory. Then he displays on this platform. He may execute circle flights from his platform, flying out and around the colony and back again. While on his platform he may give his advertising calls, which have been described as "rather soft *fra-fra* or *frawnk* calls."[57] When potential mates are attracted to an advertising male, they are first greeted with threat displays and chased away. Eventually, after the female repeatedly and passively returns to the male's platform, she may be accepted for pair bonding.

Courtship behavior in the Great Egret has been described by Wiese, following observations made in Louisiana and Florida.[64] While displaying, the male erects and fans his aigrettes or scapular plumes and may sway back and forth. During the wing stroke he runs his open bill downward, along the forward edge of a partially extended wing, passing the primary wing feathers through his bill. Wiese states that this display has become highly ritualized, since the bill often does not actually touch the primaries. A stretch display starts with an upward toss of the neck and head, followed by a bob or a bending of the legs at the heels. In the bow display, a male picks up a twig from the nest platform and performs tremble-shoving and a bob; during tremble-shoving the twig is thrust forward or downward, accompanied by a sideways head-shaking movement. In the snap display, the male thrusts his neck and head downward, bobs, erects his head and neck feathers, and snaps his mandibles. The female will often execute some of these displays while watching or accompanying the male.

The Great Egret is a fisherman but will take amphibians, reptiles, crustaceans, insects, and mammals as well. It may have feeding territories which it defends against trespass by other Great Egrets and smaller herons and egrets. However, it will give ground when approached by a Great Blue Heron. Its hunting techniques are not as involved as those of the Snowy Egret. The classic heron hunting techniques of stand and wait and wade or walk slowly are employed. In addition, it employs hopping, wing flicking, hovering, and dipping—all to be described in the Snowy Egret account.

The Great Egret uses some hunting techniques that have not yet been recorded for the Snowy Egret. During gleaning it pecks food from objects above ground or water, such as vegetation. In leapfrog fishing in a flock of egrets, an individual will fly from the rear of the foraging flock to a position at its head. In plunging, while in flight or hovering, an egret will dive headfirst out of the air into the water after prey. During swimming feeding, while swimming on the surface of the water, the egret captures prey.

Snowy Egret
Egretta thula

The Snowy Egret's breeding range is scattered over much of the Americas, south of northern United States and north of southern South America. Formerly, its scientific name was *Leucophoyx thula*; then in 1973 *Leucophoyx* was merged with *Egretta*; and it is now known as *E. thula*. The Snowy is a common breeding bird in the Great Basin at suitable marshes and other wet areas. It can be found at streams and other bodies of water throughout the Basin during migration and during its postbreeding dispersals in late summer.

The Snowy Egret has become much more widely distributed and abundant in the Great Basin since the 1860s. Despite spending over two years in the field in the Basin collecting and studying birds from 1867 to 1869, Robert Ridgway did not see a single Snowy Egret. Not a single Snowy along the Carson, Truckee, Humboldt, and Jordan rivers in three summers' time. Not a single Snowy in the Washoe Valley, Truckee Meadows, Humboldt marshes, Ruby marshes, or Ruby Valley or in the Great Salt Lake and Salt Lake City area. Today, you can find the Snowy Egret, from spring to fall, in this selfsame country, without straining your eyes or getting off the highways. Since Snowy Egrets are so conspicuous, it is difficult to believe that Ridgway would have missed them if they were present in the 1860s.

During the latter part of the nineteenth century and the early part of the twentieth century, Snowy Egrets were severely worked over by the plume hunters and were almost exterminated in the United States. In 1903 plume hunters were receiving thirty-two dollars an ounce for aigrettes, which was about twice the price of gold.[65] The exquisitely delicate aigrettes of the Snowy Egret were in great demand. The egrets had to be shot to obtain fresh plumes, since shed or molted "dead" aigrettes were soiled and worn and sold for only a fraction of the price of "live" ones. Shooting was productive only during the nesting season, when the egrets were in nuptial plumage and congregated at breeding rookeries. Plume hunters soon learned to start shooting only after a heronry was ripe. If shooting started too early in the nesting season, the egrets would abandon the rookery. The strategic time to start shooting was after the eggs had hatched, when parents were so strongly attached to their nestlings, they would remain despite the shooting. Following the slaughter of the adults, all the young and any remaining eggs in the heronry were doomed. Once given protection from the plume hunters, however, the Snowy made a remarkable comeback.

Sociality is highly developed in Snowy Egrets. Although the birds maintain feeding territories, they nest colonially, often in company with other species of egrets and herons. Although Snowy Egrets may defend feeding territories, they feed in open areas, often close to other feeding egrets and herons. During courtship displays, a circle of Snowies may gather to watch a displaying male.

An unpaired male advertising for a mate employs a stretch display, during which he extends his bill upward, pumps his head up and down, and loudly calls *a-wah-wah-wah*. Nearby Snowy Egrets may soon encircle the displaying male to watch, and from these birds a mate may emerge. Courtship flights take place—such as aerial stretch displays and aerial circle flights—and after pair formation the female may join the male in these maneuvers.[57] The tumbling flight is the most spectacular aerial display. The bird or birds mount high in the sky and then drop earthward, tumbling over and over, only to right themselves just before landing. On the ground a jumping-over display may occur, where a Snowy Egret executes a jump-flight, leapfrog fashion, over its mate. The mate may or may not respond in kind. Watch for these displays in the spring. Not only are the movements spectacular, but they are framed by the ethereal beauty of fully erected, dazzling white plumes.

Snowy Egret watching can be a stimulating outdoor pursuit in the Great Basin. There are many breeding colonies, including some at readily accessible national wildlife refuges and wildlife management areas. Often egrets can be seen from highways that closely approach streams, marshes, ponds, and lakes. Since these egrets hunt in shallow water, in open places, they will be fully exposed to your view. Their hunting behavior is varied and sophisticated—making for enjoyable watching.

The stand and wait and the wade or walk slowly techniques are employed when a Snowy Egret is still-hunting or stalking prey. During a stand and wait bout, an egret may engage in bill vibrating, which creates a disturbance in the water that supposedly attracts prey to the spot. Bill vibrating involves inserting the tip of the bill into the water and then rapidly opening and closing the mandibles.

The Snowy's disturb and chase repertoire is varied.[55] It walks quickly to flush prey but may resort to running. Hopping attacks are launched by jumping up and flying to land next to prey. A hunting Snowy Egret has foot moves that would put an ace soccer player to shame. Not only can it flush prey from cover by foot stirring and foot raking, but it can also engage in foot probing and foot paddling—by extending a foot as a probe or by rapidly paddling on the substrate with its feet. It even has an aerial modification of foot stirring called hovering stirring, in which it stirs with one foot while hovering over shallow water and striking at prey from the air. Interestingly, Meyerriecks believes that the egret's bright yellow toes act as a lure during foot stirring, when the egret "quivers its foot on the surface of the water" instead of stirring the substrate.[66]

Like some of the other herons, the Snowy Egret uses wing movements as an integral part of its disturb and chase repertoire. One such movement is wing flicking. While wading or while at a stop, the egret abruptly executes what appears to be a balancing movement; "it suddenly extends and withdraws both wings in a short, rapid flick."[66] A series of several wing flicks may be given. Meyerriecks believes that this wing flicking, which typically occurs on sunny days, casts sudden shadows on the water, flushing the egret's prey. Snowies also employ their wings in open-wing feeding. "The wading bird, feeding typically in shallow, open water, will suddenly extend one wing fully, whirl, withdraw the wing, whirl, fully extend the same or the other wing, then rapidly pursue any prey disturbed by the quick wing movements."[66] Open-wing feeding may also terminate a running bout, with one or both wings being extended. Finally, the Snowy Egret practices underwing feeding—by extending its wings while wading, putting its head under one wing, scanning the water, and capturing prey. The egret may shift its head under the wings as it turns. The wing not only shields the eyes of the heron from the sun, but its shadows startle and flush prey and reduce surface glare.[66]

There are other behavioral components to aerial hunting by Snowy Egrets besides hovering stirring. They employ dipping, during which they catch prey while flying low over water. While hovering over the water, they may capture prey in the water. Foot dragging can occur if an egret lowers one or both feet into the water while flying low. Strikes are made while the egret is in flight.

Cattle Egret
Bubulcus ibis

The Cattle Egret is not a native of the Great Basin or even of North America. It is an Old World immigrant in the process of settling in North America. Its remarkable colonizations—which have carried it from Africa to South America and from South America to North America—have conferred a touch of tachycardia upon many avid birders, as they strive for the first occurrence and nesting records from here and there.

Cattle Egrets were first reported from the eastern end of the Great Basin; the initial sightings were made near Brigham City in September 1963 and Farmington Bay in August 1964.[27] In May 1973, a Cattle Egret was seen in Ruby Valley in the interior of the Basin. In the western end, the egrets were first reported from Malheur National Wildlife Refuge in August 1974 and from Washoe Valley, Nevada, in August 1975. Cattle Egrets were first reported nesting near Bear River in the eastern end of the Great Basin in 1978[67] and at Carson Lake in the western end in 1980.[68] Since summer sightings of Cattle Egrets are now quite common, nesting is probably becoming more widespread in the Basin.

It is almost certain that the Cattle Egrets which colonized the New World came from Africa. Apparently, in South America, "the earliest written records of the bird's presence were made by a Surinam collector between 1877 and 1882" from the Courentijne River.[69] By 1916 and 1917, the egrets were in Colombia.[70] Some people have suggested that the Cattle Egret may have been introduced into South America in the early part of the twentieth century to help control ticks. However, Cattle Egrets are not tick specialists; and, when they do take them, they apparently pick them off the ground more than off of cattle.

The spread of the Cattle Egret into the Americas has all the earmarks of a natural phenomenon, just as its subsequent spread over the Americas certainly has been a natural phenomenon. The Cattle Egret is not a sedentary species but wanders widely—it is quite capable of flying across the Atlantic Ocean, from Africa to South America, on its own. On occasions, Cattle Egrets have been seen hundreds of miles out to sea. Further, what appears to be a Cattle Egret was photographed on Saint Paul's Rocks in the Atlantic Ocean in April 1963.[71] Saint Paul's Rocks, about 1,050 miles from Sierra Leone, are roughly midway between Sierra Leone in Africa and the Guyana-Surinam coast of South America; and they are located close to the axis of the southeast trade winds. Hence a crossing here, with a steady tail wind and a resting point en route, might be easier to accomplish than a crossing at the closest route between Africa and South America. The shortest route across the Atlantic is only 520 miles long but lies across wind.

"The Cattle Egret's invasion of Central and North America appears to have been two-pronged, *via* Colombia through Middle America and *via* Trinidad through the Antilles."[72] It was first recorded on the mainland of the United States in the vicinity of Lake Okeechobee, Florida, in May 1948. Since then it has spread west to California and north into Canada.

The Cattle Egret is neither a great fisherman nor a frequenter of shallow water like many of its kin. It favors dry land and prospers where herds of large mammals graze and browse. Cattle Egrets form a commensal association with large herbivorous mammals, that is, an association which benefits the egrets but neither benefits nor harms the large mammals. In wildlife movies filmed in Africa you have seen Cattle Egrets accompanying large mammals—either foraging alongside of them or perched on their backs. Although they seem to favor the hippopotamus, they often associate with the elephant, buffalo, waterbuck, zebra, giraffe, eland, and rhinoceros. Where herding occurs, they form commensal associations with livestock—especially cattle. When watching for Cattle Egrets in the Great Basin, don't expect to have hippopotamuses or elephants attract them; you'll have to rely on cattle.

Insects form the food staple of Cattle Egrets, particularly orthopterans such as grasshoppers. A herd of large-bodied mammals, moving about grazing or browsing, stirs up the insects and causes them to take evasive action. The movements of the insects betray them to the birds. Thus, the large mammals unwittingly serve as beaters and flush game before the bills of the egrets.

Field studies have shown that Cattle Egrets are much more efficient in capturing prey when following grazing cattle than when foraging alone. Grubb carried out an interesting quantitative study of the effectiveness of Cattle Egrets foraging with black Angus cattle in a pasture on Saint Catherine's Island, off the coast of Georgia. He believed he could tell when an egret had captured an insect, since it "always jerked back the head characteristically when swallowing prey." Grubb discovered that a Cattle Egret foraging with a cow captured, on the average, 2 insects per minute; if it shared a cow with another egret it averaged 0.8 insect per minute; but if it foraged alone, without a cow, it averaged only 0.6 insect per minute. It required only 13.7 steps between captures if with a cow, 44 steps between captures if sharing a cow, and 76.5 steps between captures if alone.[73]

Grubb's study showed that not only was it more productive and efficient for an egret to forage with a grazing cow, but it was best if the egret could avoid sharing its cow with another egret; and, interestingly enough, most of his resident egrets defended their feeding station with a cow against trespass by another egret. When with a grazing cow, the egret spent most of its time foraging up by the head of the cow, less time along the side of the cow, and very little time about its rear. The relative time spent in each of these three locations correlated with their productivity. The egret averaged 2 insect captures per minute when at the head of a cow but only 0.7 when along its side and 0.4 when at its rear.[73]

Cattle Egrets do not restrict their foraging to accompanying large grazing or browsing mammals. They have been observed catching insects disturbed by a flock of Sandhill Cranes, by tractors, and by mowing equipment. They will visit newly plowed fields to feed on exposed larvae and grubs.

Cattle Egrets do not restrict their diet to insects. They capture other invertebrates and some vertebrates such as fish, amphibians, and small mammals. Besides using beaters to flush insects, they employ other hunting techniques. They have been known to employ stand and wait, bill vibrating, standing flycatching, gleaning, wade or walk slowly, walk quickly, running, hopping, leapfrog feeding, and aerial flycatching.[55]

Green-backed Heron
Butorides striatus

This heron has experienced two name changes in recent years. Formerly, it was known as the Green Heron, *B. virescens*. Then in 1976 the Committee on Classification and Nomenclature of the American Ornithologists' Union decided that it was conspecific with *B. striatus*. Its common name then became the Northern Green Heron, and its scientific name became *B. striatus*. When two or more species are found to be conspecific, we mean that they really are only one species and were mistakenly separated in the past. In 1982 its common name became the Green-backed Heron.

The Green-backed Heron is not a prominent bird in the Great Basin avifauna. Following his extensive fieldwork from 1867 to 1869, Ridgway wrote, "It appeared to be entirely wanting in the Great Basin—at least we could never find it, even in localities where other species of the family were found in the usual numbers."[25] The modern checklists which have been compiled at Great Basin locales, either do not list this species or list it as being of accidental or rare occurrence. These evaluations may well be true; but, in my opinion, this species is often overlooked. The Green-backed Heron is a wary, unobtrusive bird which inhabits wet areas which are relatively unfrequented by birders. It is usually a solitary nester, and its comings and goings in the Basin are of much less concern to most birders than are those of the glamor species of water and marshbirds, such as pelicans, grebes, ibises, and egrets.

The Greenback is known to nest in the northeast corner of the Great Basin, "in the marshes near the mouth of Bear River."[56] Although Linsdale was unable to locate any nesting records for western Nevada,[74] I suspect that some nesting occurs in the western part of the Basin. I have seen the bird here in early summer, and it was listed as an "uncommon summer resident" in a privately printed pamphlet on the birds of the Carson River Basin.[75] During mild winters this heron has been recorded in December from the Salt Lake Valley.[27] Over much of the Great Basin, the few reported sightings of this species have occurred during spring or during late summer and fall.

The Green-backed Heron is not much of a social species. It is usually a solitary nester, only occasionally nesting in small groups or in large colonies. The male establishes a breeding territory in which the nest is located. The nest is usually placed in a tree or shrub in the vicinity of water; occasionally it is located on the ground. The breeding behavior of this heron has been well studied. Both individual displays and mutual displays by both the male and the female have been observed—these displays involve feather erections, mandible snapping, calling, movements, posturing, and flights.

The Greenback feeds on a variety of items, including small fish, amphibians, reptiles, crayfish, insects, and other invertebrates. It has been known to vigorously defend feeding territories. A general hunting technique common to all herons is stand and wait, a type of still-hunting. The Green-backed Heron does this by getting into a crouched position, not an upright one as does the Great Blue Heron, and remaining motionless until prey approaches within striking distance. The Green-backed Heron has been seen using a modification of stand and wait called baiting. In describing baiting, Meyerriecks wrote, "I would like to relate an amazing example of learning in relation to feeding, first seen by Hervey B. Lovell. A green heron that foraged in an area where people threw bread into the water was apparently able to associate the presence of the bread with its use by the fish as food. Lovell actually saw the green heron seize some bread, place it in the water, and wait for the fish to grasp the bait. Indeed, he saw the green heron repeatedly retrieve the bait when it began to float into unsuitable foraging areas!"[66] The Green-backed Heron's repertoire includes still another modification of the stand and wait technique, one called standing flycatching—the heron, while in a stand-and-wait position, grabs dragonflies which approach within striking distance.

A second general hunting technique of herons, also employed by the Green-backed Heron, is stalking—using a wade or walk slowly approach. In employing this technique, the heron moves very cautiously and slowly; it may alternate bouts of stalking with periods of stand and wait.

A third general technique involves flushing prey by disturb and chase methods. The Greenback has been observed practicing both foot stirring and foot raking. During foot stirring the mud, vegetation, or bottom sediments are stirred with a foot, while during foot raking they are raked or scraped with the toes.

Green-backed Herons will enter deep water after prey, and their fourth set of general hunting techniques is employed in such situations. These techniques are jumping, diving, and swimming feeding. In jumping the heron jumps, feet first, from a perch several inches above the water, while in diving it enters the water headfirst from a perch or from the shore.

Green-backed Herons, both young and old, have been seen practicing swimming feeding.[55]

Black-crowned Night-Heron
Nycticorax nycticorax

The Black-crowned Night-Heron is the most cosmopolitan species of heron in the world today, and it is probably the most abundant one. It is found over much of the world, except for the northern part of the northern hemisphere. This night-heron is a common breeding species in many marshes and other wet areas in the Great Basin. It is a common migrant and may show up as a wanderer during its postbreeding dispersal.

The Black-crowned Night-Heron is a rare to uncommon overwintering bird in most parts of the Great Basin. However, in western Nevada in the Reno-Stillwater region, winter rookeries are regularly present. In Truckee Meadows, where the cities of Reno and Sparks are positioned, there are always several winter rookeries. From the early 1930s until recently, the largest winter rookery here was located in west Sparks. For years the number of herons in this rookery approached one hundred birds; during recent years the number has been decreasing, despite the protection afforded by the people living there. Unlike the west Sparks' rookery, winter rookeries in Reno have been subjected to constant and heavy persecution. Householders have objected to the "filth" deposited by the birds on the roofs and sides of houses adjacent to roost trees. In one instance, a number of years ago, someone had the tops cut off the poplars in which the birds were roosting along a lane in southeast Reno. The decline in numbers of winter-roosting herons in the Reno-Sparks area may well be a consequence of human population growth in the nearby Washoe Valley—which once supported a very large Black-crowned Night-Heron breeding rookery before human disturbance eliminated it.

Blackcrowns nest in colonies, often in mixed colonies with other species of herons and water birds. The colonies nest in a variety of situations from ground level on up to high in trees. After arrival on the breeding grounds, males leave their flocks and perform advertising displays from a series of temporary territories, which often include old nest sites. Displays may take place on nest platforms or on boughs of shrubs or trees. During his famous snap-hiss ceremony, also known as song and dance or reversed stretch, the performing bird arches his back, lowers his head, and while raising one foot makes a click or snapping sound in his throat, followed by a prolonged hiss. A twig ceremony may take place in which he holds a twig

Black-crowned Night-Heron

in his bill and then loudly snaps his bill on the twig, while moving his head up and down. Other herons are attracted to the site of an advertising male. If an onlooker approaches too closely, it is greeted with a threat display or attacked. If a receptive female is attracted to the male, she performs an appeasement display while approaching him: she fans her head and neck feathers, spreads her long white nape plumes, and stretches and nods; then she may bend her head to wing touch herself with her bill. The male may make an overture to the female and display to her. During the overture the male's head is lowered and turned until a cheek is parallel to the ground, guttural calls are uttered, the head is raised, and the plumes are erected. The male and female may then engage in mutual displays.

Many species of birds possess special body adornments. These adornments are often exquisite and beautiful, but some may be seen as grotesque and ugly. Usually the special adornments are present only on males or are best developed on males. In many species the adornments have become incorporated into the display complex of the males in such a way that the display movements, postures, and maneuvers appear to be designed to show them off. The nape plumes are the most conspicuous special adornment of the Black-crowned Night-Heron. Usually there are two or three of these long white plumes, extending down the back from their attachment on the nape of the bird. The male's plumes are typically longer, and sometimes more numerous by one or two, than the female's. Observations have shown that the male's nape plumes are critical to his success in mating. Males with damaged nape plumes may fail to obtain mates, since females are less responsive to their overtures. When pair-bonded males have their nape plumes removed experimentally, the pair bonds are weakened, and the pairs tend to break up. However, experiments have shown that the rosy leg color of the male does not function, in the eyes of the female, to enhance his leg movements during the snap-hiss ceremony—painting the male's legs blue-green had no effect on how quickly a pair bond formed or how long it lasted.[76]

The designation of night-heron is quite apropos for this species, since it hunts by night and roosts by day—most herons are diurnal hunters and nocturnal roosters. Field observations indicate that night-herons have difficulty maintaining feeding territories and competing for prey with the larger species of herons during the daytime, but at night they have the waters much to themselves. Observations indicate that, in localities where the larger kinds of herons are scarce, night-herons tend to do more daytime feeding. Black-crowned Night-Herons hunt typical heron prey such as fish, amphibians, reptiles, insects, crustaceans, small mammals, and small birds. This heron also has a reputation of raiding the nests of other species of birds for eggs and young. Its usual hunting technique is that of stand and wait. It also has been seen employing bill vibrating, walk slowly, hovering, plunging, feet-first diving, and swimming feeding.[55]

Other Ardeids

The Little Blue Heron, *Egretta caerulea*, is of occasional occurrence in the eastern end of the Great Basin.[27] Sightings here have become more regular of late; *American Birds*, reporting on the autumn field season for 1979, stated, "For the fourth consecutive year, the Salt L. Valley reported Little Blue Herons."[77] The scientific name of this species has recently been changed from *Florida caerulea* to *Egretta caerulea*.

The Tricolored Heron, *Egretta tricolor*, is of accidental occurrence in the Great Basin. An immature specimen was "collected at Malheur National Wildlife Refuge on 31 October 1943," and this heron was photographed at Honey Lake "standing on the shore of Hartson Reservoir, Lassen County, on 24 August 1971."[78] There are several undocumented sight records for this species in the eastern end of the Basin in the Great Salt Lake area. Recently this heron experienced a change in both its common name and its scientific name. Formerly, it was known as the Louisiana Heron, *Hydranassa tricolor*.

FAMILY THRESKIORNITHIDAE

The Family Threskiornithidae contains the ibises and spoonbills. Only one member of this family, the White-faced Ibis, is a breeding species in the Great Basin.

White-faced Ibis
Plegadis chihi

The White-faced Ibis has a discontinuous distribution in western United States, Mexico, and South America. Breeding colonies of this ibis are located in suitable marshes in the Great Basin: around Great Salt Lake and Utah Lake, at Ruby Lake, in the Carson Lake–Stillwater area, at Honey Lake, and at the Malheur National Wildlife Refuge. During migration, flocks of ibises can be seen feeding and loafing at wet areas throughout the Great Basin. A few have been known to overwinter in the Basin.

The White-faced Ibis was an abundant and widespread bird in the Great

Basin when the earliest ornithological studies were made here in the 1860s and 1870s. Ridgway reported, "This bird, known locally as the 'Black Curlew,' or 'Black Snipe,' was first observed in September [1867], at the Humboldt Marshes, where it was one of the most abundant of the water-birds, since it sometimes occurred in flocks composed of hundreds of individuals."[25]

In recent years there have been declines in the number and size of breeding colonies of ibises in the Basin. However, over the years this bird's fortune seems to fluctuate. A colony may decrease in size or disappear entirely, while nearby an inactive colony becomes active again, or a new colony pops up, or another colony increases in size.

Great Basin ibises have had problems with pesticide contamination, probably acquired during migration in Mexico. This has resulted in the thin-shelled egg syndrome and reproductive failure. The August 1970 issue of *Audubon Field Notes* reported, "The colony of White-faced Ibis near Bear River Refuge in Utah used to have about 5200 birds. This year only 900 ibis were present, and, according to the Denver Wildlife Research Center, there has been almost complete failure of nesting in the last three years owing to the thin-shelled egg phenomenon that is associated with the accumulation of DDT and DDT-type residues in the birds' body tissues. In this case it is believed that much of the DDT is picked up in Mexico where the ibis winter." The next issue, that of October 1970, reported, "A colony of 600 White-faced Ibis at Stillwater Nat'l. Wildlife Refuge in western Nevada was found to contain many young as well as thin-shelled eggs; it was thought that the colony may be somewhat smaller than usual but that production has not yet been affected significantly by the thin-shelled egg syndrome."

The colony referred to in the above quotation was actually located at Carson Lake, southwest of the Stillwater Wildlife Management Area. By 1976 this same journal, under its new name of *American Birds*, was now talking about the effects of water diversion and drought as well as pesticide contamination on the nesting success of the White-faced Ibis. The October 1976 issue related that "thin egg shells caused poor hatching in Utah. Low water caused high mortality in w. Nevada. After averaging 3995 young over the past four years, Carson L. started with 500 active nests May 17, which produced only 300 live young, about one-quarter of the potential. The nesting area dried up entirely by June 29, and predators moved in freely on the young not dead from exposure, dehydration, or lack of food."

Subsequent to 1976, the White-faced Ibis experienced several poor nesting seasons as well as some improvements during 1979, 1980, and 1981. Since these are troubled times for marshbirds throughout North America and the world, and since Hugh Kingery's summaries of the nesting season reports for *American Birds* so well chronicle the story of one of these marshbirds in the Great Basin, we shall now examine the recent reports on the White-faced Ibis. In these summaries, the letters in parentheses represent the initials of the observers who contributed information.

Kingery's 1977 summary reads, "No White-faced Ibises nested in n.w. Nevada, where Carson L., in past years had one of the nation's three largest nesting colonies (LCH). Ruby L. had 300 young, compared to last year, nesting in a sump coveted by power boaters (SHB). . . . abundant at Logan, Utah, ibises frequently fed in flocks of over 200 (only a few immatures observed), but observers could find no nests. The Bear R. colony apparently moved from the refuge to an adjacent gun club, success unknown."[79]

For 1978: "In Nevada 800 pairs of White-faced Ibises nested in seven scattered sites (PL); at Ruby L., they raised 400 young. At Carson L., 400 had poor success owing to poor water conditions and trampling by cattle; 3300 pairs nested there in 1973 (LCH). Bear River's 1000 pairs produced 1500 young, and flocks up to 300 fed in the fields around Logan, Utah."[80]

For 1979: "White-faced Ibises enjoyed success throughout the Mountain West, with 1300 young at Bear R., and 500 at Ruby L. . . . and the first nesting at Stillwater Ref., since 1960—probably refugees from dried up Carson L., nearby, but also signaling the improvement in Stillwater's habitat."[81]

For 1980: "White-faced Ibis news improved. In 1978 all of Nevada had 800 nests. This year, the Ruby L. area had 255+ pairs, with good success—3 young/nest (SB), and Stillwater W.M.A.'s 1800 nests produced 4500 young, up from 1200 nests and 2500 young last year (MR). Bear R. had 1000 pairs and 800 young (RV), and Logan, Utah's colony had the same size and success as last year (1500 pairs, 10% success due to predation—KA). At Ogden Bay, 100 were present June 16 (JN)."[68]

For 1981: "White-faced Ibises erupted over the Region. Nesting reports included 6600 reaching flight stage at Fallon, Nev. (Stillwater W.M.A., and Walker L.—ML), 150 young at Ruby L. (SB), ten young at Fish Spgs. (none in 1980 but 80–100 in 1978—KF); but Bear R. had none (thought to have nested off-refuge—RV). The most striking feature of the ibis' summer, were widely scattered birds and flocks apparently not nesting. These included 518 at the Salt Lake City airport June 25 (E & RS), 375 at Springville, Utah July 25 (MW)."[82]

And for 1982: "They had good nesting success at Bear R. (1000 adults June 1—RV), Fish Springs with 15 young (JGo), and Ruby L. (75–100 pairs fledged 2–3/nest—SB). Ibises were scattered throughout the Region during the summer, with post-breeding peak counts of 322 at Ogden Bay W.M.A., Utah July 14 (JN) and 300 at Lehi, Utah July 30 (MW)."[83]

Despite our concern about the welfare of the White-faced Ibis, we know relatively little about its natural history. Field observers have not carried their studies much beyond counting adults, nests, and young. However, Chapman did witness aerial displays in the San Joaquin Valley of California in the spring. Of these he wrote:

> On several occasions, however, we were privileged to see flocks of from ten to forty of these usually dignified birds perform a surprising evolution. In close formation, they soared skyward in a broad spiral, mounting higher and higher until, in this leisurely and graceful manner, they had reached an elevation of at least 500 feet. Then, without a moment's pause and with thrilling speed, they dived earthward. Sometimes they went together as one bird, at others each bird steered its own course, when the air seemed full of plunging, darting, crazy Ibises. When about fifty feet from the ground, their reckless dash was checked and, on bowed wings, they turned abruptly and shot upward. Shortly after, like the rush of a gust of wind, we heard the humming sound caused by the swift passage through the air of their stiffened pinions.[34]

Also, a nest-relief ceremony has been witnessed; when the female is relieved by the male, mutual preening and billing may take place, accompanied by some cooing notes.

White-faced Ibises nest in emergent marsh vegetation. In the Great Basin, they seem to favor tules. They are colonial nesters and often nest in mixed colonies with herons and egrets. One study showed that, within mixed colonies, the ibises tended to nest next to each other.

This ibis uses its long, decurved bill to probe in the mud or soft ground for a variety of prey. It has been known to feed on crayfish, earthworms, insects, mollusks, and amphibians. It may catch small fish by stand and wait or wade slowly techniques.

The White-faced Ibis is gregarious and is usually encountered in flocks. In the air ibises alternate flapping with sailing on set wings. They usually fly in some kind of alignment—in a diagonal line, or abreast in a line, or in an inverted V. Hoffmann's classic *Birds of the Pacific States* carries a good description of their flight, landing, and feeding:

> One sees lines of Ibis appear, their long necks and curved bills straight out and their feet extended well beyond the tail. Their lines tend to form a V but constantly break and re-form. As they prepare to alight, they tumble and dart downward, swooping suddenly this way and that. After alighting they often stretch their wings over their backs; then they fringe the edge of the shallow water, standing in the same attitude that

the Egyptian artists portrayed thousands of years ago. When they feed they often walk steadily forward, thrusting the whole head and neck under water.[84]

The Sacred Ibis, *Threskiornis aethiopicus*, a relative of the White-faced Ibis, was worshiped by the ancient Egyptians of dynastic times. Their great god Thoth was usually represented as an ibis or as ibis-headed. Dead ibises were taken to the sacred city of Hermopolis for burial, ibises were mummified, and it was a capital offense to kill this bird. Today the Sacred Ibis breeds only in Iraq; it had already almost completely disappeared from Egypt by 1850.[85]

Other Threskiornithids

The Roseate Spoonbill, *Ajaia ajaja*, ranges from the Gulf states into South America. On July 2, 1919, a specimen of this species was collected from a flock of five birds on a ranch near Wendover, Tooele County, Utah.[27]

FAMILY CICONIIDAE

In North America this family is represented by but one species—the Wood Stork, *Mycteria americana*. The breeding range of this stork extends from southeastern United States into South America. Following nesting there is a postbreeding dispersal; wanderers will occasionally reach the Great Basin during our summertime. There are some sight records and several specimens have been collected in the Utah part of the Basin.[56] In the Nevada part of the Basin, there is a sight record from near Beowawe[74] and several from near Fallon.[86] In older bird books this species will be called a Wood Ibis, but in 1973 its common name was changed to the more fitting one of Wood Stork.

FAMILY PHOENICOPTERIDAE

This family contains the flamingos, of which there is only one species present in North America—the Greater Flamingo, *Phoenicopterus ruber*. This species occurs south of the United States in places such as the West Indies, but it has occasionally bred in the Florida Keys. There are sight records of flamingos in the Great Basin—several in the Great Salt Lake area and several in the Stillwater Wildlife Management Area, in the Fernley Wildlife Management Area, and in Reno. Flamingos in the Great Basin may well be escapees from zoos, parks, or racetracks. In 1982 this species' common name was changed from the American Flamingo to the Greater Flamingo.

FAMILY RALLIDAE

The members of the rail family are predominantly solitary, secretive birds of omnivorous feeding habits. Although many of the species are migratory, they are weak fliers and are often reluctant to fly or fly far when flushed. The common expression "thin as a rail" is an allusion to their slim bodies, which allow them to slip through dense marsh vegetation. Of the seven species recorded from the Great Basin, four, possibly five, nest here.

Virginia Rail
Rallus limicola

The Virginia Rail, a breeding species in both North and South America, is found in many parts of North America, often coast to coast. The Virginia Rail is present as a breeding species in suitable marshes throughout the Great Basin. Some Great Basin observers rate it as locally common, others as uncommon or even of occasional occurrence. It is difficult to estimate the number of rails present in a region. Not many marshes, especially ones of any size, have been systematically combed for the seldom flying, small, secretive rails. Ears are often better than eyes in detecting the presence and estimating the population density of rails during the early part of the breeding season. In his *Rails of the World*, Ripley commented, "I recently was reassured to discover that the very secretive nature of rails tends to protect them. In the marshes of a rivermouth in Litchfield, Connecticut—site of my early explorations for Virginia and Sora Rails—tests by using a tape of rail calls played from a boat showed a dense population of apparently nesting pairs, particularly of Virginia Rails, beyond the dreams of avarice of any naturalist."[87] The greatest threat facing rails is the loss or degradation of their marsh habitats. Secretiveness will afford little protection against this, with people's hunger for land and water being what it is. Although an occasional Virginia Rail can be found in the Great Basin during the winter, most rails migrate south to overwinter.

It is virtually impossible to catch more than a fleeting glimpse of the Virginia Rail in its thick world of emergent vegetation, so we do not know much about its natural history. The territorial song or calls of the male have been described as a loud, rapid, metallic *kid-ick, kid-ick, kid-ick* or as *cut, cut, cut-ah, cut-ah*, often repeated many times.[65] Although the rail is active by day, it can often be heard calling at night. Territory is defended by threatening bill thrusts, by chasing, or by bill dipping—dropping the bill into the water at territorial boundaries.[87] One of the calls of this rail has been likened to the grunting of a hungry pig; this grunting may have a threat function. At times, the female joins in the calling of the male—as in a duet. The members of a pair exchange soft calls, and they practice courtship feeding and mutual preening.

The Virginia Rail usually builds its nest over water on a pile of vegetation. The most readily available marsh vegetation is used in building the nest, such as cattails, rushes, and sedges. If the nest site permits, rushes or sedges are arched to form a living canopy above the nest. This canopy may conceal the nest, but at the same time it may direct the eye of a knowing observer to the site; Walkinshaw has located many nests by first detecting the canopy.[88] Flooding may threaten rail nests during rainstorms—the moment rails detect rising water, they add material to the nest by pulling in dead vegetation and working it under the eggs.

The Virginia Rail uses its long, curved bill to probe in the mud for earthworms or insect larvae and to thrust in the water after small fish, snails, and other aquatic life. It runs with speed and agility through thick emergent vegetation, over mud, and on fallen or bent vegetation. If need be, it can swim.

Sora
Porzana carolina

The Sora, a summer resident and breeding species over much of temperate North America, is a breeding bird at suitable marshes throughout the Great Basin. Observers rate it as an uncommon to common summer resident at various Great Basin localities. Since this rail is a secretive marsh dweller, the impression that the average observer gains of its relative abundance may not be very accurate. Ridgway found the Sora to be widespread in the Basin during the summer in the 1860s; he wrote that it was "constantly met with in all suitable localities in the Interior."[25] An occasional Sora may overwinter here, but most migrate south of the Basin.

The Sora is the most widely distributed and familiar rail in North America. In the spring it is among the more persistent marsh musicians—during the day and often by night. Its high-pitched sequence of short, clear whistles is called a whinny, and the twelve to fifteen notes of *whee-hee——hee* are given in a descending key. Also heard is a clear, whistled *kerwee* or *ter-ee*.

Female Soras are prodigious layers of eggs. Their clutches may contain as many as eighteen eggs, although the average size is closer to ten to twelve. Both males and females incubate the eggs. But they are thin—"thin as a rail"—nine-inch birds and cannot properly cover a large clutch of eggs when incubating. Thus in a large clutch the eggs are frequently arranged in two layers, so the small-bodied bird can cover them. This presents a problem of delivering sufficient body heat to the bottom layer of eggs. Soras are thought to remedy the situation somewhat by frequently rearranging the eggs in the nest. Nests often contain newly hatched young and eggs in various stages of development. This may be entirely or partly due to incubation commencing before the clutch is completed, as well as to uneven heating of the eggs.

At one time rails were hunted as game birds—particularly in the eastern United States. Soras bore the brunt of the hunting there, since they often congregated in large numbers after the breeding season and during migration. In the late summer and fall Soras feed on the seeds of aquatic plants, especially wild rice, and their flesh is quite tasty.

Common Moorhen
Gallinula chloropus

The Common Moorhen is a cosmopolitan species found in the Americas, Eurasia, and Africa and on various islands. In 1982 its name was changed from the Common Gallinule to the Common Moorhen. In the United States, this species' breeding populations lie to the west, east, and south of the Great Basin. This moorhen is of occasional occurrence at various sites in the Basin. However, the Common Moorhen is listed as an occasional breeding species on the official checklist of the Stillwater Wildlife Management Area in western Nevada. There is some evidence that this species may occasionally nest in the Utah end of the Great Basin. Hayward has witnessed males fighting and displaying at Utah Lake in early May. Thus, the Common Moorhen may be a casual member of the Great Basin avifauna.[56]

American Coot
Fulica americana

The American Coot ranges over the Americas—from Canada to South America. During the summer it can be found at virtually every pond, lake, and marsh in the Great Basin. In the colder locales in the Basin, such as in the northwest and northeast corners, where winters are more severe and there is more closed water, the number of coots present in the winter declines. In other locales, such as the Reno-Stillwater area, the coot is abundant the year around. There is a considerable migratory movement of coots through the Great Basin, and the number of coots at the Bear River National Wildlife Refuge builds up tremendously during the fall.

The American Coot stands in bold contrast to its fellow rallids—the shy, secretive rails—for it is often conspicuously present on open water or along an open shoreline. Coots are frequently found in association with ducks. At park ponds in cities, such as at Virginia Lake in Reno, birds are protected from undue disturbance, and people frequently feed wild ducks bread and other table scraps. At these parks wild ducks and their associates, the coots and gulls, become so tame that they will swim or walk to within a few feet of a human being. Such a place is ideal for striking up an acquaintance with the American Coot.

Even when in among ducks and at a distance, a coot catches the eye, since it pumps its head and neck back and forth when swimming or walking. This unique head nodding is done in unison with the movements of the feet. Gallinules and moorhens also head nod; this is thought to be an orientation movement. When walking, a coot often assumes a hunchbacked posture.

When a coot is seen on land, its feet deserve special attention. The toes are long, and along each side of a toe are several membranous flaps or lobes. These lobes increase the surface area of the feet. Thus, added propulsion is gained when swimming, and the coot can walk on mud and muck without miring. In order to live on the mud and muck of marshes, rallids need foot specializations that will spread their weight out more. Coots have lobate feet; rails have long toes; and gallinules and moorhens have long toes, with a narrow membrane along both sides of each toe. The principal reason we know relatively little about rallids is that we have no means of moving around in a large marsh without becoming mired and exhausted. If we could move on foot effectively, we could manage the problems of secretive habits, mosquitoes, and summer heat.

When disturbed, coots show the usual rallid reluctance to fly. Instead they move away by swimming or diving or by rushing over the surface of the water. When rushing on the surface they vigorously beat their wings and feet against the water, noisily splashing about. The technique of rushing or running on the surface of the water is also used to gain sufficient speed to lift off when they are going to fly. Thus, when rushing away in alarm, coots are also approaching speeds where they can quickly lift off if need be.

American Coots are highly territorial. They apparently form permanent pair bonds, and each pair establish a permanent territory. In studies Gullion carried out at various locales in California, including Honey Lake in the Great Basin, the resident pairs of coots defended a core area in the winter around their nesting site of the previous year.[89] Defending permanent territory may well help keep the members of a pair permanently bonded. At the commencement of the breeding season, the pairs attempt to expand their territories from the core areas out to the boundaries of the previous year; in doing this, they may have to contend with outside coots and yearling coots trying to wedge their way in to form new territories. Territorial defense is at a low ebb while coots are incubating eggs but rises to its greatest intensity after the eggs hatch and young are present. Coots, ducks, blackbirds, other birds, garter snakes, turtles, and others are attacked when they trespass.

Early in the breeding season the male courts the female, frequently pursuing her with beating wings over the water. In the most common display, the male swims toward the female with his head on the water. His wing tips are lifted above his raised and spread tail in such a position that the white markings on either side are conspicuous. As he approaches the female, she often assumes the same position, and he turns, presenting his prominently marked tail to her view. He may then repeat the display. At times each bird, with its head extended on the water, will swim toward the other, calling *kuk kuk kuk kuk*. They may straighten and brush together as they turn, while tossing water about by poking it with their open bills. Sometimes they engage in allopreening. The female preens the feathers on her mate's head, then presents her head to be preened by him.

American Coots forage for food both on land and in water. In water they often upend, with their heads and necks submerged like dabbling ducks, or they dive for food. Their staples are the leaves, seeds, and roots of aquatic plants, although aquatic insects and mollusks are also taken. On land these birds forage by walking about. They may graze on grass, and they have a fondness for grain.

The American Coot often masquerades under the name of Mud Hen. As such it is despised by hunters and is believed to compete with ducks for food. It is seen as having no redeeming features as a game bird, being too easy to shoot and too unpalatable to eat. In earlier times gun clubs organized Mud Hen shoots to clear the waters somewhat before the duck season opened. Also, during earlier times it was claimed that "this bird has for many years been sold by the hotels of San Francisco and other large cities as 'ducks.'"[90]

Other Rallids

The several sight records of the Yellow Rail, *Coturnicops noveboracensis*, in the Great Basin are undocumented by specimens or photographs. Ridgway may have seen one in September 1868 in Ruby Valley, Elko County, Nevada. He shot several birds while at Parley's Park in Utah during the summer of 1869, but they were "all lost in the tall grass and sedges among which they fell."[25] So Ridgway never collected a specimen for positive identification. He identified the birds which he had seen in Nevada and Utah as Black Rails but stated that their appearance did not completely agree with that of Black Rails, since they showed a conspicuous white area "along the hinder edge of the wing" when they flew. This description better fits the Yellow Rail than the Black Rail.

There are also sight records from 1947 and 1969 for Yellow Rails near Cedar City, Utah, at the southeastern edge of the Great Basin.[27] In mid May of 1932 Linsdale, a highly competent field ornithologist, saw a Yellow Rail on several occasions in a small patch of cattails near Millett in Nye County, Nevada.[91] In the past the Yellow Rail has been seen during the summer in California, along the western edge of the Basin, at Quincy in Plumas County and at Bridgeport and Long Valley in Mono County.[92] Confirmed nesting occurred in Mono County in 1922, 1939, 1947, and 1950.[93] But these are old records. Our current knowledge about the status of this rail in California is very meager. Some believe that it may have been extirpated as a breeding bird in California and that current records are all of rare migrants. It is apparent that the records are too scattered and tentative to assign the Yellow Rail a position in the Great Basin avifauna at this time.

The little-known Black Rail, *Laterallus jamaicensis*, frequents coastal marshes in California. The Black Rail is on the official checklist of the Malheur National Wildlife Refuge as being of accidental occurrence there.

The Purple Gallinule, *Porphyrula martinica*, of accidental occurrence in the Great Basin, is far out of place here. It ranges from southeastern United States into South America. A specimen was collected southwest of Salt Lake City, and one was seen near Nephi on the eastern edge of the Great Basin.[27]

FAMILY GRUIDAE

Sandhill Crane
Grus canadensis

The Sandhill Crane is the only member of the Family Gruidae found in the Great Basin. The breeding range of the Sandhill Crane includes Siberia, Cuba, and regions in North America—particularly in the West and Southwest. In the Great Basin, its main breeding concentrations are in northeastern Nevada. This crane also nests in the northwestern part of the Basin—at scattered localities from Honey Lake northeastward to Malheur

National Wildlife Refuge. It also nests just beyond the northeastern edge of the Basin "south of Bear Lake and along the Bear River and its tributaries near Randolph."[27]

At one time the Sandhill was a widespread summer resident and breeding bird in the Great Basin. Describing its status in the 1860s, Ridgway wrote that the "Sand-hill Crane was an abundant species in nearly all localities where extensive grassy marshes or wet meadows existed."[25] But the Sandhill Crane has not been very successful in adapting to the changes humans have wrought on earth: it has been declining in numbers and disappearing from many of its former breeding localities, not only in the Great Basin but also over much of its range; and two of its subspecies—the Mississippi, *G. c. pulla*, and Cuba, *G. c. nesiotes*, Sandhill Cranes—have been listed in the Federal Register as endangered subspecies.

Of course, cranes in general have fared poorly in the face of advancing civilization, and six full species have been listed in the Federal Register as endangered: the Whooping Crane of North America and five other species from the East—in Siberia, Mongolia, Korea, Japan, China, and India. The Sandhill Crane was once present as a breeding species in a number of places in the eastern part of the Great Basin from which it subsequently disappeared. It was reported as nesting in the Salt Lake Valley, at Parley's Park, in Cache Valley, along the Jordan River, at Fish Springs, and possibly in Snake Valley.[94] In the western part of the Great Basin, it disappeared from Carson Valley, where it apparently was a breeding species in the 1860s.[25]

In recent years, the Nevada Department of Wildlife has been monitoring the breeding population of Sandhills in Elko County. The cranes nest in wet meadows where there are patches of willows. In September, prior to migrating south to overwinter, the cranes assemble at staging areas, mostly located in Ruby and Lamoille valleys, to feed in areas where grain is grown. In order to measure the recruitment of young birds into the population, young to adult counts are made in September: in 1978 only 5.8 young per 100 adults were counted; in 1979 the count was much higher at 17.4 young per 100 adults; and in 1980 there were 9.3 young per 100 adults.[95] After leaving Elko County, the majority of the cranes overwinter on the Colorado River Indian Reservation in Arizona. Lesser numbers overwinter at the Cibola National Wildlife Refuge, along the Gila River, and near Brawley, California. In the late winter, the cranes begin their return journey to northeastern Nevada. On the way large flocks stop over at Lund, Nevada, in the southern end of the Great Basin, to feed and rest before completing their return trip. The cranes are counted several times at Lund: the high count in 1980 was 1,028 cranes on March 6, and in 1981 1,094 cranes were present on March 5.[95] The Elko County crane population is considered to be a stable one at this time.

Cranes are great dancers—the wild, long-legged artistry and primal choreography in a Sandhill ballet are high theater. Sandhill Cranes can be seen performing at any season. Some dancing takes place when they are in their winter and spring flocks, but the frequency increases with the advent of the breeding season. Both males and females dance. Often with some bowing, the cranes perform a series of several to many stiff-legged leaps into the air. Soon rising to a height of eight to ten feet or more, they pass beside or even over each other. While on high, they may strike forward with their feet, the wings raised or drooped, necks stretched forward and upward, and bills elevated. A wild clamor of croaking whoops of *crrr-uk-crrr-uk* accompanies the dance. As the cranes abandon themselves to the beat of the dance, their cries rise to a din. Cranes are born to dance—they may even dance when young. This urge is so volatile that captive cranes can sometimes be induced to dance by the sight of a human imitating their dance.

Sandhill Cranes do not form pairing bonds until they are almost three years old, but once paired they remain together as long as both are alive. Upon returning to their breeding grounds in late winter or early spring, they seek out their territories of previous years. At this time, yearling cranes, who have accompanied their parents since birth, are driven away. The territories are located in freshwater marshes or in wet meadows. Areas where the waters are more than slightly alkaline or saline are not inhabited by cranes. Sandhill Crane territories are usually large, since all nesting season activities, including most of the food foraging, usually take place on the territory.[96] During territorial conflicts, the antagonists fight with wings, feet, and bills. After being driven off by their parents, yearling cranes join other nonbreeders to form flocks, whose members forage and roost together. Once a pairing bond is formed, the birds leave the nonbreeding flock and seek out a territory, often in a marsh or meadow uninhabited by other cranes.[96] The new pair may spend one or several breeding seasons on their territory before attempting to nest for the first time.

When constructing a nest, the members of a pair may stand close to each other, near the nest site. Seizing dead vegetation in their bills they pitch it backward, toward the nest site, with a sideways or over-the-shoulder motion. But they engage in nest building at a leisurely pace, working but for a short period each day. After sufficient material has accumulated to work with, one of the birds—often the female—stands on or next to the nest site, putting together the nest. Once eggs are laid, both sexes incubate. The average clutch size is two eggs, but clutches of only one egg are laid. In clutches of two eggs, hatching occurs on successive days, and the second chick to hatch is often the larger and stronger of the two. On occasions, sibling aggression has been observed between the two chicks in a brood. It is not known whether this aggression is much of a mortality factor or not, but reproductive success is low in Sandhill Cranes. Few parents successfully rear two young; most parents end up with one or no young.

Sandhill Cranes are omnivorous feeders. They especially like roots and grain, and they consume small mammals, frogs and toads, snakes, crayfish, and insects of various kinds. They dig up earthworms with their bills. Sandhill Cranes are believed to prey on nestling Red-winged Blackbirds. Redwings and Sandhills frequently nest in the same areas, and Redwings manifest a healthy hatred of Sandhill Cranes. Adult Red-winged Blackbirds often attack cranes on sight, flying at them and pecking them. Redwings may even land on the back of a crane and ride a while, pecking at the crane.

The Whooping Crane, *G. americana*, a North American relative of the Sandhill Crane, is an endangered species. At the low point in 1941, there were only fifteen Whoopers left in the wild. One of the recovery techniques intended to improve the position of the Whooping Crane is a cross-fostering program involving Sandhill Cranes. All the remaining wild Whooping Cranes nest deep in Canada, in the Wood Buffalo National Park, located on the boundary between Alberta and the Northwest Territories. The Whoopers migrate about 2,500 miles to overwinter at the Aransas National Wildlife Refuge, located on the Gulf Coast of Texas. Thus this endangered species has all its eggs in one basket, so to speak. The cross-fostering program with Sandhill Cranes is an attempt to establish another wild flock of Whooping Cranes and put the species' eggs into two baskets. Some Whooper eggs are removed from nests in Canada and flown south, where they are put in the nests of Greater Sandhill Cranes at the Grays Lake National Wildlife Refuge in southeastern Idaho—just off the northeastern border of the Great Basin. Other eggs have come from the captive population at Patuxent. The Greater Sandhills incubate the eggs of the Whooping Cranes and rear the chicks as their own. The Sandhill Cranes from Grays Lake fly south to overwinter at or near the Bosque del Apache National Wildlife Refuge in southern New Mexico—a trip of only 850 miles. If the cross-fostering program is successful, a new population of Whooping Cranes may emerge, with entirely different breeding and wintering grounds and a much shorter migration route.

8

Waterfowl

Collectively, swans, geese, and ducks are referred to as waterfowl. They are all members of the Family Anatidae, one of the two families which comprise the Order Anseriformes. The other family in this order is Anhimidae, a small South American family of three species of screamers. Waterfowl have webbed front toes and are foot-propelled swimming birds. Since there is but one large family of waterfowl, I shall subdivide it into taxonomic groups which we call tribes for ease of discussion.

TRIBE DENDROCYGNINI: WHISTLING-DUCKS

Whistling-ducks are long-legged, rather shy birds. Their name was derived from the whistlelike, often multisyllabic cries emitted by both sexes when in flight. Although they are not primarily perching ducks, they were formerly called tree ducks.

Fulvous Whistling-Duck
Dendrocygna bicolor

The breeding range of the Fulvous Whistling-Duck lies below the Great Basin, in scattered areas from the southern end of the United States into South America. It also occurs in and about Africa and India. In older bird books the Fulvous Whistling-Duck will be referred to as the Fulvous Tree Duck.

The Fulvous Whistling-Duck is of occasional or even casual occurrence in the Great Basin. There are several records from the eastern end of the Basin: a specimen was collected in the Bear River marshes in November 1908; a specimen was collected at Clear Lake Refuge, Utah, in May 1959; and a pair were observed near Cedar City in May 1969.[27] In the western end of the Basin, three specimens were collected at Washoe Lake early in 1877 from several large flocks—the first ever to be seen at this lake; another skin, now in the collection of the British Museum, was subsequently collected in the winter at Washoe Lake.[91] Another specimen was shot out of a flock of about twenty birds west of Fallon in November 1940.[86] There are sight records from Minden, Nevada, and Honey Lake, California. Although I am unaware of any documented instances of Fulvous Whistling-Ducks nesting in the Great Basin, the fifth edition of the AOU checklist suggests that casual nesting may have occurred at Washoe Lake.

TRIBE CYGNINI: SWANS

The swans are the largest of all the waterfowl. They have extremely long necks, which they extend when they fly. Most species are white, although the young may have some grayish brown coloration.

Tundra Swan
Cygnus columbianus

The Tundra Swan nests on the tundra of North America and Eurasia. In the Great Basin it is a common, often abundant migrant and an occasional to common winter resident. Excellent swan watching can be found at Bear River, Ruby Lake, and Malheur National Wildlife Refuges and at Stillwater Wildlife Management Area. The fall migration, peaking in late November, can be truly spectacular at the Bear River Refuge—up to 41,000 swans have been counted at one time there. During migration and even during winter, Tundra Swans may be seen on other bodies of water in the Great Basin as well—such as Washoe Lake, Pyramid Lake, and Honey Lake. The common and scientific names of this swan were recently changed from Whistling Swan, *Olor columbianus*, to Tundra Swan, *Cygnus columbianus*.

Diagonal lines or inverted Vs of these great white birds approaching across a dimly lit sky can be inspiring—as their wild clamor of hoots and yelps precedes them. When they are on the water, you may note that there is some internal organization to the flock. Family groups of two adults and one or several immature swans may be evident. The adult male and female remain pair-bonded for life and may be accompanied by their offspring of that year during migration and overwintering.

The feeding repertoire of swans involves three basic techniques—two for feeding on submerged vegetation and one for straining food particles out of surface water.[97] When feeding in shallow water on underwater plants, a swan merely submerges its head and neck. In deeper water it extends its reach by tilting its body forward with its rear end elevated in the air. When filtering food particles out of water, it positions its bill along the surface and allows water to enter at the front end of the bill. The water is then squeezed out of the sides of the bill; the serrations along the edges of both mandibles act as a sieve, allowing the water to pass while trapping the food particles.

Swans enhance their feeding effectiveness by working the vegetation over with their feet. This may involve raking the bottom vegetation with their powerful feet; raking not only shreds vegetation but helps free roots and rhizomes of mud. Swans are not members of the "clean plate club" but leave behind a rich soup of vegetation, and other water birds may be attracted to feed in their wake. Feeding Tundras may be accompanied by Mallards, Northern Pintails, Redheads, Canvasbacks, Gadwalls, Buffleheads, and Canada Geese.[98] Swans feed mainly on the stems, seeds, and roots of aquatic plants, but some animal food is taken; sago pondweed is one of the favorite foods of Tundra Swans. Tundras are very wary at all times: while a flock feeds on submerged vegetation, one or more swans remain on guard. Sometimes these swans will graze on land.

Over the years, the Tundra Swan has had its problems, but during recent years it has slowly increased in number—from an estimated 49,000 in 1950 to 157,000 in 1971.[99] Before the trade in bird skins was eliminated, this swan, along with the Trumpeter Swan, was shot for its skin. Eskimos and Indians harvest swans and swan eggs on the breeding grounds, and "the hunting of flightless molting birds by Indians and Eskimos are significant mortality factors in some areas."[100] The Tundra Swan was never a popular game species in the United States, since it was difficult to get a decent shot at this wary bird, and its meat was not highly regarded. Two Great Basin states, Utah and Nevada, have the rather dubious honor of still regarding this beautiful bird as a game species!

We have all heard that a swan sings before it dies. D. G. Elliott actually heard a swan song. He reported that "a number of swans passed over us at a considerable height. We fired at them, and one splendid bird was mortally hurt. On receiving his wound the wings became fixed and he commenced at once his song, which was continued until the water was reached, nearly half a mile away. I am perfectly familiar with every note a swan is accustomed to utter, but never before nor since have I heard any like those sung by this stricken bird. Most plaintive in character and musical in tone, it sounded at times like the soft running of the notes in an octave."[101]

Trumpeter Swan
Cygnus buccinator

The breeding range of the Trumpeter Swan consists of scattered nesting sites in southern Alaska, western Canada, and western United States. Until 1982 the scientific name of this swan was *Olor buccinator*. Then the genus

Olor was merged with the genus *Cygnus*, and the bird became *C. buccinator*.

There are two breeding colonies of Trumpeters within the Great Basin: one at Ruby Lake National Wildlife Refuge and one at Malheur National Wildlife Refuge. Both of these colonies were established by a series of transplants of cygnets and adults during the years that the Trumpeter Swan was an endangered species. Some of the Trumpeters at these two refuges overwinter as well. In addition, overwintering has occurred at Fish Springs National Wildlife Refuge.

Since the Tundra Swan is so common here, it is difficult to reconstruct the history of the Trumpeter Swan in the Great Basin prior to the introduction of transplants—starting in 1939 at Malheur and in 1947 at Ruby Lake.[102] It is often impossible to positively differentiate between Trumpeter and Tundra swans in the field by sight; you have to look for the Trumpeter's larger size and the lack of a yellow spot before its eyes. Voice differences are somewhat better indicators, with the Trumpeter's being lower-pitched, more sonorous and hornlike. But, then, the listener must be completely familiar with the voices of both swans, and the swan under scrutiny must cooperate by sounding off. Even when the bird is in hand, dissection of its breastbone may be required. In swans a loop of the windpipe is tucked inside the keel of the breastbone. In the Tundra Swan the two arms or sides of the loop lie close together inside the keel; in the Trumpeter Swan they are separated, as one arm of the loop bends away from the other, upward over a bony partition. Therefore records of occurrence for Trumpeters in the Great Basin are of varying worth, with specimen records being more valuable than sight records and sight record values varying with the competence of the observer-listener.

It is evident that the Trumpeter Swan was in the eastern end of the Basin prior to the introduction of transplants. The documentation consists of a mounted specimen, collected at Spring Lake, Millard County, Utah, in April 1892 and now in the collection of the University of Utah, and of six immature Trumpeters captured near Salt Lake City and sent in January 1901 to the New York Zoological Park. Further, it has been suggested that the Trumpeter was "probably a nesting species in the northern part of the state [of Utah] in early days."[56]

The evidence for the pretransplant occurrence of Trumpeter Swans in the western part of the Great Basin is based on one specimen record and several sight records. Bendire reported collecting a specimen in March 1877 at Malheur Lake.[103] Ridgway listed the Trumpeter Swan for Nevada with a question mark after its name; he stated, "In December, 1867, Swans were exceedingly numerous in the vicinity of Pyramid Lake, but as no specimens were obtained, we do not know certainly whether they were the Trumpeter or Whistler (*C. americanus*). Their note was almost exactly like that of the Sand-hill Crane (*Grus canadensis*)."[25] In recent years the swans I have seen at Pyramid Lake and along the lower part of the Truckee River have been Tundra Swans. In the 1870s, Henshaw reported the presence of Trumpeters on Washoe Lake and the sink of the Carson River in the fall. Today, Tundra Swans are frequently seen on Washoe Lake in the late fall, winter, and early spring. In November 1935, McLean reported seeing a Trumpeter up close, within thirty feet, and hearing its "deep resonant call," just northeast of Eagle Lake, California, between Termo and Grasshopper Valley; he also reported that his companion, A. L. Brown, had seen Trumpeters as a boy at Honey Lake, California.[104]

For those concerned with the welfare of endangered species, the story of the Trumpeter Swan has become heartwarming of late. It is a rare bird, indeed, who survives endangerment and recovers enough to be taken off the endangered list—as did the Trumpeter Swan. During the nineteenth century this swan was abundant and widespread in North America. By 1932, however, a National Park Service survey showed only 69 Trumpeters—57 adults and 12 cygnets—in all of the United States.[102] With the handful left in Canada, there were probably fewer than 100 adult swans outside of Alaska in 1932. We are not sure of the size of the Alaskan population in 1932, but it was probably quite a few times larger than the combined U.S.-Canada populations. In 1959 an aerial census revealed 1,124 Trumpeters in Alaska. But in 1964 these birds were found breeding in the interior of Alaska, and with the addition of these interior swans the aerial count for Alaska in 1968 was 2,848.[105] Since some individuals are overlooked on aerial surveys, count values are minimal, and there could have been 3,400 or more swans in Alaska in 1968. When the revised edition of *Rare and Endangered Fish and Wildlife of the United States* appeared in December 1968, the Fish and Wildlife Service no longer listed the Trumpeter Swan as either rare or endangered. By 1971 the Survival Service Commission of the International Union for Conservation of Nature and Natural Resources had removed the bird from its *Red Data Book* as an endangered species.

The settlement of North America brought on the decline of the Trumpeter Swan. For ages this swan had lived in ecologic balance with the hunters of North America, supplying them with meat, eggs, and feathers. But with the coming of Europeans the swan's troubles began. Early in the nineteenth century, swan skins became an article of commerce. In 1806 the Hudson's Bay Company exported 396 swan skins; between 1823 and 1880 the company sold 108,000 swan skins, mainly those of Trumpeters, on the London market. The skins were used in manufacturing powder puffs and as garment ornaments. The quills made fine pens. That great painter of birds,

John James Audubon, used Trumpeter quills in his work; he wrote that these quills were "so hard, and yet so elastic, that the best steel pen of the present day might have blushed, if it could, to be compared with them."[102] With settlement and the reclamation of wetlands, the Trumpeter Swan lost most of its breeding grounds in the United States and Canada. It disappeared as a breeding species in north central conterminous United States and over most of the interior of Canada. The relatively few breeding populations left in North America either were sedentary ones or performed only limited migrations. So the bird disappeared as a migrant and overwintering species in the states south and east of its former breeding grounds. As a migrant and overwintering species the Trumpeter had been much more susceptible to hunting than the Tundra Swan, since it was less wary and flew much lower—within reach of the hunters' guns.

Although the Trumpeter Swan's recovery has not been spectacular within the conterminous United States, it is no longer tottering on the brink of extirpation. Protection against illegal hunting is efficient. In 1935 a national refuge was established at Red Rock Lakes, Montana. Marsh management and winter feeding programs were initiated. For example, the trapping of muskrats was limited, since the main nesting sites of Trumpeter Swans there have been on top of muskrat houses. A program of translocating Trumpeters was instigated in 1938.

The Great Basin has been involved in the translocation program almost from its inception; on October 16, 1939, three cygnets were transferred from the Red Rock Lakes Refuge to the Malheur Refuge. The first transplant at Ruby Lake National Wildlife Refuge was made in 1947, when twelve swans, originally sent to Malheur from Red Rock Lakes as cygnets, were translocated to Ruby Lake. Since these initial transplants, a series of other transfers of cygnets, nonbreeders, and breeding pairs has been made to both Malheur and Ruby Lake Refuges from Red Rock Lakes.

The transplant program has encountered difficulties and failures. In 1979 at Malheur, the Trumpeters probably had their best year since their first successful breeding in 1958—"with 35 cygnets on or near the refuge."[106] In 1980 "Trumpeter Swans had their best year ever at Ruby Lakes N.W.R., Nev., with 15 cygnets included in the Dec. 1 count of 50."[107] The Ruby Lake population is a stable one, but with a lot of mortality. Usually there are about ten nesting pairs of Trumpeters, which fledge somewhere between ten and fifteen young a year. Several of these pairs may nest at nearby Franklin Lake, the rest on the refuge. The Trumpeters here are late breeders, and Franklin Lake often dries up during the summer. In 1981 the last clutch did not hatch until around the beginning of July. In November 1981, *American Birds* reported that the "cygnets fledged by Trumpeter Swans during the past few years at Ruby L., have found one new probable nesting site, the Newark Valley near Eureka."[82]

Trumpeter Swans mate for life. Some young swans form pair bonds during the second winter of life, and all have paired by the third winter of life. Trumpeters use their voices and postures and movements to express a variety of emotions. Displays occur throughout the year, and mutual displays are common. "The most common display attitude is one in which the quivering wings are raised horizontally and partly extended. This posture, when accompanied by the pertinent vocal effort and extended position of the head, neck, and body, is used with some variation on greatly different occasions."[102] Precopulatory display involves "mutual head dipping," as both members of the pair engage in dipping their heads into the water. In postcopulatory display the female calls as the male spreads his wings; then both rise in the water, calling and turning in a partial circle.[108]

The Trumpeter Swan is the largest of all the waterfowl. A large male will weigh thirty pounds, often much more, and have a wingspread of ten feet.[109] An adult male swan is called a cob, an adult female a pen, and a young swan a cygnet. The adult plumage is completely white, although swans feeding in waters where the marsh muds are high in iron may have a rusty wash to the head, neck, and underparts. Before taking flight, swans may show flight-intention movements by pumping their heads and necks up and down and calling. Once in the air they are strong fliers at about two wingbeats per second. The neck is fully extended, and the feet may be thrust forward and buried out of sight under the feathers. In flight a flock may travel in an inverted V or in a line—abreast or staggered.

Trumpeter Swans tend to return to their old nest sites, year after year. Trumpeters have large territories with adequate feeding areas. A male strongly defends his territory against trespass by other swans and even geese; ducks are tolerated. Since a refuge like Malheur or Ruby Lake can contain only a limited number of Trumpeter territories, young pairs of swans find it impossible to breed in competition with established pairs, so there may be a large component of nonbreeders in a population. Further, the nonbreeders often find it difficult to feed, since that often entails trespassing on a defended territory.

A firm support for the nest is needed. A favorite site is often the top of a muskrat house. The average clutch size is five eggs, and the incubation period ranges somewhere around thirty-two to thirty-seven days. Many eggs fail to hatch, and cygnet mortality is typically very high, ranging upward from 50 percent. The cygnets initially feed on aquatic insects and crustaceans but soon are feeding on the leaves, stems, roots, and rhizomes of aquatic plants. The feeding techniques of Trumpeter Swans are like those of the

Tundra Swans. After the cygnets hatch, molt begins for the adults—since they shed all their flight feathers at once, they are flightless then. During this period, which lasts about a month, the swans avoid predators by swimming or by diving underwater when threatened. The family groups form flocks for migration and overwintering. Today, the Trumpeter Swans in the conterminous United States remain in their breeding areas to overwinter or perform very limited migration. Since these swans live at quite high latitudes, they have to rely on conditions such as hot springs to keep ponds open during the winter.

TRIBE ANSERINI: GEESE

Geese are large birds with long necks and short legs; they fly with their necks extended and are noisy in flight. Geese are gregarious most of the year. They spend much of their time out of water grazing on land.

Canada Goose
Branta canadensis

The Canada Goose is a North American species which breeds from the arctic tundra to as far south as the Great Basin. Formerly, there was an Asiatic subspecies of the Canada Goose, the Bering Canada Goose, *B. c. asiatica*, breeding on islands in the Bering Sea, but this subspecies has been extinct since 1914. During migration and winter, Canada Geese can be found over the rest of North America as far south as Mexico. The Canada Goose is a common breeding species at larger lakes and marshes in the Great Basin and is an abundant migrant here. The number of geese overwintering in the Basin varies according to the severity of the weather, but some are present every winter. The status of this goose has not changed much since the Basin was settled, for Ridgway observed in the 1860s that "this species was the only one of the genus found breeding in the Great Basin, where it remained throughout the year about all the larger lakes."[25]

Members of the various breeding populations of Canada Geese are alike in their general color pattern, although some are darker and others are paler. Therefore, it is easy to recognize a Canada Goose when you see one. But, when you see some migrants, you will note that they come in different sizes and proportions—some are not much bigger than a Mallard, others are almost as large as a swan. Delacour was able to recognize eleven subspecies of living Canada Geese and one extinct subspecies.[109] Of these eleven subspecies, four are lightweights, two are middleweights, and five are heavyweights. The subspecies breeding in the Basin is a heavyweight by the name of the Great Basin or Moffitt's Canada Goose, *B. c. moffitti.*

Several of the lightweight subspecies have shown up in the Great Basin during migration or in the winter. Of these, the smallest and darkest of all the lightweights, the Cackling Goose, *B. c. minima*, is not much larger than a big Mallard. It breeds along the western coast of Alaska. In migrating south, it pauses for a major rest stop in the Klamath Basin of northern California, before going on to overwinter in the Central Valley of California. The Cackling Goose has been seen and collected at locales throughout the Great Basin as a migrant or visitant; during recent years in western Nevada, it has been of casual occurrence. However, this was most likely the subspecies which Ridgway collected in Truckee Meadows in November of 1867, about which he wrote, "This miniature of the Canada Goose was an abundant winter visitant in western Nevada, but it was not seen anywhere in summer, when all had gone northward to breed."[25] Ridgway called his miniature goose the Hutchin's Goose. But what we recognize today as the subspecies *hutchinsii* is a slightly larger lightweight which breeds in eastern arctic America and passes east of the Great Basin during migration, to overwinter in Texas and Mexico.[109] Specimens of *hutchinsii* have been collected from several different locales in Utah.[56]

Another lightweight subspecies that has been seen and collected in the Great Basin is the Aleutian Canada Goose, *B. c. leucopareia.* This subspecies was frequently seen in the Lahontan Valley of western Nevada during the fall and winter, and specimens were collected in November and December of 1940 and January and November of 1941 by Alcorn.[86] Specimens of this subspecies have been collected in the Bear River marshes.[56]

The Aleutian Canada Goose is an endangered subspecies. Its story is a tragic one. Formerly, it was an abundant bird, breeding at many sites in the outer two-thirds of the Aleutian Islands. In the fall it migrated south to overwinter from British Columbia to California, although some went to Japan. There is a possibility that this subspecies, instead of *minima*, was Ridgway's "miniature of the Canada Goose." Although the Aleutian Canada Goose has always been hunted on its breeding grounds during migration and on its wintering grounds, its decline has been mainly due to predation by arctic foxes—which were introduced onto islands in the Aleutian chain for fur-farming purposes from about 1836 to 1930.[110] Once established on an island, the arctic foxes promptly wiped out all the ground-nesting colonies of Aleutian Canadas. Fortunately, fur farmers neglected to introduce the foxes on one breeding island—Buldir Island, a small volcanic rock in the western Aleutians, isolated and lacking a good harbor. The 1973

Canada Goose landing

edition of *Threatened Wildlife of the United States*, published by the Fish and Wildlife Service, stated the estimated world population of these geese to be about 250 to 300 individuals. The spring count in 1977 showed 1,150 birds.[110]

The Fish and Wildlife Service appointed a recovery team in 1975 to aid the Aleutian Canada Goose; a four-point recovery program is being implemented. Arctic foxes are being removed by poisoning from three former nesting islands—Agattu Island, Amchitka Island, and Kanaga Island. A program of breeding captive geese is under way at several propagation facilities. Captive-reared geese are being reintroduced on the fox-free islands. Finally, the basic biology of the subspecies is being studied, and the geese are being given some protection while migrating and on their wintering grounds.

Often it is impossible to differentiate between subspecies of geese by sight alone. There is considerable individual variation in size and coloration within a subspecies, and subspecies may interbreed when they come in contact in nature, giving rise to intermediate individuals. Nevertheless, closely scrutinize the lightweight Canada Geese you encounter in the Great Basin, where goose-watching conditions are great, particularly in the fall and winter. The Aleutian Canada Goose is slightly larger than the Cackling Goose; and it always has a white ring around the base of its black neck, followed by a dark brown band. This white ring is often quite broad in front (up to twenty millimeters). Cackling Geese may or may not have the white band, but if present it will be narrow. When viewed from above, the bill of the Aleutian Canada Goose is more pointed than the bill of any other subspecies of Canada Goose.[109]

The two middleweights are each represented by an identified specimen collected in the Great Basin. What was apparently a Lesser Canada Goose, *B. c. parvipes*, was collected in Millard County, Utah, in January 1964.[56] A specimen of the Dusky Canada Goose, *B. c. occidentalis*, was collected at Soda Lake near Fallon in January 1942.[111]

Canada Geese are monogamous and pair for life. Although the pair bond is very strong, a dead mate will be replaced. The members of a pair remain together the year around. This type of pair bond has long intrigued human observers, since it is obviously based on something beyond mere sexual attraction and a seasonal urge to reproduce. Is some of the glue which holds these birds together akin to love? Well, who knows. But we do have observational evidence that companionship and even affection for each other may be involved. Hunters have often observed that, when a single goose is shot out of a flock, another goose will leave the departing flock and circle back toward the fallen bird, calling plaintively.[112]

Not only are pairing bonds strong, but family bonds are strong as well.

Young geese remain with their parents for most of the first year of life and sometimes even well into their second year of life. During the second summer of life, the summer after they were born, some yearlings form pairing bonds; but these are usually only temporary—when the bonds break down, the yearlings rejoin their respective parents or their siblings to spend the coming fall and winter with them.[113]

The entire existence of Canada Geese is structured around family life; they are highly gregarious birds and remain in flocks outside of the nesting season. A small flock may consist of a single family. Although a large flock will contain some single geese and some pairs, its basic unit of organization will be families—parents and their offspring. The larger the family, the higher its status and the more it dominates other families within the flock. Investigations utilizing telemetry and color marking have shown that the family unit is maintained during the performance of daily activities. The subunits within a flock are not spatially recognizable when the flock is taking off or while it is in the air in steady flight. But when landing each subunit usually becomes recognizable: a single goose will veer away from the rest of the birds before landing; members of a pair will land close together; and the members of a family will all land close to each other.[114]

Some Canada Geese are pair-bonded and are nesting by the time they are two years old; the majority have mated by the time they are three. Courtship starts on the wintering grounds and may continue en route to the nesting grounds. Most young geese destined to breed that year arrive on the nesting grounds already pair-bonded. Pair bonding is probably initiated by a young gander (male) pursuing a goose (female). Ganders often fight over a goose—pinching with their bills and beating with their wings. Triumph ceremonies seem to be important in pair bond formation. Calling loudly, the gander approaches his intended goose—often after a victorious fight with another gander—his open bill close to the ground, his neck extended and weaving to and fro. If the goose is receptive, she may join in converting the display into a mutual display. If the goose follows the gander when he flies, they are considered to be pair-bonded. The triumph ceremony is not restricted to courtship—the members of a pair perform it after the gander has successfully defended his mate or family or territory.

A territory is defended with aggressive displays and calls by both the gander and the goose. The nest is located on the territory, and, ideally, the territory is large enough to include adequate food and water within its boundaries. There is a lack of unanimity among observers concerning the universality of territorialism in Canada Geese. From studies of heavyweight Canada Geese—or Honkers—Williams concluded that the ganders did not display territorialism; that is, they did not stake out a piece of the breeding grounds and defend it against trespass by other geese. Instead, the gander merely guarded "an area around his mate, wherever she may be during the preincubation period." Williams believed that situations where many nests were crowded into a small space—as on an island, sandbar, or haystack—were possible because there was no defense of real estate by ganders. He observed that "one haystack, which measured 10 by 30 by 8 yards in Oregon, had 11 goose nests on it."[112] Of course, territories come in all sizes. Territorialism is not a function of low nest densities, nor are high nest densities necessarily a function of the lack of territorialism. Colonial nesting birds actually have territories, but they may be so tiny as to extend no further out from a nest than a bird sitting on that nest can reach with its bill. Apparently, Honkers can exploit prime nesting sites by nesting colonially.

The movements of the precopulatory display of the Canada Goose are suggestive of bathing movements and have probably been derived from that mundane practice. The principal component of the display is mutual head dipping. Usually the display is begun by the gander. While the gander and goose are facing each other, rapid tipping or upending in the water, as in bathing, may precede the mutual head dipping. During the display the gander and goose dip their heads and necks in and out of the water, as in bathing. After copulation, while facing each other, the birds raise their breasts up, hold their heads high, and extend their wings out from their bodies, while one or both call out.

The goose does the nest building. She looks for a site that is isolated, close to water, and elevated with good visibility in all directions. A variety of sites are used—including the tops of muskrat and beaver houses, bars, islands, dikes, and even matted bulrushes. Occasionally, the birds will nest in a tree, in an abandoned heron or Osprey nest. Geese readily take to artificial structures, and wildlife management programs have provided a variety of types aimed at increasing productivity. Elevated structures are provided in areas subject to flooding. These can be a simple wooden platform or one with a tire, basket, or washtub on it; the platform may have sides around it like a small sandbox.

The goose does all the incubating, while the gander remains on watch nearby. Weather permitting, the goose leaves the nest several times a day on short breaks. Before she leaves, she covers the eggs with down from the nest lining. On her break, she is accompanied by her gander, who stands by while she feeds, drinks, and, maybe, preens or bathes. As in other anatids, the goose does not automatically acquire, through hormonal control, a bare incubation patch on the underside of her body. Instead, starting during egg laying and continuing through incubation, she removes down and some contour feathers from the underside of her body, with her bill, to create an incubation patch of bare skin. The plucked down and contour feathers are added to the nest. Bare skin is more efficient in transmitting body heat to the eggs than is skin covered with feathers.

After an incubation period of about four weeks, the eggs hatch. The precocial goslings are soon led away from the nest to a rearing area by their parents. Often they spend the first night in the nest and depart the next day. Long distances, requiring several days' travel by water, may be traversed to reach the rearing ground, which has open water, food, cover, and suitable roosting places. When swimming one parent leads the parade, followed by the goslings, with the other parent bringing up the rear.

Early in life the goslings cannot recognize their siblings and their parents as individuals, and their parents have difficulty recognizing them. As a consequence, confusion ensues when two broods come together. The more aggressive pair of parents may capture the brood of the more submissive pair, leaving them broodless. Or the mixed broods will attempt to unsort, with each pair of parents moving off with a mixture of their own goslings and those of the other pair. Sometimes a number of families will amalgamate to form a large rearing group or crèche. Nonbreeding adults and subadults may even join the group as escorts. One such group seen in Missouri was composed of one hundred goslings and twenty-one adult and subadult geese.

After their goslings are a few weeks old, the parents undergo their postnuptial (prebasic) molt. This is called a synchronous molt, since all the flight feathers are shed simultaneously. During the five or six weeks required for new flight feathers to complete their growth, the geese are flightless. Synchronous molting is characteristic of anatids, loons, grebes, pelicans, and certain other water birds who can escape from predators by going out into the water or underwater. The flight feather molt of land birds is spread out in time, so they are never flightless—land birds would be too vulnerable to predators if they were in a flightless stage. Of course, not even water birds have a synchronous molt of body feathers. Body feathers are molted feather by feather over such a period of time that a bird never goes through a naked stage—a naked bird would soon succumb to the elements.

The various breeding populations of Canada Geese engage in molt migrations—when some subadult geese and nonbreeders or failed breeders among the adults migrate away from the breeding grounds for the summer postnuptial molt. Molt migrations, generally performed in a northerly direction, require hundreds to several thousands of miles to reach the molting grounds. Geese from the Great Basin have been found on molting grounds in the Far North—in the Beverly and Aberdeen Lake region—lying 250 to 350 miles west of Chesterfield Inlet on Hudson Bay. Molt migrations are thought to reduce food competition back on the nesting grounds and thus ease the pressure on the breeders and their goslings.

The patterns of migratory paths or corridors to wintering grounds of the Canada Geese are bound to be complex, since these birds are so far-flung in latitude and longitude on their breeding grounds. In general, the lightweights or small Canada Geese nest out on the arctic tundra; the larger subspecies nest south of, or inland from, the tundra in wooded, prairie, or shrubland regions. Considerable leapfrog migration occurs in Canada Geese, where more northerly breeding populations will fly over more southerly breeding populations and winter further south. Some leapfrogging is characteristic of the subspecies *moffitti*, which breeds in the Great Basin as well as north of it.[100] A great many of these geese may overwinter in the Basin, particularly in the western part at places such as Malheur, Honey Lake, and Stillwater. There is also some overwintering in the eastern part of the Great Basin. Geese coming from the more northern breeding sites of this subspecies tend to leapfrog the more sedentary southern populations. Many *moffitti* collect in southern Idaho along the Snake River, and some of these move southwestwardly into western Nevada and into the Central Valley of California. Some coming out of northeastern Utah overwinter in extreme southern California and in Arizona in the vicinity of Phoenix.[99] The number of geese overwintering in the Great Basin varies according to the severity of the winter. Local movements occur if an area is plagued by snow and closed waters.

During long flights and migration, Canada Geese mainly fly in formation. A flight of birds in an inverted-V formation is almost universally thought to be a flock of geese, and many observers are surprised to learn that birds other than geese fly in this formation. During a study of over one hundred migrating Canada Geese flocks, only about 20 percent were not in a definable formation. The remaining 80 percent were in inverse-V formations, or in J formations, or in diagonal lines.[115] Since geese so very often fly in formation on long flights, it is tempting to speculate that they may be deriving some benefit from doing so.

One of the potential aerodynamic advantages of formation flying could be a decrease in drag—the force opposing forward motion. Another potential advantage could be an increase in lift—the upward force that keeps a bird airborne—for in flight an upwash of air is created off the wing tip, and each bird could theoretically gain lift by positioning its inner wing tip on the rising vortex of air left by the bird in front of it. Theoretical calculations indicate that a flock of twenty-five birds with optimal spacing in an inverse-V formation could have an increase in flight range of about 70 percent and that this formation is required to equally distribute the drag saving among the birds.[116] To gain lift or decrease drag while flying in formation, a number of conditions must prevail: each bird must beat its wings at the same frequency and in phase with the bird in front of it; the inside wing tip of each bird must be properly placed in the upwash of the bird in front of it; and all the birds must remain in the same horizontal plane.

Although camera positions and angles often leave much to be desired, analyses of films of several species of geese flying in an inverted-V formation have led to the following conclusions: each goose beats its wings at its own frequency, not in synchrony with the goose ahead of it; the angle of the V varies from time to time; the spacings between the birds vary; and the birds rise and fall in relation to each other and do not stay within a single horizontal plane.[115] So there is some controversy over whether geese gain aerodynamic efficiency by flying in formation. However, it has been suggested that there are other benefits to formation flying. By being displaced to one side of the preceding bird, a goose avoids flying in its turbulent wake, which prevents a loss in aerodynamic efficiency. At the same time, the birds can maintain visual contact without any danger of colliding with each other. The rear end of a flying Canada Goose is well designed to function as a white flag to aid visual contact, since the upper tail coverts above the base of the tail are white and contrast well with the darker tail and back feathers.

The Canada Goose feeds on aquatic plants, but it is also a terrestrial grazer and is often seen on land feeding on clover, grasses, and other plants. It readily takes agricultural crops such as grains and corn. A major tool in game management involves planting agricultural crops for geese to consume during migration and on the wintering grounds. The creation of sanctuaries and the planting of food crops have brought the Canada Goose back to a point where it is probably more abundant today than at any time in its past. However, it still remains the number 1 game bird in this country.

Other Geese

The Greater White-fronted Goose, *Anser albifrons*, is of occasional to uncommon occurrence in the Great Basin. It breeds in the Far North, and the main flows of migrants to wintering grounds pass to the east and west of the Basin. Up until recent years this goose was a regular winter visitant in western Nevada. Concentrations of fifty to three hundred or more geese would appear and loiter for a few weeks in January or February in the wet fields adjacent to Kleppe Lane on the eastern edge of Sparks. With the decline of ranching the marshes and wet fields disappeared, as industrial developments such as warehousing took over, and the geese stopped coming here.

The Snow Goose, *Chen caerulescens*, is a common to occasional migrant at marshes and lakes in the Great Basin. It breeds in the Far North, and some of the national wildlife refuges and wildlife management areas in the Basin—such as Bear River, Malheur, and Stillwater—lie on its migration corridors. It is also an occasional to rare winter visitant in the Basin. This bird's status as a winter visitant may have changed over the years, since back in the 1860s Ridgway found it to be an "abundant winter visitant to the lakes in the Great Basin."[25] At one time the Snow Goose, *C. hyperborea*, and the Blue Goose, *C. caerulescens*, were considered to be separate species. Now they are considered to be merely different color phases—or what we call morphs—of a single polymorphic species, which is called *C. caerulescens*. The white morph which occurs in the Great Basin is called the Snow Goose, *C. caerulescens hyperborea*, and the dark morph is called the Blue Goose, *C. c. caerulescens*. The Blue Goose occasionally occurs as a migrant in the Great Basin.

Ross' Goose, *Chen rossii*, is an occasional to uncommon migrant in the Great Basin, except at Honey Lake, where it has been reported during migration in fair numbers.[117] It is reported to be of rare occurrence in the fall and winter at Bear River National Wildlife Refuge.[118] This goose breeds in the Far North, and its migration corridor passes along the northwest corner of the Great Basin.[99] It may be of more regular occurrence in the Basin than the records indicate—it is easily overlooked, since it often occurs with Snow Geese and looks much like them.

The Emperor Goose, *Chen canagica*, is reported to be of accidental occurrence at Malheur National Wildlife Refuge.[61]

The Brant, *Branta bernicla*, is a coastal species, but records of its occasional occurrence in the Great Basin are scattered over more than one hundred years. There are specimen records from the Great Salt Lake area in the eastern end of the Great Basin and from the Stillwater area in the western end. In addition, there are sight records for the Basin going as far back in time as Ridgway's December 1867 record at Pyramid Lake[25] and Henshaw's October 1872 record at Rush Lake, Utah.[56] The sight records for western Nevada include Washoe Lake (spring), Virginia Lake (fall), and Paradise

Park in Reno (winter). Formerly, the West Coast Brant was considered to be a separate species called the Black Brant, *B. nigricans*. Now it is regarded as merely a subspecies of the Brant, *B. bernicla*.

TRIBE CAIRININI: MUSCOVY DUCKS AND ALLIES

Members of this tribe are often referred to as perching ducks, since they spend more time in trees than do other waterfowl. They generally nest in cavities. Many species possess iridescent coloring.

Wood Duck
Aix sponsa

The Wood Duck has a discontinuous breeding range: one part occupying eastern United States and immediately adjacent Canada, the other part occupying the western edge of the United States and immediately adjacent Canada. Only a bit of the western border of the Great Basin, in the area of Wadsworth, Carson City, and Stillwater, lies within its breeding range. The Wood Duck has been recorded nesting at the Stillwater Wildlife Management Area, and I have evidence of nesting from along the Truckee River below Wadsworth and from along the Carson River below Dayton. The Wood Duck probably has long been a breeding species in western Nevada, since Ridgway encountered a pair in July 1867 along the Truckee River.[25] It is an occasional to uncommon visitant over the rest of the Great Basin, with records scattered throughout the year—more in the fall, with fewer in the spring and winter. Since the Wood Duck barely penetrates the Basin as a breeding species, it is best regarded as a peripheral species rather than as a full-fledged member of the Great Basin avifauna.

Wood Ducks are highly arboreal—they not only perch in trees but nest in cavities in trees. In Nevada suitable tree cavities are at a premium in the Truckee and Carson River bottoms, and I was intrigued by the discovery that along the Carson River these ducks were nesting in old Black-billed Magpie nests. Then, I discovered that this is evidently a tradition of some standing among Carson River Wood Ducks, for I eventually uncovered a report in the literature that in 1949 E. C. D. Marriage found a Wood Duck nest containing nine eggs in an old magpie nest near Carson City.[119] In Wood Duck management programs, artificial nest boxes are used with success.

The male Wood Duck is the most beautiful duck in North America, if not in the entire world. Its feathers have long been treasured for making artificial flies for fishermen. The best time and place in the Great Basin to see wild Wood Ducks is probably during the winter in Reno. Often they can be seen at Virginia Lake or on the private ponds along the country lanes on the southwestern edge of Reno.

TRIBE ANATINI: DABBLING DUCKS

The dabbling ducks constitute our largest tribe of waterfowl. Worldwide, there are about forty species, mostly belonging to a single genus—*Anas*. They are called dabbling ducks because they upend in the water when feeding. An iridescent patch of feathers, called the speculum, is usually present on the wings of both sexes. When getting airborne they do not have to taxi but can spring directly into the air from the surface of the water.

Green-winged Teal
Anas crecca

Green-winged Teal are distributed throughout the northern hemisphere. Formerly, the ones in North America were considered to constitute a species of their own with the scientific name of *A. carolinensis*. Since 1973 the American Ornithologists' Union has recognized the North American teal as being merely a subspecies of a species ranging over Eurasia and North America. Our subspecies is known as the American Green-winged Teal, *A. crecca carolinensis*.

The Great Basin lies south of the prime Green-winged Teal breeding grounds in Canada and Alaska, and this teal is an uncommon breeding bird in the Basin. However, it is a breeding species at a number of refuges and wildlife management areas here. The Greenwing is an occasional to abundant wintering species at various wet places in the Great Basin. It is most abundant here during migration. Several of its migration corridors traverse the Basin; one major corridor brings teal from Saskatchewan to the Great Salt Lake marshes. From here some of these birds use a corridor leading across Nevada into California.[99]

This teal is the smallest of all the North American dabbling ducks yet one of the swiftest fliers among ducks. Flocks may dart, turn, and maneuver in the air much like shorebirds. Of all ducks, teal make the greatest use of mud flats when feeding. Green-winged Teal feed on a variety of plant seeds— including those of grasses, pondweeds, bulrushes, and other sedges. Greenwings have a highly developed form of social courtship in which a group of males will swim and display around a female while whistling loudly.[120]

Mallard
Anas platyrhynchos

The Mallard is the most widely distributed and abundant duck in the northern hemisphere, occurring in Eurasia and North America. It is a breeding species over much of North America, except for Mexico and southern and eastern parts of the conterminous United States. Mallards are permanent residents in the Great Basin. Here they are common, often abundant breeders wherever there is suitable shallow-water habitat and food. All of our national refuges and wildlife management areas list them as a common breeding species. In some areas there is a decrease in numbers during the winter. There are Mallard migration corridors traversing the Great Basin,[99] and the numbers of Mallards present may peak during the fall months at places such as Bear River, Stillwater, and Malheur.

The Mallard is the most familiar of all our ducks. Hunters recognize it as a major game species. Farmers recognize it as the source of most of our domestic ducks. Parents and grandparents recognize it as a tame bird hanging around park ponds, where they can take children and grandchildren to feed and watch waterfowl. All our domestic breeds of ducks, with the exception of the Muscovy Duck, *Cairina moschata*, were derived from the Mallard. The Mallard was probably first domesticated in Mesopotamia, but China was another early center of domestication—the Chinese bred Mallards mainly to develop breeds which would profitably produce meat and eggs for the table. Most of the ornamental breeds were developed by the Europeans; some of the domesticated breeds are pure white in color. Today, we find domesticated Mallards not only in their proper setting—the barnyard—but also in our city park ponds. Many domesticated ducks have been dumped back into the wild. Since the various breeds can readily hybridize among themselves, we see all kinds of mixed-up-looking ducks in our park ponds: ducks with color combinations between the white and the wild colorations, ornamentations such as crests of feathers on the head, and body sizes ranging from the average to the exaggerated—for large size was developed for table meat. At Virginia Lake in Reno and at other sites in the Great Basin, you can see the full gamut of hybridization.

Most drake dabbling ducks are unusual in that they wear their colorful nuptial (alternate) plumage for most of the year. During early summer, the drake Mallard undergoes a postnuptial (prebasic) molt and acquires his dull eclipse plumage. Although the eclipse plumage is worn in the summer, it is akin to the winter (basic) plumage of most birds. While in eclipse plumage the drake experiences a synchronous molt of his flight feathers and, then, his tail feathers. Consequently, he is flightless for three to four weeks, until the flight feathers grow enough to support flight. The flightless period is centered in July but may occur anytime from late June to early August.[100] The eclipse plumage is worn for only a few weeks; then the prenuptial molt converts the drake back into his colorful nuptial plumage by the start of fall.

Courtship and pair bonding get under way in the fall and continue through the fall and winter into the spring. Ducks do not pair for life, and new pair bonds must be formed in advance of each breeding season. Since Mallards are so tame and display on open water, it is easy for a curious observer to become acquainted with their display behavior.

The most common courtship behavior of the female is called inciting. This is a ritualized display in which a female incites her mate or prospective mate against another drake or elicits sexual behavior from her mate. The mate may attack the drake he has been incited against or may direct his turning-of-the-back-of-the-head display toward the female. The displays of the Mallard have been well studied by Konrad Lorenz, whose pioneering studies in comparative animal behavior under natural conditions won him a Nobel Prize in 1973. Lorenz believed that in ritualized inciting, such as that by the female Mallard, "the original significance of the instinctive behaviour, namely bringing the mate into conflict with a real adversary, is completely obscured by the secondary meaning of 'avowal of love' for the incited male."[121] When inciting the female turns and swims after her mate or prospective mate while threatening another drake over her shoulder. She repeatedly turns her head to one side back over her shoulder while uttering a series of trembling *gagg* notes. She always incites with over-the-shoulder movements regardless of the direction in which the threatened male is located. Another common display by the female which elicits sexual displays from the males is called nod-swimming. In nod-swimming the female swims rapidly in a circular course around a male, with her head and neck stretched out in front of her over the water.

Males of the various species of dabbling ducks share about ten different movements which they string together in various combinations to form courtship displays. The ten movements are initial bill shake, head flick, tail shake, grunt-whistle, head-up-tail-up, turning toward the female, nod-swimming, turning the back of the head, bridling, and down-up.[122] As you can see, most of Lorenz' names for the movements are descriptive. However, several must be explained in greater detail. The grunt-whistle, probably the drake's most common display, is initiated when he inserts the end of his bill underwater. He then completes a sideways shake of his head and bill, while rearing his body upright but keeping his neck and head in the water. As the tip of the bill breaks through the surface on the upward swing

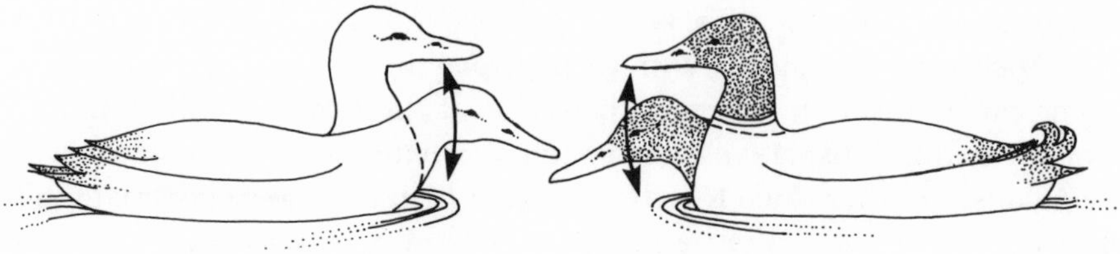

Pre-copulatory behavior: mutual head pumping.

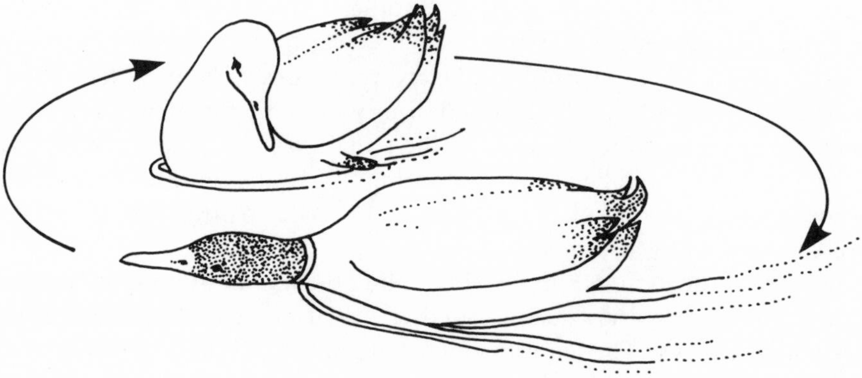

Post-copulatory behavior: male Mallard nod-swimming around bathing female.

of the head shake, it tosses up water droplets. After the shake, while still in an arched position, with his body pointing upward and his neck and head downward, the duck emits a sharp whistle followed by a grunt. Then he raises his head and settles back on the water. Nod-swimming by the male is similar to nod-swimming by the female. In bridling, the male throws his head back while raising it high.

The copulatory displays of the Mallard are highly stereotyped. The precopulatory display is mutual head pumping, during which the male and the female pump their heads up and down. During the postcopulatory display the female bathes vigorously, while the male bridles, then nod-swims around the bathing female, and finally turns the back of his head to her.

Since pioneering fieldwork by Munro in British Columbia[123] and Sowls in Manitoba,[124] waterfowl biologists have been questioning whether Northern Pintails, Mallards, and most other species of dabbling ducks actually display territorial behavior. The nature of breeding territories varies from species to species. Some are so tiny as to barely encircle the nest. Others are small and are used solely as courtship and copulation stations by the males. Still others are large enough to accommodate the male's courtship displays and contain the nest or nests of his mate or mates. Finally, there is what is called classic territory, so large that it accommodates all the courtship and nesting activities and provides cover and feeding and loafing areas. But, regardless of the type of territory, it possesses precise boundaries, and these boundaries are defended against trespass.

Territorial defense also varies from species to species. In most species the male alone defends the territory, but in some species the female alone or the male and the female both defend the territory; typically, it is the male who sets up a breeding territory. In most species the defense is directed against males of the same species, but some defend against females as well. Most territories, whether breeding, winter, feeding, or roosting territories, are defended only intraspecifically—against trespass by members of the same species. But some species defend interspecific territory—against other species as well as their own. A few species defend against trespass by other kinds of animals, such as mammals, as well as birds.

Now that we have acquired some insight into the meaning of territorialism, we are ready to consider the reason why some biologists question its existence in dabbling ducks such as Mallards: no precise boundaries are defended. Pairs of ducks may overlap in their use of real estate. They may trade loafing areas or alternate in the use of a loafing area.[124] Further, the drake primarily defends his mate, as she moves about, instead of defending rigid boundaries. The area or areas he does defend can change from day to day and vary in size and shape. There is considerable variation in the behavior of individual drakes and their hostility toward others. In three-bird flights, when the drake is chasing an intruding pair away, he directs his attack at the female, not the male, of the intruding pair. Therefore, some ornithologists view these three-bird flights as being fueled more by the male's readiness to engage in sexual promiscuity than by his drive to defend territory.

Once it was realized that breeding dabbling ducks did not defend and utilize real estate in a manner strictly compatible with the operation of territoriality, there was a tendency to back off from the concept. Real estate was now envisioned as constituting home ranges.[124] The concept of a home range, which had been developed by mammalogists, applied to the terrain a mammal traversed in carrying out its daily activities. It was the physical space within which the mammal lived day by day, without any qualifications on whether or not any or all of it was defended. If we envision physical space as constituting a home range, the opportunity still remains of recognizing any defended areas on the home range as being territory. However, there is one drawback, in that the concept of a home range was derived from studies on terrestrial mammals who lack the high mobility of birds. Hence, the home range of a mammal is a single continuous piece of real estate, whereas the home range of a breeding duck includes a number of discontinuous pieces of real estate.

The hen Mallard selects the nest site. She leads her mate out on evening reconnaissance flights until a site is found. She prefers grassy upland sites to marsh sites. Most nests are located within one hundred yards of water, but good sites over one mile from water may be utilized. Early nesting Mallards show a preference for nesting on islands in close proximity to nesting Canada Geese. Geese will allow ducks to build nests within a few feet of their own nests. Studies of island-nesting ducks have shown that the density of duck nests and nesting success are highest when duck nests are located close to goose nests. Observers have suggested that the reason for this is that the ducks are benefiting from a protective nesting association with the geese—the larger geese have been seen threatening and chasing potential nest predators such as raccoons, skunks, and crows out of the area. I shall discuss the subject of protective nesting associations in greater detail in my natural history account of the Golden Eagle.

Once the hen Mallard starts laying eggs, she plucks down from the underside of her body to add to the nest; just before completing the clutch, she plucks large amounts. This creates an incubation patch of bare skin and provides down for covering the eggs when she leaves the nest. The female alone incubates the eggs, leaving the nest twice a day to feed, drink, bathe, and preen. Once incubation is under way, the male departs from the scene to form male flocks and to undergo molt. The eggs hatch after twenty-six to twenty-nine days of incubation. The hen soon leads the precocial downy ducklings to water; she cares for them for seven to eight weeks before they fledge.

Mallards feed, as all dabbling ducks do, in shallow water by tipping up; but they will dive for food on occasions. Their principal food is the stems and seeds of emergent and submerged plants—such as bulrushes, smartweeds, and pondweeds. Their intake of animal food, which is limited, is confined mostly to the summer. The Mallard has acquired the greatest appetite for agricultural crops of all our ducks. Not only does it glean for grain in wheat and barley stubble and feed on waste corn, but in some regions of Canada and the United States it is responsible for serious crop depredations—while the grain is in swaths, ripening, prior to threshing.

We all know that ducks quack. We believe that if something waddles and quacks we can safely assume it is a duck. However, most of the duck quacking heard in nature originates from female Mallards. Not only does the hen Mallard quack, but she does it loudly, shrilly, repeatedly, and at times, such as in the fall, when most ducks are quite silent. The drake's quack is soft and reedy in nature.

Mallards are sometimes used as experimental birds in scientific investigations, since they are fairly easy to come by and take readily to captivity. Their most interesting role as experimental birds has been in investigations on orientation. Bellrose launched these studies by capturing Mallards during their fall migration in Illinois and displacing them in various directions and at various distances. He was astonished to discover that when they were released, after first being displaced, they all flew and disappeared from sight in a general northward direction. But subsequent recoveries indicated that they soon resumed their southward migration.[125]

Matthews followed up on this type of research by displacing and releasing largely nonmigratory Mallards from southwestern England. No matter in which direction they were displaced, these Mallards always flew off on a northwest initial heading. They did this even when they were released in full view of the place where they were captured. Further, the Mallards deviated from their initial heading within twenty minutes after being released. However, when Matthews displaced and released Mallards from London Park, they all flew off on an initial heading to the southeast, as did Mallards from Sweden.[126] There is still no adequate explanation of why Mallards, when captured and released, briefly adhere to an initial heading which is population-specific as to direction. Since this phenomenon does not make sense, Matthews has labeled it nonsense orientation. Nonsense orientation has been detected in other water birds besides Mallards—including some

Male Mallard dabbling

teal, pintails, and the Canada Goose. It was first detected in the Common Tern.

Because Mallards migrate more by night than by day, Bellrose decided to release some at night. He attached a penlight device to their legs so that the birds could be followed with binoculars until they disappeared from sight. His Mallards all took off on a northward initial heading, as they had by day. Independently, Bellrose and Matthews showed that their released Mallards could practice nonsense orientation only during clear weather, when they could see the sun or stars. On overcast days and nights, they were unable to orient on their population's initial heading and flew off in random directions from the release sites. Their inability to practice nonsense orientation on overcast days and nights indicated that they were probably orienting in the daytime by using the sun and at night by using the stars.

Matthews then proceeded to investigate how Mallards were using the sun and stars in their orientation—specifically, whether they were employing time-compensated orientation or not. Because of the apparent movement of the sun through the sky, one must know the local standard time of day to use the sun as a compass. In the northern hemisphere the sun is in the east at 6:00 A.M., in the south at noon, and in the west at 6:00 P.M. Matthews realized that, if the Mallards were using the sun compass, the only clock available to them would be an internal biological one. If groups of Mallards are kept under different schedules of artificial night and day, their biological clocks can be set ahead of or behind local time by a precise number of hours. Once their biological clocks have been set ahead or behind local time, birds will make predictable and constant errors in their orientation if they are employing time-compensated orientation. For example, if a Mallard is released at noon local time, with a biological clock which has been set six hours ahead of local time, it will mistakenly regard the sun's position as being west instead of south. Consequently, in its nonsense orientation it will confuse west with north; and if its population's nonsense orientation is to the northwest, it will fly off on an initial heading to the southwest.

When Mallards with reset biological clocks were released during the daytime by Matthews, they made the exact errors in initial headings which had been predicted. When released at night, they did not make any errors in their initial headings during nonsense orientation. Hence, it was concluded that Mallards orient during the daytime in a time-compensated way, by using the sun compass, but that their nonsense orientation at night is not time-compensated. They are not using the position of individual stars to orient by—unless it is Polaris, the North Star. More likely they are using the major patterns of stars, which does not require time compensation.

Northern Pintail
Anas acuta

The Northern Pintail is the most widely distributed waterfowl in the world. Although it is a breeding bird over much of the central and northern parts of North America and Eurasia, it is not present in most of eastern Canada and eastern United States. The Northern Pintail is an uncommon to abundant breeding species in the Great Basin. A number of its migration corridors cross the Basin, and during the spring and fall it may be the most abundant duck here—in the fall several hundred thousand pintails may be present at the Bear River National Wildlife Refuge. It is least abundant during the winter. The Northern Pintail is one of the most abundant species of waterfowl in the world. The Mallard is the most abundant duck in North America, but the Northern Pintail is not far behind. Some years it is in the number 2 position, other years in the number 3 position. The Northern Pintail ranks number 1 in abundance along the Pacific Flyway, where the Great Basin lies.

The concept of flyways dates back to 1935 to a biologist from the old Bureau of Biological Survey by the name of Frederick C. Lincoln. After analyzing several thousand recoveries of banded waterfowl, Lincoln decided that waterfowl recognized four great, north-south migratory pathways in North America. He named these four pathways, from east to west, the Atlantic Flyway, Mississippi Flyway, Central Flyway, and Pacific Flyway. Later, in describing a flyway, Lincoln wrote, "It is a vast geographic region with extensive breeding grounds and wintering grounds connected with each other by a more or less complicated system of migration routes." And he broadened the concept to include all North American migratory birds with the statement that "there is a growing mass of evidence in support of the belief that all populations of migratory birds adhere with more or less fidelity to their respective flyways."[127]

Today it is easier to envision flyways for waterfowl as more of an administrative concept than a biological one. Lincoln even admitted that "the breeding grounds of one or more flyways may (and usually do) overlap broadly." In fact, band returns actually indicate that ducks banded on their breeding grounds in the prairie provinces of Canada use all four flyways in migrating south. Admitting that the flyway concept had value as an administrative concept, Lincoln wrote, "Beginning in 1948, it has served as the basis for administrative action by the Fish and Wildlife Service in the annual hunting regulations."[127]

Another serious biological flaw in the concept of flyways has been noted by Bellrose: "Flyways fail to define the passage of waterfowl because they

Frederick Lincoln's map of the Pacific Flyway. This map was presented as figure 16 in Lincoln's *Migration of Birds* (Fish and Wildlife Service Circular 16, United States Government Printing Office: 1950).

cover too extensive an area and do not delineate movements of waterfowl that are lateral to a north-south direction."[99] To rectify this situation, Bellrose employed the concept of migration corridors, the actual pathways used by a migratory species. He determined the width, direction, and amount of traffic for each corridor by radar studies of migration, band recoveries, survey counts of waterfowl, and actual observations from the ground and air of migrating waterfowl. His information on migration corridors is presented in his book, *Ducks, Geese and Swans of North America.*[99]

Male and female Northern Pintails tend to migrate to their wintering grounds in segregated flocks. Once on the wintering grounds the sexes will start to intermingle, and courtship and pair formation will get under way during the winter. Thus many pintails are paired by the time they complete their spring migration and arrive back on the breeding grounds. The marshes on the east side of Great Salt Lake serve as an important way station for hundreds of thousands of pintails coming down from Alberta and Saskatchewan, on their way to overwinter in California. The majority of these pintails reach California by moving along a migration corridor that leads across the Great Basin from the Great Salt Lake marshes, through northeastern Nevada to the Stillwater–Carson Sink area, and on to California. Another major pintail corridor on the Canada to California route crosses the northwest corner of the Great Basin and passes through the Malheur area. The Basin is also involved in the molt migrations of Northern Pintails. Marshes at the north end of Great Salt Lake are visited by tens of thousands of drake pintails during their molt migration to acquire their eclipse plumage; these drakes may be coming down from Alberta.[99]

Northern Pintails, along with certain other species of ducks, show a strong propensity to nest in close association with colonially nesting larids—gulls and terns—on islands in inland lakes. By so doing the ducks gain excellent protection from crows, magpies, and other nest predators, for it is virtually impossible for a predator to approach a larid colony without being immediately mobbed. As long as the ducks are nesting in association with Common or Black Terns or with pure colonies of Ring-billed Gulls they prosper. But, if California Gulls are present, the ducks may end up fledging few or no young, since California Gulls with young to feed are quick to prey on ducklings.[128] Since the ducks apparently don't always discriminate between good larids—terns and Ring-billed Gulls—and bad larids—California Gulls—some observers question whether anatid-larid nesting associations are an ecological trap for the anatids.[129] Evidently, Northern Pintails breeding in Canada show some discrimination, since Vermeer's observations indicate that they and Lesser Scaup show a strong propensity to nest in association with Common Terns—good larids.[128] An anatid-larid nesting

Male Northern Pintail pseudo-sleeping

association in the northwest end of the Great Basin, at Hartson Reservoir in Honey Lake Valley, was studied. Here, on two islands, 2,050 California and Ring-billed gulls' nests were counted. Of the 33 pintail nests found here, 28 were successful, 3 were destroyed by predators, 1 was deserted, and 1 was lost to flooding. Other species of ducks found nesting in association with the gulls here were Mallards, Gadwalls, Cinnamon Teal, Northern Shovelers, American Wigeons, Redheads, and Ruddy Ducks; there were 107 duck nests in all. Although some predation on ducklings by California Gulls was observed, it was not considered to be serious. The ducklings enjoyed the protection afforded by levees and emergent vegetation.[130]

As in other dabbling ducks, there is little evidence of territoriality in Northern Pintails. There is virtually no hostility between males. In three-bird flights, where the resident drake chases the female of an intruding pair, the evidence indicates that his intentions are more rape than territorial defense.[131]

Burping is the most common male courtship display—with his neck up and bill pointing somewhat downward, the drake utters a *geee* call. Inciting by the female and turning the back of the head by the male appear to be instrumental in forming pairing bonds. Mutual head pumping constitutes the precopulatory behavior. In postcopulatory display, the male does a single bridling movement and may also turn the back of his head and burp. There is no nod-swimming.[108]

Blue-winged Teal
Anas discors

The Blue-winged Teal is widely distributed over the mid northern part of North America as a breeding species; it winters mainly south of the United States. It is an occasional to uncommon breeder in wetlands in the Great Basin, and a few overwinter here. It is a rare migrant along the Pacific Flyway.

Apparently, the Blue-winged Teal was formerly abundant in the Great Basin. In 1868 Ridgway found this teal to be "rather common in May at Pyramid Lake, where breeding in the meadows."[25] In 1900 Hanford found the Blue-winged, Green-winged, and Cinnamon teal to be common breeders at Washoe Lake, Nevada.[132] In August 1872 Nelson found the Bluewing to be an abundant breeder in the vicinity of Elko; he wrote, "They were so numerous that one morning I found five broods. The young were from three to ten days old."[133] In 1872, Henshaw found it to be as abundant as the Green-winged Teal in Utah.[56]

The Blue-winged Teal is probably now the fourth most abundant duck in North America. Yet the evidence indicates that it experienced a major decline in numbers during the twentieth century in the Great Basin, dwindling from abundance to rareness as a breeding species here. That the change was a real one and not just an apparent one is not too difficult to believe. The Great Basin was never prime breeding habitat for the Bluewing. Prime breeding habitat lies in the northern prairies and parklands, not in arid shrublands. Banding studies have shown that the Bluewing is not a strong migrational homing species—it does not return year after year to the same area to nest. Banded yearling teal have not shown much in the way of natal-site tenacity; they often fail to return to where they were born in order to nest. Few banded hens have returned to the same nesting area two years in a row. Blue-winged Teal appear to be highly opportunistic in selecting nesting grounds. They will often breed outside of their normal breeding range if they encounter unusually favorable breeding conditions during migration.

Cinnamon Teal
Anas cyanoptera

The Cinnamon Teal is found in both North and South America; it is a breeding bird in western North America from southern British Columbia to Mexico. This teal is an abundant breeding species in the Great Basin, and a few overwinter here. The marshes of the Great Basin are prime breeding habitat for this teal, and over half of the total North American population is said to breed in the marshes east and north of Great Salt Lake in Utah.[99]

The Cinnamon Teal is a close relative of the Blue-winged Teal; except for males in their nuptial plumage, the two species even look much alike. Cinnamon and Blue-winged teal appear to have quite similar habitat requirements—they frequent small bodies of shallow water, where the emergent plants are short and the submerged aquatic plants numerous. Both rely heavily on the seeds of aquatic plants as a food staple. The prime breeding habitat of the Cinnamon, in general, lies west and south of that of the Bluewing; but their breeding ranges do overlap—as in the Great Basin. The Cinnamon Teal heavily exploits the alkaline water habitats in the Basin, and this may give it the upper hand in competing with the Blue-winged Teal—if competition does occur between the two. In some of its displays and behavior the Cinnamon Teal is quite like the Northern Shoveler, and some ornithologists regard it as being intermediate between the Blue-winged Teal and the shovelers.[108]

The home ranges of Cinnamon Teal are small and often overlap, with little hostility occurring between pairs. In a Utah study, most of the activities of a pair were seen to be centered within a space as tiny as thirty square

Cinnamon Teal preening

yards.[134] Nesting densities are often very high in the Great Basin. In studies made in the early 1950s, a nesting density of about 244 pairs of Cinnamon Teal per square mile was obtained for Honey Lake,[135] and about 120 pairs per square mile nested in Utah.[134]

The Cinnamon Teal forms new pairing bonds each year. Courtship begins on the wintering areas, and most teal arrive on the breeding grounds already paired. Inciting by the female is probably her main display during pair bond formation; the male usually responds by turning the back of his head as he swims before her. There is little lateral movement of the head and bill during inciting, which consists mainly of alternate chin lifting and head lowering.[108] Although there are some aerial pursuits, courtship occurs mainly on the water. The male's displays are much like those of the Blue-winged Teal. He makes short jump-flights toward the female, during which he displays the blue area on the upper side of his wing. After landing near the female, he may engage in a series of ritualized displays—mock feeding, shaking, wing flapping, bathing, and preening.[136] The precopulatory behavior is mutual head pumping. The intention movement given prior to flight involves shaking the head sideways.

Northern Shoveler
Anas clypeata

The Northern Shoveler is widely distributed in North America and Eurasia. In North America it breeds mainly in the northwestern part of the continent, with only scattered nesting to the east. In the Great Basin, the Northern Shoveler is a common breeding species and an occasional to uncommon wintering species. Several shoveler migration corridors cross the Basin. Thousands of Northern Shovelers from Alberta and Saskatchewan are funneled through the Great Salt Lake marshes. This species experienced common and scientific name changes in 1973. It was formerly known as the Shoveler, *Spatula clypeata*. Then the genus *Spatula* was merged with the genus *Anas*, and its common name received the modifier of Northern.

Although some waterfowl biologists consider the Northern Shoveler to be the most territorial of all dabbling ducks, there is controversy over the nature of its territoriality. An attempt was made to experimentally answer the question of how orthodox its territoriality is—whether static real estate boundaries are the object of defense or whether merely an area around a female is defended. The experiments involved enclosing a female in a trap, letting a male pair-bond with her and set up a territory around her, and then shifting the location of the female. When the location of the female was changed to a site visible to the male but outside his territory, he would quickly abandon his territory and attempt to set up a new territory around the translocated female. If the female was shifted to a site not visible to the male, he would remain on his territory for up to two days, defending it against trespass by other males.[137] Although this last act showed that the male would defend his territory in the complete absence of a female, the experiments also clearly showed that the territory had significance to the male only because of the prior presence of the female and that the male would promptly abandon his territory to follow a female. This is not orthodox territoriality, since it arises out of defense of a female, not of territory per se.

The Northern Shoveler is quite unique among North American surface-feeding ducks in its food habits, since up to one-third of its food may consist of animals such as mollusks, crustaceans, and insects. Although it concentrates more on shallow water, it does some feeding in deep water; in shallow water it spends less time upending and more time feeding by merely submerging its head and neck. Not only does it pick up food directly, but it also filters small food items such as duckweeds and crustaceans out of the water. Water is taken in at the tip of the bill and expelled through the lamellae or "teeth" on the sides of the bill; the particulate matter is retained in the mouth. Filter feeding is the means by which this shoveler feeds in deep water, since it is not a diver. Social feeding can be observed—groups of shovelers swimming together will rotate on the water, stirring up the surface water prior to filtering it, one feeding right behind the tail of the next.

The bill of the Northern Shoveler is unique in size and design. So long that it dominates the head of the duck, it is spoon-shaped and about twice as wide near its tip as it is near its base. The lamellae along the sides of the bill are long, fine, and numerous. This is a highly specialized instrument for filter feeding in deeper surface water or off of the bottom muds in shallow water.

The courtship and pair formation displays are much like those of the Cinnamon Teal. Inciting by the female and turning the back of the head by the male are prominent displays. While inciting, the female moves her head up and down without any sideways movements. Jump-flights by the male toward the female are common. This is a hovering flight, with the wings making a noisy rattling sound. After landing near the female, the male often upends.

Gadwall
Anas strepera

The Gadwall is a breeding species over much of the mid latitude of North America and Eurasia. In North America it breeds primarily in the western parts of Canada and conterminous United States, and it winters southward through Mexico. The Gadwall is a common to abundant breeding species over the Great Basin wherever suitable habitat occurs. Bellrose reported six Gadwall migration corridors penetrating the Basin.[99] Depending on the severity of the weather, the Gadwall is an occasional to abundant overwintering species, with lesser numbers remaining in the northeastern and northwestern corners of the Great Basin. Over the years, the Gadwall has waxed and waned in numbers in North America. During the first half of the twentieth century it was in a decline—as were other species of water-dependent game birds. But, since the end of the first decade of the second half of this century, the Gadwall has experienced a highly significant increase in numbers—something very unusual for a duck.[99]

The breeding biology of the Gadwall has been of interest to wildlife biologists, since its nesting success is the highest of all the dabbling ducks. An average nesting success rate of 67.5 percent was once calculated after lumping the nesting results from 2,173 nests throughout its breeding range.[99] There has been speculation as to the reasons for this very high rate of success. It has been suggested that, since Gadwalls are very late nesters and locate their nests on islands and in tall, dense vegetation, they are more immune to nest predators than are other dabbling ducks. Predators tend to concentrate less on nests late in the season and less on nests that are relatively inaccessible or better hidden. Some observers believe that the Gadwall's propensity for nesting in dense stands of stinging nettles may also discourage predators.

Gates made a major study of the nesting of Gadwalls at Ogden Bay, Utah. These Gadwalls were indeed late nesters as waterfowl go, with pairs initiating nesting from mid May to mid July. Here, with a lack of islands, most of the nests were located on elevated sites provided by dike banks and natural levees. Gates found that the behavior of his Gadwalls was such that it made sense to envision home ranges—encompassing loafing, feeding, and nesting sites—instead of territories. At Ogden Bay the home ranges varied from thirty-four to eighty-seven acres in extent and averaged sixty-seven acres.[138] Gates believed that the hen Gadwall determined the general location of the home range of the pair, since at least 60 percent of the surviving hens marked in 1956 returned in 1957 and nested in the immediate vicinity of their 1956 nest sites. The home ranges of the pairs were shared or overlapped. Each pair had one or more feeding ponds and channel or dike sites for loafing, but two or more pairs were often feeding on the same pond or loafing in the same section of a ditch. The aggressive behavior of the drake was directed toward defending the presence of his mate, not toward a static piece of real estate.

Gates noted that the drakes engaged in two types of aerial chases. During what he called territorial chases (three-bird flights), the drake chases the female of an intruding pair, with the female's mate bringing up the rear. In what he called harrying chases, a number of drakes chased a nesting hen, usually an incubating one, with the intent of copulating with her. These harrying chases are more accurately called rape chases. A nesting female is vulnerable to rape, since pair bonds dissolve early in the nesting cycle, and the incubating female lacks the protection previously afforded by her mate.

American Wigeon
Anas americana

The American Wigeon is a breeding bird in northwestern North America; it is an occasional to uncommon breeding species in the Great Basin. Several of its migration corridors cross the Basin, and it reaches its greatest abundance here during migration. Although wigeons overwinter in the Great Basin, their abundance varies considerably from place to place.

This wigeon has experienced changes in both its scientific and common names. Its scientific name changed from *Mareca americana* to *Anas americana* when the genus *Mareca* was merged with the genus *Anas* in 1973. An obsolete spelling was then corrected, and widgeon became wigeon. Prior to being called the American Widgeon, this bird was known as the Baldpate.

The feeding habits of the wigeon set it apart from other North American surface-feeding ducks. Most dabbling ducks prefer the seeds of aquatic plants or cultivated grain as food, but the wigeon prefers the stems and leafy parts of aquatic plants. It will even go on land to graze. Unlike other dabbling ducks, it is often seen feeding out on open, deep water—in company with coots, swans, and diving ducks. The wigeon is not a diver itself but relies on acquiring aquatic plants dislodged by the divers during their underwater feeding. It also robs coots and diving ducks such as Redheads and Canvasbacks. When stealing plant food, the wigeon waits at the surface of the water for a diver to emerge with vegetation dangling from its bill and then grabs some. Coots and diving ducks are quite tolerant of this thievery.

Other Dabbling Ducks

The American Black Duck, *Anas rubripes*, occurs in eastern United States and Canada. It is of occasional occurrence in Utah and of accidental occurrence at Malheur National Wildlife Refuge. Unsuccessful attempts have been made to introduce the American Black Duck into the Bear River and Farmington Bay areas of Utah.[27] This duck is believed to be closely related to the Mallard, and in eastern North America it appears to be the ecological equivalent of the Mallard—that is, it plays the same ecological role that you would expect the Mallard to play if it was there. When the American Black Duck and the Mallard come together as breeding or wintering birds, they readily hybridize. Most of the Utah specimens studied appear to represent such hybrids.[56]

The Eurasian Wigeon, *Anas penelope*, is of rare occurrence in the Great Basin. This species was formerly called the European Wigeon. Every year some single birds, and sometimes several together, are reported from localities in North America.

TRIBE AYTHYINI: POCHARDS AND ALLIES

Pochards dive from the surface and swim underwater when feeding. They are heavy-bodied and must patter across the surface of the water to get airborne. A large flap or lobe is present on each hind toe. Pochards' legs are located far back on their bodies—walking on land is an awkward act for them.

Canvasback
Aythya valisineria

The Canvasback's breeding range lies in western North America. In the Great Basin, it is an abundant breeding species at Ruby Lake National Wildlife Refuge. At other localities such as Bear River, Fish Springs, and Malheur, it is only a rare to uncommon breeder. Unfortunately, its welfare at Ruby Lake is threatened by the demands of fishermen and boaters for unrestricted access to the refuge during the nesting season. Several migration corridors lead through the Great Basin, with a major one connecting the Great Salt Lake area with California via the Stillwater–Carson Sink area.[99] The Canvasback is an occasional to uncommon wintering species at localities in the Basin. According to Ridgway, the "Canvas-back was abundant in winter at the lakes and marshes of the Great Basin" in the 1860s.[25]

The Canvasback is one of our most desirable game birds, and its meat is highly prized for the table. It has been experiencing a population decline for many years. Although this duck has lost a lot of breeding habitat in the prairie regions of the United States and Canada, habitat loss has not been the principal reason for its decline. In May 1980 H. Albert Hochbaum, the world's authority on this species and the author of the wildlife classic *The Canvasback on a Prairie Marsh*, wrote of the Canadian nesting areas, "Despite the loss of wetlands to drainage, there still are not enough ducks returning to make full use of the fine marshes remaining."[139]

Overshooting has seriously damaged the Canvasback. Hochbaum has observed how local populations are particularly vulnerable to hunting, "especially during September and early October when many young are barely on the wing and their mothers are just recovering from the wing molt. Through successive years of early gunning, the native stock of young Canvasbacks and some other species has been depleted. This 'burning-out' of breeding stock has been going on for many years on marshes in the United States. Now the process is underway at Delta and other Canadian nesting areas."[139]

As drought years and heavy hunting pressure plagued the Canvasback near the end of the 1950s and the beginning of the 1960s, several management practices were implemented, including reducing the hunters' bag size and closing the season. Even closing the season is not too effective—since so many hunters cannot identify ducks, especially on the wing, they shoot promiscuously and discard the illegal ones. Hochbaum's telling remark on this point was, "In 1963, a year when no Canvasback shooting was allowed in Manitoba, this species nevertheless was second only to the Mallard in early season harvest on the Delta Marsh." Effective management practices have not been seriously contemplated, let alone implemented—such as stopping the shooting altogether at certain critical marshes with important local or migrant concentrations of Canvasbacks. In fact, a diametrically opposed movement developed in the United States of opening up areas within our national wildlife refuges to duck hunting. It is disturbing to see hunters at a great national wildlife refuge such as Bear River; and it is tragic that Hochbaum has had to ask the question, "Will editors redefine 'refuge' in the next edition of *Webster's Dictionary*?"[139]

We have seen how very vulnerable female and young Canvasbacks are to hunting. As the Canvasbacks dwindled in the face of overshooting, females formed a disproportionate share of the casualties. "By the early 1970s, of every 100 birds, 70 to 75 were drakes."[139] With the sex ratio so highly imbalanced in favor of males, the breeding effectiveness is much lower than the total population figure would indicate, since the majority of males would not be contributing to the process of reproduction. Only a female can lay an egg. Then, the breeding effectiveness of the females is further reduced by the extremely high level of brood parasitism imposed on them by the Redheads laying eggs in their nests. Canvasbacks and Redheads come into frequent contact since their breeding ranges and breeding habitats are so similar.

One of the major concentration points for migratory Canvasbacks in the Basin lies in the marshes around Great Salt Lake—up to ninety thousand or so ducks will concentrate there during the fall. The major migration corridor out of these marshes leads westward across the Great Basin, through the Stillwater–Carson Sink area, onto the California wintering grounds. Band recoveries indicate that many of the Canvasbacks migrating through Utah and Nevada are coming from Alberta, with a few coming from Saskatchewan and Alaska. A less populated migration corridor leads from Canada to California, across the Great Basin at Malheur.[99]

Canvasbacks form new pairing bonds each year. They have an extended courtship period. Although this starts on the wintering grounds, most birds still have not completed pair formation before reaching the breeding grounds. During pair formation displays, the female alternates threatening movements with inciting—during which she neck stretches, with head high and bill pointed at her mate or mate-to-be. The drake may respond by chin lifting. The male's display repertoire includes neck stretching; a sneak posture as he approaches the female, with his neck extended and his bill on the water; a kinked-neck call, given with his neck drawn back and his head lowered while uttering a cooing call; and a head throw, during which he snaps his head back and then forward again—the backward toss accompanied by *uk-uk*, the forward toss by a cooing call. A female often tries to escape courting drakes by diving and taking flight. They follow her, and aerial pursuits ensue. Males do not show much hostility toward each other. They may defend their females occasionally, but their home ranges overlap; they do not hold orthodox territories.

Although young ducklings feed heavily on animal food, up to 80 percent of the adults' food consists of aquatic plants; pondweeds are a major food staple. The Canvasback may dive up to thirty feet or so in procuring food. Nevertheless, it does a lot of shallow-water feeding and some surface-water feeding; it may even tip up like a dabbling duck when feeding in shallow water. Males desert their females after incubation gets under way and band together to go to a molting ground. Females abandon their young before the young can fly. Late nesters may abandon their young when they are only two or three weeks old and undergo their molt.

Redhead
Aythya americana

The Redhead has a limited breeding range in western North America. It is an abundant breeding species in the Great Basin, and the largest nesting concentration in North America is reported in the marshes around Great Salt Lake.[99] Sizable nesting concentrations also occur in the Malheur Basin and in the Carson Sink. Several migration corridors pass through the Great Basin, and the Redhead is an occasional to common wintering bird here.

The Redhead is a duck of unsavory reputation. During the mating season, hen Redheads often lay their eggs in the nests of other ducks, and when they do raise young by themselves they are poor parents. This practice of laying eggs in the nests of other birds, thereby putting young out for unsolicited adoption, is referred to as brood parasitism. Redheads not only parasitize other Redheads, but they lay eggs in the nests of other species of ducks—including the Northern Pintail, Mallard, Canvasback, Blue-winged Teal, Cinnamon Teal, Gadwall, Northern Shoveler, Lesser Scaup, and Ruddy Duck—as well as the American Coot and American Bittern. This may involve 50 to 80 percent of the nests present. Since Redheads do some nesting on their own, they are classified as nonobligate brood parasites. Some of the most detailed studies of Redhead brood parasitism have been made in the Great Basin, at Malheur and at various marshes in northern Utah.

Redheads are heavily parasitized by other Redheads, as well as occasionally by Ruddy Ducks. Each year many of the hens present on a breeding ground do not build nests of their own or incubate eggs. Often more don't than do. At Knudson marsh in northern Utah, forty-two hens were trapped while parasitizing nests. Some appeared to be adults, others yearlings, but none of them possessed an incubation patch—showing that they were not nesting and incubating eggs themselves.[140] So nonnesters contribute to the parasitism, as well as nesters, by dumping eggs into nests. Usually three or four hens will parasitize a single nest, but up to thirteen hens have been trapped parasitizing a single host nest![140] At Bear River National Wildlife Refuge, numerous Redhead nests were found containing nineteen to thirty-nine Redhead eggs.[141] Clutches with as many as eighty-seven eggs have been found.[140] This is sometimes called dump nesting.

There is a tremendous wastage of Redhead eggs. Even in successful nests, a large proportion of the eggs fail to hatch. In studies involving 661 successful nests, on the average 5.4 eggs per nest failed to hatch. This represented almost half of the eggs in the nest, since the average clutch size was 11.1 eggs.[99] Of the eggs which failed to hatch, about 25 percent were infertile, and 75 percent contained dead embryos. The eggs with dead embryos were probably not fully incubated, because they were added after the hen had begun her incubation. Many eggs are found scattered around outside the nests of Redheads. Regardless of whether they are parasite eggs, they are never retrieved and do not hatch.

Although repeated parasitic intrusions can cause a hen to desert her nest, parasitic eggs are often accepted. The Canvasback is severely affected by Redhead parasitism, since the breeding ranges and habitats of these two species overlap so much. A study by Olson revealed that about 50 percent of the Redheads produced on pothole habitat in Manitoba were being reared by Canvasbacks.[99]

Hen Redheads are not very good mothers. The drive to incubate eggs does not reach the intensity in Redheads that it does in most other species of ducks. They are prone to quickly desert the nest and eggs when threatened by flooding, repeated parasitic intrusions, or other disturbances. With newly hatched young the hens are quite solicitous, but their maternal instincts rapidly wane. They rarely offer effective distraction displays, such as injury feigning, when their young are threatened by a predator. They desert their broods quite early in life, usually before the young are old enough to fly.

Although the Redhead can dive and swim underwater, propelled by its feet, it differs considerably from most other diving ducks. Unlike diving ducks in general, it inhabits marshes, sloughs, and other shallow-water feeding grounds to a considerable extent. It commonly feeds in shallow water like a dabbling duck—by upending or by submerging its head and neck. The vegetative parts and seeds of pondweeds are a major food staple, and this bird takes much less animal food than other diving ducks. Redheads are very gregarious. There is little or no evidence that they defend either territory or parts of a home range.

The Redhead is a troubled species and has not been faring well in recent years. It has lost breeding habitat to drainage projects, has low nesting success, has been periodically hit by botulism, and is so easily shot that it is very vulnerable to hunters.

Lesser Scaup
Aythya affinis

There is only scattered nesting in the Great Basin by the Lesser Scaup—here at the southern edge of its breeding range. The Lesser has been reported breeding at Ruby Lake and Malheur National Wildlife Refuges and at Honey Lake; it is "possibly a rare breeder in northern Utah."[27] There is limited migration through the eastern and western range of the Great Basin, and this scaup is an occasional winter visitant in the Basin.

The Lesser Scaup is another diving duck with a sex ratio strongly imbalanced in favor of males, with some estimates of up to 70 percent or more males.[99] Nevertheless, it is one of the most abundant of all North American ducks. Its northern breeding grounds are more stable, and it has greater nesting effectiveness than the Canvasback.

The Lesser Scaup's diet is often heavily slanted toward animal food such as mollusks, crustaceans, and aquatic insect larvae. Although it sometimes feeds in shallow water by upending, it is usually seen out in deeper water diving for its food. In diving it springs upward and forward, entering the water headfirst, with its wings closed against its sides. Underwater it is foot-propelled, stroking with both feet simultaneously. During migration or when wintering, scaup may form huge rafts of thousands of birds. Females often desert their young long before the young can fly.

The Lesser Scaup is one of several species of ducks which employ an unusual type of behavioral response called pseudosleeping when mildly alarmed but not threatened enough to respond by flight or attack.[142] The duck stops whatever activity it is engaged in, such as feeding, and assumes the sleeping posture. It turns its head back over its shoulder and buries its bill under the scapular feathers. However, the duck remains awake and alert—keeping an eye on the potential source of danger while assuming a low profile. It may even swim slowly away while in the sleeping posture. Pseudosleeping is an example of displacement activities, which have been defined as "movements that, as regards their causation and their function, seem to occur out of context."[21] As we shall see, birds engage in many different kinds of displacement activities when under stress—as do humans. Among ducks pseudosleeping has also been recorded in the Ring-necked Duck, Canvasback, and Ruddy Duck.

Other Diving Ducks

The Ring-necked Duck, *Aythya collaris*, is not a prominent bird in the Great Basin. Its prime breeding grounds lie north of us in Canada, and there are only a few known breeding areas in the Basin. Currently, this diving duck is listed as an uncommon breeding species at Ruby Lake Refuge and as a rare breeder at Malheur Refuge. Several sets of eggs collected in 1911 and 1912 indicate that it may once have been a rare breeding species in northern Utah.[27] It is not an abundant migrant in the Great Basin, although one migration corridor cuts across the northwestern end. During the winter this duck is of occasional to common occurrence in the Basin.

The Greater Scaup, *Aythya marila*, is not a regular member of the Great Basin avifauna but is only an occasional visitant during migration or winter. Here it is far from its northern breeding grounds and coastal wintering grounds and far from its migration corridors. Greater Scaup have been seen and collected at various Basin locales. Because this scaup looks so much like the commoner Lesser Scaup, it may often pass unnoticed. There is some evidence that the Greater Scaup was once a much more common visitant. Early ornithologists reported it as being common in the fall in the late 1800s around Great Salt Lake and Utah Lake.[56] Ridgway observed it as a winter visitant at Pyramid Lake.[25]

TRIBE MERGINI: EIDERS, SCOTERS, MERGANSERS, AND ALLIES

The members of this tribe are sometimes called sea ducks. They are diving ducks whose breeding ranges mainly lie in the northern part of North America and Eurasia. Although some species visit inland freshwater ponds and lakes, they mostly overwinter in coastal areas on salt water. Sea ducks rely heavily on animal food. Their presence is mainly felt in the Great Basin during the winter. The classification of these ducks was altered considerably in 1982. Prior to this, mergansers were placed in a subfamily by themselves, and the other sea ducks were in a subfamily with the pochards.

Hooded Merganser
Lophodytes cucullatus

The Hooded Merganser is an occasional to uncommon winter visitant and migrant in the Great Basin. The prime breeding range of this species lies far to the east or to the north of us, although some nesting has been reported from the Sierra Nevada just west of the Basin.[99] In the 1860s the Hooded Merganser was evidently a rare breeder along the western edge of the Great Basin. Ridgway observed that it "was occasionally met with in summer in the wooded valleys of the Truckee and Carson Rivers, but it seemed to be very rare."[25] I believe that there still is limited nesting going on in this area, because every spring I encounter several pairs on a pond close to the Truckee River at Verdi, Nevada.

As yet a thorough search of the Truckee and Carson River bottoms for nesting Hooded Mergansers has not been made. These mergansers are quite shy and easily disturbed by any human presence, yet they show strong site tenacity in returning each year to the same site to nest, and there are still stretches along the Truckee and Carson rivers that are relatively isolated from much human traffic. Hooded Mergansers, like Wood Ducks, nest in tree cavities along wooded streams and ponds. The merganser is specialized for diving and catching fish underwater. Its bill is spikelike instead of being flat like that of other ducks. The upper mandible does not overlap the lower mandible. Instead of lamellae, there are sawtooth projections along the edges of the mandibles which aid in grasping fish.

Common Merganser
Mergus merganser

The Common Merganser is an occasional to common migrant and winter visitant in the Great Basin. Although most waterfowl books depict the

Female Common Merganser

Basin as lying outside of the breeding range of this species, breeding does occur here, especially along the western edge. Nesting has occurred at Eagle Lake in California[143] and at Lake Tahoe, Pyramid Lake, and along the Carson and Truckee rivers in Nevada. It has also been reported in the eastern part of the Great Basin at Goose Creek, Nevada.[91] The Common Merganser is a breeding species at Malheur Wildlife Refuge. It also occurs in Eurasia, where it is known as the Goosander.

Red-breasted Merganser
Mergus serrator

The Red-breasted Merganser's breeding range lies in the northern part of Eurasia and North America; the Great Basin lies far south of the prime breeding range of this species. It is an occasional to uncommon migrant and an occasional to uncommon winter visitant in the Great Basin, being more common in the eastern than the western part. There are summer occurrence records for Utah, and past nesting has been reported for the Great Salt Lake area.[56]

Other Sea Ducks

There are five species of sea ducks whose breeding ranges, migratory routes, and wintering grounds lie outside the Great Basin and whose presence here is of an accidental or occasional nature. These are the Harlequin Duck, *Histrionicus histrionicus*; Oldsquaw, *Clangula hyemalis*; Black Scoter, *Melanitta nigra*; Surf Scoter, *Melanitta perspicillata*; and White-winged Scoter, *Melanitta fusca*. The Black Scoter was formerly known as the Common Scoter, *Oidemia nigra*, before the genus *Oidemia* was merged in *Melanitta* in 1973.

The Common Goldeneye, *Bucephala clangula*, is an uncommon to common migrant and an occasional to common winter visitant on bodies of water in the Great Basin.

Barrow's Goldeneye, *Bucephala islandica*, seldom visits the Great Basin—and then only as an occasional to rare migrant and winter visitant. This goldeneye has been observed and collected at various sites in Utah[56] and at Ruby Lake Refuge in the eastern part of the Basin. In the western part it has been observed in the Stillwater area, at Virginia Lake in Reno, and at the Malheur Refuge.

The Bufflehead, *Bucephala albeola*, is an occasional to common migrant and winter visitant in the Great Basin. Although Buffleheads are occasionally seen in the Basin in the summer, there is no evidence of their breeding here. Their prime breeding range lies far to the north.

TRIBE OXYURINI: STIFF-TAILED DUCKS

Stiff-tailed ducks of various species are found over much of the world—their long, stiff tail feathers provide them with their name. As they swim on the surface, the ducks may cock their tails upright. Their small wings make it difficult for them to get airborne, and once aloft their flight is labored. Their legs are attached far back on their bodies and they walk on land with some difficulty. However, stifftails are superbly designed for diving and swimming underwater.

Ruddy Duck
Oxyura jamaicensis

The Ruddy Duck is widely distributed over western North America and occurs in the Andes of South America. It is a common breeding species in the Great Basin. In the northeastern and northwestern corners, where the weather is more severe, it is of occasional occurrence in the winter; over the rest of the Basin, it is more common during the winter. Several migration corridors funnel Ruddies through the Great Salt Lake, Stillwater–Carson Sink, and Malheur areas.[99]

Apart from the few groups of obligate brood parasites among birds, waterfowl are most prone to lay eggs in the nests of other birds. They also frequently form dump nests of eggs which are never incubated. A dump nest may contain many eggs, laid by several different species. Among North American ducks with these habits, the Ruddy Duck is second only to the Redhead. Ruddies parasitize other Ruddy Ducks; they also lay eggs in the nests of many other species of ducks and in the nests of grebes, bitterns, coots, and gallinules. In some localities Ruddy Ducks and Redheads will heavily parasitize each other.

Male Ruddy Ducks do not regain their nuptial plumage until spring. Consequently, unlike other ducks, much of their courting and pair bonding occur on the breeding grounds, not while wintering or migrating. During courtship the females do not do anything obvious to stimulate the males. They lack an inciting display and typically gape and bill threaten any closely approaching drake. In contrast, the drakes engage in complex displays. Their main display, called bubbling, apparently serves not only as a courtship display but also as a territorial display, since it is performed even when other Ruddy Ducks are not present in the area. Prior to bubbling, air is

Male Ruddy Duck

diverted into the tracheal air sac. The drake then repeatedly beats his bill against his now inflated neck, producing a drumming sound. Air forced out from beneath the feathers forms bubbles in the water around the drake's breast. The drake terminates the display by uttering a belching call. Often the drake swims just ahead of the hen, tail flicking by cocking his tail so that the white undertail feathers are displayed. The drakes execute short display flights or ringing rushes toward hens. The ringing or popping sound accompanying these rushes is probably produced by the drakes' feet repeatedly striking the water.[144] Precopulatory display consists of head shaking and bill dipping by the drake. The hen merely assumes a receptive position. Postcopulatory displays involve several bubbling displays by the drake.[136]

Observations by Joyner have revealed that a limited area around the nest is defended by a pair of Ruddy Ducks. This is generally interpreted to mean that Ruddies display behavior more akin to orthodox territoriality than do most ducks.[99]

Following implementation of a marshland improvement program at Fish Springs National Wildlife Refuge in the eastern part of the Great Basin, some interesting observations on breeding area and nesting site pioneering by the Ruddy Duck and the Redhead were made. From 1962 to 1964, following purchase of a marsh to form the Fish Springs Refuge, parts of the marsh were drained, and the entire area was subdivided to form nine large impoundments. With the development of the impoundments, the amount of emergent vegetation present was greatly reduced, but the amount of aquatic insects available for food was greatly increased. Typically, Ruddy Ducks and Redheads nest in emergent vegetation. In 1960, only two broods of Redheads were observed at Fish Springs, and the Ruddy Duck was not present as a nester until 1963. But, in the breeding season of 1968, 960 young Redheads and 180 young Ruddies were produced here! This demonstrated remarkable pioneering capabilities on the part of these two species. Not only did they quickly move in numbers to exploit a newly available major source of breeding season food, but they showed remarkable adaptability in being able to nest extensively on dryland sites, instead of in traditional sites in emergent vegetation.[145]

Ruddy Ducks feed more on plant than on animal food and have a special liking for the vegetative parts of pondweeds. However, during the summer they feed on a great variety of insect food. They regularly dive for food. On occasions they will feed on the surface or in shallow water by immersing their heads and necks. Their flattened bills allow them to readily probe in the bottom muds for midges and other insect larvae.

9

Shoreline Birds

FOR WANT of a better collective name, I am referring to the birds which constitute the Order Charadriiformes as shoreline birds. This order contains three groups of birds: the shorebirds or waders; the gulls, terns, and jaegers; and the alcids. These are birds who at one time or another frequent shorelines or the waters adjacent to shorelines. They may also occur in such places as marshes, wet meadows, and tundra.

FAMILY CHARADRIIDAE

The plovers and lapwings of this family have a worldwide distribution. They frequent open, bare habitats. They are plump birds with short, thick necks; their short, stout bills are usually swollen near the tips. When foraging they run swiftly and stop abruptly.

Snowy Plover
Charadrius alexandrinus

The Snowy Plover has a far-flung distribution over all the continental land masses and many islands of the world. In North America it is mainly found in some of the western states and along the Gulf Coast. The Snowy Plover is a summer resident and migrant in the Great Basin. It is considered to be an uncommon summer resident at some of its breeding locales in the Basin—such as in northern Utah,[27] in northeastern Nevada, and at Malheur National Wildlife Refuge.[61] But it appears to be rather common at other locales—such as in central Utah,[56] at Pyramid Lake, and along the western edge of the Great Basin at Upper and Lower Alkali lakes, Honey Lake, and Mono Lake in California.[146]

This tiny, six-inch plover inhabits the most hostile of all the Great Basin environments: the unvegetated or feebly vegetated alkali and salt flats. Although the flats are often located at permanent or ephemeral bodies of water, these are usually shallow and too alkaline or salty to serve as drinking water. The flats are so lacking in plants, boulders, and other topographic relief that there is virtually no shade from the sun or protection from the wind. Here, in a setting where heat burdens and evaporatory water losses can soon be lethal, Snowy Plovers dig their shallow nest scrapes.

The Snowy solves its problem of remaining in water balance by ecological and behavioral means. Its diet consists of insects; this wet food, even during the driest and hottest time of the year, contains much water of succulence. Thus, with each mouthful of food the plover obtains drink. This plover does not evaporate inordinate amounts of water from its respiratory system by panting during hot weather; instead, it restricts its activities to the vicinity of water and, when under heat stress, goes and stands in the brackish water.[147] As the brackish water evaporates from the surface of the plover, it subtracts heat, the so-called heat of vaporization, from the body surface.

The Snowy Plover apparently lacks the necessary physiological adaptations which would allow it to drink saline or alkaline water to replenish its evaporatory and excretory water losses. Experiments on captive plovers have shown that their kidneys and salt glands have only a limited ability to excrete salt and that the water in their immediate environment is usually too salty for them to drink.[148] But then they evidently have no real need to drink free water during the summer, since even when fresh water is available nearby they do not visit it to drink.[147]

During the heat of the day, eggs as well as birds are in danger of overheating. Since these plovers nest in a shallow scrape in the superheated ground, unshaded and uninsulated, their eggs will quickly bake if not attended. Both sexes incubate, and on a hot day they frequently change places, sitting on the eggs and insulating them with their bodies. At high air temperatures the sitting bird often gets to its feet and stands over the scrape, casting a shadow over the eggs. As air temperatures rise above 41° C, standing becomes more frequent than sitting, and the parents apply water collected in their breast feathers at nearby pools to the eggs.[149] Although the parent is probably cooling itself off by getting wet and standing instead of sitting on the eggs, both bird and eggs benefit from the enhanced evaporatory cooling. As the heat abates in the late afternoon, incubation attentiveness decreases, and both parents forage for food. Much of our knowledge of this fascinating little plover has resulted from the field studies of James Purdue.

In courtship the male Snowy Plover performs a butterfly flight—with slow, deep wingbeats—during which it utters a rolling trill of *rrai-ai-ai-ai-*

aiaiaiai.[150] Plovers often nest in loose colonies, and a female may build more than one scrape. These multiple nests are often called dummy nests or cock nests. However, only one scrape will be lined with a bit of plant material or debris and used for nesting. When intruders approach a nest, a parent will try and lure them away by performing an injury-feigning distraction display—by pretending to have a broken wing. Nests and eggs and birds at rest are often difficult to detect on a barren flat, since they blend in so well with their background. At times, when the surface is soft, a nest site can be located by noting where the tracks of the plovers converge—they approach and leave the nest by foot.

Killdeer
Charadrius vociferus

The Killdeer is widely distributed over the Americas. It is the most widely distributed shorebird in the Great Basin, occurring at wet places in the valleys and in mountain meadows up to elevations of seven thousand to eight thousand feet during the summer. Although Killdeers are abundant here during the summer and during migration, relatively few overwinter. Overwintering occurs in the valleys at localities where the snowfall is light and the waters remain open most of the time.

The Killdeer is the best-known shorebird in North America. Not only is it often seen, but it can be heard even more often—by night as well as by day. Its common name is imitative of its frequent cries of *kill-dee.* The noisiness of the Killdeer is also commemorated by the specific part of its scientific name, which is derived from the Latin for vociferous or vocal.

Since the vocal presence of the Killdeer is so obvious, its calls warrant closer scrutiny. Aretas Saunders, an early student of bird vocalizations, has described three major types of Killdeer calls.

> The first is the common call heard when one approaches one or more birds, or the vicinity of a nest: *dee dee dee dee-ee kildee dee-ee*, etc., the notes usually slurred slightly upward at the end, at least the longer ones. A second call is the long trilled *t-rrrrrrrrrrrr*, often heard when the nest or young are threatened, and when the birds are fighting or displaying. The third call is one from which the bird evidently has derived its name. It is usually indulged by birds flying about in the air in loose flocks, particularly early in the morning or toward evening. A number of observers or writers on the notes of this species seem not to have separated this call from the first one. It differs always by the fact that the notes slur downward, instead of upward, on the end. I should write it *kil-*

deeah kildeeah kildeeah, at least in those forms where the first note is lower in pitch than the second. It is often rendered, however, when the first note is highest in pitch, when it sounds more like *keedeeah keedeeah*.[151]

Some Killdeers arrive on the breeding grounds already paired, but most pair bonds are formed after arrival. The male advertises both on the ground and in the air. He may fly back and forth over his territory uttering *kill-dee* or drawing his cries out into a trill. Sometimes he may climb almost out of sight in a high circling flight. Often he flies with the slow, deep wingbeats referred to as butterfly flight. Much of the advertising is done from the ground. A Killdeer will stand and watch for the appearance of other Killdeers while calling *di-yeet* or *di-yit* every few seconds. Between bouts of calling, the Killdeer may move from one display spot to another by running or flying. When advertising on the ground, unmated males often engage in scraping. The male first assumes the scraping position—leaning forward, with breast on or close to the ground, wings folded and bent downward. Then, in this position, he kicks alternately with his two legs, digging a scrape. Scraping is accompanied by various calls ranging from *kill-dee* through *di-yit* to a trill. Sometimes the act is symbolic or displacement scraping, and no scrape is dug—the Killdeer performs the act without kicking his legs. During intense bouts of scraping, displacement nest building may take place. The male, standing on tiptoe, will pick up small objects such as pebbles and plant material from the ground and toss them back over his shoulders—as in nest building.[152]

When an advertising Killdeer is approached by another Killdeer, the advertiser may step up the tempo of calling and scraping, but more often he flies or runs to chase the intruder. In hostile chases the advertiser runs with his body held in a horizontal position, his back feathers ruffled, while uttering trilled calls. The intruder may end the chase by suddenly tilting forward in a crouch, exposing his orange rump to the chaser.[152] Sometimes hostile encounters attract other birds, and soon there is a mob of Killdeers in aerial and ground displays—calling, posturing, scraping, and picking up and tossing objects. These encounters usually end with the intruder or intruders leaving the area. Sometimes Killdeers approaching an advertising male elicit little or no hostile behavior; these birds are thought to be receptive females. Once the male and the female form a pairing bond, they remain together.

The members of a pair spend much of their time defending their territory against other Killdeers. However, they may leave the territory for a while, usually to feed. The male engages in much scraping and tossing while on his territory. Scraping is part of the precopulatory behavior of the Killdeer, and some scraping bouts end in copulation. While scraping the male may utter a stuttering trill which often attracts the female. Facing the female, the male tilts forward and repeatedly fans his upraised tail at her, conspicuously displaying his white tail tips. At times, the female may crouch under the tilted tail of the male. Sometimes she replaces him at the scrape. He may walk away, pecking and tossing, only to soon return and replace her at the scrape. Both may peck and toss, or a calling duet may ensue.[152]

One of the scrapes on the territory is used for nesting. This shallow depression in the ground may be left unlined, or it may be lined with small objects such as pebbles and bits of wood and grass. Four eggs are usually laid, and both parents incubate the eggs. Often nesting in the open, with little or no shade, the Killdeers may have difficulty in preventing overheating of the eggs. During the heat of the day, they may forego sitting on the eggs and instead crouch over them with their wings slightly extended, shading them from the sun. Evaporative cooling may also be employed. Before taking over for a session of incubation, a Killdeer may stand in water and wet down its belly feathers. Once back at the scrape, it will lower its belly feathers onto the eggs and then stand over the eggs with its belly feathers barely touching them.[153] Of course, evaporative cooling is the only effective way of cooling eggs in very hot environments. Shade by itself will not do this.

At hatching the young are downy and precocial; they are soon led away from the nest by their parents. Often moving to water, if it is nearby, the chicks feed on their own, but they are guarded and brooded by their parents. They can first fly when about forty days old.

Both parents courageously guard their eggs and young. The behavior of the parents in protecting eggs and young varies according to the circumstances of the threat menacing them. At the distant approach of a potential ground predator, the incubating parent may steal away from the nest; then both parents will fly wildly around the territory, sounding their alarm calls. At the near approach of the predator, a distraction display, involving injury feigning based on the broken-wing ruse, will be performed by one of the parents. In later stages of nesting, both parents may perform. Calling all the while, the Killdeer will lower its wings and hunker down, exposing its bright orange rump and tail. Then, with tail pressed against the ground, it will flail the ground with one wing and the air over its back with the other, half-opened wing. When the predator moves in on this seemingly injured bird, the Killdeer gets up and runs for a short distance with its wings drooping and its tail spread. Again crouching, it repeats the entire performance.

The performance may be modified by the bird simultaneously beating the ground or air with both wings.[154] Thus offering itself as bait, the displaying parent lures the predator away from the eggs or young. These fascinating displays continue until the young are about ten days old.

Sometimes, when distracting human intruders, a nesting Killdeer will employ dummy-nest-brooding behavior. After running a short way, the Killdeer will abruptly stop at a depression in the ground or behind a plant. It will settle down in a manner reminiscent of adjusting and sitting on eggs.[154]

Distraction displays will not work against nonpredatory intruders, such as cows or horses. Here the eggs may be in danger of getting trampled underfoot. The Killdeer will wait until the intruder closely approaches the nest; then it will slip off of the eggs and stand in front of the nest, calling loudly, with its feathers ruffled and its wings beating the ground or air. If the intruder continues to approach, the Killdeer will run toward it and suddenly fly up in its face—the bird may even hit the intruder on its muzzle.[154] The intruding beast is usually so startled that it bolts away.

The two narrow black bands on the neck of the Killdeer are thought to have a disruptive function—they tend to blur or break up the outline of the bird when it is standing motionless in open terrain. Our two other ringed plovers, the Semipalmated and the Snowy, have only one ring or partial ring on the neck; however, these plovers are much smaller than the Killdeer. One ring would not effectively disrupt the outline of the Killdeer but, rather, would render it more conspicuous, as would one thick ring. You can see this size effect at work in young Killdeers. The Killdeer chick is very small and has only a single black collar mark, which functions effectively as disruptive coloration.

When feeding, the Killdeer employs the plover technique of alternately running and standing motionless. A Killdeer will stop abruptly and stand with head up, then suddenly snatch its prey off the ground. It will occasionally bob or teeter. It is one of our most beneficial shorebirds, since it feeds almost entirely on insects and other invertebrates. What little plant food it takes is mainly weed seeds.

Other Charadriids

The Black-bellied Plover, *Pluvialis squatarola*, is an occasional to uncommon migrant in the Great Basin. It nests on the arctic tundra of Siberia and North America. Some overwinter on the Pacific Coast of North America, but most overwinter in the southern continents. This bird was formerly assigned to the genus *Squatarola*—as *Squatarola squatarola*. Then in 1973 the genus *Squatarola* was merged with the genus *Pluvialis*.

The Lesser Golden-Plover, *Pluvialis dominica*, is an uncommon migrant in the eastern part of the Great Basin[27] and a rare one in the western part, where it has been recorded at Malheur National Wildlife Refuge.[61] It breeds on the arctic tundra of Siberia and North America and overwinters on Pacific Islands and in South America, Australia, and southern Asia. In earlier times, this plover migrated through the United States in vast waves during the spring. But it was slaughtered during migration and while on its wintering grounds by both game and market hunters, and its ranks became greatly depleted. Since receiving protection during migration, it has made somewhat of a recovery. Prior to 1982 its common name was the American Golden Plover.

The Semipalmated Plover, *Charadrius semipalmatus*, nests on the arctic tundra of North America. It overwinters from the coast of central California southward to Patagonia in South America. The Semipalmated is a regular migrant in the Great Basin—on its way to and from its breeding grounds. This plover is considered to be more of a coastal migrant, with much less traffic passing through the interior of North America, particularly in the Intermountain West. At least one major bird guide has a distribution map which indicates that there is no migration to speak of through the Great Basin and the rest of the Intermountain West. This is simply not true. Each spring I see Semipalmated Plovers during migration at the various wet areas I visit in western Nevada. They can be seen quite regularly at the Fernley and Stillwater Wildlife Management Areas.

The Mountain Plover, *Charadrius montanus*, is a rare migrant in the Great Basin. There are a handful of sight records for the Basin. Several specimens have been collected in the Brigham City–Bear River Refuge area in Utah[56] and at Carson Lake in Nevada.[155] The closest Mountain Plovers nest in the high plains region of Wyoming and Colorado to the east of the Great Basin. Formerly, this plover was in the genus *Eupoda*. Then in 1973 *Eupoda* was merged with the genus *Charadrius*.

FAMILY RECURVIROSTRIDAE

The avocets and stilts of this family are distributed over much of the world. They are very long-legged wading birds, with slender necks, small heads, and long slender bills—which may be straight, downcurved, or upcurved.

Black-necked Stilt
Himantopus mexicanus

The Black-necked Stilt is a breeding species in western United States, and, to a limited extent, along the Atlantic Coast. Its breeding range extends into South America. The Blackneck is an uncommon to common nesting species in the Great Basin, reaching its greatest abundance in the eastern end. Black-necked Stilts are often found in association with American Avocets but usually are not as abundant as the avocets. However, at some wet locales in the Basin—the Bear River and Fish Springs National Wildlife Refuges and Stillwater Wildlife Management Area—this stilt is reported to be an abundant summer resident.

Although stilts and avocets are often seen together, may nest in the same area, and even occasionally lay eggs in each other's nests, their habitat preferences are not identical. Stilts are found more often in marshes, prefer fresher water than avocets, and may occur around fresh water where there are no avocets. They will even frequent grassy areas that have been temporarily flooded. When feeding they do not venture as far away from shore on mud flats as do avocets.[156]

The stilt bears a descriptive name, since it stands on what appears to be legs of unnatural length for a bird. Outside of flamingos, stilts have the longest legs relative to body size of any bird. They often rest by standing on one leg in the water, but on land they rest sitting down. Displacement pecking and displacement sleeping have been witnessed during interactions between birds. Hamilton had the intriguing thought that possibly stilts and avocets dream. This is somewhat disquieting, since dreaming is such a prized human attribute. Hamilton wrote, "Possibly recurvirostrids may dream while sleeping. I have observed recurvirostrids moving their bills rapidly, as when swallowing, when the bills were under their wings and their eyes were closed. Calling has also been observed being performed by birds which appeared to be sleeping."[156]

Although this stilt is a member of the Family Recurvirostridae, it lacks a bill, as the family name suggests, that is bent back on itself. The stilt bill is long, straight, and needlelike, and the bird's hunting techniques are less varied than those of the avocet, with its recurved bill. The stilt's principal hunting technique is pecking—during which it seizes insects on or near the surface of the water or on land while standing still or walking slowly. Occasionally, it will plunge its head and neck underwater or run, with upstretched wings, after a rapidly moving insect. Most of the stilt's food is animal material, although some plant seeds are harvested.

The long legs of the stilt are well designed for wading, but this requires practice—a newly hatched chick, like a newly foaled colt, is wobbly on its legs. During takeoff, the legs provide sufficient spring and air speed to get the stilt airborne. In normal flight the legs are brought up in line with the body, trailing behind the tail. Just before touching down, the legs are lowered again. Stilts tend to alight on land more frequently than avocets. When they do alight on water, they do it carefully—they first hover over the water with dangling legs before gently touching down.

In some display flights, the bright red legs are conspicuously dangled. The most unusual of these display flights, the butterfly flight, occurs early in the year and may be a territorial display. While performing a butterfly flight a stilt hovers at elevations of fifteen to thirty feet or so above the water. Its body is positioned at about a forty-five-degree angle above the horizontal, with neck pulled in, tail spread, and long red legs dangling.[156] The hovering bird then sinks toward the surface of the water, only to immediately arise again before touching down. After performing these up-and-down maneuvers, the stilt may move a short distance on and repeat the performance.

The stilt nests in a lined ground scrape; sometimes quite a mound of material is accumulated. Like avocets, stilts will shove additional nesting material under their eggs if the nest is threatened by rising floodwaters. Both sexes incubate the eggs, and within a day after all the chicks have hatched the parents lead them away to food and cover. The parents brood their chicks to protect them from the elements. A parent kneels on its lower legs, with its body well above but parallel to the ground, and the chicks stand underneath its elevated body. This is the same brooding posture the avocet uses.

Stilts seldom perform distraction displays inside the nesting colony. Along with avocets, who often nest in loose colonies with stilts, the stilts fly out to intercept a ground intruder before it reaches the colony. If an intruder does enter the colony, the stilts fly around overhead, screaming at it. But they do not dive at or buzz intruders, as do avocets. Upon intercepting an approaching ground intruder outside of the colony, stilts and avocets engage in communal distraction displays. Both species will perform distraction incubation, running around and suddenly crouching as if they were settling down on eggs. Stilts have a wing flagging display, during which they flag one wing at a time. A tremendous outcry accompanies the displays. There is much crouching, screaming, and quivering of outstretched wings. Displaying stilts may vigorously bob up and down in shallow water, striking at it and splashing it around with their breasts. Injury feigning has been reported, during which stilts raise one wing as if it were smashed and even collapse one leg.[157] At the approach of winged predators, such as hawks, stilts will fly up, along with the avocets, to scream and dive at them and drive them away.

American Avocet
Recurvirostra americana

The breeding range of the American Avocet lies in western United States and in the southern prairie region of Canada. This avocet reaches its highest numbers in the Great Basin, where it is a common and often abundant summer resident at wet locales. Stragglers are occasionally present in the winter. The American Avocet enlivens the bleakest yet most characteristic of Great Basin summer landscapes—the shallow, foul bodies of alkaline or brackish water and their fringing flats of mud, alkali, or salt. Here, where the earth is overpowered by heat and light, this strangely beautiful bird nests and gains its summer livelihood.

The avocet is well designed for wading and swimming and for traversing mud flats. It has very long legs, and the three front toes on each foot are enclosed in a web. The long legs allow the avocet to wade in fairly deep water. The webbed feet provide a swimming bird with powerful propulsion and prevent it from miring when wading or walking ashore.

The avocet's physical equipment for swimming, wading, and mud walking in its quest for food is impressive. But it does not seem to be equally well designed for capturing food—its long, upcurved, awllike bill would seem to make for slim pickings. Surprisingly, the odd-shaped bill is a handy, all-around tool; and the avocet wields it effectively in a variety of fashions.

During the most characteristic movements of feeding, the avocet wields its bill like a scythe—sweeping it from side to side over the surface of the mud flat or underwater. Pausing after each forward step, the avocet rests the upcurved bill tip flat on the mud or bottom, just to one side of its body and line of movement. The slightly open bill is then swung to the other side, scraping the mud or bottom as it goes. This is a tactile method of feeding, in which the avocet captures prey by touch and not by sight. Usually after a single side-to-side sweep, the avocet raises its head and swallows its prey, before taking another step forward to repeat the process. Sometimes an avocet will sweep from side to side several times before raising its head to swallow prey. On occasions, avocets will forage cooperatively—a group will form a phalanx, wading forward, shoulder to shoulder, pausing to scythe after each step.

Filtering and scraping are the bird's two other methods of tactile feeding. Filtering occurs on mud flats where there are shallow pools of water.

American Avocets

While the bill is randomly moved over the mud for a few seconds, its tip is rapidly opened and closed. In scraping, the neck is employed to push the bill tip ahead over the surface of the mud and then to pull it back.

Not all foraging is touch foraging—visual foraging also takes place. While walking slowly and pausing between steps or while standing still, the avocet may locate prey by sight and capture it by pecking at it. In deeper water the head and neck of the avocet may disappear underwater as the bird jabs at its prey. An avocet may run or flutter after a flying insect and snatch it out of the air with its bill.

More animal than plant food is consumed. Both adult and larval aquatic insects are taken, including brine flies; brine shrimp and other crustaceans are also eaten. The plant food of avocets consists mainly of the seeds of marsh and aquatic vegetation. Since much of their food is located by tactile techniques, the birds probably scavenge to some extent and pick up dead along with living prey.

During the nesting season, pairs of American Avocets may join in group displays. The avocets form a rough circle with all the birds facing inward, each bird standing next to its mate. While displaying the avocets lean forward, with their necks extended and their bills just above and parallel to the surface of the water. Sometimes the group will circle first in one direction, then in the other direction.

The American Avocet is not known to have formal pair-bonding displays. The female initiates pair formation by approaching a male to associate with him. He drives her away, but she persistently returns. She never actively resists the male but adopts nonaggressive postures—she never looks directly at him or engages in display pecking. Looking directly at another bird is often an aggressive posture, whereas looking away or turning the head is an appeasement posture. Eventually, the male may accept the female and freely associate with her.

Avocets have fascinating copulatory displays. The precopulatory display choreography is designed around breast-preening movements. When breast-preening, an avocet lowers its bill into the water, lifts its bill and head upward by fully extending its neck, and then by depressing its head brings its bill into contact with its breast feathers. Copulation is initiated by the female adopting a posture with her legs slightly apart, neck extended, and neck, head, and bill parallel and close to the surface of the water. Taking a position close to the female, the male begins to breast-preen. The speed of his preening movements increases until they become frenzied and the bill dipping is so violent that water is splashed high in the air. From time to time the male will change his position, from one side to the other of the female; he will always walk behind her in doing so. Following copulation the male and female cross bills and run forward together for a short distance. The male may hold a partly outspread wing over the back of the female as they run.[156]

Avocets nest in colonies. The distance between two nests may vary from several feet on up to over one hundred feet. The nest is a scrape, built by rotating the breast on the ground. Scraping often occurs after copulation and may be accompanied by nest material tossing. Sometimes very little nesting material—sticks, pebbles, weeds, and feathers—is added to the scrape; at other times a mound of material is added. Both male and female incubate the eggs, and often a nest-relief ceremony occurs at the change-over. The relief merely walks up to the nest and covers the eggs. But, as the departing bird walks away from the nest, it picks up small objects from the ground and tosses them over its back in the direction of the nest.

Avocets have a repertoire of distraction displays to use when a colony is threatened by an intruder. In the immediate vicinity of nests, the birds will fly about, diving and screaming at a ground intruder. To deal with more distant threats, an avocet will walk toward the intruder, tipping its fully extended wings from side to side and uttering low nasal calls. Distraction incubation may occur. Avocets may alight on water and call and behave as if they were injured. Since avocets often nest in loose colonies with stilts, stilts may participate in communal distraction displays.

FAMILY SCOLOPACIDAE

The classification and nomenclature of shorebirds were changed considerably by the AOU Committee on Classification and Nomenclature in 1973 and again in 1982. In 1982 the Family Scolopacidae was enlarged by merging the phalarope family, Phalaropodidae, into it as the Subfamily Phalaropodinae. Then, a second subfamily was formed by merging all the subfamilies previously present in Scolopacidae into a single subfamily—Scolopacinae. The Subfamily Scolopacinae was then divided into eight tribes, seven of which are represented in the Basin. So that you can use bird books published prior to the 1982 and 1973 changes without complete confusion, I shall note all the name changes for Great Basin species of scolopacids as I proceed through my natural history accounts.

SUBFAMILY SCOLOPACINAE

TRIBE TRINGINI: TRINGINE SANDPIPERS

The tringine sandpipers occur as breeding birds over much of the northern hemisphere—often in the Far North. They are migratory, and some overwinter far south of the equator. Their presence is often loudly announced by their piercing call notes.

Willet
Catoptrophorus semipalmatus

The Willet has an unusual breeding distribution in North America. There is a coastal breeding population along the Atlantic and Gulf coasts and an inland breeding population in western United States and adjacent Canada. The Willet is a common breeding species and migrant in the Great Basin. On rare occasions, a straggler is seen in the Basin during the winter.

When Willets arrive on the breeding grounds in the spring, some behave as if they were already paired. For a time, before egg laying gets under way, Willets gather in open places in marshes and elsewhere to engage in communal displays. Birds' display movements are often choreographed to conspicuously feature some unusual color pattern or peculiar adornment of each species. Willets have a striking wing pattern—with a broad band of white partially bordered on two sides by black—and this species' displays are designed around wing waving, in which the boldly colored wings are repeatedly flashed at the viewer. In aerial displays the wings are vibrated through a narrow arc; in ground displays they are raised above the back and vibrated. The male's precopulatory display consists of wing waving, and he will even wing wave during copulation.[158]

Willets do not breed until they are two years old. Usually four eggs are laid in a scrape in the ground. The female incubates the eggs by day, with the male probably incubating at night. Willets are not the best of parents, and their deficiencies may offset their reproductive success. Incubation may start before the clutch is completed, and then the eggs do not hatch all at the same time. After one or two chicks hatch, the parents may depart with them, abandoning the rest of the eggs.

Although Willets are not colonial nesters, a number of pairs may nest within sight and sound of each other. If a ground intruder enters the nesting area, a number of adults may fly around in alarm and escort the interloper out of the area. When a nest is threatened, the distraction display staged by an adult Willet is an unconvincing one—the wings may be drooped or beaten feebly, but there is no sincere pretense of being crippled. The parents escort the chicks about for a while but then abandon them before they are old enough to fly.

Spotted Sandpiper
Actitis macularia

The Spotted Sandpiper's breeding range encompasses most of North America, extending from the northern tree line in Alaska and Canada as far south as southern United States. It is a common summer resident and migrant in the Great Basin, and next to the Killdeer it is the most abundant and widespread shorebird here. These sandpipers rarely overwinter in the Basin.

Spotted Sandpipers frequent the margins of lakes and ponds where there are sandy or pebbly stretches of shoreline and streams with sandbars and gravel bars. They range from the valleys on up to altitudes of nine thousand feet or so.

Bird species often have their own individual ways of doing things—often ways unintelligible to us. The Spotted Sandpiper has several such traits. According to its geometry, the shortest distance between two points is not a straight line but an arc. When flushed from a point along a shoreline, the bird does not fly straight up or down the bank to escape danger. Rather, it flies out over the water on a semicircular course that eventually intercepts the same bank further along from where it was flushed. If flushed again, it will again fly along a semicircular course. Even when the danger is past and the sandpiper is returning to its original location, it will do so by following semicircular courses out over water, instead of flying straight back along the shoreline.

Not only does the Spotted describe semicircles in flying from one point to the next, but on foot it is perpetually teetering up and down like a perpetual motion machine. As Stearns and Coues described the teetering, "the forepart of the body is lowered a little, the head drawn in, the legs slightly bent, whilst the hinder parts and tail are alternately hoisted with a peculiar jerk, and drawn down again, with the regularity of clock-work. The movement is more conspicuous in the upward than in the downward part of the performance; as if the tail were spring-hinged, in constant danger of flying up, and needed constant presence of mind to keep it down."[90] Now what can a bird possibly accomplish by spending its waking hours teetering up and down? It seems such a waste of energy and such an awkward position from which to view the world. Theorists have not been overly successful in coming to grips with the riddle of this perpetual teetering. The best hypothesis they have to offer is not very intellectually satisfying and has yet to be tested in the field. It is based on the observation that certain other kinds of birds who also feed along the margins of streams or lakes—like dippers, waterthrushes, and wagtails—also teeter. And that "this tipping to and fro causes the bird to blend into the lapping wavelets and the play of light and shadow they create on the shore, effectively camouflaging it."[159]

In most species of birds, the male plays the dominant role in courtship and the female plays the dominant role in the care of eggs and young. Shorebirds show a strong tendency to reverse this procedure—the phalaropes have even evolved to a point where they show a complete reversal in sex roles. Although the Spotted Sandpiper has not evolved to the point reached by the phalaropes, it still shows a considerable amount of reversal in sex roles. The female Spotted plays the main role in courtship strutting displays, whereas the male does all or most of the incubating and provides most of the care to the young. Female Spotted Sandpipers are on the average slightly larger and more heavily spotted than the males. Since at best this is subtle sexual dimorphism, earlier observers did not suspect sex role reversals in this species but automatically described the dominant bird in courtship as being the male. Eventually, studies in which birds incubating or tending young were internally sexed showed them to be males.

Other Tringine Sandpipers

The breeding range of the Greater Yellowlegs, *Tringa melanoleuca*, lies in the boreal coniferous forests of Alaska and Canada. This bird is an uncommon to common migrant at wet places in the Great Basin. Occasionally, a nonbreeding straggler summers in the Basin, or a straggler overwinters here. The Greater's scientific name was *Totanus melanoleucus*, until the genus *Totanus* was merged with the genus *Tringa* in 1973.

The Lesser Yellowlegs, *Tringa flavipes*, has a breeding range that lies in the far northern forests of Alaska and Canada. It is an occasional to common migrant at wet places in the Great Basin. Stragglers rarely overwinter or summer here; the summer stragglers are nonbreeders. Before the genus *Totanus* was merged with *Tringa*, this bird's scientific name was *Totanus flavipes*.

The Solitary Sandpiper, *Tringa solitaria*, breeds in the vast boreal coniferous forests of Canada and Alaska, with nesting sites commonly located near lakes or ponds in muskeg or bog areas. This sandpiper is a rare to uncommon migrant in the Great Basin. Occasionally, Solitary Sandpipers are seen during the summer in the Basin, but there is no evidence that they ever nest here. It is common practice among shorebirds for some nonbreeding individuals to spend the breeding season on their wintering grounds or somewhere along their migration route. When Solitary Sandpipers occur in the summer in the Great Basin, they may be seen in what appears to be aerial courtship activities, or they may appear to be maintaining territories in which they feed. But these activities can occur in the absence of nesting—proof of nesting demands eggs or chicks. Since the Solitary does not form migratory flocks, and individuals migrate mostly at night, we just do not have much information about its status in the Great Basin. It is easily overlooked in the field, in the presence of flocks of other shorebirds, and may be commoner than the records indicate.

The Wandering Tattler, *Heteroscelus incanus*, is of accidental occurrence in the Great Basin. There are several sight records from Farmington Bay on the east side of Great Salt Lake.[27]

TRIBE NUMENIINI: CURLEWS

Curlews are long-legged and long-billed birds. Their bills are downcurved and are used for probing for invertebrates, such as worms. Curlews breed in the northern parts of North America and Eurasia and migrate southward to overwinter—often crossing the equator. The Upland Sandpiper is also a member of this tribe.

Long-billed Curlew
Numenius americanus

The Long-billed Curlew is a breeding species in western Canada and United States; it was extirpated as a breeding species from its range in midwestern United States. This curlew is a breeding species and migrant in the Great Basin. In recent years it has been declining in number in the Great Basin and the rest of the West, losing much breeding habitat and migratory way stations to agriculture and other land development. During the past ten years, I have witnessed a striking reduction in the number of curlews during spring migration in western Nevada.

Some Longbills are paired, and some are not, when they arrive on the breeding grounds. Upon arrival some paired curlews and unpaired males immediately occupy territory. Other curlews remain in their migratory flocks for several weeks, before the flocks break up and territorial behavior ensues. Most territorial defense is by the male. Some of the hostile interactions between territorial curlews involve displays, while others are more violent and lead to fighting.

Long-billed Curlews use their feet, wings, and bills in fighting. They mainly fight from a head-to-head position in which the bills, feet, and wings can all be brought into action. They may flop several feet off the ground while hitting with their wings. Sometimes they fight for a moment side by side, at which time their feet are their main weapons. At other times one will jump on the back of another and deliver a kicking and pecking attack.[160]

The less violent territorial encounters involve postures, movements, and even chasing. Frequently, during hostile encounters, one or both curlews will assume the upright posture. This is a threat display in which the back is tilted upward, the neck fully extended, and the head held high. Trespassers are usually evicted from a territory by the threat display of supplanting, during which the defender charges to meet the intruder on foot or in the air. When a curlew is closing in or chasing a trespasser on foot, the upright run is employed if the distance separating the two birds is greater than about ten meters. At distances of less than ten meters, the crouch run is employed.[160] In the first instance the defender runs in an upright posture; in the second the defender runs with his back tilted downward, his head low, his body feathers raised, and his tail spread. The crouch run may terminate in an aerial chase. Other territorial displays or movements are wing raising and concealment. Wing raising is a threat display in which the bird raises his wings over his back while in an upright posture. Concealment is used to bewilder the intruder while getting the defender close enough to launch a crouch-run attack. The defender, while approaching or being approached by the intruder, suddenly throws himself down and hides in the grass. The curlew may repeat the performance while moving forward, until he is close enough to suddenly launch a surprise attack on the intruder.[160]

During territorial boundary disputes between curlews with contiguous territories, the out-of-context displacement behavior is often seen. Violent grass pulling or displacement pecking may occur, as the curlews peck at the ground, picking up large pieces of litter and breaking them into pieces in their bills. Boundary disputes often end abruptly as the birds drift apart and begin displacement feeding.[160]

Unmated males advertise for females by bounding-SKK flights and ground calling. Julia Allen's excellent descriptions of these two displays are as follows. "In the Bounding-SKK Flight display, the bird rapidly flutters upward, rising almost perpendicularly, then sets its wings in a downward curved arch resembling an umbrella . . . In that position, it slowly glides back down, pinions motionless on the breeze, sometimes coming to within 0.3m of the ground before rising again. The Soft 'Kerr Kerr' call consists of a high pitched and very melodious 'kerr' note that tapers off at the end and is repeated in a series." In ground calling the "basic note is a long drawnout 'whee' that fades in, rises in the middle, and then fades out. It is uttered with the bill only slightly opened and is repeated in a series. The following are variations on the basic 'Whee' call: 'whee,' 'whee-a-ee-a-ee,' 'wheer,' 'whee-a-ee-a-eer,' 'ee-a-ee-a-ee,' and 'eee.'" These two advertising displays are also important in helping establish a pairing bond once a receptive female is attracted to the male, as is scraping. The scrape ceremony produces a number of nest scrapes on the male's territory. Following a ceremony, the male engages in tossing, in which he stands with his back or side to the scrape and throws nesting material toward it.[160]

This curlew nests in grassy areas close to marshes or lakes as well as in dry, upland situations, at a distance from water, even at the edges of alkali flats. One of the scrapes made during a scrape ceremony is lined and used as a nest. The usual clutch size is four eggs. Both parents incubate, with the female sitting by day and the male by night.[160]

The Long-billed Curlew is the largest of the North American shorebirds, with females being up to one-third larger than males. This curlew is well named, since its appearance is dominated by a very long and downcurved bill. The bills of females are noticeably longer than those of males. The male's bill is about three times as long as his head or less, whereas the female's is more than three times as long as her head. Also, the female's bill is straight along part of its length, with most of the curvature occurring near the tip; the bill of the male is symmetrically curved. Of course, the curlew chick must come out of the confines of the eggshell with a short bill. But not only is the chick's bill short, it is also straight and stubby. The slimness, curvature, and length develop gradually after hatching.

In feeding, the curlew employs two different techniques. It pecks and removes food from the ground or from vegetation, and it probes by inserting its bill partially or fully into crevices, burrows, and soft ground. Insects and other invertebrates constitute the curlew's principal prey. Some berries are taken.

Other Numeniines

The Upland Sandpiper, *Bartramia longicauda*, known as the Upland Plover prior to 1973, is of occasional occurrence in the Great Basin. Its breeding range lies to the north and to the east of the Basin. There are several sight records for this species from along the eastern edge of the Great Basin in Utah.[56] It has been suggested that this species possibly was once a breeder in Utah.[27]

The Whimbrel, *Numenius phaeopus*, breeds on the tundra of North America and Eurasia. It is of accidental occurrence in the Great Basin during migration, having been recorded from several sites in Utah and from Malheur National Wildlife Refuge.

TRIBE LIMOSINI: GODWITS

Godwits are large, long-legged shorebirds. Their long bills are straight or slightly upcurved. Godwits breed in the northern parts of North America and Eurasia and migrate south to overwinter.

The Hudsonian Godwit, *Limosa haemastica*, which breeds on the Canadian tundra, migrates through the United States to the east of the Great Basin. There is a sight record for this species from Cedar City on the southeastern edge of the Basin.[27]

The Bar-tailed Godwit, *Limosa lapponica*, breeds in Eurasia and western Alaska. It migrates out over the Pacific Ocean to overwinter on Pacific islands. A few strays wander down the Pacific Coast in North America. There is a sight record for this species from the Bear River National Wildlife Refuge.[27]

The Marbled Godwit, *Limosa fedoa*, breeds in the prairie regions of northern United States and southern Canada. It migrates mainly to the Pacific Coast. The Marbled Godwit is an uncommon migrant at wet places in the Great Basin. Stragglers may be present here during the summer and winter.

TRIBE ARENARIINI: TURNSTONES

Turnstones are plump, short-necked birds; their short, thin, flattened bills are slightly upturned at the tips. Turnstones possess features which are suggestive of both plovers and snipes. There are two species of turnstones, both of whom breed in the Far North—one in North America and the other in both North America and Eurasia.

The Ruddy Turnstone, *Arenaria interpres*, is a breeding bird on the arctic shores of Eurasia and North America. It is of accidental or occasional occurrence in the Great Basin during migration. Specimens have been collected at the Bear River National Wildlife Refuge and at Soda Lake near Fallon, Nevada. There are a number of sight records from various locations in the Basin. Formerly included in the Family Charadriidae, turnstones were transferred to the Family Scolopacidae in 1973.

TRIBE CALIDRIDINI: CALIDRIDINE SANDPIPERS

Calidridine sandpipers are small shorebirds of worldwide distribution. Countless flocks pass through the Great Basin to and from their northern breeding grounds. These sandpipers are usually seen on the move—running over mud flats and along shorelines as they forage or swiftly twisting and turning through the air in tight flocks as they fly from one foraging station to another nearby.

The Red Knot, *Calidris canutus*, breeds far to the north—on the arctic tundra. Its migration occurs mainly along the Pacific Coast. However, there is some inland migration, particularly east of the Rocky Mountains. There is apparently some erratic migration along the eastern edge of the Great Basin in Utah, and flocks of up to 1,500 birds have been seen.[56] Specimens have been collected from sites in Utah. There is a specimen record for the western end of the Great Basin from the Carson Sink, and this species is of accidental occurrence at the Malheur National Wildlife Refuge.

The Sanderling, *Calidris alba*, has a breeding range that lies far to the north—on the tundra of Canada and Eurasia. It is a rare to uncommon migrant at wet places in the Great Basin. An occasional straggler is seen here in the winter and summer. Prior to 1973 this bird's scientific name was *Crocethia alba*, but then the genus *Crocethia* was merged into *Calidris*.

The breeding range of the Semipalmated Sandpiper, *Calidris pusilla*, lies on the low arctic tundra of North America. Migration mainly flows to the east of the Rocky Mountains. During migration this sandpiper is a rare transient in Utah in the eastern part of the Great Basin, where a specimen was collected and a handful of sight records have been compiled.[27] There is but a single record for the Semipalmated in the western part of the Basin. Ridgway reported collecting three juveniles from a flock of sandpipers at the Humboldt marshes on August 26, 1867.[25] Later, Linsdale reported that these three birds were Western Sandpipers and that one of them was in the collection of the United States National Museum.[91] As of now, we just do not have much information on the status of the Semipalmated Sandpiper in the Great Basin. It resembles the Western Sandpiper in appearance, travels with other sandpipers, and can be easily overlooked. Formerly, its scientific name was *Ereunetes pusillus*, but in 1973, when the genus *Ereunetes* was merged into the genus *Calidris*, it became *C. pusilla*.

The breeding range of the Western Sandpiper, *Calidris mauri*, lies on the coastal tundra of Alaska. This species migrates mainly along the Pacific Coast. The Western Sandpiper is a common and often abundant migrant in the Great Basin; stragglers are occasionally seen here during the winter and summer. Formerly, its scientific name was *Ereunetes mauri*.

The breeding range of the Least Sandpiper, *Calidris minutilla*, lies in the low (southern) arctic tundra of North America. During migration this sandpiper is a common, often abundant transient in the Great Basin. Stragglers may be present here during the summer and winter. Formerly, the scientific

Western Sandpiper

name of this species was *Erolia minutilla*. After *Erolia* was merged into *Calidris* in 1973, the Least became *C. minutilla*.

The breeding range of the White-rumped Sandpiper, *Calidris fuscicollis*, lies on the North American tundra. This species migrates through the interior of North America to the east of the Great Basin. There is but a single sight record for the Basin, from Fish Springs National Wildlife Refuge.[56] Formerly, the Whiterump's scientific name was *Erolia fuscicollis*.

The Baird's Sandpiper, *Calidris bairdii*, breeds on the high arctic tundra of Siberia and North America. The heaviest migration passes through the plains region of North America. The Baird's Sandpiper is a rare to uncommon migrant in the Great Basin. Stragglers are of rare occurrence here during the summer and winter. Formerly, this sandpiper's scientific name was *Erolia bairdii*.

The breeding range of the Pectoral Sandpiper, *Calidris melanotos*, lies on the tundra of Siberia and North America. Most of its migration passes through the interior of North America, barely brushing the Great Basin. The Pectoral Sandpiper is an uncommon migrant through the eastern end of the Great Basin, and stragglers are seen on rare occasions during the winter.[27] This sandpiper is a rare spring migrant and an occasional fall migrant at Malheur National Wildlife Refuge.[61] There is only a single sight record for western Nevada—a September record for the Stillwater Wildlife Management Area.[161] Formerly, this bird's scientific name was *Erolia melanotos*.

The Dunlin, *Calidris alpina*, breeds on the coastal tundra of North America and in northern parts of Siberia. The Dunlin is an occasional to uncommon migrant in the Great Basin. Occasional stragglers may be seen here in the summer and winter. Formerly, this species' scientific name was *Erolia alpina*.

The Stilt Sandpiper, *Calidris himantopus*, nests on the arctic tundra of North America and migrates through eastern United States; it is of accidental occurrence in the Great Basin. There are several sight records for the Stilt Sandpiper in the eastern part of the Basin and one from near Hazen, Nevada, in the western part. Prior to 1982 its scientific name was *Micropalama himantopus*.

TRIBE LIMNODROMINI: DOWITCHERS

The dowitchers are plump birds with fairly short necks and legs and long bills. They breed in the northern ends of North America and Eurasia and migrate south to overwinter.

Short-billed Dowitcher
Limnodromus griseus

The breeding range of the Short-billed Dowitcher is discontinuous, with several breeding populations in Canada and one in southern Alaska. The population in Alaska migrates along the West Coast—with a limited inland flow. Hence, this dowitcher is present in the Great Basin during migration. Since it is often difficult to differentiate between Short-billed and Long-billed dowitchers in the field, observers usually lump sightings as simply being dowitchers; and this is what appears in checklists and reports. However, we do have some specific sightings of Shortbills in the Great Basin by competent observers. Several specimens have been collected in the eastern part of the Basin.[56] Specimens have also been collected in the western part at the Fernley Wildlife Management Area, Washoe Lake, Lahontan Valley, and Goldfield.[161] Although we are as yet uncertain about the status of the Short-billed Dowitcher in the Great Basin, at best it is but an uncommon migrant here.

Long-billed Dowitcher
Limnodromus scolopaceus

The Long-billed Dowitcher breeds on the tundra of Siberia and North America. During winter and while migrating this species frequents freshwater ponds, in contrast to the coastal mud flats preferred by the Short-billed Dowitcher. The Longbill is a common, often abundant migrant in the Great Basin. Stragglers are occasionally seen here during the summer and winter.

TRIBE GALLINAGONINI: SNIPE

Snipe are small to medium-sized birds. They possess relatively short legs and very long bills—with which they probe in the soft ground when foraging. Snipe have elaborate courtship flights. They are practically worldwide in their breeding and migrate to the south to overwinter.

Common Snipe
Gallinago gallinago

The Common Snipe is a breeding bird over much of the subarctic and temperate regions of Eurasia and North America. It is an uncommon to common summer resident and breeding species in the Great Basin. Most snipe migrate out of the Great Basin to overwinter. However, a few overwinter here in areas in the valleys where the waters remain open and the wet ground in marshes, along streams, and at seepage areas does not freeze or remain covered with snow. Prior to 1982, the scientific name of this species was *Capella gallinago*.

The Common Snipe is a creature of wet, soft ground, where it can obtain food by probing deeply with its long bill. In the Great Basin it usually announces its presence by exploding into the air when an intruder enters a wet meadow, marsh, or seepage area or walks along an irrigation ditch where snipe are feeding. This explosion from ground to air is often the first indication that snipe are present, although sometimes you may first note areas where the ground is riddled with their probe holes. When danger approaches a snipe freezes motionless; on the nearer approach of danger, the bird explodes into the air in an erratic, zigzag flight before straightening out. The explosion, the erratic flight, the grating *scaip* alarm notes, and the neckless, long-billed appearance all proclaim the bird to be a snipe.

To the gullible participants in snipe hunts, the snipe is but a mythical bird. Would-be snipe hunters are trusting souls who get taken on a one-way ride to a snipe hunt. The hunter is taken to some lonely, remote station at night, given a gunnysack and a flashlight, and told to wait while the rest of the group departs to station other snipe hunters, before commencing to drive snipe their way. Once snipe start coming their way, the hunters are to blind the victims in the beam of their flashlights and clap them into their gunnysacks. After a long, lonely wait, without the appearance of a single snipe or the reappearance of the rest of the party, the would-be snipe hunter finally realizes that something is amiss and that it's a long way home on foot!

To those who have overheard the winnowing *huhuhuhuhuhuhu* cascading down without ever actually seeing the distant bird racing through the sky, the snipe is but an eerie, disembodied voice out of the spring sky.

To hunters the Common Snipe is a prized antagonist—an elusive, swiftly flying, challenging target, gracing the table with rich dark meat of excellent flavor. The world's record for shooting snipe belongs to James Pringle, a hunter who shot 366 birds in six hours on December 11, 1877, in Louisiana. In thirty years of hunting snipe, Pringle's score was a depressing 78,602 birds.[162]

Greater sport can be realized from watching snipe than from shooting them. Winnowing, the aerial courtship flight of the male, can be readily witnessed in the Great Basin during the spring. This occurs often at dusk but also during the daytime, especially on overcast days or with the approach of a storm. Winnowing occupies a lot of space, more than just the sky above the male's territory. Males can often be seen conspicuously perched on fence posts and utility poles around their winnowing grounds. Some of the suburban lanes on the outskirts of Reno-Sparks are particularly good places from which to watch the performance.

The winnowing male mounts in the air to a height of several hundred feet. He then races along on rapidly beating wings, often describing a great circle in the sky. He follows an undulatory path, performing swooping dives at the rate of about eight per minute; and he terminates each dive by climbing steeply back up, on rapidly beating wings, to his original flight level. During the swoop the wings are partly closed, and the tail is spread with the outer tail feathers clearly separated from the others; the swoop is not a straight descent but a side-slipping one. As the bird descends, the eerie *huhuhuhuhuhu* or *who-who-who-who-who-who* is produced. The winnowing sound—a nonvocal tremolo—is believed to be produced by the humming vibrations of the stiff outer tail feathers as they interrupt or damp the airflow from the wings. When ascending from the ground to winnow and when descending from winnowing, the snipe utters a yakking call of *cut-a cut-a*. Sometimes the female will also winnow.

There are other courtship displays besides winnowing. During pair formation the male, while yakking, may mount a short distance into the air, only to hold his wings above his body and drop along a downward curve toward the ground. Just before touching down, he flies up again to repeat the performance.[150] He may go from this performance into winnowing. Ground displays, which involve both sexes, also take place.

The Common Snipe nests in boggy areas. Usually four eggs are laid, and the female does all the incubating. After an incubation period of about eighteen days, the eggs start hatching. The male usually commandeers the first two chicks to hatch and leads them off to raise by himself; the female raises the last young to hatch. Hence, the family separates into two subfamilies. At first the parent will probe for food for the chicks, since their bills are slow in developing. By the time the young are fifteen to eighteen days old, they are beginning to fly and feed themselves. At about six weeks of age, the young begin to form groups or wisps with other young snipe.[163] Eventually, the young snipe will migrate south together, in advance of the adults. Then, apparently, the adult females may migrate south ahead of the adult males.

Common Snipe eat predominantly animal food of the likes of insects, earthworms, and crustaceans. The seeds of marsh plants such as sedges, smartweeds, and bulrushes constitute their principal vegetable food. Although they pick up some food at the surface, much of their feeding is based on probing in soft ground for burrowing animals such as earthworms.

A bird will thrust its bill perpendicularly into the ground, sometimes to the hilt, and work it around. The outer surfaces of both mandibles near their tips are provided with nerve endings housed in tiny pits. Hence, the snipe can feel with the tip of its bill. Further, the tip of the upper mandible is independently movable, so that a burrowing animal can be grasped and withdrawn from the ground.

SUBFAMILY PHALAROPODINAE

This subfamily contains the phalaropes. Prior to 1982 the phalaropes were accorded familial rank—as the Family Phalaropodidae. They are best known for their sex role reversals. Their breeding ranges lie in the northern parts of North America and Eurasia. They migrate south to overwinter—often at sea.

Wilson's Phalarope
Phalaropus tricolor

The Wilson's Phalarope breeds in western North America, from the western interior of the United States northward into southern Canada. This phalarope is a common, often abundant breeding species in the Great Basin. It is most numerous during migration. Prior to 1982 the scientific name of this bird was *Steganopus tricolor.*

Phalaropes are odd birds—they are one of the few which have experienced a major reversal in sex roles. The female Wilson's is larger, has the more colorful nuptial plumage, and actively pursues and courts males. The only contribution she makes to family life is laying the eggs. Once she accomplishes this, she departs. The male has the sole responsibility for incubating the eggs and rearing the young.

There are many species of birds in which the males play little or no role in family life beyond inseminating the females. Quite often these males also practice polygyny, mating with more than one female. The question of whether the female Wilson's Phalarope not only has no family responsibilities but practices polyandry as well is unresolved. Polyandry, in which the female has more than one mate, is of rare occurrence in the bird world, as it is among humans. The present-day claims that female Wilson's Phalaropes do or do not practice polyandry are founded not on any positive proof but merely on circumstantial evidence and speculation. There is a possibility that female Wilson's Phalaropes may engage in some polyandry, since a relative, the Red-necked Phalarope, does so. The female Red-necked Phalarope practices serial polyandry by mating with a second male after she has laid eggs for a previous male.

While phalaropes are still in spring flocks, females initiate courtship by pursuing the males. A female does not establish and defend a static territory. Rather, she selects a male to court and defends a mobile territory around his person. If another female attempts to approach her male, she interposes her body between the male and the interloper and assumes the head retraction threat posture—the neck is retracted, the bill is pointed forward or slightly downward, and the back and crown feathers are ruffled. If the interloping female does not retreat before this threat display, the defending female lunges by running or swimming toward her. If the lunge does not work, the defending female will attempt to land on her back and peck her head.[164]

Initially, a male phalarope is not very receptive to being selected as a mate. He generally ignores the female unless she tries to get close to him, especially when she is facing him. Then, he repulses her with the head retraction threat display—he seldom has to get more violent than this. Despite the male's hostility, the female persistently stays with him. But, even after the two are further along in their relationship and have begun investigating nesting sites together and engaging in mutual scraping, the male may threaten the female when she presses closer than a foot or two. The female may respond with several bill-down movements. Pointing the bill down is apparently an appeasement display, since it turns off the male's hostility. The male's opposition to close approach by the female gradually wanes as they search for a nest site.[165] But this hostility emerges again soon after the last egg is laid. Then the pair bond comes to an end, and the male has nothing further to do with his mate.

Courtship begins in flocks of Wilson's Phalaropes during spring migration. Not only do females pursue males on the water, but aerial chases also take place. These aerial chases, which apparently involve unpaired phalaropes, begin when several females chase after a male who has taken flight. Once an aerial chase is under way, the male may augment the number of females chasing him by swooping low over swimming females and enticing them to join the chase. At times, as the chase progresses, the male will dangle his legs and hover low over the water, while holding his wings high and beating them in a narrow arc above his body. While hovering the male utters *ernt* calls. Females will hover around a hovering male; sometimes a hovering female will drop down and contact a female below her. A female may approach an aerial chase uttering *wa* calls, then fly ahead of the male in a loon flight—in which her hunchbacked posture is quite like that of a flying loon. Before landing, the male may flutter his wings in shallow beats below his body. When the male lands, the females also alight and attempt to follow him. Threats and attacks then occur between the females.[164]

A display called chugging, which a female directs at a male at close range, may be involved in the formation of the pair bond. The female approaches within a few feet of the male—with her neck feathers elevated, her neck stretched upright, and her bill pointing straight ahead—and utters several froglike calls or chugs. Early on, during the social phase of courtship within the flocks, the male may respond aggressively by threatening the female. Later on, chugging may occur during precopulatory behavior.[164]

The sex role reversal in Wilson's Phalaropes involves structural and behavioral reversals between males and females. The female develops a bright nuptial plumage, is the aggressor in courtship, and doesn't develop brood patches or the urge to incubate eggs or rear young. The male has a drab nuptial plumage, is more passive during early courtship, and develops paired brood patches and an urge to incubate eggs and rear young. Experimental studies have indicated that these structural and behavioral reversals are controlled by reversals in the endocrine systems of the female and male. During molting and early on in the breeding season, the female's ovaries produce high levels of the male sex hormone called testosterone. Although the male has the innate capacity to develop bright nuptial plumage and courtship aggressiveness, he does not because his testes do not secrete high levels of testosterone early enough in the season.

The development of paired brood patches on the abdomen of the male phalarope is controlled dually by the testosterone produced by the testes and by the hormone prolactin produced by the hypophysis (the pituitary gland). The female's hypophysis produces little prolactin during the breeding season; consequently, she doesn't develop the paired brood patches. High levels of prolactin are also responsible for the development of broodiness in birds. Since the female phalarope possesses little prolactin during the breeding season, she has no urge to incubate eggs or rear young. The male, having high levels of both testosterone and prolactin once the breeding season is under way, develops both the paired brood patches and a high drive to incubate eggs and rear young.[166] During the formation of the paired brood patches some down feathers are shed, the skin becomes thickened and edematous in the patch areas, and the number of blood vessels there increases. This gives the male two "hot-water bottles" with which to apply heat when incubating eggs.

Several groups of birds, including waterfowl, do not develop brood patches. Unlike phalaropes and some of their relatives, other kinds of birds may develop a single median incubation patch or even three incubation patches on their abdomens. In most birds the development of a brood patch or patches is controlled dually by the female sex hormone estrogen and by prolactin. In many passerine or perching birds, the male will often do some incubating although only the female develops a brood patch.

The Wilson's Phalarope is a versatile feeder, wielding its needlelike bill through a variety of motions while walking, wading, or swimming. It feeds on an array of animal life, emphasizing insects and crustaceans. Most observers first note its side-sweeping or swishing technique, in which it moves its bill through the water from side to side, straining out food items. This phalarope also pecks or jabs with its bill and at times practices bill pushing—lowering its open bill into the water while swimming forward.[167] Phalaropes will also upend in the water, submerging their heads and necks while paddling along.

The most unusual feeding involves spinning. The phalarope spins around in the water like a top—it may rotate at rates approaching sixty revolutions a minute, dabbling in the water for food about once per revolution.[168] This is thought to stir up prey from the bottom of shallow water or activate motionless invertebrates in the water so they may be better seen. Spinning has been shown to be innate in young Wilson's Phalaropes. Incubator-hatched chicks two to twenty-eight days old who have never witnessed spinning will spin on dry ground before pecking bits of egg yolk from a shallow dish of water. Wilson's Phalaropes will form commensal associations with other birds whose feeding activity involves stirring up invertebrates. Phalaropes have been seen feeding alongside of wading American Avocets and Northern Shovelers feeding in shallow water. An average of five phalaropes accompanied each shoveler; and the phalaropes' pecking rates increased threefold when accompanying a duck over their rates when feeding alone.[169]

There are many locales in the Great Basin where one can sit in the comfort of an automobile and watch phalaropes during the spring and early summer—such as the boat-launching area on the southeastern shore of Washoe Lake or the shallow ponds along Interstate 80 north of Fernley, Nevada. The single most vital locale for Wilson's Phalaropes in the Basin, and for that matter in the entire West, is Mono Lake, California. Mono Lake is a major staging area for Wilson's Phalaropes preparing for the long migratory flight to South America. Vast numbers of phalaropes come here to molt and fatten up on brine shrimp and brine flies in July and August before attempting the long southward flight. In the safety of this isolated lake and in the presence of unlimited food, the molting phalaropes may double their body weight before moving on.[170] But the intertwined futures of the birds and this lake are in doubt. The insatiable and unconscionable water greed of the city of Los Angeles has diverted the freshwater streams feeding Mono

Lake. As the lake continues to shrink and to increase in salinity, it may no longer afford sustenance to the birds.

Red-necked Phalarope
Phalaropus lobatus

The Red-necked Phalarope has a circumboreal distribution, nesting on the southern part of the tundra and in forest openings in northern Eurasia and North America. It winters at sea. During migration this phalarope is abundant off the Pacific Coast, but some inland migration occurs. The Red-necked Phalarope is a common, often abundant migrant in the Great Basin. Prior to 1982 this phalarope was called the Northern Phalarope, *Lobipes lobatus*.

The overcast night sky of May 14, 1964, rained Red-necked Phalaropes on Reno and its environs. Untold thousands of migrating Red-necked Phalaropes, trying to set down, were attracted to ground lights. Flock after flock was attracted to the neon glare of downtown Reno. Phalaropes were wildly flying around and between buildings and lights and colliding with obstacles by the scores. All the dead phalaropes I examined were Red-necked Phalaropes. The next morning dead phalaropes were seen as far afield as Mount Rose, a peak in the Carson Range to the southwest of Reno. The sudden appearance of hordes of wildly flying birds in downtown Reno during the dead of night created an uneasy stir. Since gambling is a twenty-four-hour-a-day sport in Reno, the dead of night is not very dead. There is constant street traffic between casinos, and crowds of people were attracted from the casinos to witness the strange happenings. There were quite a few nervous Nellies in the crowds, since the audience had been preconditioned to fear bird invasions by the recently released Hitchcock movie *The Birds*!

Red Phalarope
Phalaropus fulicaria

The Red Phalarope breeds on the arctic tundra of North America and Eurasia and winters at sea in both hemispheres. It migrates offshore and is of accidental occurrence in the Great Basin during migration. A partially paralyzed bird was collected at Bear River National Wildlife Refuge in September 1951 by botulism workers.[56] The Red Phalarope is listed as being of accidental occurrence at the Malheur National Wildlife Refuge.[61]

FAMILY LARIDAE

The Family Laridae contains the skuas, gulls, terns, and skimmers. Larids are birds with long wings and webbed feet. They have a worldwide distribution. Prior to 1982, the skuas were in a separate family—Stercorariidae—and the skimmers were in a separate family—Rynchopidae.

Ring-billed Gull
Larus delawarensis

The breeding range of the Ring-billed Gull lies mainly to the north of the Great Basin, extending through the northwestern states into the prairie country of Canada. It also breeds in the midwestern states. The Ringbill is the common resident gull in the Great Basin during the winter. Small numbers occur here during the summer, but this gull is known to nest in the Basin only in northeastern California and southeastern Oregon. Ringbills have nested at Honey Lake, California, for many years.[171] In describing its past nestings in southeastern Oregon, Gabrielson and Jewett wrote, "Each of the great colonies of California Gulls in Warner Valley, Malheur Lake, and the Klamath Basin contains its quota of Ring-billed Gulls nesting in a separate and distinct area that is usually almost entirely surrounded by the nests of the more abundant California Gulls."[103] Since it is not always easy to differentiate between California and Ring-billed gulls at a distance, other great California Gull nesting colonies in the Basin should be closely examined to see if they too have a quota of nesting Ring-billed Gulls.

California Gull
Larus californicus

The breeding range of the California Gull lies in the interior of western United States and Canada; its wintering grounds are along the Pacific Coast. The California Gull is the commonest summer gull in the Great Basin, breeding in colonies on islands in many lakes. A substantial part of the world's population of this gull breeds in the Basin. Most of our California Gulls migrate across or around the Sierra Nevada to overwinter on the West Coast. However, a few remain to overwinter in the Great Basin. In the Reno area the number of overwintering California Gulls has increased substantially since the 1960s. This may be due in part to the wholesale duck-feeding programs which take place at Virginia Lake, Idlewild Park, and Paradise Park. Townspeople flock to these bodies of water to feed the ducks.

The competing gulls get more than their proper share—often chasing ducks and taking food away from them.

Gulls become quite tame at sites where birds are protected and fed and will allow a near approach without flushing. Under these ideal conditions, a bird-watcher can study gulls from a range of five to ten feet and learn to identify the various species. Further, it takes a gull several years to acquire its definitive adult plumage, and intermediate as well as adult plumages can be studied close at hand at a feeding site.

The California Gull is one of about forty-four species of gulls in the world—all of which are called sea gulls. Even in the Great Basin, far removed from any semblance of a sea, this catchall misnomer is applied to California Gulls. To the human mind gulls are birds of beauty and grace which figure prominently in poetry and art. To the early Mormon settlers in Utah they were heaven-sent. In June of 1848, President Brigham Young, then back in Missouri, was informed of a Mormon cricket plague and the intervention of the gulls with the words, "Sea gulls have come in large flocks from the lake and sweep the crickets as they go." The following report was sent to the *St. Louis Republican* on July 16, 1849, "Hundreds and thousands of gulls made their appearance early in the Spring, and as soon as the crickets appeared, the gulls made war on them, and they have swept them clean, so that there is scarce a cricket to be found in the valley."[172] In 1913, in grateful recognition of the contribution of the "sea gulls" to the successful settlement of Salt Lake Valley, the Mormons erected a monument to them in Temple Square in downtown Salt Lake City.

Early on, before the avifauna of the Great Salt Lake islands was well known, there was some question of which species of "sea gulls" had come to the rescue of the Mormon settlers. Ornithologists originally identified the gulls as probably being either Herring Gulls or Franklin's Gulls. There was even some suspicion that Ring-billed Gulls nested on the Great Salt Lake islands. But others, particularly Behle, showed that only California Gulls nested on the islands of Great Salt Lake, and they did so in vast numbers. The Herring Gull doesn't nest in the Great Basin and is at best an uncommon transient here. The Ring-billed Gull is mainly a winter resident. The Franklin's Gull does nest in limited numbers in the area, but not on any of the islands of Great Salt Lake. Therefore, there is little doubt but that the "sea gulls" of 1848 and 1849 were California Gulls.[45]

From the human viewpoint, the role of the California Gull in the economy of nature is not entirely beneficial. On the credit side, California Gulls do feed on harmful insects and are useful scavengers. Spring flocks prey on pests such as cutworms, white grubs, and wireworms from freshly plowed fields. In the spring and summer, gulls assemble to gorge on crickets and grasshoppers—often in their greed eating and disgorging food several times a day. During highs in black-tailed jackrabbit population cycles, I have encountered California Gulls from Anaho Island up to twenty miles away from Pyramid Lake feeding on roadkills. On the debit side, the California Gull is a serious predator on the eggs and young of American White Pelicans, Double-crested Cormorants, and waterfowl. In Utah these gulls have an appetite for fruit and may damage the cherry crop.[45]

The locations and sizes of the California Gull colonies in the Great Basin have fluctuated greatly over the years. These changes have probably been influenced by alterations in nesting conditions, availability of food, and human disturbance. After a fairly complete census in 1931, Behle estimated that about eighty thousand adult gulls were present in the Great Salt Lake region.[45] The large nesting concentration at Mono Lake has been estimated to have numbered as many as fifty thousand gulls.[173] With the progressive desiccation of the lake, brought about by the pirating of its freshwater supplies by Los Angeles, the breeding gull population and its reproductive success there have been declining. Since 1979, Negit Island, the main nesting site at Mono Lake, has lost its insularity. Dry lake bottom now connects Negit to the mainland, and this will soon be the fate of other nesting islets as the lake continues to desiccate. The future of the California Gulls of Mono Lake is very bleak.

We actually know very little about the behavior of California Gulls. Feeding gulls usually alight and feed from the surface of the water. At times, they will plunge headfirst from the air into the water after fish. One recent April day, I witnessed a small group of about ten gulls engage in plunging at a pond in a park in northwest Reno. Most of the plunging originated from a swimming position on the water. A gull would fly two to eight feet straight up above the spot where it had been swimming, hover briefly, and then with open wings plunge headfirst into the water. Upon resurfacing, the gull would shake off the water. At times, a gull would be holding a string of algae in its bill upon resurfacing and, occasionally, a small item which could have been a fish. A few of these gulls were plunging ternlike: flying, hovering, and plunging from here and there, about six feet off the water, without taking up a swimming position between plunges.

Caspian Tern
Sterna caspia

The Caspian Tern is a cosmopolitan species in that it occurs in various parts of the world, but it has a highly discontinuous distribution. In the Great

Basin there have been breeding colonies reported from Great Salt Lake, Utah Lake, Bear River Refuge, Lahontan Reservoir, Pyramid Lake, Honey Lake, Summer Lake, Warner Lake, and Malheur Lake and its vicinity. But breeding colonies come and go, as local breeding conditions and disturbances by humans and gulls fluctuate, and there may have been or will be more breeding colonies here. The sizes of the breeding colonies also fluctuate over time. The Caspian Tern has experienced a recent name change; prior to 1982, it was *Hydroprogne caspia*.

Caspian Terns usually nest on islands, often in the midst of a California Gull colony. At Pyramid Lake the Caspian colony has been located in the very center of a California Gull colony on a sandy beach at the south end of Anaho Island. Sometimes the tern and gull nests are located only a few feet apart. On Rock Island at Utah Lake, the tern nests have been "completely surrounded by the nests of the gulls and even mixed in among them to a large degree."[174] On both Anaho and Rock islands the terns nested somewhat later than the gulls, and tern eggs and newly hatched young arrived at a propitious time for parent gulls. In commenting on the situation at Rock Island, Hayward wrote, "The terns were more shy than the gulls. When we approached near enough to frighten the terns from their nests the less fearful gulls would immediately pounce upon their eggs and attempt to destroy them. This would happen even within a few feet of us."[174]

Early ornithologists did not always differentiate between the Caspian Tern and the Royal Tern. Audubon referred to both as being Cayenne Terns.[175] Then Ridgway in the 1860s and Henshaw in the 1870s encountered Caspian Terns in the Great Basin: Ridgway at Washoe Lake, near Pyramid Lake, at the Humboldt marshes, and near Salt Lake City and Henshaw at Washoe Lake. Both referred to the Caspian Terns they saw as Royal Terns. However, these constitute two separate species—the Royal Tern is found on sea coasts and has never been recorded from the Great Basin. Undoubtedly, what Ridgway and Henshaw saw were Caspian Terns.

Forster's Tern
Sterna forsteri

The breeding range of the Forster's Tern is discontinuous. It lies partly in south central Canada and western United States and partly along the southeastern Atlantic Coast and Gulf Coast. The Forster's is a common, at times abundant summer resident and breeding bird at Great Basin marshes.

Forster's Terns will be most frequently encountered fishing over open, shallow water. Unlike gulls, who feed mainly from the surface of the water or on the ground, terns are aerial high-divers who plunge from high in the air into the water to grab their prey. When searching for small fish and other aquatic prey, Forster's Terns beat back and forth over the water, sometimes hovering, with their heads and bills pointed somewhat downward, scrutinizing the water below. When prey is sighted, a tern will bring its wings in close to its body or fold them and plunge downward, headfirst, into the water. Sometimes a tern will briefly disappear beneath the surface, before mounting into the air again with a few splashing strokes of its wings. When catching prey close to the surface, a Forster's may scoop it up without plunging into the water. Sometimes flying insects are captured.

Our knowledge of the natural history of the Forster's Tern is fragmentary—evidently, no ornithologist has been sufficiently turned on by terns to do an in-depth study. Hoffmann briefly described their mating behavior:

> On the breeding ground a Forster Tern keeps up a constant *kit, kit, kit*, like the cluck of a driver to a horse, or a hoarse high *kee-dee-dee-dee*, but if we approach too near the nest the Tern breaks into a low hoarse *kyarr* or *kerr*. In the mating season two birds fly with long sweeping strokes till they have reached a considerable height and then scale down side by side with wings stiffly set, or one, with head and tail depressed and wings arched, weaves back and forth over the other.[84]

Back in the days when it was legal to use the feathers of wild birds for millinery purposes, the Forster's Tern suffered heavy casualties. Possessing a beautiful and luxurious plumage and long, graceful wing and tail feathers, it was in great demand for adorning women's hats. Forster's were particularly vulnerable to the guns of the plume hunters, since, as soon as one bird lay wounded in the water, the remaining terns would hover solicitously over it and fall one by one to the hunters' guns.

Black Tern
Chlidonias niger

The Black Tern inhabits Eurasia and North America. In North America, it nests in the southern half of Canada and the northern half of the United States and overwinters to the south of the United States. The Black Tern is an uncommon to common nesting species at Great Basin marshes.

Although this beautiful bird is as graceful as any tern and is truly a tern, in some ways it is not exactly ternlike. Unlike that of other terns, its plumage is predominantly dark. Then, instead of feeding mainly on fish captured by plunging from the air into the water, the Black Tern feeds mainly on insects. It captures flying insects on the wing, or it swoops by or hovers and picks resting insects off of marsh vegetation. At times, a Black Tern will hover above the surface of the water and obtain prey, but seldom does the tern dive into the water after aquatic prey.

Other Larids

The Parasitic Jaeger, *Stercorarius parasiticus*, breeds on the arctic tundra of North America and Eurasia and winters at sea. This jaeger is of occasional occurrence in the Great Basin. The greatest number of sightings have been made in the vicinity of Great Salt Lake, particularly at Bear River Refuge, where two specimens have been collected.[27] A juvenile was picked up, photographed, and released in Spring Valley, White Pine County, Nevada.[176] The Parasitic Jaeger has been sighted several times at both Pyramid Lake and Malheur Refuge.

The Long-tailed Jaeger, *Stercorarius longicaudus*, breeds on the tundra of North America and Eurasia and winters at sea in the Pacific and Atlantic oceans. It migrates over the oceans and is seldom seen on shore during migration. This jaeger is of accidental occurrence in the Great Basin. There are several records for Utah—at Bear River Refuge, at Farmington Bay, and near Ogden. Two specimens, one from Bear River and one from near Ogden, have been taken. I am aware of only one other record for the Great Basin, a sight record at Pyramid Lake in October 1979.

The breeding range of the Franklin's Gull, *Larus pipixcan*, is located in the prairie country of western North America, with some scattered nesting to the west. There are now several breeding colonies of this gull in the Great Basin, which reflect the westward expansion of its breeding range. Ornithologists and collectors visiting the Basin in the 1800s and early 1900s did not encounter this species.[56] Wetmore discovered a breeding colony at the Bear River marshes in 1916, and since then this gull has prospered greatly at the Bear River National Wildlife Refuge. Around 1957, a second breeding colony appeared in the Great Basin at Malheur Lake, and this gull is now a common breeding species there.[177] Finally, in the summer of 1981, for the first time, this gull was reported nesting in the Fallon area.[82] It is apparent that the Franklin's is a fairly recent immigrant in the Basin and a new member of its avifauna. Since it is still expanding its range and is occasionally seen at other Great Basin sites, it behooves us all to keep watching for new nesting colonies.

The breeding range of the Bonaparte's Gull, *Larus philadelphia*, is in the boreal forests of Alaska and Canada. It winters off the Pacific, Atlantic, and Gulf coasts. The Bonaparte's is an occasional to uncommon transient during migration in the Great Basin, and an occasional bird may be seen during the summer or winter here.

The Heermann's Gull, *Larus heermanni*, breeds on islands in the Gulf of California and off the west coast of Baja California. Some of these gulls migrate northward to overwinter along the West Coast of the United States. This gull is of accidental occurrence in the Great Basin. A single adult was identified and photographed on Anaho Island at Pyramid Lake in June 1961.[178]

The Mew Gull, *Larus canus*, breeds in the Far North—in Alaska, western Canada, and Eurasia. In North America it winters along the Pacific Coast. The Mew is of accidental occurrence in the Great Basin. It was recorded at Pyramid Lake in March 1981.[179]

The breeding range of the Herring Gull, *Larus argentatus*, lies north of the Great Basin in Alaska and Canada and in Eurasia. There is some nesting in the eastern part of the United States. This gull is an uncommon to rare transient in the Great Basin; specimens have been collected in both the eastern and the western ends. The Herring Gull may be more common here than the limited records indicate, since few observers look for it, and it is easily overlooked among California Gulls.

In 1973 the Thayer's subspecies of the Herring Gull, *Larus argentatus thayeri*, was split away and elevated to full species status as the Thayer's Gull, *L. thayeri*, by the Committee on Classification and Nomenclature of the American Ornithologists' Union. Since then, there have been a few winter sight records for Thayer's Gulls from Reno and Honey Lake. Much more work is needed to ascertain the status of this species in the Great Basin.

The Western Gull, *Larus occidentalis*, is a coastal gull and is of accidental occurrence inland from the Pacific Coast. Howard reported a bone found in prehistoric Lovelock Cave in western Nevada as being suggestive of this species.[74] There is a sight record of a Western Gull from Virginia Lake in Reno.

The Glaucous Gull, *Larus hyperboreus*, breeds on the arctic shores of North America and Eurasia, and only a few overwinter very far south. The records for this gull in the Great Basin are from north central Utah—from the Bear River National Wildlife Refuge, Farmington Bay, and the mouth of the Provo River and Utah Lake.[56] Several specimens have been collected. The Glaucous Gull is classified as a "rare transient or vagrant" in Utah.[27]

The Black-legged Kittiwake, *Rissa tridactyla*, has a circumpolar distribution, nesting on the coasts and islands in northern waters. Kittiwakes migrate southward to overwinter at sea, seldom appearing on land. The

Black-legged Kittiwake is of accidental occurrence in the Great Basin. A specimen was found dead at Fish Springs National Wildlife Refuge in March 1972.[27]

The Sabine's Gull, *Xema sabini*, nests on the arctic tundra of North America and Siberia. It migrates off-coast, winters on the open oceans, and is rarely encountered inland. The Sabine's is of accidental occurrence in the Great Basin; it has been seen at various sites in both the eastern and the western ends. Some years are better than others for Sabine's Gulls in the Basin, and they will show up at several different sites.

The Common Tern, *Sterna hirundo*, is a breeding species in both Eurasia and North America. In North America its breeding range lies in Canada from the prairie provinces eastward, as well as along the eastern coast of the United States. Apparently, there is a considerable migratory flow of these terns off the coast of California. The records indicate that this tern is more common in the eastern part of the Great Basin than in the western part. In the eastern part, it is said to be an "uncommon transient, especially in autumn."[27] In the western part, it is considered to be of accidental occurrence at Pyramid Lake, Stillwater, and Malheur. However, Common Terns not only look much like Forster's Terns but often associate with them; thus they may be cursorily dismissed as being Forster's Terns.

The Least Tern, *Sterna antillarum*, has a fairly cosmopolitan distribution, occurring in many regions of the world. Its distribution in the United States is patchy—it occurs on the East and West coasts and along some of the rivers in the interior of the country. There is a single sight record for this tern in the Great Basin: one was seen in June 1981 fishing at an impoundment at Fish Springs National Wildlife Refuge.[180] Prior to 1982, the scientific name of the Least Tern was *S. albifrons*.

FAMILY HAEMATOPODIDAE

The American Black Oystercatcher, *Haematopus bachmani*, of the Pacific Coast is of accidental occurrence in the Great Basin. There is one sight record for Farmington Bay in August 1949.[27]

FAMILY ALCIDAE

The Ancient Murrelet, *Synthliboramphus antiquus*, is strictly an oceanic bird, nesting on islets and sea cliffs. Land occurrences are of an accidental nature. Yet three specimens have been taken far inland, along the eastern edge of the Great Basin—near Lehi, at Logan, and on Gunnison Island in Great Salt

Lake.[56] Specimens have also been taken in Nevada—at Elko, in the Lahontan Valley, and at Carson City.

The Parakeet Auklet, *Cyclorrhynchus psittacula*, is another oceanic bird of accidental occurrence at inland sites. There is a single sight record for this species in the Great Basin—at Farmington Bay in August 1962.[27]

10

Diurnal Birds of Prey

THE ORDER Falconiformes contains the diurnal birds of prey—raptors who hunt during the daytime, such as vultures, eagles, hawks, and falcons. These raptors are armed for hunting with strong, hooked bills and powerful feet with sharp, curved talons. These birds command the skyways. Many are fast and powerful in level flight and faster yet in a swoop or stoop. Some have mastered the art and science of soaring. Many glide or sail. A few can hang in still air or hover without the support of even a head wind. All three families of North American falconiforms—Cathartidae, Accipitridae, and Falconidae—are represented by species in the Great Basin.

FAMILY CATHARTIDAE

The New World or American vultures belong to the Family Cathartidae. The fossil record shows this to be an ancient family. Some ornithologists judge the New World vultures to be more closely related to the storks (Order Ciconiiformes) than to the falconiforms with which they are presently classified. It has even been suggested that the New World vultures and the storks be split away from the orders with which they are presently associated and combined in a new order by themselves. New World vultures differ considerably from Old World vultures—they have weakly hooked feet and slitlike instead of round nasal openings, and they lack a syrinx or voice box. The Old World vultures are related to the eagles and clearly belong in the same family in the Order Falconiformes as the eagles.

New World vultures are carrion eaters; their weak bills can only tear apart soft, rotten carcasses. These vultures are the most skillful of all land soaring birds. Two of the world's largest flying birds belong to this family. The California and Andean condors possess wingspans of up to ten feet.

The largest flying birds of all times were vulturine birds known as teratorns, who are usually placed in the same family or at least the same superfamily as the New World vultures. For some years the largest known flying bird of all times was a Great Basin species named *Teratornis incredibilis*. *T. incredibilis* was named and described by Hildegarde Howard on the basis of a unique, extremely large, prehistoric carpal bone found in cave deposits in Smith Creek Canyon in the Snake Range of White Pine County, Nevada. Howard estimated that this species had a wingspread of as much as sixteen to seventeen feet—hence the bird's specific name![181] Recently, the Great Basin lost the distinction of having the largest known flying bird of all times, when an Argentine fossil was found with an estimated wingspread of about twenty-four feet.

Turkey Vulture
Cathartes aura

The Turkey Vulture ranges from southern Canada through North, Central, and South America. In the Great Basin it is a common breeding species and migrant. It moves out of the Basin to overwinter. However, the Turkey Vulture has been reported to be of rare occurrence during the winter at Malheur National Wildlife Refuge.[61]

Although the Turkey Vulture is still widespread in the Great Basin, there has been an alarming decrease in abundance in Nevada during recent years. I now frequently spend an entire day moving around in the field during the summer without seeing a single Turkey Vulture; often I see just a few. Back in the 1950s and 1960s this seldom occurred. It is noteworthy that some observers in Utah also sense the Turkey Vulture as being less common there than formerly.[56] The observations of early ornithologists clearly indicate that this bird was common and widespread in the Great Basin before ranching and settlement were well under way. Back in the 1860s, Ridgway was impressed by the ubiquity of the Turkey Vulture in the Basin; he wrote, "It was abundant throughout the summer, when it was found in nearly all localities . . . the distribution of the Turkey-Buzzard was so general that it might be met with in any sort of locality . . . both in the valleys and on the mountains, and at all elevations."[25]

Since Turkey Vultures feed mainly on carrion, their presence is obvious as they glide along searching the ground below or alight on highways to feed on roadkills. Less obvious is the answer to the question of how they locate carrion—by sight, or by smell, or by both? Ornithologists have long searched for the answer to this question. Way back in 1834, Darwin presented meat wrapped in white paper to Andean Condors, but with inconclusive results. Later, Audubon's experiments with Turkey Vultures—he used a stuffed deer skin and hid a hog carcass—seemed to favor sight to the exclusion of smell. Bachman's experiments on Turkey Vultures also favored sight over smell—the birds did not respond to the odor of offal but were attracted to a painting on canvas of a skinned and cut-open sheep. At last, in 1909, an indication that Turkey Vultures could use smell to locate carrion emerged from experiments conducted by Beebe. He exposed Turkey Vultures and Black Vultures to tainted meat placed under one of three inverted boxes. The Turkey Vultures were attracted to the box containing the meat, but the Black Vultures were not. However, after a while the Black Vultures imitated the Turkey Vultures and came to join them at the box. Today, there is good evidence both that—among the New World and Old World vultures tested—at least one, the Turkey Vulture, has a keen sense of smell and uses it along with vision to find carrion and that another New World vulture, the King Vulture, may also rely partly on smell to locate carrion.

Controlled field experiments conducted by Kenneth Stager conclusively showed that Turkey Vultures can be attracted by odors. Stager employed a forced-air unit to dispense odors from fresh or decaying animal carcasses and a compressed-air unit to dispense the fumes of ethyl mercaptan, an extremely odorous and volatile chemical compound. His Turkey Vultures also detected animal bait concealed in chambers made from perforated cardboard cartons. Stager's field observations showed that the Turkey Vulture utilizes "a low-level searching flight that brings it close to the ground and in range of low-lying olfactory cues."[182] Prior to these experiments, troubleshooters had located leaks in a natural-gas line in southern California by introducing a high concentration of ethyl mercaptan into the line and then searching along its length for concentrations of Turkey Vultures, either circling in the air or sitting on the ground close to the line.

The next time you encounter Turkey Vultures in the field, see if you can tell whether they are trying to locate carrion scent trails. Are they searching for carrion at low levels? Olfactory cues are strongest close to the ground, but scanning vision is more restricted here. Do the birds ever overlook carrion which should be clearly visible from the air? Do they ever overshoot carrion until downwind of it, then promptly circle in a downwind pattern where they are obviously receiving olfactory cues? Are they searching for carrion by using the energy-inefficient method of frequently gliding into the wind instead of with the wind? All these types of behavior would suggest that the vultures were relying more on smell than on vision in searching for carrion. Of course, once the general location of carrion is detected by smell, the precise location could be determined with the aid of vision.

The skies of the Great Basin are clear and bright and wide open—with but few trees to screen them from view. Here is an ideal setting for viewing the flight of birds. When you see Turkey Vultures in the skies, note how seldom they are in flapping flight. Their wings are designed for soaring and gliding; they lack the massive pectoral musculature required for strong power flying. Turkey Vultures are masters of static soaring, the type of soaring done over land. During static soaring altitude is gained by sailing or gliding in a mass of rising air. Static soarers exploit two general types of updrafts—thermals and deflection (slope or obstruction) currents.

Thermals result from unequal heating of the surface of the earth by the sun. The lay of the land varies from place to place, as do the kinds of exposed soils, rocks, vegetation, bodies of water, and artificial structures, with their different heating and cooling properties. As rising columns of hot air or thermals are formed above heating land surfaces, a Turkey Vulture can gain altitude by sailing in circles inside a thermal and being carried upward by it. Since the vulture is not lighter than air, it cannot remain stationary within the thermal—if it did, it would stall and plunge earthward in response to the pull of gravity. The Turkey Vulture has to keep moving so that air will flow over its set wings, creating lift, an upward force, to counterbalance gravity. The vulture keeps moving on set wings by gliding downward as it circles within the thermal—sacrificing altitude for momentum.

In order for a bird to glide downward within a thermal and still show a net gain in altitude, the thermal has to be rising faster than the vulture is sacrificing altitude for momentum within it. A static soarer has to be able to gain enough lift to avoid stalling while maintaining only a slight downward-gliding angle. Measurements made from a sailplane showed that the sinking speed of a Turkey Vulture is only about 0.6 meter per second. Therefore, if the thermal is rising faster than 0.6 meter per second, it will carry the vulture upward. The Turkey Vulture can maintain such a low sinking speed because its wings are designed for static soaring. The wings are fairly long and narrow and deeply slotted at the tips—this provides the vulture with a high lift-to-drag ratio. Further, the Turkey Vulture has very light wing loading; a vulture has about 1.8 square centimeters of wing surface per gram of body weight. A Golden Eagle, which has higher wing loading, at about 1.4 square centimeters per gram of body weight, would have a higher sinking speed and require a stronger thermal to carry it upward.

Deflection currents are updrafts created as wind is deflected upward by obstacles in its pathway. The Great Basin is rich in deflection currents. Here there are hundreds of mountain ranges with innumerable ridges, slopes, and peaks, and in this desert-mountain country winds are constantly stirring. By gliding within a deflection current that is moving upward at a speed greater than 0.6 meter per second, the Turkey Vulture can stay aloft and move around on set wings. It can adjust the steepness of its angle of glide or its sinking speed within the deflection current so that it approaches the ground, remains at one level, or gains altitude as it sails along. Riding a deflection current upward provides a Turkey Vulture with the gift of altitude, which can then be expended by gliding in any direction in or out of the deflection current. Traffic flows best northward and southward in deflection currents in the Great Basin, since most of the mountain ranges trend in a north-south direction, and the prevailing wind is from the west. However, ridges associated with side canyons run at angles to the main trend of a range, providing complex patterns of deflection currents.

When a Turkey Vulture has gained sufficient altitude, it can leave a thermal or deflection current and glide in any direction. As we have seen, in gliding the vulture moves forward by sacrificing altitude for momentum: it coasts downward in the air. The lower its sinking speed, the further it can glide with a given sacrifice of altitude. If it wants to increase its sinking speed and sacrifice altitude more rapidly, it can steepen its glide angle by partially folding its wings. When the wind is blowing, the vulture can gain back some altitude by turning and gliding head-on into the wind. Since lift is proportional to the square of the speed of the air moving over the wings, the lift on the wings will increase exponentially, shooting the vulture upward in the air—thus the wind's speed adds to the bird's speed in developing lift. During this maneuver to increase lift, the bird can also increase the lift developed by its wings by increasing the angle of attack of its wings. This is done by tilting the leading edge of its wings slightly upward. However, the leading edge of the wings cannot be tilted upward very much, because so much turbulence is created that lift is destroyed, not gained, and the bird stalls.

Most birds soar or glide on flat wings—on wings that are extended horizontally, straight out from the body. But not the Turkey Vulture, for it holds its wings in a dihedral when soaring or gliding. The wing tips are held well above the horizontal plane and when viewed from the front or back form a V, not a straight line, with the body. The Turkey Vulture does much rocking and tilting when it glides through erratically moving air, but instead of floundering in rough air the vulture automatically comes back on an even keel each time. Since its center of gravity is suspended pendulum-like in its body below the wings, and its wings are held in a dihedral, it automatically returns to its normal gliding position after being displaced. When wings which are held in a dihedral are tilted, the down-tilted wing comes into a more horizontal position than the wing on the opposite side. The more horizontal the position of a wing, the greater the lift component developed by the wing which is directed straight upward. The strong upward lift developing on the down-tilted side of a dihedral automatically rights the dihedral. As we have seen, the Turkey Vulture has very low wing loading and a low sinking speed and can glide far from a given point in the air, if it glides slowly along a shallow angle. However, at slow speeds it really gets tossed about in rough air. The automatic repositioning conferred by holding its wings in a dihedral is a definite asset to the vulture.

The wings of the Turkey Vulture and those of other birds designed for static soaring are highly slotted. When a Turkey Vulture glides directly overhead, note that most of its flight feathers at the tips of the wings are widely separated from each other and stand out individually. These flight feathers at the tip of the wing beyond its bend are called primaries; the slots between some of the primaries are wider at the tips than at the bases. The highly slotted wings of a Turkey Vulture allow it to adopt a shallower glide angle and to slow down considerably when gliding without stalling. As we have seen, much of the turbulence created by a wing cleaving the air is created out at the tip of the wing, and the turbulence of the tip vortices destroys lift. Primary slotting greatly reduces the amount of turbulence associated with the tip vortices and allows the vulture to develop sufficient lift to prevent stalling at slow speeds. Further, the fairly long and narrow wings of the Turkey Vulture have a moderately high aspect ratio—the ratio of wing length to width. High aspect ratios lessen the effects of tip vortices, since on a long wing only a relatively small part of the total wing length lies close enough to the tip to be affected by the tip vortex. Another design feature of the wings of static soarers such as the Turkey Vulture which increases the lift-generating efficiency of their wings is cambering. A cambered wing is concave on its undersurface, convex on its upper one.

In the afternoon, Turkey Vultures may congregate at roosts to spend the night. Even during the breeding season in areas of high population densities, nonbreeding vultures may roost together at night. Many vulture roosts are traditional ones, in the sense that they are frequented year after year. Rick Stetter and I found such a roost in the Great Basin on a late September afternoon, along Soldier Creek, just off the northwest tip of the Ruby Mountains. At least thirty Turkey Vultures were soaring and gliding over and around a grove of tall cottonwood trees. Other vultures were perched, here and there, in the tops of the trees, and some were landing and others taking off. A solitary Red-tailed Hawk was gliding and soaring in company with these vultures.

Early in the morning when the sun first comes up, Turkey Vultures will frequently be seen perched in the tops of their roost trees, with their wings fully extended sideways and their bodies oriented to the sun. Some may hold a spread-wing posture for up to thirty minutes or so. They may turn their backs to the sun and elevate their back feathers to expose the underlying skin to the sun's rays. The spread-wing posture is usually interpreted as being a sunbathing posture. Since the Turkey Vulture's body temperature may drop several degrees C during the night, the bird reheats itself in the morning economically with solar heat, instead of creating additional heat by shivering and expending metabolic fuel. The vultures remain at the roost until later in the morning, after the sun has been out long enough to generate thermals to support soaring.

The Turkey Vulture's favorite nesting site is a cliffside cave or recess; other types of sites may be used, including hollow logs. I once found a pair of vultures nesting along the lower end of the Truckee River in a hawk's abandoned stick nest in a tall cottonwood tree. Both sexes incubate eggs. When cornered on a nest with young present, an adult vulture may engage in death feigning—it goes limp, with no muscle movement, and will allow itself to be picked up while playing dead. Death feigning has been observed in several species of birds and mammals. Because of the opossum's propensity for doing this, it is sometimes referred to as playing possum.

Turkey Vultures feed almost exclusively on carrion; they have unusual resistance to any ill effects from eating diseased carrion. They can feed with impunity on waterfowl killed by botulism, although some of the other raptors are not that fortunate. In the past Turkey Vultures have been accused of feeding on livestock killed by cholera or anthrax and then spreading these diseases in their droppings. Experiments based on mixing these droppings with food given to healthy hogs have shown that this is not true—the disease organisms apparently do not survive their sojourn in the digestive tract of the vulture. The Turkey Vulture and its digestive tract live up to the genus name of *Cathartes* and the family name of Cathartidae. These names represent the latinized form of the Greek *kathartes*, meaning purifier or purger; and, indeed, this is what Turkey Vultures do.

California Condor
Gymnogyps californianus

Today, the California Condor is an endangered species, limited to a handful of individuals in southern California. This condor formerly occurred in the Great Basin. There is a prehistoric cave-deposit record from Smith Creek Canyon in White Pine County, Nevada.[181] During historic times, in the spring of 1854, a California Condor was shot by Grizzly Adams along the Walker River in Nevada.[183] There is a sight record from the southeastern end of the Great Basin; Henshaw apparently saw a condor in 1872 near Beaver, Utah, in the company of Turkey Vultures.[56]

Extinct Cathartids

There are prehistoric cave-deposit records of extinct species of cathartids from the Great Basin. Howard reported that the bones of at least six La Brea Condors, *Breagyps clarki*, including those of one immature specimen, were found in Smith Creek Canyon in northeastern Nevada and that "possibly *Breagyps*, too, used the cave as a nesting site at intervals." She also reported that the bones of two specimens of *Coragyps occidentalis* were found there.[181] *C. occidentalis* is an extinct relative of the Black Vulture, *C. atratus*, which resides in the Americas to the south of the Great Basin.

FAMILY ACCIPITRIDAE

The Family Accipitridae contains the kites, eagles, hawks, and their allies. Accipitrids have a worldwide distribution. Prior to 1982, the Ospreys were in a separate family—Pandionidae.

Osprey
Pandion haliaetus

The Osprey is a cosmopolitan species whose breeding range extends over much of North America, Eurasia, and the Australian region. There is some migration into the southern hemisphere for overwintering. The Osprey is not a widespread member of the Great Basin breeding avifauna; its known nesting sites are confined to the eastern and western rims of the Basin. However, at various wet areas within the Basin, one or several Ospreys are occasionally seen during migration. In forty-one days of hawk watches at the Wellsville Mountains in Utah in the fall of 1979, thirteen migratory Ospreys were counted; in fourteen days at the Goshute Mountains in Nevada, four Ospreys were counted.[77] Ospreys are rarely found in the Great Basin during wintertime.

Breeding pairs of Ospreys have been recorded from Fish and Panguitch lakes in the eastern rim of the Great Basin.[184] On the western rim, there is a sizable breeding population at Eagle Lake, California, and since 1976 a pair nesting on the Nevada side of Lake Tahoe have fledged one or two young a year. Since the adults fish at Marlette Lake, in the Carson Range two miles east of Lake Tahoe, two nesting platforms were constructed there in 1980 by Gary Herron of the Nevada Department of Wildlife. By the end of the 1983 breeding season there was still no nesting activity at Marlette Lake.

The large nesting population at Eagle Lake has been there for some time, since back in 1921 an estimated thirty to thirty-five pairs nested along the southwestern shore.[185]

Following World War II, massive and widespread environmental contamination by pesticides like DDT and industrial pollutants like the PCBs (polychlorinated biphenyls) caused the Osprey to virtually face extirpation in the conterminous United States. The number of nesting pairs decreased drastically, and the few still nesting had very poor success—particularly in eastern United States. Ospreys nesting along the Connecticut River dropped from around two hundred pairs in 1940 to five pairs in 1969. Studies showed that Osprey eggs in some areas were heavily contaminated with DDE, a stable breakdown product of DDT, and with PCBs. Many eggs contained dead embryos, and the hatching success was less than 20 percent in some populations.[186] Now, as our environments receive better protection against pesticides and industrial pollutants, the Osprey is recovering somewhat. However, it is still vulnerable to contamination during migration in Central and South America, where DDT continues to be used to control malaria.

Ospreys are consummate fishermen, and their economy is based almost totally on fish. The aerial plunges of Ospreys after fish are spectacular to watch and easy to view, since they take place out over open water. A fishing Osprey flies at a height of forty to one hundred feet above the surface of the water, head down, scanning the water below. Its attack range extends only a few feet below the surface. When it encounters a fish close to the surface, it will plunge downward as soon as it is directly above its prey. However, it may have to hover above the fish and then even circle around and return to hover some more, waiting for the fish to venture close enough to the surface to be vulnerable. While hovering its body is arched, tail spread, legs lowered, with its wings beating forward and backward. In plunging from a hovering position, the Osprey turns over and dives straight down—almost headfirst, with its wings only slightly spread. Just before reaching the surface of the water, it swings its feet forward, beyond its head. Its wings are extended upward and backward as its body enters the water, and its body may disappear beneath the surface. Whether it is successful in grabbing a fish or not, the Osprey breaks out of the water with a single powerful beat of its wings.

The Osprey grabs the fish in its talons, with one foot behind the other on the body of the fish and with the head of its prey pointing forward. If the head of the fish is not pointing forward when captured, the Osprey will turn the fish around into this position once it gets airborne. Like that of most birds, the Osprey's foot bears three forward-directed toes and one backward-directed toe. However, the outer toe on each foot of an Osprey is reversible. Hence, it can grab and carry a fish by placing four toes on each side of the fish's body. Further, the bottom surface of each toe is covered with sharp spicules, another adaptation for holding on to slippery fish.

Adult Ospreys arrive back on their breeding grounds in the Great Basin in late March or early April. Pair bonding evidently persists from one year to the next, and the same nest sites may be used for many years. Two types of display flights have been described for this species. One consists of undulating flight—a series of downward dives and upward swoops is performed, and the upward climb may be aided by wingbeats. The male may do this by himself or as a type of ritualized pursuit of the female. The other display flight is almost a hovering one, performed slowly, at close to stalling speeds, during which the tail is down, the head up, and the body tilted. A fish may be carried during this display, and both sexes may vocalize.[187]

A number of different kinds of birds have been observed forming protective nesting associations with Ospreys by building their nests in the sides or bottoms of the stick nests of Ospreys—these include House Wrens, House Sparrows, European Starlings, Common Grackles, night-herons, swallows, and jays. In eastern United States the Common Grackle is particularly prone to do so, and in some instances six or seven pairs of grackles have been observed nesting in the sides of a single Osprey nest.[188]

Black-shouldered Kite
Elanus caeruleus

The Black-shouldered Kite is of occasional occurrence along the western edge of the Great Basin. It was first reported in the summer of 1971 from the north end of Walker Lake. Then two were recorded on the Mason Valley Christmas Bird Count on December 30, 1971, from a site near Yerington, Nevada.[189] Since then winter and spring sightings have been made on a number of occasions in the Reno, Hazen-Fernley, Pyramid Lake, and Honey Lake areas. This kite was first sighted on the Malheur National Wildlife Refuge in October 1977.[190] These sightings from along the western edge of the Great Basin reflect, no doubt, the remarkable population recovery and growth that this bird has made in California in recent years.

Until 1982 this bird was known as the White-tailed Kite, *E. leucurus*. Then its common and scientific names were changed, since it was now considered to be conspecific with the Black-shouldered Kite of the Old World.

Bald Eagle
Haliaeetus leucocephalus

At one time the breeding range of the Bald Eagle included most of North America. However, unlike the rugged and remote mountainous terrain inhabited by the Golden Eagle, the coastal and inland water sites inhabited by the Bald Eagle were attractive to settlers and became densely populated. Persecution, habitat destruction and degradation, and pesticide pollution followed. The breeding range of the southern subspecies was more severely affected than that of the northern subspecies, which extended from the Great Lakes northward into Canada and Alaska. After disappearing as a breeding species over much of its former breeding range in the conterminous forty-eight states and having poor reproductive success where it still occurred, the Southern Bald Eagle, *H. l. leucocephalus*, was listed in the Federal Register on March 11, 1967, as an endangered subspecies.

The Bald Eagle has disappeared from the sites in the Great Basin where it was known to nest in the late 1800s and early 1900s. In recent years the only documented breeding site in the Basin of which I am aware is at Eagle Lake, California.[117] In the early 1900s, the Bald Eagle was recorded nesting at sites in Utah on the eastern edge of the Basin, but "breeding records at the present time are not well confirmed" for any Utah sites.[56] Ridgway reported nesting eagles in the western end of the Basin. Visiting a nest on Anaho Island in Pyramid Lake in August 1867, he was told that this nest had been used the previous year by a pair of Bald Eagles.[25] Hardly a summer passes without someone reporting either adult Bald Eagles or a possible Bald Eagle nest in the Great Basin. But more fieldwork is needed before we can determine whether there is any Bald Eagle nesting here beyond Eagle Lake.

The Bald Eagle is an occasional to common winter visitant in the Great Basin. Field observations during recent years indicate that there is a much heavier influx into the eastern part than into the western part. January counts in Utah revealed 658 Bald Eagles in 1979[191] and 742 in 1981,[192] whereas seven years of aerial wetland surveys in Nevada revealed a low of 25 in 1980 and a high of 53 in 1977.[95] Of course, not all of Utah and Nevada lie inside the Great Basin, and part of any state count may include some non–Great Basin birds. But, then, the western part of the Basin has never witnessed anything comparable to the concentration of 259 eagles on March 14, 1979, feeding on the heavy winter kill of carp at the Bear River National Wildlife Refuge. Of these eagles, 142 were adult Bald Eagles, and the remaining 117 were a mixture of Golden Eagles and immature Bald Eagles.[191]

Bald Eagle

We commonly envision Bald Eagles as being invariably tied to water, because that is where the fish upon which they prey are found. However, although they do nest near water and feed mainly on fish, the eagles can live away from water in the absence of fish. There they feed on carrion or hunt land prey. When overwintering in the Great Basin, Bald Eagles mainly frequent waters where they can procure fish, but some will be found in desert valleys lacking fishable waters. I have seen small concentrations of Bald Eagles in dry valleys in west central Nevada, and they have been found in dry valleys in northeastern Nevada. At the eastern edge of the Basin, a population of over one hundred Bald Eagles has been overwintering in two desert valleys west of Provo—in Cedar and Rush valleys and on the slopes of the foothills dividing the two valleys. These eagles roost communally in four separate roosts. Pellet studies showed that black-tailed jackrabbits were their food staple, but most of the jackrabbits are evidently taken as carrion. This population has been known to overwinter in these valleys since 1960.[193]

Bald Eagles scavenge shorelines for dead or dying fish. They also do some active fishing for live fish. They catch fish that are at the surface of the water by approaching on a shallow glide or flat run, then lowering their feet and grabbing the fish in their talons. Bald Eagles will wade into shallow water after fish.

These eagles are fish pirates who systematically rob Ospreys. A Bald Eagle will waylay an Osprey making its way back to its nest with a fish. By diving at the Osprey, the eagle forces it to drop its prey, which the eagle then tries to grab before it hits the water or ground below. The eagle will occasionally grab a fish out of the talons of an Osprey before the Osprey drops the fish. Bald Eagles have also been known to rob other raptors, such as the Peregrine Falcon and the Red-tailed Hawk, of their prey.

Bald Eagles are attracted to other kinds of carrion besides fish. As we have seen, they utilize jackrabbit roadkills in the Great Basin. They also gather at lambing grounds to feed on the afterbirths and on stillborn lambs. Bald Eagles will actively hunt birds and mammals. They have been known to force diving birds to plunge underwater. Then, by following the birds' underwater progress from the air, the eagles are in a position to grab them when they return to the surface to breathe.

Aerial courtship is one of the most spectacular events to witness when watching Bald Eagles. A male and a female will grab each other's feet while in the air. Then, with their talons locked together, they tumble earthward for several hundred feet before pulling apart.

Northern Harrier
Circus cyaneus

The breeding range of the Northern Harrier extends over much of North America and Eurasia. It is a common permanent resident at many locales in the Great Basin. Hawk watches at the Wellsville and Goshute mountains show that there is some harrier migration in the eastern part of the Great Basin.[77] At some of the northern locales in the Basin, this harrier is less numerous in winter than during the rest of the year. Checklists from both Malheur and Sheldon National Wildlife Refuges record it as being uncommon during the winter. In western Nevada, however, it is often more abundant during the winter than during the summer. The status of the Northern Harrier apparently did not change much with settlement of the Great Basin. Ridgway, commenting on its presence here in the 1860s, wrote, "No marsh of any extent was visited, either in winter or summer, where this Hawk could not be seen at almost any time during the day skimming over the tules in search of its prey."[25] And it has been written that "all the early ornithologists who visited Utah soon after its settlement by the Mormons recorded the Marsh Hawk as being common to abundant, particularly in the Salt Lake and Utah Lake valleys."[56] Prior to 1982, the common name of the Northern Harrier was the Marsh Hawk.

Given its former name, this hawk should frequent marshes, and it does. But it is far from being limited to marshes. The Northern Harrier can be seen hunting not only in marshes in the Great Basin but also in meadows, agricultural fields, and even arid sagebrush or shadscale shrublands. Although this hawk usually nests and hunts in the valleys and foothills, on occasions it can be found nesting and hunting in montane meadows. I have even seen it hunting by soaring in circles, high over large chaparral areas among the yellow pine forests in the Carson Range.

The Northern Harrier belongs to a genus of hawks commonly called harriers. These hawks harry prey by beating back and forth in open country, low over the ground. When hunting the Northern flies so low that it often has to hedge-hop over fence lines and other ground obstacles. It flies slowly, alternately beating its wings and gliding. During the glide its wings are held above its body in a dihedral, forming an angle of 120 degrees or so—much like the Turkey Vulture. And, much like the Turkey Vulture, its flight posture is unstable, and the tilted dihedral automatically rights itself.

This hawk's hearing as well as its sight are thought to be very acute. The Northern Harrier has unusually large ear openings in its skull and has an owllike facial disk of specialized feathers. The facial disk in the Common Barn-Owl has been shown to aid in collecting sound. Hence, the hunting harrier is probably able not only to see movement but also to hear the rustling and squeaking noises made by mice and other small animals. When prey is detected, the harrier may hover briefly and then drop straight down or make a corkscrew drop to the ground. Small mammals, mainly rodents, comprise the bulk of its prey. However, small birds, reptiles, and even crickets and other large insects are taken. Ridgway collected several Northern Harriers in 1867 and noted, "The stomachs and crops of those killed at Pyramid Lake were filled to their utmost capacity with the remains of small lizards, and nothing else."[25]

The spectacular courtship flight of the male Northern Harrier can be seen during early spring in the Great Basin. These flights begin even before he acquires a mate. Up to twenty-five or more U-shaped dives are performed, one after another. The performance usually takes place within a hundred feet or so of the ground. The male describes a U as he dives downward, then pulls out of the dive on a rounded curving course, swooping upward. At the peak of his upward swoop he almost stalls, then executes a somersault or sideways turn and dives back downward to execute his next U. After the male is pair-bonded to a female, he will direct these courtship dives and swoops at her—performing close to where she is hunting or perched. The female may participate in the diving and swooping. Or the male may dive at the female, who turns over on her back in the air, presenting her talons to him—but they never lock talons. The pair may also engage in mutual soaring.

In some instances a male Northern Harrier will engage in polygyny and have two, rarely three, mates. The female alone incubates the eggs and takes care of the young. The male hunts and brings food to the nest for the female and young. He passes the food to the female while both are in flight—behavior characteristic of harriers. As the male approaches and flies by the nest, he calls to his mate. She leaves the nest to join him in the air. The male then maneuvers into a position ten to twenty feet above the female and drops his prey. The female either turns partly over on her back or swings her feet out to the side to grab the falling prey in her talons.

Sharp-shinned Hawk
Accipiter striatus

The Sharp-shinned Hawk inhabits wooded or forested areas, often nesting in conifers. In Utah it is said to prefer "wooded areas along canyon and valley streams."[56] In the western part of the Great Basin, it is largely confined to the foothills and mountains during the summer but shows up in the valleys during the winter. Some appear in Reno during the winter, at times staking out feeding stations in residential sections of the city where householders are feeding birds. When birds visit the feeder, the hawk has easy hunting.

The Sharpshin is one of three species of accipiters residing in the Great Basin. The other two are the Cooper's Hawk and Northern Goshawk. Accipiters are highly specialized hunters of birds; they are often referred to as bird hawks. However, they will prey on other types of vertebrates, particularly on small mammals such as chipmunks and tree squirrels. The Northern Goshawk and Cooper's Hawk are more inclined to hunt mammals than is the Sharp-shinned Hawk.

The anatomy and hunting techniques of accipiters are designed for hunting in wooded or forested areas or around thickets. Very seldom will you see these birds hunting in open stretches of Great Basin shrublands. Often they will try to ambush prey—they will hide back in the foliage of a tree and wait for a bird to fly close by or for a mammal to wander away from cover before pouncing on it. Sometimes they will cruise through the woodlands or along thickets in an attempt to suddenly come upon prey. They strike prey with their feet and kill it with their viselike grip.

Accipiters seldom soar when hunting, for they are not built for this—their short, broad, rounded wings and long tails are designed for sudden bursts of speed and high maneuverability. They twist, turn, and cut as they dash headlong between trunks or among limbs in a forest or through thickets, all without breaking a wing or neck. The short wings not only occupy a narrow flight path but have low inertia and allow rapid maneuvers; the long tail is an effective rudder for executing quick turns. Accipiters seldom engage in static soaring. Their short, broad wings do not effectively develop lift at low speeds—the tip vortices are too close together when the wingspan is short. Accipiters have a characteristic way of flying which enables an observer to instantly recognize them in the field. They flap-glide as they dash through the air: they flap their wings briefly, then set their wings and sail or glide for a way, before starting another short series of wingbeats.

Our three Great Basin accipiters are so very much alike in their makeup and way of life that one may be moved to question the wisdom of nature in having three almost identical versions of about the same thing. Wouldn't one species of bird hawk in the Great Basin be sufficient? The logical answer to this question is that, since avian prey comes in an array of sizes, the economy of nature will operate most efficiently with an array of predator sizes to exploit this fast-moving prey base. Our accipiters come in various sizes: Sharp-shinned Hawks average about ten to fourteen inches in length; Cooper's Hawks are in the neighborhood of fourteen to twenty inches long; and Northern Goshawks range from about twenty to twenty-six inches overall. Sexual size dimorphism is very pronounced, and the females are typically much larger than males of the same species. Hence, instead of three different sizes of accipiters available to exploit the avian prey, there are actually six different sizes of accipiters to do so. The following data, compiled by Ian Newton, vividly illustrate how the six different classes of accipiters to a considerable extent divide the available prey by weight.[194]

North American Accipiters	Mean Body Weight of Accipiter	Mean Body Weight of Prey
Sharp-shinned Hawk, male	99 grams	18 grams
Sharp-shinned Hawk, female	171	28
Cooper's Hawk, male	295	38
Cooper's Hawk, female	441	51
Northern Goshawk, male	818	397
Northern Goshawk, female	1,137	522

In raptors the female is often somewhat larger than the male. However, the degree of sexual size dimorphism illustrated by the above data is exceptionally high. Newton found a strong positive correlation between the magnitude of sexual size dimorphism in raptors and the "speed and agility of the prey" they were concentrating on. In carrion eaters such as vultures there is little or no sexual size dimorphism; if there is any, it is in favor of males being larger than females—as in most nonraptorial birds. The accipiters who specialize in bird prey show the greatest amount of sexual size dimorphism.[194]

Accipiters renew their pairing bonds at the beginning of each breeding season. There is evidence, at least in Northern Goshawks, that mates of the previous year will get back together again if both are still alive. Courtship consists of flights and calling. There is nothing spectacular about accipiters' aerial courtship performances, although the male may dive and then swoop upward. After the female and male are pair-bonded, dawn screeching duets may take place.[195] The male usually initiates the duet, sometimes even before sunrise, by starting to scream. Then the female joins in. Her deeper voice can be distinguished from the higher-pitched voice of the male. Dawn screeching duets subside as the incubation period gets under way.

The females do most of the incubating of eggs, although they may be briefly relieved by the males while they feed. Male accipiters do practically all the hunting during the nesting season. They bring prey back to the vicinity of the nest for the females to feed on and to feed to the nestlings. Prey is decapitated and plucked before feeding gets under way; often discrete plucking perches are used for this purpose. Both sexes may be very aggressive in defense of the nest—they may even strike human intruders. Their battle cry is a cackling *ca ca ca ca.*[195]

Northern populations of accipiters are migratory. Not only is there migration into the Great Basin from the north, but there is altitudinal migration, with accipiters coming down out of the mountains to overwinter in the foothills and valleys. In recent years hawk watches have shown that major migratory movements of Sharp-shinned and Cooper's hawks occur in the eastern part of the Great Basin along the Goshute Mountains in Nevada and the Wellsville Mountains in Utah.[77] There is only limited Northern Goshawk migration here. But then goshawks show up in numbers below the Canadian border only during invasion years, after there has been a major die-off in the north among prey populations such as Ruffed Grouse and snowshoe rabbits.

Birds usually move on broad fronts or along wide corridors during migration, but certain geographic, topographic, or plant features can concentrate migrants into narrow channels by attraction or repulsion. Features which funnel migrants into narrow channels are called guiding lines or leading lines; sometimes ornithologists use the German term leitlinie. Examples of leitlinie which concentrate migrants by attraction are north-south-trending mountain ranges, which can provide slope or deflection currents for migrants to glide on, and north-south-trending rivers with bottomlands well provided with food and shelter. Examples of leitlinie which concentrate migrants by repulsion are the shorelines of large bodies of water and the edges of large deserts. Birds, including raptors, hesitate to cross large expanses of water or desert. When they encounter them they may turn and follow the shoreline or desert edge.

The large Great Salt Lake Desert in the eastern part of the Basin apparently functions as a leitlinien; migratory raptors are funneled along its western and eastern edges. Fall hawk watches maintained at the Goshute Mountains on the western edge of this desert show a funneling of migratory Sharp-shinned Hawks, Cooper's Hawks, Red-tailed Hawks, Northern Harriers, American Kestrels, and Golden Eagles here. There is also funneling to the east of the Salt Lake Basin along the Wellsville Mountains in Utah.[77]

Cooper's Hawk
Accipiter cooperii

The Cooper's Hawk is widely distributed in North America from southern Canada to northwestern Mexico. It is a common summer resident and migrant in the Great Basin. It is rare in the winter in the eastern part of the Basin but fairly common in the western part. Each winter some Cooper's Hawks appear in the city of Reno and, like their smaller relatives the Sharp-shinned Hawks, may loiter and hunt in the vicinity of bird-feeding stations. The Cooper's seems to prefer riparian habitat for nesting purposes—it often nests in aspens or conifers around streams. Although the Great Basin population is in good health, this species has had problems in other regions of North America.

Northern Goshawk
Accipiter gentilis

The Northern Goshawk is widely distributed throughout North America and Eurasia. It is an uncommon permanent resident in the montane forests of the Great Basin, but there is some altitudinal migration into the foothills and even valleys during the winter. There is also some migration into the Basin by goshawks coming from farther north. Although the montane forests in the Basin are mainly coniferous ones, aspens often grow along streams, around springs, and in seepage areas on the mountainsides. These aspen groves are the preferred nesting habitat of Great Basin goshawks. Nesting has been recorded from many of the major mountain ranges in the Basin.

Swainson's Hawk
Buteo swainsoni

The Swainson's Hawk, a breeding species in western North America, migrates far south to overwinter on the pampas of Argentina. Once it was a very common breeding species in the Great Basin. Ridgway reported it to be "one of the most abundant of the large Hawks of the Interior" in the 1860s.[25] Since the 1950s the Swainson's has been declining in numbers in the Basin, and it is now a rare to uncommon breeding species here. It is in trouble over most of its range in North America.

There are scattered sight records of the occurrence of Swainson's Hawks in the Great Basin during the winter. I believe these winter sight records are probably based on misidentifications. However, there is always the possibility of the presence of a straggler or of an individual physically unfit to migrate. After thirty years of fieldwork in the Great Basin, I have yet to see my first overwintering Swainson's. Observations show that this hawk is migratory throughout its range. The only known regular overwintering in the United States occurs in southern Florida. Here, up to several hundred hawks, mostly immature ones, have been overwintering since 1950.[195]

A truly intriguing report of Swainson's Hawks overwintering in the Great Basin was that of Ridgway. After describing how common this hawk was, he went on to say that it "seemed to be less common in winter than in summer." Since Ridgway was acquainted with the Swainson's and had collected specimens of it while in the Great Basin, it is difficult to summarily dismiss his observations as being misidentifications. Then, too, Bendire, another competent field ornithologist, observed overwintering by Swainson's Hawks in the late 1800s. He reported, "On the eastern slopes of the Rocky Mountains it winters from about latitude 39° southward, a few remaining in favorable localities still farther north. On the Pacific coast I have observed a few wintering in southeastern Oregon in about latitude 42°, the majority passing southward."[188] But, no matter what conditions were in the 1800s, today's Swainson's Hawks are highly migratory.

The migration of Swainson's Hawks is the grandest among North American raptors. It spans much of the length of two continents. Due to funneling in Central America, most of the world's population of Swainson's may pass overhead there in several huge flocks. Earlier, when the species was in better health, flocks containing hundreds of birds or even several thousand could be seen during migration in western United States.

The wintering habits of the Swainson's Hawk have been in a state of flux in recent years. Formerly, while in Argentina, the birds foraged in large flocks. These flocks wandered in search of the clouds of migratory locusts which plagued the pampas. Now, with more efficient pest control practices, winter food for the hawk is not so readily available.[195] Obviously, this has required some adjustments in the winter behavior of Swainson's Hawks and may have affected their winter range. Also, deterioration of their winter range could be involved in their population decline.

Although Swainson's Hawks are fairly sizable birds, their favorite prey is on the diminutive side—being large insects, especially orthopterans such as locusts, grasshoppers, and crickets. The birds locate prey by watching from low perches or by coursing low, in sluggish flight, over open country. They will chase grasshoppers or crickets on foot, jumping with raised wings in pursuit. Large flocks of Swainson's may be attracted to infestations of locusts, grasshoppers, crickets, or armyworms. In Texas and Florida they have been seen following farm machinery, such as plows, to capture newly exposed insects and rodents. In California they have been seen standing silently at the entrances to gopher burrows and grabbing gophers as they came above ground. In addition to insects and rodents, Swainson's

Hawks feed on a variety of bats, birds, amphibians, and reptiles. In New Mexico one was seen capturing a rattlesnake—it killed the snake but in the process was bitten and soon died.

The Swainson's Hawk is a late breeder; it may still be sitting on eggs in early June, when most Great Basin raptors are fledging young. Swainson's Hawks tend to return to the same nesting locality each year, but, arriving after their long migration, they often discover that their nests of the previous year have been appropriated by ravens or by other raptors. Swainson's Hawks nesting in the hunting ranges of Great Horned Owls often suffer severely from predation by the owls. And, when nesting near each other, both the hawks and crows raid each other's nests.[196]

Occasionally, small, defenseless birds nest close to the nests of aggressive birds such as hawks and eagles or close to the nests of stinging or biting colonial insects. These small birds are usually tolerated, not molested, at the nest sites of hawks and eagles. But these same nest sites are vigorously defended against intrusion by potential nest predators such as snakes, crows, and ravens. Such passive associations between defenseless nesters and formidable defenders are referred to as protective nesting associations. A number of species of small birds form protective nesting associations with Swainson's Hawks by nesting in or under a tree in which a hawk nest is located. These include the Gray Flycatcher, Chipping Sparrow, and Lazuli Bunting. Others, such as the Bullock's Oriole, House Sparrow, House Finch, and Mourning Dove, sometimes build their nests either on or in the sides of the hawk nest.

Birds who are aggressive and better able to defend their own nests may sometimes nest in the same trees as Swainson's Hawks. These associations are often not harmonious, and hostile interactions may continually occur between the species. This may result in unsuccessful nesting on the part of one or both species. Aggressive nesting associations have been observed between Swainson's Hawks and Black-billed Magpies, Western Kingbirds, and Scissor-tailed Flycatchers.

Red-tailed Hawk
Buteo jamaicensis

The Red-tailed Hawk is widely distributed over North America. It is a common permanent resident at many Great Basin locales. There is a heavy influx of migrants and winter visitants into the Basin from the north; in many Great Basin valleys, Red-tailed Hawks reach their peak numbers during the winter months. The status of the Redtail apparently has not changed much with settlement of the Basin. In the 1860s, Ridgway found it to be a "very common species in all wooded localities of the Interior."[25]

The Red-tailed Hawk is one of a number of species of the genus *Buteo* found in the Great Basin. Buteos are consummate soarers and spend little time in the air unless conditions are suitable for static soaring. You will seldom see one in flapping flight, unless it is making short flights from one hunting perch to another.

Buteos are conspicuous wherever they occur. Large in body and robust in form, they spend much time on high perches out in the open—near the top of a tall tree, on a utility pole or transmission tower, or on top of a rocky outcrop. From these high hunting perches they survey the ground for movement by mammalian or reptilian prey. In the air they soar within thermals in wide lazy circles or glide in deflection currents along ridge lines. When soaring the buteos' broad tails are fanned. Their long, broad wings are held flat, with the primaries at the tip of the wing widely separated from each other. Their long, highly cambered, and widely slotted wings make them masters of slow soaring. Buteos soar in a much steadier fashion than Turkey Vultures—without all that rocking and tilting.

The Red-tailed Hawk locates prey while perched or soaring, then swoops or dives to make the attack. In flapping flight the hawk goes into a glide when ten to fifteen feet from its prey, then lowers and fully extends its legs. Just before contact with the prey, all four toes on each foot are fully extended; upon contact they close upon the prey. Most of the velocity of the feet at the moment of contact occurs in a horizontal, not a vertical, direction. If the prey is positioned with the long axis of its body perpendicular to the glide path of the hawk, the hawk strikes with both feet simultaneously, side by side. However, if the prey has its body more parallel to the glide path, the hawk strikes with the lead foot on the head of the prey and the second foot on its back. Sometimes a strike is made with one foot.[197]

Redtails feed mainly on mammals from mouse to rabbit size. In summer, ground squirrels may be a major food staple. Often various kinds of snakes and lizards are captured. Birds and large insects, such as grasshoppers and Jerusalem crickets, are occasionally taken. Snakes are often fed to nestlings. Red-tailed Hawks have been seen capturing rattlesnakes in the Great Basin, and the remains of Great Basin rattlesnakes (*Crotalus viridis*) were found in two Redtail nests just outside of Reno on Peavine Mountain.[198] Capturing rattlesnakes is dangerous business—hawks are sometimes bitten and killed doing so. Field observations show that the Red-tailed Hawk captures and often mortally wounds rattlesnakes by sinking its talons into the head and neck of the snake. Sometimes a hawk will grasp a rattlesnake in its talons and fly off with the snake dangling, alive and loudly rattling. In an attempt to subdue a rattlesnake, a hawk may try to beat the snake with its wings or fly up and drop the snake to the ground from a height of several hundred feet. Decapitated rattlesnakes have been found in the nests of Red-tailed Hawks, but the question remains, is the snake decapitated and the head discarded before the body is eaten or given to the nestlings? It would make sense to get rid of the fangs and poison glands by discarding the head before starting to eat.

The Red-tailed Hawk has a penchant for building its stick nest in the top of a tall tree. In the Great Basin tall trees are mostly confined to the mountains. Since so much of the foothills and valleys are shrublands, the tall trees utilized by hawks here are found mainly in streamside situations, on ranches, or near human habitations. A considerable amount of nesting occurs on ledges on the sides of canyons or on rocky outcrops. Then, too, the vertical dimensions of the Great Basin shrublands have been increased in many places by utility poles and transmission towers, the cross arms and cross braces of which are used by Red-tailed Hawks and other raptors and ravens as elevated nesting sites and as hunting perches for winter survival. So the alterations made in shrublands by telephone and power companies are not as unsightly to the eyes of raptors as they are to our eyes.

Manifesting a trait common to many raptors, the Redtail brings fresh greenery to the nest during the entire nesting season, although the frequency of this act drops off as the nestlings approach their time of fledging. This habit is a puzzling one, since these green twigs are not added to the structural fabric and thus do not contribute to the physical strength of the nest. Also, early in the breeding season, before selecting the actual nest they are going to use, the birds may carry greenery to several different old nests. Ornithologists have long cast about for an explanation of this behavior. Do the birds derive aesthetic pleasure from the act? Does the greenery aid in nest sanitation by contributing some fresh lining material? Does it afford a bit of shade to the nestlings? None of these explanations appears to be very plausible. It may be that the nest-building drive does not end abruptly but gradually diminishes. The nesting birds may derive some sort of drive reduction or emotional satisfaction out of performing the act.

Watching Red-tailed Hawks in the Great Basin can be an exciting and challenging sport. Identifications of hawks are sometimes difficult to arrive at, because there is so much individual variation in coloration—one has to obtain a good look at the upperpart of the tail and at the underside of the wings and body. The resident subspecies of the Red-tailed Hawk in the

Great Basin is the Western Red-tailed Hawk, *B. j. calurus*. Not only are light-colored, typical forms present, but there are many rufous and melanistic ones as well. During the winter and migration the subspecies Harlan's Hawk, *B. j. harlani*, with the dirty-white tail marked with black, can be found in the Basin. Harlan's Hawk nests in the northwestern corner of Canada and adjacent Alaska; it was formerly considered to be a separate species, *B. harlani*. In the eastern end of the Basin the pale prairie subspecies, Krider's Red-tailed Hawk, *B. j. kriderii*, can be seen on occasions in Utah.[27]

Ferruginous Hawk
Buteo regalis

The Ferruginous Hawk is a breeding species in the arid West, ranging from the southern end of the Canadian prairie provinces through the United States. The Ferruginous is a common breeding species in the eastern part of the Great Basin, but there is only limited nesting in the western part. Some Ferruginous Hawks overwinter in the Great Basin.

In the past the Ferruginous Hawk was severely persecuted by humans, although it preys almost entirely on pest mammals such as rodents and rabbits. Also, the increasing settlement and uses of its range for agricultural purposes have had deleterious effects on the bird.

Field observations indicate that Ferruginous Hawks prefer to nest in trees; in Utah and Nevada, their favorite nesting sites appear to be the junipers which edge into many of the valleys. However, in the Great Basin and the rest of the arid West, these hawks are often reduced to nesting in shrubs or on rocks or on the ground. But the preference for height is still there, and the hawks readily nest on elevated structures such as utility poles, transmission towers, old windmills, and abandoned buildings. This preference for height by Ferruginous Hawks has been exploited in several management programs. Artificial nesting platforms and nests have been erected in an attempt to increase the size of breeding populations and their level of reproductive success. While these programs may have some local impact, less disturbance by humans and their agricultural practices is needed to bring the species back to a healthy population level. Ferruginous Hawks thrive in the presence of high densities of rodents and rabbits, but agriculture does not. The virtual elimination of the prairie dog by settlers was probably the first serious blow delivered to this hawk.

The availability of prey, more than of nesting sites, seems to influence the Ferruginous Hawk's reproductive success. Studies carried out, off and on, from 1966 to 1974 in central Utah clearly demonstrated this. Here, the black-tailed jackrabbit, the main prey of the Ferruginous Hawk, forms 85 to 90 percent of the total food biomass during some years. However, the abundance of jackrabbits fluctuates tremendously within a period of several years. During the years of jackrabbit abundance, the total number of Ferruginous Hawks in the area greatly increased. The number of breeding pairs of hawks and the number of young hawks fledged also increased.[199]

The question has repeatedly arisen, how vulnerable are adult jackrabbits to Ferruginous Hawk predation? Does the hawk have enough unused wing-loading capacity to carry off an adult jackrabbit, which weighs seven to eight pounds? During a field study in west central Utah, the ages of jackrabbits carried back to the nest were estimated. Based on the sizes of the hind limbs, over 90 percent of the remains at the nests were of jackrabbits less than thirteen weeks of age. But the observers pointed out that not only were young jackrabbits easier to carry than adults, they were more abundant and easier to kill as well.[200] During a study in Curlew Valley on the Utah-Idaho border, an observer twice saw Ferruginous Hawks carrying large jackrabbits. Since he occasionally found the entire remains of adult jackrabbits at Ferruginous nests, he concluded that "it can be assumed ferruginous hawks are capable of regularly killing and carrying adult jack rabbits."[201]

The hunting techniques of the Ferruginous Hawk are much like those of its close relative the Rough-legged Hawk. Much of the hunting is still-hunting, with the hawk waiting and watching from a low perch or on the ground. After detecting prey, the hawk launches itself with a few wingbeats and glides up to its victim. Much of the ground hunting is done by waiting at the entrance of a rodent burrow or near the fresh workings of a pocket gopher. Pocket gophers are captured when pushing up their molelike, underground tunnels: the hawks "wait for the gopher to push close to the surface. They will then spread their wings, and, rising a few feet in the air, come down stiff-legged into the loose earth when the gopher is transfixed and brought out."[188] Aerial hunting is done by quartering back and forth fairly close to the ground, utilizing flapping flight, gliding, and hovering, or by soaring at higher altitudes to detect prey.

In the Great Basin as elsewhere, Ferruginous Hawks manifest a strong propensity for incorporating dry cow manure or cow chips into their nest structure. The significance of this act is obscure, as is their occasional habit of incorporating cow or antelope bones into the nest structure.

Rough-legged Hawk
Buteo lagopus

The Rough-legged Hawk's breeding range lies in the northern part of North America and Eurasia. This hawk is a common, often abundant migrant and winter resident in the Great Basin. The numbers coming here to overwinter vary considerably from year to year.

The August 1981 checklist of the Ruby Lake National Wildlife Refuge records the Rough-legged Hawk as being of occasional occurrence there during the summer, but not as a known breeder. However, I have never seen a Roughleg during my thirty summers of fieldwork in the Great Basin. Nor do I know of any other summer records for the Basin in recent years. Curiously, however, back in the late 1800s ornithologists recorded the presence of Rough-legged Hawks in the Basin during the summer. Ridgway found them in 1867 in the western end, and Ridgway in 1869 and Henshaw in 1872 found them in the eastern end.

Recently, some ornithologists have stated that Henshaw's and Ridgway's summer records for this hawk in Utah "may have resulted from misidentification since there are, to our knowledge, no nesting records or specimens taken in summer."[56] I find it difficult to agree with a misidentification hypothesis, since both Ridgway and Henshaw had collected December specimens and were familiar with the species. The Roughlegs' flight and hovering are quite recognizable, and birds of the common or light color phase are readily identifiable in the field. Further, Ridgway sounded as if he knew what he was talking about when he wrote:

> This common species was observed nearly everywhere in the vicinity of the fertile valleys. It appears to be resident in western Nevada, for it was extremely abundant in July at the Truckee Meadows [Reno-Sparks], where during the day half a dozen or more were often noticed at one time sailing in broad circles over the meadows. The flight of this Hawk is extremely similar to that of the Golden Eagle, a fact which probably explains why the Indians class it with the Eagles instead of with the Hen Hawks (*Buteo*). Most of those seen were in the light-colored, or normal, phase of plumage.[25]

We shall never know whether the summer hawks Ridgway and Henshaw saw were breeders or just nonbreeding stragglers. But I do believe that they were properly identified as being present. It is relevant that a pair of Rough-legged Hawks may have nested at Lake Okeechobee, Florida, in the summer of 1937.[202] This is far south even of their normal winter range.

As we have just noted in the quotation from Ridgway, Indians recognized the Rough-legged Hawk as being more closely related to Golden Eagles than to other species in the genus *Buteo*. Ridgway's conclusion was based on the names used by some of the Great Basin Indians for the two species. The Paiutes called the Golden Eagle Queh-nah', the Rough-legged Hawk Assut'te-Queh-nah'. The Shoshones called the Golden Eagle Gueh'-nah, the Rough-legged Hawk Pe'-ah-Gueh-nah'.[25]

The Roughleg feeds mainly on small vertebrates, with its food staple being rodents. During the winter in the Great Basin, this hawk can often be seen still-hunting, perched erect on a low fence post or even on the ground, watching for prey movement. Some hunting is done by quartering. The hawk flies back and forth, low over open fields or marshy areas, where mice and other prey can be surprised and dropped upon. Although it is a large, robust hawk, the Roughleg can hover. It will hang in the air on rapidly beating wings as it closely scrutinizes the ground below—it may even momentarily hang motionless on set wings in a head wind. When prey is seen below, it lowers its legs, folds its wings, and drops downward. In the spring, the Rough-legged Hawks follow the melting snow northward, back onto the arctic tundra to nest. Only after the snow has melted are rodent runways and rodent traffic fully exposed to the views of hunting hawks.

Other Buteos

The Red-shouldered hawk, *Buteo lineatus*, is not a regular member of the Great Basin avifauna but is of accidental occurrence here. There are several fall to spring sight records for the vicinity of Reno and for the Carson Range. This hawk is listed as being of accidental occurrence on the Malheur National Wildlife Refuge checklist and of rare occurrence on the checklist for the Honey Lake wildlife area. There are several sight records from Utah in the eastern end of the Basin.[56]

The Broad-winged Hawk, *Buteo platypterus*, is not a regular member of the Great Basin avifauna. It breeds in the eastern part of North America and is highly migratory. Broadwings winter from the Florida Keys as far south as Brazil and Peru. Fall hawk watches in the eastern end of the Basin show this species to be a rare but apparently regular migrant through the Basin.[77]

The White-tailed Hawk, *Buteo albicaudatus*, has a range which lies far to the south of the Great Basin, extending from southern Texas to central Argentina. This species is not a member of the Great Basin avifauna. Its occurrence here, at best, is accidental. There are several records, undocumented by specimens or photographs, of its occurrence in Tooele and Juab counties in Utah.[27]

Golden Eagle
Aquila chrysaetos

The Golden Eagle is widely distributed over Eurasia and North America. It ranges over the northern part of North America and, in the West, extends into Mexico. The Golden Eagle is a common permanent resident in the Great Basin—Ridgway's statement of the 1860s that the "magnificent Golden Eagle is an almost daily sight in the mountain-regions of the Interior" still rings true.[25] Fall hawk watches in the eastern part of the Great Basin at the Goshute and Wellsville mountains show that there is a migratory movement of Golden Eagles into the Basin from the north.[77] Golden Eagles can frequently be seen perched on utility poles along busy highways in the Basin, such as Interstate 80 and U.S. 50, especially early in the morning. Just west of Elko they may nest in plain view of I-80.

The Golden Eagle is a creature of the mountains but often hunts in the intermontane valleys. Its favorite nesting sites in the Great Basin are on ledges along canyon walls or on cliffsides in the foothills. A tree nest in the valley will occasionally be used. Also, following the erection of transmission lines in some of the wide, flat, shrubland valleys, the Golden Eagle has taken to nesting on transmission towers. While making a nesting survey by helicopter along the transmission line from Tracy Station east of Sparks to Valmy near Battle Mountain, Nevada, in the spring of 1980, we found eight active Golden Eagle nests on transmission towers. On the modern aluminum or self-weathering steel towers, the eagles usually build their nests inside of the framework of the lower bridge, a cross-brace-like structure. Here, surrounded by a framework of metal, the stick nest is better protected from the strong winds which are often blowing. Also, the young receive partial shelter from the sun.

Shrubland utilization is enhanced by the transmission towers not only during the nesting season but also during the rest of the year. The high towers serve the year around as excellent elevated hunting perches for eagles, hawks, falcons, and ravens. They are particularly important during the winter, when good soaring conditions are limited. During the highs in the jackrabbit population cycles, the shrublands traversed by the transmission lines are rich in jackrabbits. We have counted as many as fifty Golden Eagles perched on transmission towers during a single winter morning's flyover of the line from Tracy Station to Valmy. Along with the eagles there were

Golden Eagle

many Red-tailed and Rough-legged hawks and Common Ravens and some Prairie Falcons perched on the towers.

Transmission lines are not a complete boon for raptors. In the past it was possible for birds with long wingspreads like eagles to electrocute themselves by making contact between two conductors or between conductors and the supporting structure. On modern transmission lines the conductors are so spaced that this cannot occur. However, Sierra Pacific Power Company has had some problems on its intertie with the Idaho Power Company system, reminiscent of the insulator flashovers that Southern California Edison Company started having when it built its Big Creek lines in 1913. Eagles and other raptors often void just as they launch into the air. The extremely long string of semiliquid excrement, as it falls away, can establish an arc across an insulator, resulting in a flashover.[203] This can kill the raptor or badly burn it.

Some environmentalists view transmission lines solely as flight hazards for raptors. It is true that a strong gust of wind could hurl an eagle against a transmission tower, particularly if the bird was taking off or landing. Landing is probably the most dangerous flight maneuver, since the bird must slow down to reduce its impact—eagles and other large raptors drop below the level of the perch and glide up to it to lose air speed. Getting hit by a sudden, powerful gust when their air speed is minimal could be disastrous. But then all birds, large and small, in the city or in the wild, have flight accidents. Houses, trees, and canyon walls are flight hazards, just as transmission towers are—it's something birds must contend with as long as they fly. No environment is completely lacking in flight hazards. Innumerable birds are killed or injured when they collide with windows in towns and cities. Windows are certainly among the most serious flight hazards confronting birds.

Most Golden Eagle nests in the Great Basin are located on ledges along canyon walls or on cliffsides. A pair of Golden Eagles usually have several different nest sites fairly close together. From year to year the birds may alternate sites, although one may be favored over the others. Very often, a Golden Eagle nest will be located in the same canyon or on the same cliffside as a Prairie Falcon nest and a Common Raven nest. These close associations often appear to proceed with very little friction between the birds. Small, defenseless birds occasionally form protective nesting associations with Golden Eagles by building their nests in the sides of an active eagle nest. During our raptor studies in Nevada, David Worley found a European Starling nest in the side of a Golden Eagle nest along Pie Creek.

In both diurnal and nocturnal raptors, sibling aggression between nestlings may lead to fratricide and the death of the smallest or smaller members of the brood. Fratricide occurs most often in times of food scarcity, and the victim is usually cannibalized by the survivor or survivors. It is commonly believed that fratricide followed by cannibalism enhances survival of the species. During years when prey is scarce, at least one nestling may survive to fledge.

Golden Eagles lay two, rarely one or three, eggs at intervals of from three to four days. The young hatch at intervals of several days, since incubation begins with the first egg. The first-hatched young often gets a jump over its younger nest mate or mates in developing size and strength.

Some observers have suggested that sibling aggression in species who usually lay two eggs often goes far beyond simple fratricide and cannibalism during years of prey scarcity. They believe, regardless of whether prey is scarce or abundant, that the nestlings engage in Cain and Abel battles and that, "instead of younger chicks passively starving to death, the older chick attacks and can cause the death of its sibling(s) shortly after they hatch."[204] The term Cainism is often used for this postulated inherent behavior trait. In their classic two-volume tome, *Eagles, Hawks and Falcons of the World*, Leslie Brown and Dean Amadon state that in Golden Eagles "in about 80% of the cases where two young hatch the elder kills the younger" and "it is possible to calculate that the potential productivity is reduced by this fratricidal struggle from about 1.6 young/pair/annum to about 0.9."[195]

The above statements are certainly not true of Great Basin Golden Eagles. In my experience, Cainism is not of any consequence here. Most successful eagle nests fledge two young per year; a few fledge three (a successful nest being a nest from which at least one young fledges).

During the breeding seasons of 1979 and 1980, David Worley and I studied the reproductive success of raptors along the Sierra Pacific Power Company's transmission line corridor from Valmy to the Idaho border near Jackpot, Nevada. Our ground and helicopter surveys showed that, out of a total of sixteen successful nests during the two seasons, Golden Eagles fledged an average of 1.75 young per nest per year. Out of eleven additional successful nests which we followed elsewhere in northern Nevada during our two-year study, an average of 1.73 young per nest per year were fledged. Our values of 1.75 and 1.73 were not unique for the Great Basin. During the same two-year period, the production values for successful Golden Eagle nests studied by the Nevada Department of Wildlife were 1.7 both in 1979 and in 1980.[95]

Cainism is evidently not of any consequence in this part of western North America. Just north of the Great Basin, in southwestern Idaho, Beecham and Kochert followed the reproductive success of Golden Eagles from 1969 through 1971. During these three years 1.4, 1.7, and 1.8 young were fledged per successful nest.[205]

All the evidence I have seen indicates that Great Basin Golden Eagles are not inherently cainistic raptors. Some nestlings may be lost to fratricide, and fratricide is probably more of a problem during years of prey scarcity than years of abundance. But sibling aggression does not inevitably lead to the demise of the younger nestling soon after hatching.

When carrying out field studies on birds, it is often impossible to ascertain the cause of much of the nestling mortality. The young may disappear, often without a trace, between your visits to the nest. Sometimes you may find one that has fallen out of the nest or evidence that a predator has visited the nest. During the three-year Golden Eagle study in southwestern Idaho, the greatest single nestling loss, seventeen out of forty-one deaths, was attributed to heat prostration.[205] On a sunny afternoon, nest sites exposed to the sun may become ovenlike, especially on cliffsides where the surrounding surfaces heat to temperatures well above air temperature. The parents may try to shade the young with their bodies. In our studies in the Great Basin, we witnessed only one possible heat death of a nestling eagle. One of the advantages gained by eagles and other birds nesting within a transmission tower bridge is that the openings in the framework allow air circulation to occur, yet the framework offers some shade from the sun.

Golden Eagles probably pair for life. Their interest in each other and in their nesting sites, while often in evidence the year around, is most highly manifested in the winter and early spring. The Great Basin—with its richness of eagles in unobstructed skies—is an unexcelled stage upon which to view eagle courtship. The best viewing occurs in late winter and early spring. Much circular soaring occurs over the nest site. While soaring, one bird or the other may engage in a series of undulations—diving downward and swooping upward. Sometimes when the male dives toward the female, she will roll over on her back at his approach.

The clutch of two, rarely one or three, eggs is laid in the latter part of February or in March. The female does most of the incubating, and the incubation period has been reported as lasting from thirty-five to forty-two days. In the later stages of incubation, the female is sometimes reluctant to flush from the eggs when disturbed. On occasions, we have had to hover close to eagle nests in helicopters and practically blow the female off the eggs. We encountered one female in April of 1979 who refused to be intimidated by the hovering helicopter and never did leave her nest. After hatching the young are cared for by the parents. They usually fledge and attempt their first flights from the nest when about nine to ten weeks old.

Although the Golden Eagle is a superb hunter, capable of killing large mammals and birds, it often feeds on carrion. During the days when wolves hunted in the West, Golden Eagles were often attracted to their kills to feed on the leftover carcasses. Today I sometimes see them on Great Basin highways feeding on roadkills or in fields feeding on livestock carcasses. Some observers are of the opinion that, when both carrion and live game are readily available, Golden Eagles often prefer the carrion.

Mammals are the main prey of Golden Eagles. In the Great Basin jackrabbits are their principal food staple, which is heavily supplemented by various species of ground squirrels when they are not in hibernation. Golden Eagles have been known to prey on adult as well as young antelope, deer, coyotes, bobcats, and foxes.

Birds can often outmaneuver an eagle in flight. The eagle's best tactic is to surprise its potential prey from above—in a stoop or dive an eagle may approach speeds of one hundred fifty to two hundred miles per hour. Golden Eagles have been known to capture larger birds such as herons, swans, geese, ducks, grouse, pheasants, and ravens as well as smaller birds such as magpies and pigeons. When the occasion arises Golden Eagles may subject other raptors to role reversals; they will capture and kill vultures, accipiters, buteos, falcons, harriers, and owls.

A hunting Golden Eagle can really shake up its potential prey. On a crisp, sunny winter day John M. Davis and I were in the vicinity of Honey Lake, alternately raptor watching and searching for longspurs in flocks of Horned Larks. Suddenly a mighty disturbance swept across the marshes, as flocks of Canada Geese bolted into the air. The cause of this irruption was soon apparent in the form of a Golden Eagle trailing one of the flocks. As far as we could tell, the eagle never did make a kill, but the flocks of geese continued to fly, long and far, reluctant to settle earthward again.

Golden Eagles capture some reptiles but very few amphibians. Snakes, including rattlesnakes, are fed to the nestlings; before carrying a rattlesnake back to the nest, the eagle decapitates it. When both poisonous and nonpoisonous snakes are available as prey, the nonpoisonous snakes such as gopher snakes constitute most of the kill. The eagle does not always win a battle with a rattlesnake—the raptor may receive a fatal bite.[198]

Sometimes Golden Eagles hunt in pairs. One bird may be able to drive game from cover into the talons of the other. Recently, I sat at my kitchen table watching a ridge on Peavine Mountain in back of my house. A single adult Golden Eagle was attempting to catch a black-tailed jackrabbit by dropping down on it from heights of twenty to thirty feet in the air. The jackrabbit escaped each attack by seeking the shelter of a shrub. After several unsuccessful swoops, the eagle would alight and flush the jackrabbit from its cover of the moment. But the eagle could never get airborne again in an attack position before the jackrabbit had found new cover. This unsuccessful chase continued for over an hour before the hunter and the hunted disappeared from view over the ridgeline. If two eagles had been cooperating in this chase, it would soon have ended.

Ridgway was the first observer to describe a pair chase prey in the Great Basin. While encamped on July 29, 1868, in the foothills of the Ruby Mountains, he wrote:

> We were standing a few yards in the rear of a tent, when our attention was arrested by a rushing noise, and upon looking up the slope of the mountain we saw flying down its wooded side, with the rapidity of an arrow, a Sage-Hen, pursued by two Eagles. The Hen was about twenty yards in advance of her pursuers, exerting herself to the utmost to escape; her wings, from their rapid motion, being scarcely visible. The Eagles in hot pursuit (the larger of the two leading), followed every undulation of the fugitive's course, steadily lessening the distance between them and the object of their pursuit; their wings not moving, except when a slight inclination was necessary to enable them to follow a curve in the course of the fugitive. So intent were they in the chase that they passed within twenty yards of us. They had scarcely gone by, however, when the Sage-Hen, wearied by her continued exertion, and hoping, probably, to conceal herself among the bushes, dropped to the ground; but no sooner had she touched it than she was immediately snatched up by the foremost of her relentless pursuers, who, not stopping in its flight, bore the prize rapidly toward the rocky summits of the higher peaks, accompanied by its mate.[25]

Being a predator, the Golden Eagle has many human enemies. Human hunters often begrudge nonhuman hunters any share of the game birds and mammals. Sheepmen attribute all of their lamb losses, real and imaginary, to the Golden Eagle and coyote—factors such as parental sterility and low fecundity, congenital illness, diseases, accidents, and exposure are assumed to be nonoperative. Of course, there is some predation by Golden Eagles on lambs. But lambs are most vulnerable to eagle predation right after birth; they are relatively safe after reaching an age of seven to ten days.[206] The level of Golden Eagle predation on lambs varies. It is complexly determined by the local lambing and husbandry conditions and practices, the amount of carrion and live prey available, and the population density of eagles. Since the Golden Eagle only occasionally kills a calf and seldom kills an adult cow, cattlemen as a class are not dedicated enemies of eagles. Ranchers can be favorably impressed by the large numbers of rabbits and rodents that eagles remove from the cattle ranges.

Field observations indicate that when a Golden Eagle dines on lamb it usually feeds on carrion, not on an animal killed for the occasion. This can be determined by wound characteristics. When a Golden Eagle kills a lamb there are talon punctures and massive subcutaneous hemorrhages on the back, but there are seldom any head or facial wounds.[207]

FAMILY FALCONIDAE

This family of falcons consists of trim-bodied raptors, bearing long, pointed wings and fairly long tails. They fly swiftly on rapid wingbeats and seldom soar. Their short, strongly hooked bills usually bear a "tooth" along each side of the upper mandible. Falcons are distributed worldwide.

American Kestrel
Falco sparverius

The American Kestrel is widely distributed over much of North and South America. This kestrel is a common permanent resident in the Great Basin; it is the most abundant of all the raptors during the summer. Its status apparently did not change with settlement of the Basin. Ridgway found in the 1860s that "it occurs *everywhere*, in suitable places; at the same time, we may remark that it is by far the most abundant of all the birds of prey, although its numbers vary greatly with the locality."[25]

This falcon is much less common in the Great Basin during the winter than during the rest of the year. Although widespread in the mountains during the summer, it moves down out of the mountains, in altitudinal migration, to overwinter in the foothills and valleys. Fall hawk watches have shown that there is a heavy migratory movement of kestrels through the Basin.[77]

In 1973 the common name of this falcon was changed from that of Sparrow Hawk to American Kestrel. The old name of Sparrow Hawk was a misnomer, since the European Sparrow Hawk, from which our bird's name was derived, is an accipiter, not a falcon. Our bird is closely related to the Eurasian Kestrel, a falcon.

The American Kestrel is the easiest of all raptors to get to know. Kestrels are well adapted to civilization. They are often found living and even nesting alongside of human habitations in cities. They tolerate a close approach—even the least adept of bird-watchers can focus on a kestrel through a pair of binoculars and follow its actions.

The kestrel is our smallest diurnal bird of prey. Although a falcon, it prefers large insects instead of birds. It also captures mice, bats, small reptiles, and small birds. The kestrel hunts by watching for prey from elevated perches or by hovering in the air—it hovers on rapidly beating wings, with tail spread, heading into the wind if it is blowing. When prey is detected on the ground, the kestrel swoops down from its hunting perch or drops down from its hovering position to make the capture. When forced to hunt birds due to the unavailability of other prey, the kestrel will pursue them in rapid flight, falconlike, if unable to strike them before they fly.

American Kestrels are loners, except during the breeding season. During winter individuals may maintain feeding territories. Pair bonds are quickly formed at the onset of the breeding season on the breeding grounds. The male performs spectacular aerial dives and climbs over the female—as she remains perched or flies. Pursuit flights may take place. The male is very noisy during this season. His high-pitched cries have been variously interpreted as, for example, *killy-killy-killy* and *ki-ki, ki-ki, ki-ki.* The cries of the female are lower-pitched. The precopulatory behavior consists of head bobbing, often initiated by the female.

The kestrel's preferred nesting site is a natural cavity in a tree; it is the only cavity nester among our diurnal raptors. Lacking natural cavities, kestrels will nest in any type of cavity they can find, including old flicker holes and even niches in buildings. An abandoned magpie nest is occasionally used. In the Great Basin tree cavities are at a premium, and kestrels have to be flexible in their nesting habits. As Ridgway observed in the 1860s, "We found them adapting their nesting-habits to the character of the surroundings. Thus, in the precipitous cañons of the Ruby Mountains, they built among the crevices of the limestone cliffs, in company with the Prairie Falcon (*F. polyagrus*), the Violet-green and Cliff Swallows, and the White-throated Swift; while in some portions of Utah they took possession of the holes dug by the Kingfishers and Red-shafted Flickers in the earthy banks of the ravines."[25] This last observation of Ridgway's is doubly interesting. As yet, I have never found flicker holes anywhere but in trees.

Merlin
Falco columbarius

The breeding range of the Merlin lies in the northern end of Eurasia and North America. There are no recent breeding records for the Merlin in the Great Basin. Several sets of eggs collected in the Wasatch Mountains document the occurrence of nesting in the eastern rim of the Basin in 1868 and 1869.[56] The reported occurrence of present-day breeding in Utah is unconfirmed by any hard evidence. The Merlin is a rare to uncommon migrant and winter visitant in the Great Basin.

The Merlin was formerly called the Pigeon Hawk in the United States and is so designated in the older bird guides. The British call this species the Merlin. The common name of our bird was changed from Pigeon Hawk to Merlin in 1973 to bring the name into conformity with international usage. The Merlin, like its relative the Peregrine Falcon, is very sensitive to chlorinated hydrocarbon pollutants and is often heavily contaminated with DDE. Eggshell thinning of 10 to 25 percent has been reported.[208]

Peregrine Falcon
Falco peregrinus

The Peregrine Falcon has a truly cosmopolitan distribution; some eighteen to twenty subspecies are spread out over the continents and many of the islands of the world. During historic times this falcon has not been a prominent figure in the Great Basin. Breeding records for the western part of the Basin are few and far between. Prior to the early 1950s there were three known aeries at Malheur National Wildlife Refuge, but they have been deserted since then.[209] The only known or suspected nestings reported for western Nevada were at Pyramid Lake[25] and Walker Lake.[210] The last reported nesting in the state occurred in 1949 in northeastern Nevada.[211]

The Peregrine has been more common in the eastern part of the Great Basin than in the western part. In the neighborhood of Great Salt Lake and in the region to the south, Nelson found three active aeries in 1939 during three weeks of searching in his spare time.[209] In a 1973 publication, Porter and White stated that "although sparsely distributed throughout Utah, the species apparently found conditions especially suitable for nesting in the environs of the Great Salt Lake and Utah Lake valleys, where its nesting sites in the adjacent mountains were within flying distance of a plentitude of preferred prey species which inhabited the marshes and shorelines surrounding the two lakes. Despite the aridity of the environment, the 20 eyries that occurred there, when and if they all were active simultaneously, comprised a population comparable to some populations elsewhere in North America where the environment is considered to be more congenial to the peregrine."[212] In 1973 Eyre and Paul stated, "In Utah about 40 eyries were known or suspected, although now possibly less than ten exist."[184] Of course, the eastern part and the southern tip of Utah lie outside the Great Basin. One must remember that much of the country in the Basin is inaccessible even to four-wheel-drive vehicles—there may be a few aeries here of which we are still unaware. Peregrine Falcons are occasionally seen in Nevada during the breeding season, and, as the rumors float, the search for active aeries continues.

At most Great Basin locales, Peregrine Falcons are seen only during migration or as winter visitants. The resident subspecies in the Basin is the American Peregrine Falcon, *F. p. anatum*. Northern populations of *anatum* are migratory, and some falcons may enter the Great Basin during migration. Some Arctic Peregrine Falcons, *F. p. tundrius*, may pass through the Basin on their way to and from South America, although their main migratory flow is through eastern and middle North America.

Two of the three North American subspecies have been officially listed as endangered in the Federal Register. Within a few years after the end of World War II, *anatum* was experiencing serious reproductive difficulties. By the mid 1960s it had disappeared as a breeding bird in eastern United States. The 1973 edition of *Threatened Wildlife of the United States* stated that *anatum* had been "extirpated as a breeding bird east of the Rocky Mountains in the United States, in Ontario, southern Quebec, and the Maritimes" and that only limited numbers still bred in the western states, Canada, and Alaska.[213] By 1970 the subspecies *tundrius*, breeding in arctic Alaska, Canada, and western Greenland, had joined *anatum* on the endangered list. While all this was taking place in North America, across the Atlantic Ocean in Europe and the Soviet Union the nominate subspecies of Peregrine, *peregrinus*, was experiencing reproductive difficulties. Then, *peregrinus* joined *anatum* and *tundrius* in the Federal Register and on our official list of endangered birds.

Widespread use of the insecticide DDT has been implicated as a major factor in the decline and demise of falcon populations. At many locales in North America and Europe, Peregrines experienced little or no reproductive success during their decline. Chemical analyses showed that both the birds and their eggs were heavily contaminated with DDE, one of the breakdown products of DDT. Adult birds had reduced sexual drives, were late in getting reproduction under way, formed weak pair bonds, and functioned poorly as parents—often eating their eggs instead of incubating them.[214] Females often laid eggs that did not hatch—the shells were so thin that the eggs would break under the weight of the incubating adult.

Experimental studies have shown that DDE causes eggshell thinning by inhibiting the action of an enzyme called carbonic anhydrase. The structure of the eggshell consists mainly of calcium carbonate, and carbonic anhydrase action is necessary to allow the delivery of calcium to take place in the oviduct, where the eggshell is being produced.[36] *Anatum* disappeared from its eastern breeding grounds after this region became heavily polluted with

DDT. In contrast, *tundrius* is not exposed to DDT on its tundra breeding grounds. It is exposed, however, during migration—in South America, DDT is still widely used. Different species of birds show different degrees of sensitivity to DDE. The Peregrine is quite sensitive. Then, Peregrines feed mainly on birds. Thus, they are at the top of complex food webs in which DDE levels can become highly magnified.

Studies indicate that DDE has less to do with the other adverse reproductive ailments manifested by Peregrines—such as delay of the onset of breeding, low sexual drive, and feeble parental care. These ill effects are more strongly elicited by other kinds of environmental pollutants: by the PCBs (polychlorinated biphenyls) and by the insecticide called dieldrin. The PCBs have a variety of industrial applications, including wide usage as plasticizers. Today, we live in a plastic world, and every time part of it burns PCBs are released. Global circulation has polluted many of the world's environments with PCBs by now, even including the bottom of the Antarctic Ocean. PCBs and dieldrin do their dirty work by stimulating the liver to produce enzymes which abet a chemical process called hydroxylation, resulting in the rapid breakdown of certain normal products in the body, including the sex hormones.[36] With low levels of these hormones a bird shows little interest in sex, in incubating eggs, and in brooding and caring for young.

Contamination by DDT and the PCBs has not been solely responsible for the present plight of the Peregrine Falcon. Its way of life was more compatible with wilderness than with urbanization, farming, and the heavy hand of settlers. Not satisfied with merely disturbing or displacing the falcon, people often shot it as "vermin" or for "sport" and robbed its nests of eggs and young.

With its coast-to-coast distribution, the Peregrine Falcon overlapped the Prairie Falcon in western North America. Where Peregrine and Prairie falcons live alongside each other, that is, in areas of sympatry, Prairie Falcons have been displacing Peregrine Falcons and occupying their former nesting sites since early in the present century. One of the first field observers to record the concurrent waning of the Peregrine and the waxing of the Prairie was the noted bird artist and ornithologist Allan Brooks. In 1937 Brooks wrote a letter to Joseph J. Hickey relating what had happened in the vicinity of Okanagan Lake, British Columbia, where he lived. Brooks stated that from "1897 to 1907 the duck hawk [Peregrine Falcon] was a regular nester in this region. I knew of 5 eyries on Okanagan Lake and about 10 in the region between the south end of the lake and the international boundary . . . The last duck hawk to nest near my home here was in 1929 . . . the following year this eyrie was occupied by prairie falcons."[209]

In Utah in 1939, in the eastern part of the Great Basin, Nelson witnessed spectacular aerial battles between Peregrine and Prairie falcons near a Peregrine aerie. He wrote, "The aerial battles were spectacular with a fierce and awesome beauty. The prairie falcons seemed to win these battles much to my surprise. The battles were not definite and always ended in a sort of draw, with observers deciding that the prairie falcons won. They had command of the air, but when the birds parted they went back to their respective sites. . . . By 1941 the peregrines at U1 [the aerie number] did not return to nest, although they were seen near the nesting sites as late as 1946." Nelson went on to summarize the situation in Utah by writing, "During the period 1939-42, I observed from 9 to 14 peregrine eyries depending upon whether some sites were alternate nests of the same pair . . . The prairie falcons were steadily taking over the nesting sites of these birds, often without a fight of any sort. The peregrines did not seem to come back or did for only a short time. About 50% of these sites were taken over by prairie falcons before 1942. The fighting between the two species had no significance other than being spectacular to watch."[209]

The field observations of Brooks, Nelson, and others imply that the decline of the Peregrine Falcon in western North America started before the days of DDT and the PCBs. Nelson's observations have also shown that the takeover of Peregrine nesting cliffs by Prairie Falcons was a symptom, not the cause, of the decline—Prairie Falcons did not wrestle away control of the skies over nesting cliffs in bloody and decisive battles with Peregrines. Finally, since the major decline in Peregrine numbers started forty years or so earlier in the thinly populated West than in the densely populated East, the loss of wilderness with concomitant human disturbance could not have been the prime factor in precipitating the western decline.

An interesting and plausible hypothesis concerning the decline of the Peregrine and its replacement at western breeding cliffs by the Prairie has been advanced by Morlan Nelson. Nelson suggested that a change in climate in the West, involving decreasing amounts of precipitation and increasing temperatures, may have created unfavorable breeding conditions for Peregrine Falcons. As the aridity increased, many bodies of water and marshes would have shrunk or disappeared, and the accompanying decrease in nesting birds such as shorebirds would have affected the food supply of the Peregrine Falcons. Also, as the temperatures increased, heat death, often the major reason for nesting mortality in cliffside-breeding raptors, may have become a serious decimating factor. Nestling Peregrine Falcons would be more susceptible to heat death than nestling Prairie Falcons. Peregrines have a later, consequently hotter, spring nesting season than Prairies. Heat exposure would be accentuated by the less cloudy skies associated with hotter, drier days and by the disinclination of Peregrine Falcons, compared to Prairie Falcons, to nest inside cliffside potholes, where overhead shade helps protect the nestlings.[209]

The Peregrine Falcon is not being allowed to slip from endangerment into extinction with only political skirmishes being fought. Raptor biologists and falconers are at work trying to increase the productivity of the species. They have succeeded in getting Peregrines to breed in captivity, supplying young to introduce into areas where the species once occurred. They have also succeeded in increasing the egg production of wild-breeding Peregrines. When the eggs are removed from the nest of a pair of Peregrines, the birds may renest and lay another clutch. The first clutch can be hatched in an incubator and a supply of young Peregrines obtained for introduction programs.

The three techniques which have been employed in introducing young, captive-produced Peregrines into the wild are cross-fostering, augmentation, and hacking.[215] In cross-fostering, the young are placed in the nest of a closely related species who may act as foster parents—young, captive-produced Peregrines have been successfully introduced into the wild in Idaho by cross-fostering with Prairie Falcons.[216] In augmentation, captive-produced Peregrines are placed in the nests of wild-breeding Peregrines. In hacking, captive-produced Peregrines are taken to hack sites before they are old enough to hunt. Human attendants unobtrusively watch over the young and provide them with meat, until they are able to support themselves by their own hunting.

The term and technique of hacking are both derived from falconry. Falconers feed young falcons at a board or post called the hack board. The hack board provides the falconers with a hold on the young falcon only as long as it cannot support itself by hunting; they must gain active control of the young bird before it becomes a successful hunter if they are not to lose it. Of course, if the objective is to introduce young, captive-produced falcons into the wild, this is accomplished by having the birds escape. The Peregrine Fund of Cornell University, which has been developing a major hacking program, has some hack sites in Utah. In 1980 eleven young Peregrines were released in northern Utah, eight on hack towers and three on cliffs.[217]

Prairie Falcon
Falco mexicanus

The Prairie Falcon is a resident of western North America, ranging from southern Canada to northern Mexico. It is a common and widespread resident in the Great Basin. Some Prairie Falcons remain in the vicinity of their nesting sites the year around. However, considerable wandering takes place in the fall and winter. The number of falcons will increase in some areas in the Great Basin, and falcons will appear in other areas where there are no nesting sites. Limited breeding records indicate that there is a strong tendency for first-year birds to wander eastward—into the plains provinces and states of Canada and the United States. There is no orientation or rigidity in these movements suggestive of migration.[218] Most observers believe that there is a correlation between the fall and winter movements of Prairie Falcons and the occurrence of their main prey species, especially Horned Larks.

Prairie Falcons nest on cliffsides overlooking the shrublands in which they hunt; a pothole or a ledge sheltered from above by an overhang is preferred. Prairies often have alternate nest sites, sometimes on the same cliffside or canyon wall, which they use every other year. The eggs are laid in a shallow scrape on the floor of the pothole or on the ledge. An old raven or hawk nest is occasionally used. The cliffside below the nest site may be conspicuously whitewashed with excrement.

Toward the end of winter, the birds return to their nesting site. Sometimes the female arrives first, sometimes the male. Noisy aerial displays take place—the dives of the male can be quite spectacular. The normal clutch size is four to five eggs. Since the female usually begins to incubate with the laying of the second egg, the young hatch over a period of two or three days. Although the female does most of the incubating, the male may cover the eggs while she is feeding on the prey he has brought her. The male does the hunting for the family, from the time the pair rejoin at the nesting site until the young are about two weeks old. Then the young are left unattended while both parents hunt. The incubation period is around thirty days long, and the young can fly when they are about forty days old.

The Prairie Falcon preys mainly on small to medium-size birds and small mammals. Although a falcon is capable of overtaking swift birds, it prefers to attack birds when they are on the ground or have flushed and are just taking off. The Prairie Falcon differs from the Peregrine Falcon, who ignores the ground and looks for birds awing. The Prairie often hunts from perches, especially during the winter. Upon detecting unsuspecting prey from the vantage point of a hunting perch, the falcon launches a swift attack, flying close to the ground to avoid early detection. During aerial hunting the Prairie Falcon stays within several hundred feet of the ground. When it detects a suitable target, it rapidly closes for the kill on a low, slanting approach. Small prey may be snatched off the ground or seized in the air, but the falcon deals larger birds and mammals a blow with its feet in passing. At the time of impact, the bird's toes are extended and most of the velocity of the feet is in a horizontal direction—prey that is seized can be dispatched in the powerful, talon-rimmed grip, which can crush the skull of a mammal. The falcon usually bites through the neck vertebrae of a bird, severing its backbone. Bird prey is usually headless, mammals and lizards frequently so, when they are brought back to the nest. During the early days of the nesting season, the male may subsist to a considerable extent upon the heads of the prey he captures. Birds are plucked and mammals and lizards are skinned before they are presented, piece by piece, to the nestlings.

Prairie Falcons prey on many different kinds of mammals and birds, but their favorite prey includes ground squirrels, Horned Larks, Western Meadowlarks, and Mourning Doves. In the Great Basin ground squirrels are unavailable when in hibernation or aestivation, and the Mourning Dove and, to some extent, the Western Meadowlark are migratory, but Horned Larks are available in numbers the year around. They are most vulnerable when surprised on the ground. Once aloft a Horned Lark flock may employ an interesting defense tactic in the presence of a hunting Prairie Falcon. The flock, in loose formation, climbs above the falcon and then follows closely behind it. From this position the larks are not subject to surprise attack and can rely on their greater maneuverability to avoid being captured. The flock tags after the Prairie until it departs from the scene.[187]

The Prairie Falcon often single-mindedly concentrates on one prey species. Observers have reported finding nest areas strewn exclusively or almost exclusively with the remains of meadowlarks, or ground squirrels, or Burrowing Owls, or Rock Doves, or quail and jays, or other prey species. Seldom in nature does a predator base its entire economy on a single prey species. The exception which comes readily to mind is the Snail Kite, *Rostrhamus sociabilis*, whose entire diet consists of the apple or moon snail, a species of freshwater snail. There is a general belief among ecologists that the composition of a predator's kill merely reflects the relative abundances and vulnerabilities of its various prey species. However, Bond has some interesting ideas about the eating habits of falcons in general. He observed that it was quite difficult for young falcons to gain enough insight to see that live prey could be converted into food and that they must relearn this relationship for each type of prey of different appearances. Then, he writes, "It is perhaps as a result of this, at any rate I have found it true, that falcons, both wild and trained are prone to form prey habits, and to confine the food taken to a single species or group as exclusively and so long as may be. When the chosen food gives out, through migration or whatever cause, a new item is selected and followed with the same single-mindedness."[219] So the question arises, is the Prairie Falcon's tendency to concentrate on a single prey species at a given time reinforced by whatever search images it has formed?

The Prairie Falcon is not judged to be the most likable of falcons. It has the reputation of having a "savage demeanor" and of being "irascible," "excitable," and "vindictive." It has been said that "one of the most disagreeable features of the Prairie Falcon is a vile and unpredictable temper. One day a falcon is as calm as can be and the next day, wilder than a hurricane."[220] "A MOODY creature at all times, peevish and whimsical, the Prairie Falcon is a bird of extremes."[221]

Some of the most flagrant cases of the "vile and unpredictable temper" of the Prairie have been witnessed by human observers after they have disturbed falcons at their nests. Ravens, Common Barn-Owls, and Great Horned Owls that take to the air when Prairie Falcons are disturbed are often attacked and may be stunned or killed. Common Barn-Owls are easy targets. Bond watched a pair of Prairie Falcons he had disturbed at their nest stoop on a pair of barn-owls who flushed from the same cliff. In a single pass the male falcon killed one of the owls, and the female falcon stunned the other.[222] Ogden watched Prairie Falcons stun or kill several Common Barn-Owls; and during the course of a field study the carcasses of seven barn-owls found within about sixty-five yards of falcon aeries had probably been killed by the falcons.[223] Prairie Falcons are quick to challenge other raptors who enter the airspace around their aerie—but they usually just harass them and only occasionally strike them. Given the Prairie Falcon's nasty disposition, it is difficult to understand why so very often Common Ravens nest close to falcon aeries. However, I do not know of any instances of small, defenseless birds forming protective nesting associations with Prairie Falcons. Those that do nest near an active falcon nest often end up as food for young falcons.

On occasions Prairie Falcons have been observed practicing piracy on Northern Harriers. Ridgway was probably the first to describe this habit. He initially encountered the falcon in the Great Basin at the Humboldt marshes in October 1867. Then, "late in November, of the same year, it was noticed again among the marshes along the Carson River, near Genoa, where it was observed to watch and follow the Marsh Hawks (*Circus hudsonius*), compelling them to give up their game, which was caught by the Falcon before it reached the ground; this piracy being not an occasional, but a systematic habit."[25] Sometimes a Prairie Falcon is on the receiving end of piracy. Large, strong raptors like the Golden Eagle and Ferruginous Hawk, if they are persistent, may take prey away from a Prairie Falcon.

The Prairie is not a grim, unhappy bird at all times. Life does have its lighter moments. Munro watched a Prairie Falcon playing with pieces of dry cow manure for about twenty minutes. The falcon would drop a piece of manure from a height of fifty feet or so, then try to catch it before it hit the ground. Alighting on the ground, the falcon tossed a piece of manure ahead for several feet and fluttered after it, pouncing on it. Finally, rising only several feet off the ground, the falcon would toss the manure up in front and then try to catch it before it hit the ground.[224]

The Prairie Falcon has been much used in the sport of falconry in North America, as it is a fairly large falcon and is readily available in western North America. Beebe has made some interesting observations in reference to the use of the Prairie in falconry and its general inferiority in this sport to the Peregrine Falcon. Often a falconer training a Prairie Falcon has used methods developed for Peregrine Falcons, who hunt by climbing high in the air and stooping on flying prey. So, when a Prairie Falcon is trained to hunt by climbing aloft over the falconer, there is a conflict with its inherent nature to practice ground-oriented hunting and to stay close to the ground. Further, although the female Prairie Falcon is large enough to fly against grouse-size prey, she is "fractious and irascible" and more difficult to train than the considerably smaller, more tractable male.[187]

There is a rather interesting story to be told about Prairie Falcons at Pyramid Lake and Louis Agassiz Fuertes, one of the greatest bird painters of all times. During July 1903, in the company of several other ornithologists, Fuertes visited Pyramid Lake "to make a study of the great white pelican rookery" on Anaho Island. Among the other members of the party were C. Hart Merriam, the formulator of Merriam's life zone system for North America, and Frank M. Chapman, an outstanding ornithologist and author of the classic *Color Key to North American Birds*. A pair of Prairie Falcons and their three grown young were present on Anaho Island. Fuertes was later to write that a Prairie Falcon "was killed (almost in self-defence) well toward the crest of the cliff-like southern acclivity. All about this point, which I took to be near the eyry, were strewn the feathers of quails and jays." Fuertes posed the dead falcon and made a drawing of a Prairie Falcon perched on a point overlooking Pyramid Lake. This drawing, dated July 1903, was later used to illustrate an article he wrote describing his experiences on Anaho Island. Fuertes was deeply touched by his day here; he wrote:

> As I looked down from my position at a height on the wall-like face of the cliff, the yellow rock merged into the chalky levels below, where the huddling herds of young pelicans crowded together; then came the white alkali beach, which lost itself in the wonderful blue of Pyramid

> Lake—the most glorious color water ever had. And against this marvellous color, the blistering sun gleaming on their broad snowy backs and wings, the old pelicans soared magnificently below me, while the falcons screamed in the clear air around my head. I think this was one of the most striking experiences I ever had, and I stood a long time imbibing the varied new sensations of sound and color before I at last turned my steps downward to join the census bureau on the lower levels, where Dr. Merriam and Mr. Chapman were diligently counting the young pelicans in the rookeries.[225]

That glorious day at Pyramid Lake stayed with Fuertes for the rest of his life. In 1926, a year before he died, he created a magnificent painting, *The Prairie Falcons of Pyramid Lake*. In this 1926 painting, the falcon in the 1903 drawing was reoriented and new details were added. In the painting a Prairie Falcon is standing on a point overlooking Pyramid Lake, with one foot on its prey—a quail—peering upward at another Prairie Falcon flying overhead. The painting was reproduced for the first time in color in the September–October 1966 issue of *Audubon Magazine*. The caption accompanying the reproduction stated, "Never before reproduced in color, this masterpiece by Louis Agassiz Fuertes was the last painting sold by the great bird artist. It was received by T. Gilbert Pearson, president of the National Association of Audubon Societies, just two days before Fuertes' death on August 22, 1927. The scene is Nevada's vast Pyramid Lake, the prey a Gambel's quail."

When I saw the reproduction of *The Prairie Falcons of Pyramid Lake*, it became my unpleasant duty to write to the editor of *Audubon Magazine*, stating that Fuertes had misrepresented nature in this fine painting. When Fuertes decided to convert his 1903 drawing into a painting, he probably recalled seeing "the feathers of quails and jays" strewn on the ground around the point on Anaho Island. But he added the wrong species of quail when he added a Gambel's. The only kinds of quail present in northern Nevada are California and Mountain quail; the range of the Gambel's lies to the south of the Great Basin.

Other Falconids

The Gyrfalcon, *Falco rusticolus*, is a creature of the arctic regions of Eurasia and North America. At best, it is of accidental occurrence as far south as the Great Basin. There is a handful of undocumented sight records for the Gyrfalcon in Utah, particularly in Salt Lake County. Some of the sight records were made by falconers.[56]

11

Nocturnal Birds of Prey

THE ORDER Strigiformes contains the raptors known as owls. Most species of owls hunt by night, although a few species hunt by day. Like the members of the other great order of raptors, the falconiforms, the strigiforms are armed with hooked bills and curved talons. Owls swallow their prey whole and regurgitate pellets of indigestible materials such as bones, teeth, fur, and feathers. They are big-headed, with large eyes set in facial disks and directed forward. Their hearing, as well as their vision, is very acute; and some species detect their prey more by sound than by sight. Owls are soft-feathered and, consequently, are rigged for silent flight and sound detection. Both families of North American strigiforms—Tytonidae and Strigidae—are represented in the Great Basin.

FAMILY TYTONIDAE

The barn-owls constitute the Family Tytonidae. They are characterized by heart-shaped facial disks and relatively small eyes. The claws of their middle toes are pectinate or comblike. Barn-owls are distributed over most of the world.

Common Barn-Owl
Tyto alba

The Common Barn-Owl is quite a cosmopolitan species, ranging over most of the warmer parts of the world. It is an occasional to uncommon resident in the northwestern and eastern parts of the Great Basin, but it is common in the west central part. Prior to 1982 its common name was the Barn Owl.

The Common Barn-Owl is completely nocturnal in its activities—it is seldom seen during the day unless it is flushed from its roosting place. The apparent uncommonness of the barn-owl may be due in part to the lack of human observers abroad at night in the Great Basin. In Nevada a considerable number of dead or crippled barn-owls are picked up along the highways each year. Common Barn-Owls are frequently found nesting in holes in banks or in potholes and crevices on cliffsides, as well as in old buildings; they are frequently found in cities. In the past barn-owls often roosted on the University of Nevada campus in Reno during the colder part of the year. These owls had the temerity to roost on ledges directly above the main entrances to some of the older buildings. They were then subjected to severe persecution by buildings and grounds people objecting to the pellets and droppings beneath the roosts.

Common Barn-Owls are among the most nocturnal of owls. They often hunt at such low light intensities that they must rely on their sense of hearing to locate and capture prey. Mice, especially voles of the genus *Microtus*, form over 95 percent of their prey. *Microtus* live in wet, grassy areas and construct elaborate systems of runways through the grass—even at high light intensities Common Barn-Owls would have difficulty seeing them in their grass-shielded runways. Instead, the barn-owls locate and capture the voles in their runways by the rustling and squeaking noises they make. Experiments conducted in total darkness in lightproof rooms have shown that Common Barn-Owls will freely hunt and efficiently capture mice by sound alone.[226]

The name barn-owl is descriptive, for this owl lives in the company of humans; in agricultural regions it nests and roosts in barns and old buildings. Recognizing that a barn-owl is the very best rat- and mousetrap on the market, farmers in Europe may build a small "owl door" opening into their barns to entice barn-owls to come and nest or roost there.

Common Barn-Owls hunt in open areas away from woodlands and forests. They may still-hunt from perches but usually course back and forth, low over the ground, alternating flapping flight and gliding. An owl may follow the same hunting route night after night. Barn-owls are rigged for silent running with loose, soft body feathers and modified flight feathers. Their noiseless flight allows them to hear their prey without being heard in return.

When the Common Barn-Owl hears the sound of potential prey from the air, it is confronted with the problem of localizing that sound within the framework of two-dimensional space. The owl has to determine the proper horizontal or compass direction to the prey. Then it must determine what its swooping angle should be, so that it intercepts the ground at the precise spot where the prey is located.

Hunting at night by means of acoustics poses some interesting technical questions. Several investigators have been intrigued enough to do some sophisticated research on acoustical hunting by the Common Barn-Owl. The owl apparently locates the source of the sound by comparing minute differences between the sounds arriving at the right and the left ears. Since the two ears of the owl are at slightly different distances from the sound source, there is a slight difference in the time of arrival of the sound. Then, too, the sound will be slightly louder in the ear nearest the source. The owl uses these subtle differences in timing and loudness in determining direction to the sound source.[227]

Determination of the vertical angle to the sound source involves anatomical asymmetry of the outer ears of the Common Barn-Owl. This asymmetry is not apparent on the surface. The outer ear structures are located under fine, acoustically transparent feathers in the heart-shaped facial disk of the owl. Two troughs run through the facial disk, one on each side, extending from the forehead down to join under the beak. Within each trough is an ear opening. Since the right trough and right ear opening are tilted slightly upward, the right ear is more sensitive to sound coming from above. The left trough and left ear opening are tilted slightly downward, and the left ear is therefore more sensitive to sound coming from below. Comparison of sound arriving at the right and left ears allows the owl to determine the angle to the sound source on the ground below. The stiff, densely packed feathers outlining and forming the two troughs reflect sound frequencies above three thousand hertz. Therefore the two troughs collect and funnel sound into the ear openings in the skull, just like hands cupped behind human ears do.

The differential tilt of the two troughs and ear openings is not the only asymmetry present in the outer ears of Common Barn-Owls. Each ear opening is located behind a preaural flap of skin, which projects to the side next to the eye. The left preaural flap and left ear opening are located higher up in the facial disk of the owl than the right ones. A line drawn through the two eyes would find the left preaural flap and ear opening above the line, the right ones below the line.

After getting a sound fix on its prey, the Common Barn-Owl swoops down on relatively shallow wingbeats. Before striking its prey, the owl often closes its eyes. It then extends its legs and feet ahead of its wings and head so that they are positioned on the intercept line to the prey. The owl has also determined the direction of movement and, hence, the body orientation of its prey by sound. Before striking, the talons are opened and the owl maneuvers so that both feet will strike, side by side, on the long axis of the prey's body. If the owl misses on the first strike, it will remain on the ground listening for prey movement.

FAMILY STRIGIDAE

Members of the Family Strigidae are called typical owls; they represent all the owls except the barn-owls. Their facial disks are rounded and their eyes are very large. Strigids are present in most regions of the world.

Flammulated Owl
Otus flammeolus

The breeding range of the Flammulated Owl is in the montane conifer forests of the interior of the West, extending from southern British Columbia through Mexico into Guatemala. The Flammulated is considered to be a rare to uncommon resident in the mountains of the Great Basin and in its mountainous eastern and western rims.

Random field observations do not tell us much about the status of inconspicuous species in an avifauna. Species can be inconspicuous because of a variety or combination of factors beyond rareness: small size, cryptic coloration, general noiselessness, remote or inaccessible habitat, nocturnal activity, shy and retiring habits, and so on. Field studies must be designed specifically to reveal the presence or absence of target species. Research on the Flammulated Owl by Johnson and Russell has revealed how enlightening such an approach can be. Prior to their fieldwork along the western rim of the Great Basin in northeastern California and adjacent Nevada, there were only two records of Flammulated Owls in northeastern California, none from western Nevada. The California records were for Fort Crook, Shasta County, in 1860, and Quincy, Plumas County, in 1907.[26]

Johnson and Russell used recordings of the voice of the Flammulated Owl in an attempt to call in owls. They played these recordings at various locales on a number of evenings, scattered between mid May and early July from 1958 to 1962. They succeeded in calling in anywhere from one to ten Flammulated Owls an evening, mainly males, at a number of sites in northeastern California in Modoc, Lassen, Sierra, and Placer counties. They obtained the first record of this owl in western Nevada, when they called in eight individuals on June 23, 1961, at a locale one and one-half miles north of Crystal Bay, Lake Tahoe, Washoe County.[228] This record was not entirely unexpected, since prior to this the Flammulated had been reported from the California side of Lake Tahoe at Meeks Bay and at Rowland's marsh.[26]

During their fieldwork Johnson and Russell collected thirteen Flammulated Owls as museum specimens. The stomach contents of ten of these owls were analyzed. One stomach was empty; the other nine contained the remains of various insects. Although there are only two other published records of the Flammulated Owl in the Great Basin part of Nevada—there is one record from the eastern base of the Toiyabe Mountains, Nye County, and nesting has been reported from Ruby Lake National Wildlife Refuge—the research of Johnson and Russell indicates that the paucity of records may not be indicative of this owl's true status.

Very little is known about the natural history of this tiny (six to seven inches) owl. Most of its food consists of insects and spiders. The territorial call of the male is a single soft *hoo*, repeated every few seconds. The preferred nesting site is a tree cavity. There are still unresolved questions concerning its winter behavior—as to what extent it engages in latitudinal and/or altitudinal migrations. The sixth edition of the American Ornithologists' Union checklist states that the Flammulated Owl "*winters* from central Mexico (Sinaloa, Jalisco, Michoacán and Distrito Federal) south in the highlands to Guatemala and El Salvador, casually north to southern California."[24] According to this statement, Flammulated Owls migrate to the south to overwinter.

The wintering habits of the Flammulated Owl attracted the attention of Johnson, who proposed that this owl may be "a permanent resident on or near the breeding grounds."[229] Johnson pointed out that most sightings are made when Flammulateds respond to territorial calls and that this type of response obviously does not work for winter studies. He offered two nonmigratory explanations of the owls' overwintering. One was the possibility that these owls may perform altitudinal migrations, down from their conifer forest breeding grounds, when nocturnal insects become difficult to obtain. Or the owls may spend the winter in cold torpidity, since they are related to birds such as Common Poorwills who do hibernate in the winter. Obviously, additional research is needed on the Flammulated Owl.

Western Screech-Owl
Otus kennicottii

The breeding range of the Western Screech-Owl lies west of the Rocky Mountains; it extends from southeastern Alaska through Mexico. The Western Screech-Owl, a permanent resident in the Great Basin, is apparently more common in the eastern than the western part. It is reported as being rare to common at various locales in the Basin. Prior to 1982 the Western Screech-Owl and the Eastern Screech-Owl were regarded as a single, continent-wide species called the Screech Owl, *O. asio*.

One of the surest ways to detect the presence of owls is by listening for their calls during the breeding season. Playing recorded call notes in likely habitat may stimulate owls to respond—they may even closely approach

Western Screech-Owl owlet

the tape recorder. In listening for screech-owls, don't expect to hear screeches since their name is a misnomer. The common call of the screech-owl is a series of mellow, hollow, tremulous whistles, on one pitch or dropping in pitch at the end, starting slowly and running together as *oo, oo, oo-oo-oo-oo-oo-oo-oo-oo*.

The preferred nesting site for a screech-owl is a natural cavity or flicker-built cavity in a tree. Look for this in cities as well as in wooded localities. Screech-owls are very aggressive at night toward human intruders in the vicinity of their nesting sites—on more than one occasion I have first become aware of the close proximity of nesting screech-owls by suddenly receiving a sharp rap on the head from out of the dark of the night. When searching for screech-owls, use the listening technique. There is much less wear and tear on the head in using sound rather than touch.

The Western Screech-Owl bears a conspicuous ear or hornlike tuft of feathers on each side of its head. The Long-eared, Great Horned, and many other species of nocturnal owls also have ear tufts, but the tufts are absent from diurnal or predominantly diurnal owls. Ornithologists have long pondered the significance of such head ornamentation. All this pondering has given rise to several hypotheses. Since the ear tufts of various species come in various sizes and positions on the head, it has been hypothesized that a species' ear tufts give that species a distinctive night silhouette in the eyes of other owls. Others have speculated that ear tufts mimic the ears of certain mammalian predators during threat displays and thus have protective value when the owl is approached by a mammalian predator who may yield to a threat display. Currently, the most meritorious hypothesis suggests that ear tufts have camouflage functions. Nocturnal owls are usually grayish or brownish in color and roost on branches during the day. When disturbed they often stretch, thinning their silhouettes. Then, the ear tufts make them look much like a broken vertical limb, since jagged breaks are more typical in nature than straight ones.

Great Horned Owl
Bubo virginianus

The Great Horned Owl ranges over all of the Americas, from the northern tree line in the arctic to the Strait of Magellan at the tip of South America. The Great Horned is a common and widespread resident of the Great Basin, inhabiting every conceivable type of desert and montane habitat, including cities and towns. In Reno it nests in residential districts where groups of taller trees are standing. It is certainly the most widely distributed raptor in the Great Basin and one of the most abundant.

The Great Horned Owl has been abundant in the Basin for some time, for Ridgway reported that it "was found by us in all wooded districts" in the 1860s. Then, he goes on to say, "In the lower Truckee Valley, near Pyramid Lake, it was abundant in December, and its nocturnal hootings were heard from among the cotton-wood groves every moonlight night, while its feathers, more than those of any other bird, adorned the arrows of the Indians on the reservation."[25]

Outside of the nesting season Great Horned Owls are solitary creatures. But by December or January the males have their territorial hooting under way, just as Ridgway heard it coming out of cottonwood groves along the Truckee River in December of 1867. Once a male has formed a pair bond with a female, the two remain together and roost together during the day until egg laying is underway—usually by February or March.

The Great Horned nests in a variety of situations. It usually does not build a nest of its own but appropriates old stick nests of hawks, ravens, crows, and magpies. In the Great Basin nesting frequently occurs on ledges and in niches on cliffsides and canyon walls. I have even found nests on the ground on hillsides overlooking shrubland valleys. Great Horned Owls do virtually no repair work on old nests, so they frequently nest in dilapidated structures. A lining of bark or down feathers may be added to the nest. Usually two or three eggs are laid in a clutch, rarely four or five. Although both sexes incubate, the female probably does most of it. The incubation period has been reported in the literature as ranging from twenty-six to thirty-five days in length; after hatching the young remain in the nest for about five weeks. Upon leaving the nest they cannot fly well. They remain in the vicinity of the nest for another four or five weeks, closely attended by their parents.

Active Great Horned Owl nests can be readily located in the valleys of the Great Basin once incubation is under way, with February and March being good months for such a search. Examine stands of taller trees. Watch for large stick nests well up in the trees. Even the outskirts of cities such as Reno may be productive. When you spot a large stick nest, stop at a distance, so that you have a shallow viewing angle from which to scrutinize the nest. At a shallow angle, the large head and horns—the ear tufts—of the incubating parent will show up above the rim of the nest. You may even find active hawk nests when you are looking for owl nests, although hawks typically nest later in the year than the owls. Occasionally you may find two Great Horned Owls perched close to one another, roosting in a tree. This will be a pair still in the engagement period. Once the eggs are laid, one parent may be seen roosting near the nest, but the other parent will be on the nest covering the eggs.

Great Horned Owls are ferocious hunters, real "flying tigers." With good size and great strength of body, bill, and talon, they can attack and eat other raptors, including hawks and other owls, as well as some carnivorous mammals. They are not reluctant to attack skunks and do not seem to be inconvenienced by the spraying they receive from the anal scent glands of the skunk. The owl's nictitating membrane, the so-called third eyelid, located under the outer eyelid, can be drawn horizontally across the cornea of the eye. This membrane probably protects the eyes from the caustic, blinding effects of the spray. The principal components of the diet of the Great Horned Owl in the Basin are jackrabbits, cottontails, kangaroo rats, pack rats, and pocket gophers.

The Great Horned Owl is a nocturnal hunter, although it may be on the wing by twilight, long before full darkness is upon the land. Most of its hunting is still-hunting, employing a perch, wait, and pounce technique. It flies to some elevated perch on a cliffside, tree, utility pole, or embankment where it can watch and listen for prey. If nothing develops at that perch, it flies on within a few minutes' time to a new site. When prey is detected, the owl plunges down from its perch and flies rapidly and silently, close to the ground, to launch a sudden, unexpected attack with its talons. Sometimes the owl will hunt using a harrier techique, flapping and gliding close to the ground and immediately dropping on any prey encountered.

Northern Pygmy-Owl
Glaucidium gnoma

The Northern Pygmy-Owl is a resident of western North America, ranging from southeastern Alaska into Mexico and Guatemala. In Utah, in the eastern end of the Great Basin, it is an uncommon permanent resident in the montane coniferous forests, but it may be found in lesser numbers in densely wooded riparian sites in the lower valley as well as sometimes in cities.[56] It is also of common occurrence in the Snake Range of eastern Nevada.[230] The Northern Pygmy-Owl is a common resident in the Carson Range and in the mountainous western rim of the Great Basin. Prior to 1982 the common name of this owl was that of Pygmy Owl.

Although this owl is a permanent resident in western North America, there is evidence from throughout its range that it may perform altitudinal migrations during severe winters. In winter pygmy-owls may show up in valleys adjacent to the mountains in which they reside. This is true of Truckee Meadows, the valley in which the cities of Reno and Sparks are located. Just to the west of Truckee Meadows lies the Carson Range, where the Northern Pygmy-Owl is a common resident and breeding bird. All our Truckee Meadows records for pygmy-owls are winter records.

The Northern Pygmy-Owl has some attributes which are not typical owlish ones. Owls are commonly associated with the night, yet this owl has a diurnal activity pattern. It hunts during the daytime—with bursts of activity around sunrise and sunset. Unlike most owls, which fly silently, there are noises associated with the flight of the pygmy-owl. And it undulates as it flies, much like a shrike. When it is perched its long tail may be tilted up, instead of extending straight backward in typical owl fashion.

The pygmy-owl is well named, ranging in total length from barely over six inches up to seven and one-half inches, tail included. The females are larger than the males. In the subspecies *californicum*, the subspecies found in the Carson Range and the adjacent Sierra Nevada, the average weight of the males is about 2.17 ounces versus 2.56 ounces for the females.[231]

Now, you wouldn't think that a raptor weighing only a couple of ounces would cause much of a commotion among montane birds and mammals, although you could envision grasshoppers and other insects being panicked by such a diminutive predator. However, the Northern Pygmy-Owl does not hesitate to attack animals much bigger than itself, such as quail and ground squirrels. It has even been known to attack that ferocious mammalian predator, the weasel.

The Northern Pygmy-Owl does cause a mighty commotion among the small birds of a montane forest when its presence is detected by sight or sound. Mobbing parties of warblers, finches, vireos, wrens, juncos, sparrows, chickadees, nuthatches, woodpeckers, and others quickly form to dance around the perched owl, screaming and feinting at it from all sides.

The voice of the Northern Pygmy-Owl is frequently heard during the day or early evening wherever this owl is present. The common call is a series of mellow, whistled notes, starting out as a slow trill and then, after a pause, continuing as two or three notes, each separated by a pause. This song has been variously interpreted by observers as *o-o-o-o-o-o-o-o-o-o——oo——oo——oo* and as *too-too-too-too-too-too-too-too; whoot; whoot; whoot*, for example.

The call of the Northern Pygmy-Owl can be readily imitated by a human observer. Pygmy-owls can be located in the field by getting them to respond to imitated call notes. However, in the process, particularly if you are hidden and motionless when calling, you may stir up mobbing parties of small birds before any owl responds to your calls. Small birds must learn to recognize the call notes of this owl as a source of potential danger, since, in localities where Northern Pygmy-Owls do not occur, small birds will ignore imitated pygmy-owl call notes. For example, small birds in the Carson Range will quickly form mobbing parties in response to imitated owl calls. But in the adjacent Virginia Range, where as yet we have not located any pygmy-owls, the small birds ignore imitated call notes and do not form mobbing parties.

Burrowing Owl
Athene cunicularia

The breeding range of the Burrowing Owl is located in open, treeless, dry regions, ranging from Arizona and Texas northward to southern Canada. This owl also inhabits suitable areas in central and southern Florida, Central America, and South America. The Burrowing Owl is a widespread summer resident in the Great Basin, where it is of uncommon to common occurrence. The impression that the Burrowing Owl is rather uncommon in most Great Basin localities goes back as far as Ridgway's descriptions in the 1860s; he wrote, "Although the 'Ground Owl' was found at widely-separated places along our entire route, it was abundant at very few localities. . . . Eastward of the Sierra Nevada we found it only at wide intervals."[25] Prior to 1982 this owl's scientific name was *Speotyto cunicularia.*

I am of the opinion that the Burrowing Owl is much more common in the Great Basin shrublands than is generally credited. Sagebrush shrublands, which I have censused in the very early morning hours at the beginning of June, have revealed the presence of many more owls than I was aware of from scrutinizing these areas at other times. Often several Burrowing Owls would be in sight at one time, perched on top of shrubs, basking in the early sunlight. The Burrowing Owl is not a noisy, up in the air, conspicuous bird—it is easily overlooked. Then, too, most birders do not spend much time in the Great Basin shrublands during the heat of summer.

Burrowing Owls are often seen in the vicinity of their burrows during the daytime. When encountered in broad daylight, they do not seem to be discomforted by the strong light or reluctant to move about. However, some studies have shown that most of their hunting may be crepuscular, at dusk and dawn, or nocturnal. In a study carried out at the Oakland Municipal Airport, Thomsen discovered that Burrowing Owls she captured "before sunrise always had distended stomachs, whereas those caught in the early evening did not, indicating that they foraged at night."[232] Laboratory experiments have shown that ambient temperatures may affect activity patterns of Burrowing Owls. There is an increase in crepuscular and nocturnal activity at ambient temperatures above 40° C, with a concomitant decrease in diurnal activity. Field observations show that young and adult owls spend little time outside their burrows during midday hours in the heat of the summer, toward the end of the breeding season. As the days become cooler, the owls spend more time outside during daylight hours.[233]

Owls swallow prey whole or in large chunks, then regurgitate the undigested feathers, fur, bones, teeth, and insect exoskeletons as pellets. In an Idaho study, 421 Burrowing Owl pellets were analyzed. Most of the prey

Burrowing Owl

remains in the pellets were from nocturnal species of animals, mainly nocturnal rodents such as kangaroo rats and voles and nocturnal insects such as Jerusalem crickets.[234]

It is generally agreed that Burrowing Owls are migratory in the northern part of their range. They are certainly migratory in the Great Basin. Burrowing Owls disappear from the Basin in late summer and early fall, when food is still abundant and the weather is mild; they reappear at the beginning of spring. During the spring migration Burrowing Owls may appear briefly in unlikely places, such as residential neighborhoods in Reno. An occasional individual may overwinter in the Great Basin. Wintering Burrowing Owls are reported to be of rare occurrence at Malheur National Wildlife Refuge[61] and uncommon at Ruby Lake National Wildlife Refuge.[235]

There are some apparent anomalies in the migratory behavior of the Burrowing Owl. Migration was not detected in the population of owls at the Oakland Airport in northern California.[232] But during a study in southern California at Imperial Valley the owls showed partial migration, with only 20 to 25 percent of the breeding population remaining to overwinter.[233] In a study of a breeding population of Burrowing Owls in a prairie dog colony in the Oklahoma Panhandle, less than 1 percent of the population remained to overwinter.[236] In a study in central New Mexico, it appeared as if some Burrowing Owls wandered about in the winter. Others migrated, with a sharp decrease in population size on the breeding grounds starting in early August, when food was still abundant, and with a sharp population increase starting in mid March.[237]

Some observers have speculated that these owls may remain underground in their burrows during severe winter weather, thus explaining their apparent disappearance from some localities during the winter. At the Oakland Airport some Burrowing Owls disappeared for a few days up to a few weeks at a time; perhaps they remained in their burrows during these times.[232] In his book on the birds of New Mexico, Ligon stated that he had never found any evidence that Burrowing Owls migrated in the Southwest. He believed that they remained underground during severe weather and that they even cached food such as mice to tide them over during inactive periods. On one occasion, when Ligon excavated what appeared to be an active winter burrow in an old prairie dog town, he found a Burrowing Owl snugly ensconced about six feet from the entrance.[238]

Ligon and others have suggested that the possibility of winter torpidity should also be considered in this owl. So the Burrowing Owl has joined the Flammulated Owl on the list of owls in which doubt over the pervasiveness of migration has evoked the suggestion of hibernation as an alternative.

The name Burrowing Owl suggests that this owl digs burrows. However, although it is capable of digging its own burrow, the owl usually enlarges or renovates the existing burrows of ground squirrels, prairie dogs, badgers, and other burrowing mammals. The same nest burrow may be used for a number of years. Burrowing Owls have been observed digging by supporting their weight on one leg while kicking dirt backward with the other leg. Since dirt shows up in the pellets when burrow work is being done, owls may do some of the work with their bills. Burrowing Owls have the peculiar habit of collecting horse and cow manure and lining the floor of the nest chamber, tunnel entrance, and tunnel with chips of this dung. If the shredded dung is removed from the entrance of the tunnel, the owl replaces it within a day's time. Observers have suggested that the odor of the dung may serve to mask the scent of the owl. This may protect the owl from detection by ground predators like badgers.[237] Tales from prairie dog country have the Burrowing Owl, prairie dog, and rattlesnake living peacefully together in a prairie dog burrow. But this is folklore. The three are natural enemies—they are not about to take up communal housekeeping.

As we have seen, the principal prey items of Burrowing Owls are nocturnal species of rodents and insects. When spadefoot toads come above ground to breed, they may figure heavily in the diet of Great Basin Burrowing Owls. The owls employ a number of techniques to capture prey. During the early evening they may locate prey by flying about and hovering while they search the ground below. They also still-hunt and locate prey while on an elevated perch. Burrowing Owls may descend to the ground and walk or run about after prey. On occasions, they pluck flying insects out of the air.

The voice of the Burrowing Owl is of some interest. The primary song of the male, functioning in pair formation and territorial display, is a two-note *coo coo*, almost on the same pitch but with the second note longer than the first.[239] When an intruder approaches a nest burrow, the owl outside on guard gives a six-note alarm call of *chip-chip-chi chi chip-chip*. Upon closer approach the owl comically bobs up and down, uttering a higher-pitched, harsher *cheed*. With continuous approach the owl will fly, in decoy behavior, to a nearby perch to continue its alarm behavior there.[233] When young Burrowing Owls are threatened, they give a rattlesnake rasp. This call, emanating from a burrow, closely resembles the rattle of a disturbed rattlesnake.

Long-eared Owl
Asio otus

The breeding range of the Long-eared Owl extends in a belt around the northern hemisphere in temperate Eurasia and North America. This owl is also found in Africa. The Longear is a common resident in the Great Basin, being one of our most abundant owls. Its status has apparently not changed much with settlement in the Basin. In describing its abundance in the 1860s, Ridgway wrote:

> Seldom, if ever, did we enter a willow-copse of any extent, during our explorations in the West, without starting one or more specimens of this Owl from the depths of the thicket. This was the case both near Sacramento and in the Interior, and in summer as in winter. In these thickets they find many deserted nests of the Magpie, and selecting the most dilapidated of these, deposit their eggs on a scant additional lining. This practice is so general, so far as the birds of the Interior are concerned, that we never found the eggs or young of this species except as described above.[25]

The Long-eared Owl is found in forested or wooded areas and in riparian stretches of willows and cottonwoods. Although the owl nests and roosts in dense cover, it hunts by night in open areas, so it requires open areas adjacent to its nesting and roosting habitat. Virtually all its prey consists of mice and other small mammals. Over much of its range its prey is predominantly voles of various species. The Longear has long wings and light wing loading for an owl—it flies with great buoyancy. During the day Long-eared Owls usually roost in small trees, perched close to the trunk. When an intruder approaches a roosting owl, the owl stretches itself into a thin, upright position, erects its ear tufts, and trusts to camouflage rather than flight.

Short-eared Owl
Asio flammeus

The Short-eared Owl is a cosmopolitan species, with its main breeding range lying between forty and seventy degrees north latitude in Eurasia and North America; it also breeds in South America. The Short-eared Owl is an uncommon to common resident and breeding species in the Great Basin. The number of owls present locally changes with seasonal movements and migrations. In west central Nevada the Shortear often reaches its peak abundance during the winter. This owl has evidently increased greatly in number in the Basin since the 1860s, for Ridgway did not encounter even one during his extensive fieldwork here.

The Shortear is one of the most frequently encountered owls, since it usually gets its hunting under way several hours before sunset and often hunts on overcast days. Short-eared Owls hunt over open country, favoring marshy and grassy areas. Rodents, particularly voles, are their principal prey, although an occasional bird may be taken. These owls employ several hunting techniques but seem to rely most on coursing back and forth, low over the ground. When coursing the owl alternates flapping flight with gliding. Upon detecting prey on the ground below, it may hover briefly before pouncing. If the wind is blowing the bird usually hunts into the wind in order to utilize it in hovering. Another technique involves searching an area by repeatedly hovering and examining the ground below for prey. Sometimes a Short-eared Owl will still-hunt from an elevated perch or descend to the ground to wait for the appearance of prey.

The wing-clapping flight of the male Short-eared Owl is among the most spectacular courtship behavior recorded for owls. The male climbs above his territory, sometimes reaching heights of up to 1,500 feet. He spirals his way upward on rowing wingbeats of moderate depth. While climbing the male engages in wing-clapping bouts, by striking his wing tips together under his body. Wing clapping produces a sound that has been compared to the flutter of a flag in a very high wind.[240] As the male gains altitude, the wing clapping lessens and some soaring may ensue. At times the male hovers and utters his courtship song of *toot-toot-toot-toot-toot* or *voo-hoo-hoo-hoo-hoo*, in a low-pitched monotone of about four toots per second. He may alternate hovering with shallow swoops that terminate with a wing clap. Then he climbs back up to his original height again. The performance ends with a spectacular descent. The male rocks back and forth with his wings held high above his back. He may descend all the way down or level off once or twice on the way down and then glide down with wing claps. If a female is present on the male's territory, she may watch the performance and give an occasional *keeee-you* call.[241]

The Short-eared Owl is a migratory species. Its migratory pattern is poorly known, since there has been only a limited number of recoveries of banded Shortears. This owl is especially migratory in the northern part of its range. The deep snow there makes it extremely difficult for owls to obtain voles, which during the winter lead a highly subnivean existence under the deep snow blanket. Short-eared Owls appear to be somewhat nomadic in their seasonal movements and often congregate in areas where population densities of voles are high. The number of Shortears breeding in a locality can vary greatly from year to year, depending on the food supply. On their wintering grounds, the owls may defend feeding territories and roost communally.

Northern Saw-whet Owl
Aegolius acadicus

The range of the Northern Saw-whet Owl extends from northeastern Alaska, down over much of Canada and the United States, on into Mexico. Over this range the owl's local occurrence is quite patchy. The saw-whet is an uncommon resident and breeding bird in the Great Basin, inhabiting montane forests and canyons and riparian thickets. During the cold part of the year, this owl wanders widely and may show up in the valleys below the level of deep snow. Its name is derived from its call notes of *skreigh-áw skreigh-áw*, which resemble the sound of a saw being sharpened by a file. Prior to 1982 its common name was merely that of Saw-whet Owl.

The Northern Saw-whet Owl is a nocturnal hunter of mice and other rodents, with an occasional bird or insect included in its diet. When roosting during the daytime this owl is quite tame and allows a close approach—even by a university class of binocular-wielding ornithology students. Ridgway's party encountered only a single saw-whet in the Great Basin, at Thousand Spring Valley, Elko County, Nevada, in late September of 1868. This owl was captured alive, by hand, when found roosting on the edge of an old robin nest in a dense willow thicket—O. L. Palmer merely walked up to it and placed his hat over it.[25]

Other Strigids

The Snowy Owl, *Nyctea scandiaca*, lives on the arctic tundra of North America and Eurasia. Its food staple is lemmings. During the years when lemming populations crash, this owl shows up, often in large numbers, during the winter in temperate North America. During irruptions, Snowy Owls have been seen in northern and central Utah, at the Malheur National Wildlife Refuge, and in west central Nevada.

The Spotted Owl, *Strix occidentalis*, ranges through the dense coniferous forests and wooded ravines from northwestern British Columbia southward to central Mexico. The Spotted Owl is not known to be a member of the breeding avifauna of the Great Basin. However, two summer specimens have been collected in the northwestern rim of the Great Basin—near Martis Peak, north of Lake Tahoe in Placer County, near the California-Nevada boundary.[228]

The Great Gray Owl, *Strix nebulosa*, inhabits the taiga or great circumboreal coniferous forests of Eurasia and North America. It extends southward in the United States in the mountains of the West. This owl is not known to be a member of the breeding avifauna of the Great Basin. There are several records of its occurrence during the winter in the northeastern corner of the Basin at Logan and the south end of Bear Lake in Utah. There is a summer record for Little Valley in the Carson Range on the western rim of the Basin.

12

Upland Game Birds

BIRDS BELONGING to the Orders Galliformes and Columbiformes have been heavily utilized by us as food, both by hunting as game birds and by raising under domestication. Just as the Orders Anseriformes and Charadriiformes furnish us with most of the game birds associated with water, the Orders Galliformes and Columbiformes provide us with most of the upland game birds. In the Great Basin, the galliforms are represented by the Family Phasianidae and the columbiforms by the Family Columbidae.

FAMILY PHASIANIDAE

Phasianidae is a large family of gallinaceous or fowllike birds present over much of the world. The pheasants, partridges, grouse, turkeys, and quail number among its rank. Phasianids possess short, stout, chickenlike bills and stout, often spurred legs. Their wings are strong, short, and rounded. Some species bear crests or wattles, and some such as the pheasants have beautiful plumage. Phasianids are mainly ground birds, although some roost in trees. They forage by scratching on the ground with their feet or by digging in the litter with their bills. Their call notes are a variety of cackles, crows, and clucks. Formerly, grouse were in their own family—Tetraonidae—and turkeys were in their own family—Meleagrididae. Then, in 1982, these two families were reduced to subfamily rank and placed in the Family Phasianidae.

Blue Grouse
Dendragapus obscurus

The Blue Grouse is a resident of western North America, ranging from the Yukon Territory of Canada southward into California and New Mexico. In the Great Basin it is mainly confined to fir and multineedled pine forests on the higher mountain ranges.[242] Its distribution is patchy. The major gaps occur in southeastern Oregon, in the northwestern quarter of Nevada, and over much of the southern end of the Basin.[243] This grouse is generally considered to be an uncommon bird in the Great Basin, but it may be locally abundant in some years.

At the onset of the breeding season in the spring, the male grouse establishes a hooting territory in which several display sites are located. There are a number of subspecies of Blue Grouse, three of which occur in the Great Basin: *sierrae* on the western edge, *oreinus* in the interior, and *obscurus* on the eastern edge.[244] The hooting displays of the subspecies vary somewhat, including the location of the display sites and the loudness of the hoots. The display sites of several subspecies such as *sierrae* are often on horizontal limbs far up in fir and pine trees, whereas those of other subspecies tend to be on the ground. Hooting displays consist of vocalizations, postures, and movements. While hooting the male stands with tail raised and spread, wings slightly drooped, and feathers on both sides of the neck parted to expose the oval gular air sacs, which are partially inflated. A series of up to five or seven owllike hoots is sounded, and this series is frequently repeated. As the displays progress, the male may move from one display site to another.

In addition to producing vocal sounds to advertise his presence on his territory, the male also produces wing sounds during flight displays. One type of flight display is called wing drumming, during which the male makes a short jump while vigorously beating his wings. He often makes an incomplete turn in the air before landing, because he is beating one wing more vigorously than the other.[243] Another type of flight display is a drumming flight, in which the male takes off on a short, circular flight, alighting near where he took off. The wing sound has been interpreted as a *burr-r-r-urrp*.[245] Hooting displays occur most frequently in the early morning and late afternoon; the sounds are among the most distinctive sounds of our montane springs. Listen for hooting when you are afield.

When his territorial displays attract another grouse, the displaying male assumes an erect posture, with his tail tilted toward the visitor; some vertical head jerking may ensue. Then he advances on the intruder, with his tail cocked and spread, his wings drooping so low as to drag on the ground. After several jerks he lowers his head and dashes forward with short, quick steps. He pulls up with a loud, deep *oop* or *whoot* note and resumes his normal posture.[243] If his visitor is a receptive female, he may display around her for several minutes before attempting copulation.

Following copulation, the male's family responsibilities end. Really, they never began. The female does all the rest by herself, without the security of a pair bond or any further contact with the male—the male merely functions to inseminate the female. Departing from her mate of the moment, the hen constructs a shallow scrape on the ground and commences egg laying. After an incubation period of about twenty-six days, the eggs hatch. The female now has a brood of six to nine or so precocial young to attend. The downy young can forage for themselves but need brooding and guidance from their mother. At first, they feed mainly on insects and other animal prey. They begin to fly within a week's time and are flying well by the end of the second week. By late summer the broods have broken up, and the juveniles are on their own.

Over much of their range, Blue Grouse perform a reverse altitudinal migration in the fall. They work their way higher upon the mountainsides to overwinter, instead of descending into the foothills or valleys like so many other montane species do. The winter economy of Blue Grouse is closely tied to the occurrence of true firs (*Abies*). Depending on the region of the country, the Douglas fir (*Pseudotsuga*) and certain species of pines (*Pinus*) also provide food and cover. The grouse feed on the needles of these trees and on buds, seeds, and twigs. Adult grouse feed almost entirely on plant food, of which two-thirds or so is conifer needles. At the end of winter, Blue Grouse move back down onto their breeding grounds, where the forests are less dense. Here dry, open spaces are available.

Ruffed Grouse
Bonasa umbellus

The Ruffed Grouse inhabits the northern part of North America. Within the Great Basin, only in Utah is it part of the native avifauna.[244] There is no hard evidence that it ever occurred naturally within Nevada before its 1963 introduction into northeastern Nevada by the state Fish and Game Commission. The Ruffed is an "uncommon permanent resident in mountains of northern and central Utah."[27] At one time it was apparently more common than it is today.

The territorial advertisements of male Ruffed Grouse are drumming displays. One or several logs are an integral part of a male's territory, since he uses a fallen log as a stage for his drumming. Drumming is a vigorous activity. The male holds his position by bracing with his tail and anchoring his claws in the log. With a flurry of strong, rapid wingbeats, he produces a dull drumming sound—caused by his wings beating against the air, not against anything solid.

Ruffed Grouse fluctuate greatly in numbers over the years. There is considerable evidence that on many parts of the grouse's range the fluctuations are cyclic in nature. The populations build up and then crash about every ten years.

Sage Grouse
Centrocercus urophasianus

In the past the Sage Grouse's range included about fourteen of our western states and extended northward to barely enter the three westernmost provinces of Canada. This grouse, which inhabits sagebrush shrublands or sagebrush-grass areas, is found in association with both tall and short species of sagebrush. The Sage Grouse is widely distributed, along with sagebrush, in the Great Basin. It normally does not occur in the pinyon-juniper woodlands or conifer forests. Nor does it occur in the valleys and lower foothills of the Basin where shadscale replaces sagebrush on the drier and more saline soils. There are two major shadscale regions in the Great Basin, which represent gaps in the distribution of the Sage Grouse. One is a tongue-shaped region extending in a northeast direction out of southwestern Nevada, through the Walker Lake depression and on into the Lahontan Basin of western Nevada. The other large shadscale region in the Basin occupies the lower valleys and foothills on the western side of Utah.[1] The Washo Indians' name for the Sage Grouse was See-yuh'.[25]

The West was settled to the detriment of the Sage Grouse. Ranches, farms, towns, cities, roads, and other human developments were centered mainly in the sagebrush zone inhabited by this grouse. Springs, seepage areas, small streams, and wet meadows were often severely damaged by livestock and human usage. Much of the sagebrush was on public lands, and federal programs to eradicate the "worthless" brush and improve the range for grazing by livestock were initiated. Hunting pressures mounted on the Sage Grouse as a game bird. There was even considerable wanton shooting of grouse outside of the hunting season. The Sage Grouse decreased in number and disappeared from parts of its former range; by the 1930s it was in serious difficulty over much of its range. Since then it has come back somewhat, as the abuse of public lands by livestock has been alleviated a bit.

This grouse is not a Sage Grouse in name only. Sagebrush gives shape and structure to the grouse's world; it shelters, protects, and provides sustenance and nesting cover for the bird. And the sagebrush world is an edible world. Most of the food of the grouse, practically all of it during the winter, consists of the leaves of sagebrush. Since the sagebrush is an evergreen plant, its leaves are there the year around as palatable and nutritious food for the grouse. So a grouse is seldom more than a neck stretch removed

Male Sage Grouse strutting

from a full stomach. In the late spring and during the summer, the Sage Grouse adds the leaves of legumes and other herbaceous plants and grasses to its menu. A limited amount of insect food is consumed during the warm part of the year. During a grasshopper irruption, the grouse may concentrate mainly on grasshoppers as a source of food. Early in life young grouse consume lots of insects.

The relationship between grouse and shrub is certainly a fundamental one: the bird derives not only name, shelter, cover, and sustenance from sagebrush but flavor as well. This was even noticeable back at a time when our palate was more attuned to wild game. While in the Great Basin in the 1860s Ridgway noted, "As an article of food the Sage-Hen cannot be recommended, unless the precaution is taken to flay it immediately, for its flesh soon becomes permeated with the disagreeable odor of the sage-brush, the leaves of which form its principal food. In fact, it is often found necessary to soak the carcase in salt-water over night before the flesh becomes palatable."[25] So the very flesh of the Sage Grouse has the essence of sagebrush. When hunting these grouse, experienced hunters aim at the smallest bird in the flock, not the largest. Young Sage Grouse, before they have started to feed heavily on sagebrush, can be quite tasty.

Whatever the food of Sage Grouse may be, it is sure to be soft. This grouse is not built to handle hard food such as seeds and nuts; it lacks the thickly muscular compartment of the stomach called the gizzard, which is necessary for grinding hard food. Most birds possess a two-part stomach. First in line is the membranous proventriculus or glandular stomach, which produces hydrochloric acid and protein-digesting enzymes. Next in line is a thick-walled, muscular gizzard or ventriculus. Birds often swallow small pieces of grit which lodge inside the gizzard and function as millstones in grinding hard food. Since the Sage Grouse lacks a functional gizzard, it is unable to grind up and digest hard food as so many of its close relatives do.

As winter wanes, male Sage Grouse assemble on their ancestral strutting grounds to reenact one of the most stirring and colorful natural history pageants in the Great Basin. Sage Grouse are lek birds or arena birds. Each year the males assemble on ancestral leks or arenas where they display communally. Each cock establishes a territory or court on the lek or arena on which he displays. The leks of Sage Grouse are called strutting grounds, since the principal display movement of the cock is reminiscent of strutting. Located in openings or clearings in sagebrush or in areas where the sagebrush is low and scattered, the territories or courts on a lek are used only for display and copulation, never for nesting or feeding purposes. Strutting grounds come in all sizes, from those with only a handful of cocks to those with four hundred or more cocks.

There are about eighty-five known species of lek birds in the world, including some grouse such as the Sage. Lek species are highly polygynous, and a male may copulate with a number of females. They show strong sexual dimorphism, with the sexes often strikingly different in size, plumage color, and ornamentation. The male has no family role to play—he merely displays on the lek and inseminates visiting, receptive females. The females are strictly on their own in nest building, incubating eggs, and raising young birds. The males' displays consist of postures, ritualized movements, and ritualistic combat; the displays are often noisy.

The social organization of male Sage Grouse on their lek is more complex than that of most lek species. In most species the organization of the lek is based solely on territorialism, and each male has a chance of being visited by females for copulatory purposes. However, the female Sage Grouse visiting a strutting ground congregate at one or several traditional locations called mating centers. Copulation is virtually restricted to these mating centers, so a male whose territory includes all or part of a center does all or much of the copulating. Such a male is called a master cock; only when the master cock is overwhelmed by females do the subcocks with territories next to the mating center mate. Hence, male Sage Grouse have a copulatory dominance hierarchy as well as territories. The location of the mating center or centers polarizes the copulatory worth of the territories on the strutting ground. Therefore the social organization of male Sage Grouse on a strutting ground has been referred to as polarized territoriality, in contrast to the simple territoriality shown by most species.[246]

Male Sage Grouse perform at the strutting grounds early in the morning, frequently again in the evening, and on some moonlit nights all night long. Their principal territorial activity is strutting. Before and after strutting the cock is in an upright strutting posture: with neck and head held high, yellow comb over each eye expanded, white neck feathers elevated, chest sac partly filled with air and sagging, wings slightly drooping, and tail feathers erected and fanned, so that the undertail region of the body is exposed and rimmed by the spikelike tail feathers.[246]

Strutting blends movement, color, and sound. The performance does not occupy much time or space. During a typical strutting bout, which lasts several seconds, the cock moves forward only a few steps. During a strut he twice raises and lowers his chest sac, which is an expansion of the esophagus. Before the sac is elevated, the cock moves his wings forward. He then draws them backward across the stiff feathers on the sides of his breast as the sac rises, producing a swishing noise. Raising and lowering the chest sac fills it with air, and two bare, yellowish skin patches on the bird's chest expand. Following the second lowering of the sac, the "air in the sac is released into the bare patches of skin, which suddenly balloon forward and then instantly collapse. As the bare skin patches distend and collapse, two sharp snaps are produced approximately 0.1 [second] apart. A peculiar resonant quality accompanies the snaps, and several brief, soft coos precede them. Thus, the entire acoustic output from the display is 'swish-swish-coo-oo-poink.'"[246]

Another type of behavior on the strutting ground is the facing-past encounter. These encounters occur along a boundary separating two territories. The two males involved take a position alongside of each other, shoulder to shoulder, facing in opposite directions. They maneuver forward and backward, as each tries to get into position to land a few blows with his wing. Harsh cries are sounded. Wing fighting may ensue, during which the males flail each other with their near-side wing. At times the birds remain motionless, side by side. They may even close their eyes as if sleeping. This behavior has been interpreted as being the out-of-context displacement sleeping.[243] All it means is that birds may perform irrelevant acts when they are thwarted in accomplishing something they are highly motivated to do or when they are in ambivalent situations where two antagonistic drives are simultaneously aroused—such as whether to fight or to flee.

Female behavior on the lek is quite simple. Hens begin arriving on the strutting ground before daybreak and generally leave before midmorning. Some return in late afternoon to attend the males' evening displays. Upon arrival hens may fly over the strutting ground and utter their "quacking call"—a "harsh, single call."[246] Landing at the edge of the strutting ground, the hens walk across it, to assemble at the mating center or centers. A female may visit the strutting ground for a number of days before copulation occurs. Once she is ready to mate, she signals her intention to the master cock by assuming the crouched solicitation posture. Following successful copulation, she bolts out from under the male, vigorously shaking her wings and tail in the so-called postcopulatory ruffling, before settling down to preen. Once inseminated, she no longer visits the strutting ground but strikes out by herself.

Sharp-tailed Grouse
Tympanuchus phasianellus

At one time the Sharp-tailed Grouse was distributed over North America from Alaska and Canada southward into western and midwestern United States. By now it has disappeared from much of the southern end of its range. Its range formerly extended across the northern end and down the eastern side of the Great Basin. Now it is confined to a few small areas in northeastern Nevada and extreme northern Utah; it no longer occurs in northeastern California, southeastern Oregon, northwestern Nevada, and central Utah.[244] Prior to 1982, this grouse's scientific name was *Pedioecetes phasianellus*.

Of all the subspecies of Sharp-tailed Grouse, the southern two have experienced the greatest reduction in range and numbers and are in serious trouble. Of the two, the western race, *columbianus*, which occurs in the Great Basin, has been the hardest hit. "It is gone from California"; occurs in one small area in New Mexico; is rare in Utah, Nevada, and Oregon; "rare or uncommon in Idaho and Washington"; and generally uncommon in the western parts of Colorado, Wyoming, and Montana.[247] Habitat loss associated with the advance of civilization is usually considered the reason for this bird's decline. Hunting pressure may also be a factor.

California Quail
Callipepla californica

The original range of the California Quail extended along the West Coast from southern Oregon into Baja California. Today the California is found inland, throughout the Great Basin. We do not know whether it was an original member of the Great Basin avifauna or not. The question of how far eastward the quail's original range extended is still unanswered, and it is likely to remain so. As far back as 1862, quail were introduced into the western end of the Basin.[248] Introductions erase old boundaries—even the quail's local distributions within its original range are uncertain. In 1944 Grinnell and Miller observed, "The distributional picture is now extraordinarily confused by reason of the planting of quail stocks in new places and transplanting them from place to place within original range."[26] Prior to 1982 the scientific name of this quail was *Lophortyx californicus*.

In reconstructing the past we must depend on the words of early observers. Unfortunately, early natural history reports are few and far between. During his fieldwork in the Great Basin from July 1867 to August 1869, Ridgway did not see a single California Quail; he wrote, "The 'Valley Quail' of California was met with only among the western foot-hills of the Sierra Nevada."[25] Although Ridgway was in the field around and about Carson City, he was apparently unaware of the early introductions of California Quail into this region and did not see any of the introduced birds. Shortly after Ridgway, Henshaw observed, "This quail is nowhere indigenous along the eastern slope, as the high mountains [of the Sierra Nevada] offer a complete barrier to its extension. Those introduced about Carson appear to just hold their own."[91] So Henshaw was aware of the introductions into west central Nevada but concurred with Ridgway that the California Quail was not native to the eastern slope of the Sierra Nevada and the western edge of the Great Basin.

Observations made in the Carson City–Reno area do not necessarily reflect the situation elsewhere along the western edge of the Basin. The subspecies *canfieldae* is a resident in Owens Valley, California, on the southeastern side of the Sierra Nevada; it may have extended northward into Nevada and the extreme southwestern edge of the Great Basin. Then, too, the subspecies *californicus* may have originally extended across northeastern California into northwestern Nevada, around the north end of the Sierra Nevada.[248] The California Quail may have originally ranged into the northwestern edge of the Great Basin in southeastern Oregon. It is believed to have been a native in Lake County, Oregon.[103]

In retrospect, it appears as if the original range of the California Quail may have barely entered the western edge of the Basin, at the northern and southern ends of the Sierra Nevada. These mountains were evidently an effective barrier in preventing the quail from moving directly eastward from California into Nevada. Nevertheless, even if this quail was an original member of the Great Basin avifauna, it was not a member of any real consequence.

Today, the California Quail is found across the breadth of the Great Basin. People have been highly motivated to establish new populations of this quail, since it is not only a game bird but also a very handsome and pleasing bird. The original introductions into west central Nevada, using California stock, were made by Lance Nightingale and Robert Lowery in the spring and summer of 1862. First twenty-two birds and then eighteen were released east of Carson City in the vicinity of what eventually became the town of Empire.[248] Subsequent introductions spread this quail into valleys throughout central and northern Nevada. The initial introduction of Californias into the eastern part of the Great Basin probably occurred in November of 1869, when General Gibbon released fourteen pairs into the Salt Lake Valley. "During the 1870s and 1880s several more introductions were made, mostly in the northern and central valleys of the state."[56]

California Quail are highly social birds—loafing, foraging, and roosting together in coveys during most of the year. Although the human hunter is their principal enemy, they thrive in many of our towns and cities. Here they frequent residential districts, parks, cemeteries, university campuses, and other places not completely abandoned to concrete and asphalt. During the colder part of the year, scores of coveys are present in Reno-Sparks. Everywhere, small to large coveys are seen, often afoot, running across city streets or filtering across lawns. The coveys repeatedly visit feeding stations where bird seeds are available.

When foraging or moving about in situations where it is difficult to preserve visual contact, the covey members continually utter their *ut-ut* call to maintain auditory contact. If a covey becomes scattered, the quail sound their loud assembly call of *cu-ca-cow* to direct reassembly. When a covey member is out of contact with the group for more than a few moments, it becomes uneasy. It may soon become panicky as it rushes about, loudly uttering the assembly call.

Quail maintain vigil against surprise attacks by predators. Their alarm call of *pit-pit* is frequently heard. In case of extreme danger, especially when under air attack by an approaching accipiter, the *kurr* call is sounded. To prevent surprise attacks, the covey relies on adult males performing sentinel duty. When you watch a covey foraging or loafing in dense cover or moving from one location to another, you will often see a male perched on a high vantage point surveying the scene; the vantage point can be a tree limb, post, or roof of a garage or house. During the hunting season, experienced hunters may try and locate well-hidden coveys by first finding the sentinel on duty, since he will be quite conspicuously perched above ground. Other hunters try to lure a covey in close by imitating the assembly call of *cu-ca-cow*.

The California Quail is a friend of ours—although quail do not eat many insects even during the summer, they consume enormous quantities of annual weed seeds. Their other major dietary item is leafy herbaceous plant material. However, although the California Quail is the state bird of California, the attention this bird has received has not always been of such a kindly nature. The quail is not as abundant in the state today as it once was. During the past century, hundreds of thousands of quail were shot and trapped yearly by market hunters. "Daily bags of 200, made by sporting men shooting the birds singly on the wing, were not unusual." The number of birds killed for shipment to large markets such as San Francisco was great. "In 1881 and 1882 over 32,000 dozen quail were shipped to San Francisco from Los Angeles and San Bernardino counties, and brought to the hunters engaged in the business one dollar a dozen. In those days restaurants charged thirty cents for quail-on-toast. By 1885 hunting had become unprofitable because of the reduction in the numbers of quail."[90]

Mountain Quail
Oreortyx pictus

The Mountain Quail is a resident of western United States. It ranges southward from southern Washington into Baja California and eastward into western Nevada and western Idaho. The Mountain Quail occurs in a number of ranges in the western end of the Great Basin in Nevada, California, and Oregon. It is also present in the Toiyabe Mountains of central Nevada, but it is unknown whether it is a native there or whether it was introduced from elsewhere.[91]

The Mountain Quail frequents dense vegetation on steep mountainsides. It has a strong aversion to open areas, and this aversion is apparent in the bird's behavior. Gutiérrez has written:

> I have recorded obvious excitement on the part of these quail when they are about to cross a modest opening in the forest canopy (15–30 meters) or chaparral . . . Their reactions can be characterized by alarm calls, alert position of the plume (vertical or forward), intention movements (bobbing of the head and/or starting and returning several times before crossing), running across a clearing, and using virtually all available cover in the form of shrubs or dead vegetation when crossing an opening.[249]

If Mountain Quail are set upon by a ground predator, they run to escape and only fly at the last moment to avoid capture. They are tremendous uphill runners. Although these quail are highly regarded as game birds, hunters rarely specialize in them. The combination of mountainous terrain, dense vegetation, uphill running, and reluctance to get up in the air makes the Mountain Quail difficult game. They are usually shot by hunters after other game as opportunity arises.

Mountain Quail are gregarious; outside of the nesting season, they group together in small coveys—generally numbering fewer than fifteen birds. When quail become separated from the covey, their assembly call is a loud *cle-cle-cle* or *kow-kow-kow*. In the fall the coveys perform altitudinal migrations, slowly drifting down out of the deep snow to overwinter. However, they are able to live in snowy habitats, since they can feed above the snow in trees and tall shrubs. Most of the migration route is traversed on foot, although the quail may fly across a canyon or other major obstacle. In the spring they return to higher country to breed. The unmated males select elevated sites as whistling posts and from these posts advertise for mates. The advertisement or location call of an unmated male is a whistle, variously described, including *plu-ark* or *que-ark*.

The Mountain Quail feeds mainly on plant food. Its major food item is seeds, followed by green vegetation such as leaves, buds, and flowers. Fruit and fungi are also taken. Acorns, when available, may become an important food item in the fall and winter, as may pine seeds. During the summer and fall, Mountain Quail may dig for underground bulbs.

Other Phasianids

A number of species of galliforms have been introduced into the Great Basin in an attempt to establish new populations of game birds. State fish and game departments have been very active in this respect. Some of the introductions have involved exotic species—species native to other regions of the world. Only four of the introduced species shall be considered.

The Gray Partridge, *Perdix perdix*, sometimes called the Hungarian Partridge, is a Eurasian species. It has been successfully introduced into many localities in southern Canada and northern United States. This species was initially introduced into Oregon in 1900, Utah in 1912, and Nevada around 1923. Some of the Oregon and Nevada introductions were successful, and the partridge spread into the extreme northeastern corner of California.[250] The introductions into Utah are not believed to have been successful; nevertheless, established populations of this partridge are now found in the western and northern counties of Utah. "These individuals are believed to have spread from the neighboring states of Idaho and Nevada where introductions have been successful."[56] Today, the Gray Partridge is established in scattered populations in the Great Basin.

The Chukar, *Alectoris chukar*, is an Asiatic species of partridge which was introduced into western United States. There are a number of subspecies of this partridge. Several subspecies were drawn upon for stock for introductions, most notably the Indian subspecies, *A. c. chukar*. Introductions into the sagebrush-grass areas of the Great Basin have been highly successful—the Chukar is now the chief upland game bird in the Basin. The initial introductions were made in Nevada in 1935, Utah in 1936, and Oregon in 1951. Nevada had its first hunting season for Chukar in 1947, Utah and Oregon in 1956. Nevada was the first state to have an open hunting season on the Chukar, followed by Washington in 1949.[251]

The Chukar found a home away from home in the Great Basin—the terrain, climate, and food were all to its liking. The numerous mountains and foothills provide rock-covered slopes up which the bird can run when disturbed and along which it can find shade and roosting sites. The hot, short summers and moderately cold, long winters are ideal, as is the aridity. Partridge need water in the summer, and the guzzlers built in the many waterless areas help provide this. A guzzler is an underground cistern from which summer evaporation losses are slight. Runoff water from rain and snow is collected over a sizable surface area and diverted into this cistern. An opening and a ramp leading underground allow the Chukar to reach the water in the cistern.

This partridge lives off the fat of the land, since cheatgrass, a fellow exotic, has taken over much of the Basin and bountifully provides seeds and leaves as food. In his monograph on the Chukar, Christensen wrote, "Where chukars are successfully established in North America it is obvious that exotic plants such as cheatgrass, red-stem filaree and Russian thistle, in seed and in leafage form, play an important part in the food requirements of the Chukar partridge. The widespread and year-around use of cheatgrass by the Chukar in North America unquestionably makes this plant the priority food species."[251] The Chukar derives its common name from its assembly or rally call. At low intensity the call resembles *chuck-chuck-chuck*. As the intensity increases the call becomes *per-chuck*, then *chukar*, and then, at highest intensity, *chuckara*.

The Ring-necked Pheasant, *Phasianus colchicus*, is an Asian species which has been introduced throughout the United States and southern Canada as a game bird. It has been successfully introduced into various areas in the Great Basin in Utah, Nevada, and Oregon. This pheasant thrives best in agricultural areas.

The Wild Turkey, *Meleagris gallopavo*, is not a native of the Great Basin. It was introduced into the Carson Range in the western rim of the Basin in 1963. Several introductions were made into the eastern end of the Basin in Utah, starting back in 1925. There are now several established populations in Utah, including one in the Beaver Mountains of Beaver County.[56]

FAMILY COLUMBIDAE

The pigeons and doves of the Family Columbidae are distributed over much of the world, but the greatest diversity of species is found in the Australasian and Oriental regions. Columbids are compactly built birds with relatively small heads on short necks. Their feathers are quite loosely anchored and are easily shed. The external nostrils open through a fleshy cere located at the base of the bill. Columbids are unique in being able to drink by lowering their bills into water and sucking it up. Young columbids are fed pigeon milk secreted by their parents' crops.

Rock Dove
Columba livia

The Rock Dove is a native not of North America but of Eurasia and North Africa. Thousands of years ago it was domesticated, and several hundred different kinds of domestic pigeons were developed from the Rock Dove by selective breeding. In time humans spread the Rock Dove over much of the world. Feral or free-living populations of Rock Doves became established in cities and agricultural areas by pigeons escaping from dovecotes. Today, the Rock Dove is a familiar figure in the Great Basin. The cities and towns of the Basin have their resident populations, as do most ranches and farms. There are also sites where Rock Doves, often nesting in fissures and crevices on cliffsides, live away from the presence of humans.

It is impossible to reconstruct the history of the spread of the Rock Dove over North America and into the Great Basin. During times of settlement domestic pigeons were transported about unnoticed. Over the years ornithologists have mainly ignored this dove. Ridgway does not mention seeing a single Rock Dove, caged or feral, in the Great Basin in the 1860s. It was not until the seventy-fourth Christmas Bird Count in the winter of 1973 that the Audubon Society allowed Rock Doves to be counted and recorded. Today, the Rock Dove is appreciated not only by pigeon fanciers and pigeon racers but also by some experimental scientists as ideal subjects for certain types of research.

Rock Doves were being domesticated in the Mediterranean-Mesopotamian world as far back as Neolithic times. Not only were domestic pigeons used as food, but they also figured prominently in religious ceremonies and divinations. Early peoples soon recognized the extraordinary homing abilities of translocated pigeons and promptly put them to work. The ancient Egyptians, Greeks, and Romans used homing pigeons to carry back messages proclaiming the results of chariot races. The armies of the Egyptians, Persians, Phoenicians, and Assyrians carried pigeons with them on military campaigns to send home messages. Homing pigeons have even figured prominently in more modern military campaigns, playing an important role in the siege of Paris of 1870 to 1871 and in both World War I and II. There are two monuments in Europe, one in Brussels and one in Lille, commemorating the patriotic contributions of the homing pigeon. Thus, the pigeon is one up on the California Gull with its single monument in Salt Lake City.

We have long pondered the migratory passage of birds and their uncanny abilities to navigate unerringly over vast distances. In recent years the

Rock Dove has become the principal subject in the experimental investigation of how birds practice orientation and navigation. There are a number of compelling reasons for their popularity as research subjects. Pigeons take readily to captivity and tolerate handling and experimental manipulation. They have highly developed capacities for orientation and navigation which have been enhanced by many years of selective breeding—they show an almost constant tendency to home, under a variety of conditions, the year around.

Both directional orientation and goal-directed orientation have been studied using pigeons as research subjects. In directional orientation a bird selects and holds a given compass direction or azimuth. Goal-directed orientation or navigation involves much more complex behavior. The bird must be able to determine where it is in relation to where it wants to go and then get there—such as pigeons do when homing and many birds do during migration. In some species young birds perform their first migration by themselves and navigate without any guidance from experienced adult birds.

Experiments have shown that pigeons have redundancy in their guidance system and have two mechanisms by which they can carry out directional orientation: they can use the sun compass, or they can use a magnetic compass. The sun compass employs the apparent movement of the sun along an arc across the sky—the ecliptic—during the day. At 6:00 A.M. local standard time the sun is due east; then, moving at an apparent speed of fifteen degrees per hour along the ecliptic, it is due south at noon and due west at 6:00 P.M. Thus, if you have an accurate clock set on local time, you can use the sun as a compass. Pigeons and many other birds possess accurate internal clocks and use the sun compass in directional orientation. Much of the sun compass research has involved experimentally resetting the internal clocks of pigeons ahead of or behind local time; this is done by keeping the birds in a lightproof building on an artificial schedule of day and night. If the pigeons' internal clocks are set six hours ahead of local time, when shown the sun again and required to use the sun compass they will make predictable errors. For example, they will mistake east for south when shown the sun at 6:00 A.M. and mistake south for west at noon.

The sun compass of pigeons has also been studied in large, circular orientation cages. Here the pigeons are taught to locate a dish containing food, when surrounded by empty dishes, by using the sun compass while viewing the sun's position through open windows. Then, by closing the window on the sun's side, the apparent position of the sun is altered, by reflecting it in mirrors through an open window on another side. Now when attempting to locate the dish containing food, the pigeons make predictable errors based on the altered apparent position of the sun.

On overcast days or when its capacity to see the sun is eliminated by putting frosted contact lenses on the bird, the pigeon practices directional orientation by using geomagnetism. Experiments have shown that birds use the inclination of the earth's magnetic field in orienting, not magnetic north. In using a compass, we rely on a pivoting or floating magnetic needle to point toward the magnetic north pole. But pigeons and other birds which have been studied do not orient on the magnetic north pole. The magnetic lines of force are parallel to the earth at the equator but progressively tilt downward at higher latitudes, until at the north pole the lines of force are perpendicular to the earth. The pigeon's magnetic compass somehow senses this downward inclination in the magnetic field, and the brain interprets it as north. Predictable errors are made by pigeons when they are outfitted with frosted contact lenses with battery-operated coils which reverse the tilt or polarity of the magnetic field around their heads: they now mistake south for north. As yet we are not sure where the magnetic receptor or magnetic compass mechanism is located in birds. However, studies have shown that homing pigeons have some permanently magnetized material in innervated tissue between the brain and the skull and in the neck muscles.

Research has not yet revealed much about how pigeons and other birds accomplish goal-directed orientation or navigation. We still do not know how a translocated homing pigeon knows where its home is or how a migrant determines the spatial relationship between its present location and its natal home site on the breeding grounds. We navigate with delicate instruments to measure latitude and longitude and with maps and charts to give meaning to our measurements. But do birds determine latitude and longitude? Do they have maps? If they do have maps, where are their maps located in the brain, and how do they determine where they are and where home is located on the map?

At one time G. V. T. Matthews proposed that homing pigeons did not require maps and that they navigated by observing the movement of the sun along the ecliptic. They would compute their north-south displacement, if any, by comparing the noon height of the sun at the release site with their memory of its noon height back home. If the noon sun was higher in the sky at the release site than back home, the release site was south of home. If the noon sun was lower in the sky at the release site than back home, the release site was north of home. The homing pigeon could then determine its east-west displacement, if any, by relying on its internal clock, which had been set on home time. If the sun at the release site was too far west along the ecliptic for the time of day recorded by the pigeon's internal clock, the release site was east of home. If the sun at the release site was not far enough west along the ecliptic for the time of day recorded by the pigeon's internal clock, the release site was west of home.[126] Matthews' sun-arc hypothesis was tested on homing pigeons and concluded to be invalid, since pigeons with artificially shifted internal clocks could home just as accurately as pigeons with accurate internal clocks. This showed that navigation did not depend on comparing the position of the sun in the sky at a precise time of day with where it would be at that time back home. With the sun-arc hypothesis invalidated, we are left, at the moment, with only the map hypothesis. But where are the maps, and how do the birds use them?

Band-tailed Pigeon
Columba fasciata

The breeding range of the Band-tailed Pigeon extends from southwestern British Columbia, through western United States, into Mexico and Central America. The Bandtail is not known to nest in the Great Basin, but in several places its known breeding range closely approaches the western and eastern edges of the Basin. There is Band-tailed Pigeon traffic within the Great Basin, however. Not only are the pigeons in the northern part of their breeding range migratory but, following nesting, flocks wander in search of food. These movements are most pronounced when the acorn crop is poor.

Usually, by late summer, small flocks of Band-tailed Pigeons can be seen in the Carson Range on the western rim of the Great Basin. Some years Bandtails are present by the hundreds in the Carson Range during late winter. The largest recorded winter invasion occurred in early 1972; flocks totaling well over a thousand birds roosted at several sites in the Carson Range, including forested areas to the west of Washoe Valley and along the southeast shore of Lake Tahoe. At their roosts in conifer forests, the ground was littered with feathers and droppings. During the day flocks could be seen flying to the east and southeast to forage in the pinyon-juniper woodlands in the Virginia and Pine Nut ranges—the pigeons which the late Pete Herlan and I collected as museum specimens had their crops loaded with pinyon pine seeds. Another invasion occurred in the Carson Range in early 1978.

Mourning Dove
Zenaida macroura

The breeding range of the Mourning Dove extends from southern Canada through Mexico. This dove is a common, often abundant breeding species and migrant in the Great Basin. A few overwinter here, particularly during mild, open winters. Prior to 1973 its scientific name was *Zenaidura macroura*. The Paiute name for the Mourning Dove was We-ho'-pe, the Washo name Hung'-o-ho'-ah.[25]

The Mourning Dove finds the Great Basin much to its liking. Ridgway noted this in the 1860s, when he wrote:

> Perhaps no bird, not even the Raven, is more universally distributed through the Interior, without regard to the nature of the country, than the common Mourning Dove, and certainly none is more abundant. It occurred about the corrals of the stage-stations in the midst of the most extensive deserts, many miles from any cultivated or wooded district, or natural water-courses, while it was also met with on the equally barren mountains and plains far from the abode of man. In the arid portions of the country, however, it is far less common than in the fertile localities, where it sometimes literally abounds.[25]

The Mourning Dove adapts well to desert situations and often nests miles away from any source of free surface water. Practically all desert birds who live and nest away from surface water feed on either insects or on meat—both of which contain much water of succulence. However, the Mourning Dove feeds almost entirely on plant seeds, and plant seeds are very dry food. Thus, in order to remain in water balance, doves must have drinking water. They are swift, powerful fliers and can quickly fly many miles during the early morning and evening to water holes to drink. Most seedeaters feed their young on insects. Mourning Doves are an exception, and even the nestlings are fed seeds. But the young are also fed pigeon milk. Produced within the crop of the adult, pigeon milk is a highly succulent food, since its principal component is water.

The drinking behavior of the Mourning Dove at a water hole is quite unusual for a bird. Most birds can drink only by taking a mouthful of water and then raising their heads in order to swallow it. However, pigeons and doves and a few other birds do not have to raise their heads in order to swallow water. They merely insert their bills into the water and suck it up in a continuous draft. This greatly accelerates drinking and reduces vulnerability to predators who also frequent water holes.

Experimental studies have shown that caged Mourning Doves kept at 39° C without any drinking water lost on the average 11.6 percent of their original body weight in twenty-four hours; dehydrated Mourning Doves drank enough water in one minute to equal 10.8 ± 2.7 percent of their body weight.[252] Therefore, within a minute or so, a dehydrated dove can take in enough water to rehydrate. When watching drinking doves, you will note that often they are satisfied after one continuous draft of water. Sometimes, they will take a deep breath and then take a second, occasionally a third, shorter draft of water.

The voice which gives this dove its name is a frequent sound in late spring and early summer. The mournful cooing consists of a soft, two-syllable coo followed by two or three loud coos—*cŏo-ŏo,* ŌŌ, ŌŌŌ or *cŏo-ōō,* ŌŌ, ŌŌ, ŌŌ. On occasions a fourth coo is sounded.[253]

Perch cooing is used as an advertising display by males. So unmated males coo much more frequently and for longer periods in the morning than do mated males, particularly mated males involved in nesting. Sometimes females will engage in weak advertising cooing. The flapping-gliding flight is another advertising display. The male initiates this by launching himself from a cooing perch on strong, noisy wingbeats—the wingbeats are so exaggerated that the tips of the two wings may meet beneath the body. The male climbs one hundred feet or so into the air and then begins a long, spiraling downward glide. While gliding the wings are extended and held in a plane lower than that of the body. The displaying male may engage in a sequence of flaps and glides before he finally alights either at his original cooing perch or at another cooing perch. Flapping-gliding flights are associated with high rates of perch cooing. Once the male is pair-bonded, these display flights cease or are greatly curtailed.[254]

Courtship and pair formation involve two displays—the charge and the bow-coo. The male charges up to a female with his body held high and horizontally, his head directed straight forward, and his tail straight backward. He may use a leap or a number of leaps in charging. When close to the female, he bows very low. The bow may be repeated several times, before the male assumes an erect position and directs a coo at the female.[254]

The male takes the initiative in selecting a nesting site and getting nest building under way. After locating a suitable site, the male will call to his mate. He may have to call from several sites and for several mornings before she will respond to his nest calls of *cōō-ōō.* If she accepts the selected site, the pair will sit together, preen each other on the head and neck, and utter soft nest calls. This preening establishes the pair's nest site attachment, and copulation may follow. In the Great Basin, Mourning Doves nest not only in trees but also in shrubs and often on the ground beneath a shrub or on the bank of a wash or erosion gully. In building their stick nest, the male brings material to the female. He usually stands on her back when presenting a twig but occasionally will place the stick himself. Sometimes he may replace the female on the nest and rearrange the nesting material. Much sitting together, preening, and soft nest calls may interrupt the actual construction work.

Although the nest of the Mourning Dove is a flimsy one, nest building is a drawn-out process. The work is restricted mainly to the early morning, and the male carries only one stick at a time to the nest site. If he happens to drop the stick along the way, he doesn't stop to retrieve it but continues on to the nest empty-billed, before starting out again for another stick. Once the nest is completed, the eggs are laid, almost invariably two in number. Both parents incubate the eggs, the male sitting by day and the female from evening until the next morning. Neither parent develops a brood patch for the incubation period of about fourteen days.

The young doves are altricial, being both naked and helpless. For the first ten days or so, the nestlings are brooded—by the male during the day and by the female from evening until the next morning. For the first four or five days of life, the nestlings are fed nothing but pigeon milk. Pigeon milk is produced by both parents. Under the influence of the hormone prolactin, produced by the pituitary gland, the lining of the crop of an incubating dove becomes greatly enlarged and honeycombed in appearance. The lining cells then begin to slough off. Each cell contains pigeon milk. This is about 65 to 81 percent water, lacks any carbohydrates, but is rich in proteins (13.3 to 18.6 percent) and fats (6.9 to 12.7 percent).[21] Nestlings put their bills inside that of a parent, and the parent regurgitates the pigeon milk. Sometimes a few seeds are mixed in with the milk. On about the fourth day of life, nestlings begin to receive increasing amounts of plant seeds along with the pigeon milk, but they are not fed insects or animal food. By the time they are adults, over 98 percent of their food will consist of seeds.

Other Columbids

The White-winged Dove, *Zenaida asiatica*, is a more southern species; it is of accidental occurrence in the Great Basin. There are a number of sight records, and several specimens have been collected in the eastern part of the Basin in Utah.[56] The only record I am aware of for the western part is for Lahontan Valley, Nevada.[255]

The Passenger Pigeon, *Ectopistes migratorius*, once the most abundant bird in North America, has long been extinct. It formerly inhabited eastern North America. On September 10, 1867, Ridgway collected a juvenile female Passenger Pigeon in the West Humboldt Mountains of western Nevada. Of this he wrote, "Only a stray individual of this species was met by us, and it cannot be considered as more than an occasional straggler in the country west of the Rocky Mountains. The specimen obtained flew rapidly past one morning, and alighted a short distance from us, upon a stick by the edge of a stream, whither it had probably come for water. Upon dissection it was found to have been feeding upon the berries of a small cornel (*Cornus pubescens*), which grew abundantly in the mountains."[25]

The breeding range of the Inca Dove, *Columbina inca*, lies far to the south of the Great Basin. There is a sight record from the southeast edge of the Basin at Parowan, Utah, for this species.[27] Prior to 1982 its scientific name was *Scardafella inca*.

The Common Ground-Dove, *Columbina passerina*, breeds far to the south of the Great Basin. There is a sight record for the southeast rim of the Basin at Cedar City, Utah.[27] Prior to 1982, this species was known as the Ground Dove.

13
Cuckoos

THE CUCKOO FAMILY—Cuculidae—belongs in the Order Cuculiformes. This order contains not only cuckoos but also the African touracos of the Family Musophagidae. Cuculiforms are characterized by having two toes on each foot pointing forward and two pointing backward. The Family Cuculidae is represented by species in the Great Basin.

The Family Cuculidae is a worldwide family. About fifty species of cuckoos are known to practice obligate brood parasitism. Brood parasites do not build nests or take care of eggs or young; they lay their eggs in the nests of other species and abandon them to their subsequent fate with foster parents. Often the foster parents rear their uninvited guests to the detriment of their own eggs and young. The best known of all obligate brood parasites is the Common Cuckoo, *Cuculus canorus*, of Europe. None of the North American cuckoos are obligate brood parasites.

FAMILY CUCULIDAE

Yellow-billed Cuckoo
Coccyzus americanus

The breeding range of the Yellow-billed Cuckoo extends over much of the United States. However, there are large areas in the Far West from which it is missing or where it is extremely rare. The Yellow-billed Cuckoo is an occasional to uncommon summer resident at a few localities in the Great Basin. Its status has changed for the worst since Ridgway's days here. Of the 1860s Ridgway wrote, "The Yellow-billed Cuckoo was so often seen or heard during our sojourn in the West, that we cannot regard it as a particularly rare bird in certain portions of that country."[25] The Yellowbill has also decreased drastically west of the Great Basin. It has been officially classified by the California Fish and Game Commission as a rare species under that state's Endangered Species Act.

The rarity of the Yellow-billed Cuckoo in the Great Basin is related to the destruction and alteration of vegetation along the larger streams—the main habitat of the cuckoo was in this dense vegetation. With settlement the riparian vegetation has been badly abused by cattle, clearing, burning, stream channelization, and flood control projects. California has initiated a recovery program for the cuckoo which includes the protection and restoration of riparian vegetation and the acquisition of suitable sites to serve as natural reserves for the bird.

Although the Yellow-billed Cuckoo is not an obligate brood parasite, its nesting habits are erratic and flexible enough to suggest that it may be an incipient brood parasite. Sometimes it will lay eggs in the nests of other species. Yellow-billed Cuckoo eggs have been found in the nests of its close relative the Black-billed Cuckoo and also in the nests of the American Robin, Black-throated Sparrow, Mourning Dove, House Finch, Red-winged Blackbird, and others.

The Yellowbill typically arrives late at the breeding grounds and often doesn't start nesting until late June or early July. Egg laying may not peak until mid July. This species' nests are poorly built, being small, flat, shallow, flimsy structures of twigs, vines, and rootlets. The eggs are relatively large compared to the body size of the female, and both embryo and nestling show rapid rates of development.[256] Incubation requires only ten to eleven days, and the young are fledged after seven to eight days. Both parents incubate, and incubation commences with the first egg. Since egg laying is spread out over a number of days, eggs are often present in the same nest with nestlings. In a California study involving four nests, the average spacing between consecutively laid eggs was 2.2 days, with a range of 1 to 4 days. On occasions a freshly laid egg has been found in a nest containing young.[257]

Some observers have concluded that a pair of Yellow-billed Cuckoos have considerable difficulty finding sufficient food to simultaneously support their own needs and the high energy demands of a growing brood of nestlings. Yellowbills have a fondness for hairy or spiny caterpillars, which are often locally available for only short periods of time. The late and often flexible nesting schedule takes advantage of peaks of abundance in the local food supply. By spreading out the egg laying within a clutch and incubating with the first egg, the adults spread out the age of their nestlings. Consequently, the young birds pass through their most energy-demanding period of growth at different times.

There is some evidence that, when parents are having difficulty procuring enough food for their entire brood, they may abandon the smallest nestling. In a nest containing three young, a noticeably smaller nestling was getting little food. After persistent begging by this nestling, one of the parents pecked at it lightly and then carried it away from the nest in its bill.[258] In another nest of three, the runt of the brood, getting little food, begged vigorously. Once again, one of the parents picked up the hapless nestling and flew away with it, only to return in a few minutes without it.[257]

The feeding activities of the Yellow-billed Cuckoo are highly beneficial to us, since they consume so many highly injurious insect pests. They are unique among birds in their ability to consume toxic hairy and spiny caterpillars—they feed on cankerworms, tent caterpillars, fall webworms, and the caterpillars of the io and gypsy moths. Sometimes a Yellowbill will work a caterpillar back and forth several times through its bill, shearing off most of the hairs or spines before swallowing its prey. At other times it does not prepare its food very well. Often the stomach of a cuckoo will be lined with a felt mass of caterpillar hairs and its intestines will be pierced by caterpillar spines.

The voice of the Yellow-billed Cuckoo is distinctive but much like that of the Black-billed Cuckoo. Its full song has been described as beginning fast and then becoming retarded and slurred at the end, as in *kakakakakakaka ka ka ka ka ka ka kow kow kow kow kow kow kow kow*.

Other Cuculids

The Black-billed Cuckoo, *Coccyzus erythropthalmus*, breeds to the east of the Great Basin in southern Canada and eastern United States. The Blackbill is of accidental occurrence along the northeastern edge of the Basin. A specimen was collected at Bountiful, Utah.[27] There are sight records from Salt Lake City, Brigham City, and Logan.

The breeding range of the Greater Roadrunner, *Geococcyx californianus*, lies to the south of the Great Basin. Roadrunners are common residents in the Mohave Desert, which abuts the southern edge of the Great Basin Desert. They are uncommon permanent residents as far north as the Cedar City–Parowan area.[27] A specimen in the University of Nevada Museum of Biology was collected at Goldfield. Deep penetrations into the Great Basin Desert by roadrunners are accidental in nature. In 1932 a decomposed carcass of a roadrunner was found in the foothills east of Provo.[56] Prior to 1982, the common name of this species was the Roadrunner.

14

Hibernators

TWO CLOSELY related orders of birds in the Great Basin contain some species which hibernate in the winter or, at least, experience nocturnal cold torpidity. The Order Caprimulgiformes contains the Family Caprimulgidae, to which the nighthawks and poorwills belong. The Order Apodiformes contains the Family Apodidae, to which the swifts belong, and the Family Trochilidae, to which the hummingbirds belong.

FAMILY CAPRIMULGIDAE

Nighthawks, poorwills, and other goatsuckers belong to the Family Caprimulgidae. The goatsucker family is distributed over much of the world. Caprimulgids are mainly crepuscular or nocturnal birds. Their soft plumages are cryptically colored in grays and browns—which help conceal the birds by day as they sleep. Contrary to legend, goatsuckers do not suck milk from goats but, rather, scoop insects out of the air while in flight. Although they possess small bills, their gapes are commodious, and their mouths are surrounded by rictal bristles. Caprimulgids bear comblike claws on their middle toes. At rest they perch lengthwise, instead of crosswise, on limbs.

Lesser Nighthawk
Chordeiles acutipennis

The Lesser Nighthawk is of accidental occurrence in the Great Basin. Its breeding range extends from southwestern United States into South America. The Lesser is a member of the breeding avifauna of the Mohave Desert, due south of the Great Basin Desert. On occasions, Lesser Nighthawks wander northward into the Great Basin. In the eastern end of the Basin, there are sight records for the Bear River Refuge, Vernon, and Elko County.

In the western end Jack and Ella Knoll saw six Lessers skimming over the vegetation, hawking insects in the Fernley marshes, on a cloudy morning in early July of 1982.

Common Nighthawk
Chordeiles minor

The breeding range of the Common Nighthawk encompasses much of North America. In the Great Basin this nighthawk is a common summer resident and breeding bird and a common, often abundant migrant. The Common Nighthawk is a familiar bird throughout its range. With the settlement of North America it took to nesting in cities—even in the center of large cities. Here it substitutes the gravel-covered, flat roofs of buildings for the sparsely vegetated, open ground sites it uses for nesting in the wild.

Although the Common Nighthawk is a caprimulgid, cold torpidity and hibernation do not appear to play the same role in its life as they do for its two relatives, the Common Poorwill and Lesser Nighthawk. Common Nighthawks can be forced into torpidity with difficulty through food deprivation and low temperatures, but they often fail to arouse from torpidity.[259] This indicates that cold torpidity is probably not a normal event for the Common Nighthawk. The Common Nighthawk is highly migratory and is one of the very latest, if not the latest, migrants to return in the spring. In the Great Basin it arrives in late May or early June, thus getting minimal exposure to our fickle spring weather.

The evening and early morning courtship displays of the male Common Nighthawk take place in the sky for all to witness. The male flies about over his territory, uttering his sharp *peent* call and booming. Sometimes he soars or even hovers as he circles about, only to plunge earthward in a steep swoop. If the female is on the ground below, the male may plunge to within a few feet of the ground, before pulling out of his swoop by turning sharply upward a moment before collision. At the bottom of his swoop the male turns his wings downward, and the loud boom is heard—this is believed to be produced by vibrations of the flight feathers, probably the primaries. Males continue to boom all during the nesting season, but booming does taper off as the season progresses. At times the male may alight on the ground and display before the female with spread tail and fluffed-out throat feathers. He wags and rocks his body, while uttering guttural croaking calls.[260]

In the wild, the two eggs of the Common Nighthawk are laid on the ground without the benefit of a nest. The female does all or most of the incubating; the incubation period is nineteen days long. After hatching, the young are fed regurgitated insects. The male roosts near the incubating female, in a tree if available or on the ground. When perched or roosting on a limb a nighthawk assumes a lengthwise position, seldom perching crosswise as do most birds. Cock roosts have been reported from southern Idaho—fifty to one hundred nighthawks congregate to roost in close proximity to each other in trees during the nesting season. It appears as if mated males, as well as unmated males and even a few unmated females, are present in a cock roost.[261]

Although nighthawks will hawk insects during the day, especially on cloudy days, their activity is generally crepuscular—a caprimulgid characteristic. In the evening males start calling from their roosts; they then take off and, perhaps, dive several times over their nesting sites, before flying on to feeding areas. Here they assemble, coming in from all directions, to hawk insects together—they capture flying insects by chasing them and using their large caprimulgid mouths to scoop them out of the air. After the male feeding assemblies have formed, incubating females leave their eggs and join the males to feed.

The name nighthawk is a misnomer, since these birds have more of a crepuscular than a nocturnal activity pattern and are often awing during the day. Neither are they hawks, nor do they look like hawks. The old vernacular name of bullbat was a better one, since the birds fly around erratically, hawking insects in the evening, like bats. Their booming does sound somewhat like the bellowing of a bull. Among the Indians of the Great Basin, the Washo called the nighthawk Kow'a-look, and the Shoshones called it Wy'-e-up-ah'-oh.[25]

Common Poorwill
Phalaenoptilus nuttallii

The breeding range of the Common Poorwill extends from southeastern British Columbia, through western United States, into Mexico. It is an uncommon to common summer resident at various locales in the Great Basin, but there are no winter records of occurrence here. As yet, no poorwill has been found in winter hibernation in the Basin. The winters are harsher here than at the sites in California and Arizona where it has been found in winter hibernation. Prior to 1982 this bird's common name was Poor-will.

The Common Poorwill will always occupy a unique position among birds, since it has been detected spending an entire winter in hibernation in the field. Swifts, hummingbirds, colies, and a few other birds are known to enter cold torpidity for briefer periods of time. Most of the studies on cold torpidity and hibernation have been done on captive birds.

The first reported discovery of a torpid poorwill goes way back to 1879, when on January 22 workmen found a Common Poorwill in cold torpidity in dense chaparral at a site east of Stockton, California. Subsequently, several others were found in cold torpidity here and elsewhere in California. Then, on December 29, 1946, Edmund Jaeger found a torpid Common Poorwill in a narrow, slotlike canyon in the Chuckawalla Mountains of southern California. When the poorwill returned to the same crevice the following winter, Jaeger was able to follow it through hibernation from November 26 to February 22. It even returned on November 24, 1948, to hibernate for the third winter, but it disappeared soon afterward. The hibernaculum or hibernating chamber occupied by Jaeger's poorwill was a shallow, vertical crevice in a rock, located about two and a half feet above the ground. The back of the poorwill was exposed, flush with the surface of the rock. During the second winter, Jaeger measured the body temperature of the hibernating poorwill on five different occasions; it ranged from 18 to 19.8° C.[262]

Jaeger's fascinating discovery has prompted various investigators to carry out physiological experiments on captive Common Poorwills. Not only have their responses to low ambient temperatures been studied but also their responses to heat stress—the poorwill nests and roosts on the ground during the summer, often in full sunlight, even in desert areas. Captive poorwills have been induced to enter cold torpidity by exposure to low ambient temperatures and food deprivation. At a body temperature of 10° C, the metabolic rate of a torpid Common Poorwill will slow down to about one-tenth of its normal resting value. This demonstrates the survival value of hibernation. During the winter when normal insect food is not available, one gram of body fat will support a hibernating poorwill for about ten days, whereas the same amount would support it only for one day at normal body temperatures. Ten grams of stored body fat would then support the hibernating poorwill for one hundred days or more or for the entire winter season.[263]

It has also been shown experimentally that the poorwill is able to enter cold torpidity not just in the winter season but in the spring as well. Poorwills captured in southeastern Idaho just after returning from migration were able to enter cold torpidity.[264] This ability could have survival value during an unseasonably cold or wet northern spring, when insects are difficult to obtain.

Another interesting facet of cold torpidity came to light during the experiments of Austin and Bradley. They discovered that torpid Common Poorwills are able to attain "good flight" at body temperatures between 27.4 and 30.8° C, the lowest reported flight temperatures for a bird. The poorwill was able to launch itself into the air from the ground at a body temperature as low as 28.3° C.[265] Being able to initiate flight at low body temperatures may enable the Common Poorwill to better exploit brief warm spells during the winter by searching for insects at twilight. A poorwill exposed to the sun and air, as Jaeger's bird was in its hibernaculum, may rewarm enough during a sunny winter's day to reach minimal flight temperatures.

Several physiological features of the Common Poorwill better adapt it for survival under heat stress and exposure to direct sunlight on the ground during the day. Studies have shown that the Common Poorwill has an extremely low standard metabolic rate for a forty- to sixty-gram bird. At rest in the dark at an ambient temperature of 35° C, its rate of oxygen consumption averaged only 46 percent of the rate expected for a typical fifty-gram nonpasserine bird. So at rest the poorwill generates much less metabolic heat to be dissipated in warm to hot environments than do most birds of its size. Furthermore, in an extremely hot environment, at an ambient temperature of 47° C, its metabolic rate increased only by 0.81 milliliter of oxygen per gram of body weight per hour—a very modest increase.[266]

In conjunction with its low production of metabolic heat when at rest, the Common Poorwill is very effective at dissipating heat by gular fluttering when under heat stress. As the ambient temperature increased from 35 to 47° C, the poorwill's rate of evaporative water loss increased from 2.9 to 23 milligrams per gram of body weight per hour. Since the poorwill has an extra-large mouth and gape, it can expose an unusual amount of moist tissue when gular fluttering. It has been estimated that, with a maximal effort at gular fluttering, a poorwill should be able to dissipate metabolic heat at over five times the rate generated when at rest at an ambient temperature of 47° C.[266]

The interesting pre-1982 name of Poor-will evokes thoughts of the meaning in a bird's name. Poor-will was an echoic name, since the oft repeated, seemingly endless calling of this bird sounds like *poor-will poor-will———poor-will*. However, if you are close to the calling bird, you may hear a faint third syllable as *poor-will-low* or *poor-will-ee*. As for the scientific name, *Phalaenoptilus* is derived from the Greek for moth-feathered and refers to the "powdery, velvety plumage" of the poorwill.[267] It is fascinating to note that among the North American Indians the Hopi name for the poorwill was Hölchko, the Sleeping One, suggesting that the Hopi may have greatly predated Jaeger in their field observations.[262] The Paiute name for the Common Poorwill was Koo-ta-gueh', and the Shoshone name was Toet-sa-gueh'.[25]

The Common Poorwill feeds on night-flying insects. Most active around twilight, the poorwill announces the beginning of its evening activities by calling before starting to hunt. If moths, beetles, and other insects are abundant, the bird usually captures a bellyful in about thirty minutes, often before full darkness sets in. A poorwill may forage from the ground by jumping up or flying up after passing insects; it also hunts on the wing, cutting here and there, in silent, mothlike flight, close to the ground. The poorwill has a tremendous gape and nets insects with its open mouth. In a study carried out in southern California, poorwills took an average of three to seventeen minutes after beginning their twilight hunting to capture twenty to sixty-one large beetles and noctuid moths. After they had filled their stomachs, they retreated to their hillside roosts to remain until the dawn hunting period. Once its stomach is filled, a poorwill is reluctant to move about, although it may call off and on during the night. On evenings following rainy periods, because of the scarcity of moths, it took the poorwills over three hours to fill their stomachs.[268]

The Common Poorwill nests on the ground; it makes little or no pretense of building an actual nest, beyond sometimes forming a shallow depression. The nest may be out in the open or, more often, in partial shade. Two eggs are laid, and both sexes incubate. An unusual event occurred at a poorwill nest in the Carson Range. The nest was originally in a shallow depression in pine needles on an east-facing slope. After the two eggs hatched, an observer visited the nest site fairly regularly to weigh the young. On five different occasions after the nest was visited, the parents moved their young to a new nest site. The distance between consecutive nest sites ranged from seven to thirty-five feet, averaging about nineteen feet per move.[269]

FAMILY APODIDAE

Birds of the Family Apodidae are called swifts. As swifts they are well named, for in level flight they are the fastest birds in the world. As apodids they are poorly named, since they are not footless as the name maintains. Swifts are present throughout much of the world. They have long, narrow, pointed wings and small legs and feet. They are the most aerial of all the birds. Swifts never alight on branches or on the ground. They rest by clinging to vertical surfaces with their strong claws.

Black Swift
Cypseloides niger

The Black Swift breeds along the Pacific Coast, in the Sierra Nevada, and in the Rocky Mountains. Al Knorr found this swift nesting on the eastern rim of the Great Basin in Provo Canyon in the Wasatch Mountains of Utah.[270] I have seen Black Swifts, just beyond the northwestern edge of the Basin, at a reservoir northwest of Susanville, and at MacArthur–Burney Falls State Park in California. The Black Swift is known to nest at MacArthur–Burney Falls State Park.[92] To date, Knorr's fieldwork has failed to reveal nesting Black Swifts inside the Great Basin, but his search for them continues.

While Ridgway was in the western end of the Basin, he twice observed Black Swifts. On May 31, 1868, he found the remains (wings, tail, feet, and sternum) of a Black Swift on a log in a cottonwood grove along the Truckee River on the Paiute Reservation. Then on June 23, at a site along the Carson River above Fort Churchill, Ridgway encountered living Black Swifts; he wrote:

> They were observed early in the morning, hovering over the cottonwood groves in a large swarm, after the manner of Night Hawks (*Chordeiles*), but in flight resembling the Chimney Swifts (*Chaetura*), as they also did in their uniform dusky color, the chief apparent difference being their much larger size. They were evidently breeding in the locality, but whether their nests were in the hollow cotton-wood trees of the extensive groves along the river, or in crevices on the face of a high cliff which fronted the river near by, we were unable to determine on account of the shortness of our stay. They were perfectly silent during the whole time they were observed.[25]

About one hundred years after Ridgway saw Black Swifts along the Carson River, I decided to see if I could locate his observation site. On a mild day at the beginning of February, I worked my way upriver from Fort Churchill. Eventually I reached a point where a huge cliffside closely fronted the Carson River and its cottonwood groves. This had to be Ridgway's 1868 Black Swift site. I raised my binoculars to scrutinize the cliffside, and there in the cool February sky were swifts, not Black Swifts but White-throated Swifts. It was a shock to see swifts flying about in the middle of winter in the Great Basin—even White-throated Swifts. Subsequent visits to the cliffside during the spring and summer showed that this was a White-throated Swift breeding site, as it still is today.

On June 24, 1982, exactly one hundred and fourteen years and one day after Ridgway found his Black Swifts, I took Al Knorr, the Black Swift authority, to visit the Carson River site. We watched the White-throated Swifts in aerial maneuvers and aerial copulation, and Al analyzed the area in terms of Black Swift nesting habitat. He concluded that Black Swifts had never nested here or in the general area. On the basis of his field studies at Black Swift nesting sites, Al believes that a potential site must meet five requirements before swifts will nest there.[271] Of these five requirements, the Ridgway site totally lacks three. It is a dry cliffside, lacking water sprays or waterfalls and the roar of falling water. It lacks high relief above an adjacent valley—it sits in a valley, not above one. And it is well-lit, lacking complete protection from sunlight.

It is obvious that Ridgway did see Black Swifts here. Al believes that Ridgway might have seen a late migratory flock, possibly retarded in its northward movement by an unseasonable spring. The flock was unlikely to be a foraging party of nesting swifts, since there does not appear to be an adequate nesting site anywhere in the vicinity.

White-throated Swift
Aeronautes saxatalis

The breeding range of the White-throated Swift is in the West, extending from southern Canada to Central America. The White-throated Swift is a fairly common breeding species in the Great Basin, occurring where there are suitable nesting and roosting cliffs. However, it is not known as a breeding species in the northwestern corner of the Basin. Although this swift is considered to be migratory in this part of its range, I have seen foraging flocks in the Carson City–Fort Churchill area in late December and early February. Does the Whitethroat hibernate in this region or alternate cold torpidity with foraging during mild winter weather?

Cold torpidity has been observed in various swifts, and both laboratory and field observations have been made on the White-throated Swift. During laboratory experiments, hypothermia did not develop until after losses in body weight had occurred. The swifts could repeatedly arouse themselves from torpidity but did not survive body temperatures below 20° C. They could crawl about with a body temperature as low as 22° C and were fully active at body temperatures as low as 35° C.[263]

For almost a hundred years, field observations have been made on the winter behavior of White-throated Swifts. In 1895 Bendire quoted a letter he had received from a Mr. F. Stephens to the effect that "the White-throated Swift is a rather common resident in southern California. In winter it is somewhat less common and disappears in stormy weather."[272]

In 1917, writing of his field studies on the White-throated Swifts of the Slover Mountains in southern California, Hanna noted that "many days often pass by during the fall and winter when no swifts are seen, and then at some unexpected time they appear in large numbers." Then he goes on to say:

> During the extremely cold wave of early January, 1913, eight, to me perfectly healthy, swifts were taken out of a crevice where they, with many others, seemed to be roosting in a dazed or numb state. They were kept in a room for about six hours and then turned loose, one at a time, a few hundred feet from the point where they were captured. All flew away in a dazed fashion and nearer the ground than usual and none were observed to return to the place where they were captured. I had hitherto thought that they were numb from the cold, or possibly from the jar of a blast in their immediate vicinity; but it has been suggested to me that possibly they were hibernating. This raises a very interesting question, as it seems possible that these birds have intermittent hibernation periods. The facts are that these birds are not observed for many days in the coldest weather, yet are found to be plentiful within the rocks, in a dormant state.[273]

Our experiences here in western Nevada, in the Carson City–Fort Churchill area, also suggest intermittent hibernation. One time, at the very end of December, a flock of White-throated Swifts was seen below McClellan Peak in the Washoe Mountains just northeast of Carson City. The weather had been very cold, bitterly cold at night, for about a month prior to the sighting. The sighting was made on an off-and-on stormy day, after the sun had broken through the clouds and was shining brightly. The swifts were hawking insects along the face of a small cliff.

In the Great Basin, Whitethroats nest and roost in crevices in sheer cliffsides—ranging from the valleys to high up in the tallest mountains. Both vertical and horizontal crevices are utilized; the openings are always very deep and only two or three inches wide. The nests consist of feathers and grasses cemented together, and each nest may be cemented to the rock wall of the crevice—swifts possess large salivary glands and produce their own "glue." At nesting cliffs White-throated Swifts are often associated with Violet-green Swallows. Cliff Swallows may be present as well.

White-throated Swifts are spectacular fliers, and their most electrifying aerial feat is their free-falling copulation act. A male and a female fly toward each other, collide, clutch each other in their claws, and while joined together tumble earthward, end over end, separating only moments before hitting the ground. Some pairs will pinwheel downward for a distance of several hundred feet. When Ridgway witnessed this aerial tumbling in the skies over the Great Basin, he interpreted it as aggressive behavior. But eventually Frederick C. Lincoln shot two pairs in the act, and each pair consisted of one male and one female.

Their common name is a good one for these birds, who are among the swiftest birds in the world in level flight. *Aeronautes saxatalis* is also a good scientific name for the White-throated Swift. *Aeronautes* is from the Greek for air sailor, *saxatalis* from the Latin for rock-inhabiting.

Other Apodids

The Chimney Swift, *Chaetura pelagica*, is a summer resident in eastern Canada and eastern United States. It is of accidental occurrence in the eastern part of the Great Basin. A specimen was taken at Kaysville, and there is a sight record from the mouth of the Provo River in Utah.[27]

The breeding range of the Vaux's Swift, *Chaetura vauxi*, lies along the Pacific Coast, extending discontinuously from southeastern Alaska to South America. The Vaux's Swift is a rare migrant in the Great Basin. A few specimens have been collected and scattered sightings have been recorded across the entire breadth of the Basin. A few summer sightings have been made in the Eagle Lake district of northeastern California.[185]

FAMILY TROCHILIDAE

The Family Trochilidae contains the hummingbirds. Hummingbirds are found only in the Americas; they display their greatest species diversity in South America. Hummingbirds are the greatest living aerial artists. Their flight repertoire includes vertical takeoffs, hovering, pivoting on a stationary axis, flying backward, flying sideward, flying on their backs, rolls, and backward somersaults. About the only flight maneuver they cannot execute is soaring.

The flight anatomy of a hummingbird differs from that of other birds. All the movement of the wings occurs at the shoulder joint; the elbow and wrist joints are rigid. In the wing skeleton, the upper arm (the humerus) and the lower arm (the ulna and radius) are permanently flexed in a V position, with the humerus forming one side of the V and the ulna and radius the other side. Further, these upper and lower arm skeletons are very short, and the hand skeleton is very long. Most of the flight surface of a wing is formed by the ten large primary feathers, which articulate with the hand. There are only six or seven secondaries, which articulate with the ulna.

The hummingbird wing is unique among birds, since it functions like a helicopter rotor—although it moves forward and backward or downward and upward instead of whirling in a circular path like a helicopter rotor. The hummingbird wing develops power on both the downward (forward) and upward (backward) strokes. Each time the direction of wing movement is reversed, the wing pivots through 180 degrees at the shoulder joint, so the front edge of the wing is always the leading edge. However, this pivots the wing so that the undersurfaces of the flight feathers are uppermost on the backward stroke.

Hummingbirds frequently hover when feeding and moving about. This is the most energetically demanding type of flight. To support the high energy costs of hovering, the flight musculature of a hummingbird constitutes 25 to 30 percent of its total body weight, compared to a figure of only 15 to 25 percent for other types of birds. Most birds develop power only on the forward stroke of the wing, and their flight muscles are mainly depressors, which move the wings downward; the elevators, which move the wings upward, are feebly developed and constitute only 5 to 10 percent of the flight musculature. Since they must develop power on the upstroke as well as on the downstroke, hummingbirds possess well-developed elevators as well as depressors—the elevator muscles comprise as much as 33 percent of the flight muscle mass.

The wings of a hovering hummingbird, like the rotor of a helicopter, move in a plane parallel to the ground. The wing tips describe a horizontal figure 8, while the head end of the body is elevated at a forty-five-degree angle above the horizontal. In this position the wings develop lift on both the forward and backward strokes to keep the hummingbird aloft. But the thrusts developed on the forward and backward strokes act in opposite directions and cancel each other out. Therefore, the hummingbird hovers with no horizontal displacement.

When a hummingbird wants to fly forward or backward, it tilts its wings in the proper direction, just like a pilot tilts the rotor of a helicopter. If it wants to fly forward, the hummingbird lowers the front end of its body toward a horizontal position, bringing the plane in which the wing tips move into a more vertical position. The more vertical the plane in which the wing tips beat, the faster the forward movement of the bird. To fly backward, the hummingbird raises the front end of its body into a vertical or upright position. In this position the wing tips beat in a plane over the bird's head—but in a plane that is tilted slightly backward, away from the horizontal.[274]

In flight maneuvers the position of the tail and the relative position of the wings are important. Take a few moments to contemplate the flight performance of the next hummingbird you encounter nectaring at a flower or at a feeder. Note the position of the body, tail, and wings and the plane of the wingbeats—you will quickly come to appreciate the rapidity with which a hummingbird beats its wings. Birds have what Greenewalt called a constant-speed motor. Except for brief bursts, each individual beats its wings at a near constant rate, regardless of the type of flight maneuver or its power demands. In the Black-chinned Hummingbird the natural wingbeat frequency is around fifty beats per second. This represents the frequency at which the flight mechanism—the bone, muscle, and other tissues—of that bird will oscillate with the least expenditure of energy. Natural rates of wingbeats are related to several properties of the oscillating system, including dimensions and weights. In general, the smaller the bird, the faster the wingbeats. The hummingbird family includes the smallest birds in the world, and these species possess the highest rates of wingbeats.

Hummingbirds capture small insects and spiders and drink much nectar. The relationship between hummingbirds and their flowering plants is one of mutual survival. Hummingbirds live out their time on the nectar of flowering plants, and flowering plants live through time on the pollination accomplished by nectaring hummingbirds. This association is not a casual or haphazard one. It is the product of untold millennia of coevolution, and hummingbirds and their flowers have become mutually adapted in their anatomy, physiology, and ecology.

Cross-pollination of flowering plants is accomplished by various means. Some plants are adapted for wind pollination, and some rely on insects, especially bees, as pollinators. Hummingbird-pollinated flowers are recognizable as a type. They have features which make them both attractive to hummingbirds and less attractive—and their nectar less accessible—to bees and other nectaring insects. To be attractive to hummingbirds, a flower must produce a more copious flow of nectar, since hummingbirds are bigger and energetically more demanding than bees. It must guard its nectar against bees, so that a hummingbird's tendency to repeat its visit to similar flowers will be reinforced by a sizable reward of nectar.

To be attractive to a hummingbird, a flowering plant's nectar must have a suitable sugar content. Hummingbirds prefer sweeter nectar over more dilute nectar, and they prefer a sugar content high in sucrose and/or glucose but not high in fructose.[275] And, if flowering plants are to exploit hummingbirds as pollinators, timeliness is essential—the plants' peak blooming periods must coincide with the peak buildups of residential or migrant hummingbirds.

The story of hummingbirds and their flowers has been charmingly told in a delightful book by Karen and Verne Grant. Hummingbird flowers have a number of structural features which tend to deny nectar to insects and to proffer it to hummingbirds. These flowers tend to be tubular and pendent or horizontally placed; they have recurved petals or lack a protruding lower lip which can be used as a landing platform by insects. Thereby their nectar is readily accessible only to hovering forms such as hummingbirds. Since the nectary is located deep within the floral tube, hovering has to be accompanied by a long tongue or proboscis. The tiny Calliope Hummingbird has a bill length of fifteen to seventeen millimeters, and the rest of our western hummingbirds have bills ranging from seventeen to twenty-one millimeters in length.[276] Then, the tongue of a hummingbird can protrude beyond the tip of the bill. If need be, the hummingbird can probe deeper still by hovering with its head and neck inside the floral tube.

It is commonly stated that our western hummingbirds have a color preference and that their favorite color is red. Hummingbird flowers are apparently proof of this preference. The Grants have written that "the colors of western American bird flowers are predominantly some shade of red, or red combined with yellow," and they state that the red color is usually found in the corollas of the flowers but sometimes in the bracts or stamens.[276]

Hummingbird feeders are commonly constructed of red glass or red plastic; the sugar water used in the feeders may be colored red. I recall a summer field trip to the Snake Range during which the late Peter Herlan and I made a stop at a café-tavern in the tiny community of Baker, Nevada. There was a hummingbird feeder just outside the front window of the bar. When discussing feeding formulas with the bartender, I asked her what solution they were using in their hummingbird feeder. She replied that it had been impossible of late to obtain all the orthodox ingredients on the market but that her hummingbirds seemed to appreciate grenadine in the feeder as much as the bar regulars appreciated it in their mixed drinks.

There are many tales of hummingbirds being attracted to red—to the red of blankets, hats, and other items. My favorite butterfly net has a five-foot tubular aluminum handle, which I have wrapped with red plastic tape to protect my hands from the metal. On occasions, I have been resting along a mountain trail with my net tossed to one side, when a Calliope or Broad-tailed Hummingbird has come to hover over the handle of the net.

Hummingbirds are not restricted to red flowers—copious, tasteful nectar is preferred to color. Also, seasonal changes in color preference may occur, as flowers of various color become important nectar sources. There is evidence that, when a hummingbird is highly rewarded by nectar or sugar water associated with a particular color, it will quickly learn to direct its search for nectar toward that color. However, it is true that the colors of nearly all our western hummingbird flowers lie toward the long wavelengths or red end of the spectrum of visible light. It has been suggested that contrast or conspicuousness of color is important to a flowering plant advertising for hummingbird guests and that red is especially conspicuous against sunlit green foliage, while yellow is conspicuous against bluish foliage.[275] There is also some evidence that the red end of the spectrum of visible light is unattractive to insects, especially bees.

Hummingbirds consume animal food, especially small insects and spiders, as well as nectar. They cannot survive on nectar or sugar water alone—they also need proteins, fats, minerals, vitamins, roughage, and other essentials. Sometimes hummingbirds visit the wells drilled by sapsuckers to drink the sap and capture insects attracted to the oozing fluid.

The feeding techniques of hummingbirds have commanded some attention. One question has centered on how they remove nectar from deep within a tubular flower. The hummingbird's tongue—a flat, tubular structure divided over much of its length—can be extended beyond the tip of the long bill. Near the end of the tongue the two divisions are troughlike, not tubelike, and diverge from each other to form a bifurcated tip. In the past writers have suggested that the hummingbird may use its "hollow tongue" like a straw to suck up nectar. However, since the internal chambers in the tongue are not hollow and do not open to the outside into the terminal troughs, there is no way that the tongue could be used as a soda straw. It may be that nectar enters the troughs and is then transported back into the mouth by a retraction of the tongue, but there is little evidence that the tongue is used in lapping up nectar. When nectaring or drinking sugar water, the bird moves its tongue in and out of the tip of its bill a number of times a second. Scheithauer has suggested that the tongue and bill may form a suction pump, with the rapid pumping movements of the tongue creating the suction.[277]

Another question involves how a hummingbird transfers insects it has captured in the tip of its long bill back into its mouth. Observations indicate that, since insects do not stick to the tip of the tongue, they subsequently do not get transferred mouthward by the tongue. Rather, when flycatching in flight, the hummingbird flies the insect into its mouth by utilizing the movement of flight to carry its mouth up to the insect. When hovering and picking up stationary insects or spiders, the hummingbird tosses its prey back into its mouth by swinging its body and head upward in flight.[277]

Black-chinned Hummingbird
Archilochus alexandri

The Black-chinned Hummingbird's breeding range extends from southwestern British Columbia, through western United States, into Mexico.

The Blackchin is an uncommon to common summer resident in many parts of the Great Basin. It nests in the valleys and on up to intermediate altitudes in the mountains. Nesting most commonly occurs in riparian vegetation, wet canyons, and towns and cities.

Except for females with dependent young, hummingbirds are solitary birds and repulse any close approach by other hummingbirds. During the breeding season the male and female associate together only long enough to accomplish copulation. Then the female builds the nest, incubates the eggs, and raises the young by herself. Even outside of the nesting season, when most birds are gregarious, hummingbirds may defend individual feeding territories.

Despite the briefness of courtship, the territorial displays of male hummingbirds are often lively and arduous. The pendulum flight of the male consists of taking off from a territorial perch and executing a series of power dives or swoops. The male Blackchin's pendulum flight is a series of shallow, U-shaped power swoops of an amplitude of five to fifteen feet or so. The pendulum flight takes place in both the presence and the absence of other birds. On the downside of the swoop, the male's wings make a whirring sound. Not only may a male Black-chinned Hummingbird perform a pendulum flight before a perched female, but he may also perform shuttling flights. In a shuttling flight the male approaches to within several feet of the female and then shuttles sidewise, back and forth, following a flight path which is shaped like a horizontal figure 8.

The female Black-chinned Hummingbird builds a tiny nest of plant down laced together with threads from spiderwebs. Often the nest is attached to a downward-slanting branch within ten feet of the ground; sometimes the new nest is built on top of an old nest. Usually two eggs are laid. The incubation period is sixteen days long and is followed by a twenty-one-day nestling period. The female is a zealous mother. And, without any male nest help, she sometimes overlaps two broods—she will have young in one nest to feed and eggs in another nest to incubate.

Like many other hummingbirds, the Blackchin can tolerate cold torpidity. Experiments indicate that, if the ambient temperature drops too low (below 14° C) while the hummingbird is in cold torpidity, the bird will increase its rate of oxygen consumption and produce additional heat. The body temperature will not continue to drop along with the ambient temperature. Thus, body temperature regulation, at lower than normal levels, may be practiced by a Black-chinned Hummingbird while in cold torpidity at low ambient temperatures.[278]

Calliope Hummingbird
Stellula calliope

The breeding range of the Calliope Hummingbird lies in western North America, extending from southern Canada into Baja California. The Calliope is a common breeding species in mountain ranges throughout the Great Basin.

The Calliope Hummingbird is the smallest bird in North America, weighing in at about three grams or barely one-tenth of an ounce. Despite its tiny size, which makes temperature regulation in cold weather difficult, particularly at night, this hummingbird nests in montane environments where the night temperatures are very low during the summer. Studies made in Wyoming indicate that, despite the low temperatures, the female Calliope Hummingbird does not enter cold torpidity when incubating eggs at night.[279] To get through the cold nights without running out of fuel, she must minimize her heat losses. Building a well-insulated nest helps her accomplish this. Further, she minimizes her radiation losses to the night sky by building her nest under overhead cover, which shields her exposed back. The Calliope Hummingbird's tendency to build her nest under overhead cover has been known for many years, at least since Mailliard studied nine nests in 1920 around Lake Tahoe. Of these nests he wrote, "Whether in a tamarack (lodge-pole pine) or in an aspen, the only two kinds of trees in which nests were found, every nest discovered by us was upon an under limb near where it forked from a larger one above, the latter giving good protection, not only from the sun, snow and rain, but, as well, from the too inquisitive eyes of would-be marauders."[280]

The territorial displays of the male Calliope Hummingbird are conspicuous in May and June in the Great Basin. From the pinyon-juniper woodlands and the lower tree line in the foothills to near the upper tree line or timberline in the mountains, males return each year to territories on the edge of open areas—montane meadows, forest and woodland clearings, mountain parks, or streamsides. Territorial males may be readily located by scanning the tops of tall shrubs or small trees in such situations. Pay close attention to the topmost spray on a tall shrub or the peak, bare projection on a small willow or conifer. There, in bright sunlight, the shrub or tree may bear a tiny hummingbird looking much like a swollen twig top. If you are not conspicuously intrusive, the male may tolerate an approach to within ten feet of his display perch. Select a vantage point and remain motionless. While perched the male will indulge in sunbathing and preening. Off and on, he will lift into flight to visit a secondary or tertiary display perch, to flycatch a passing insect, or to chase another hummingbird. Frequently,

Calliope Hummingbird nest under overhead limb

with the approach of a female or male Calliope or even when alone, he will initiate a series of pendulum flights. Mounting swiftly straight up from his perch to a height of forty to one hundred feet or more, he will pause briefly at the apex of his climb. He will then power-dive down and climb back up, each time following a narrow, steep, U-shaped pathway. At the bottom of the U, a slight wing sound may be detected.

Broad-tailed Hummingbird
Selasphorus platycercus

The Broad-tailed Hummingbird's breeding range lies in western United States and Mexico. This bird is a common breeding species in the eastern and central parts of the Great Basin. It is common throughout Utah and across eastern and central Nevada to as far west as the Pine Forest Range in Humboldt County, the Toiyabe Range in Lander and Nye counties, and the White Mountains in Esmeralda County and adjacent California. It is of accidental occurrence in the northwestern part of the Great Basin. There are no modern records for the west central part of the Basin.

The Broad-tailed Hummingbird is the most abundant hummingbird in the eastern and central parts of the Great Basin. It is most commonly found in mountains near streams. The Broadtail maintains a loud presence and is a most conspicuous bird wherever it occurs. The wings of the male make a very loud, shrill whirring or buzzing noise in flight, and he can be heard coming and going from afar. The pendulum flight of the male is a series of U-shaped power dives. He climbs almost vertically, to a height of thirty to forty feet or so, before diving. Some sharp *click-click* notes may be heard at the low point in his dive.

The female Broad-tailed Hummingbird builds a tiny, lichen-covered nest three to thirty feet up in a shrub or tree, often overhanging a stream. At times, females tend toward philopatry, returning year after year to nest in the very same place. The typical clutch consists of two eggs, and the incubation period is fourteen to sixteen days long.

Sometimes female Broadtails experience problems in finding enough food for themselves and their nestlings. The flower crop may be sparse during dry years or during midsummer. Hummingbirds defend feeding territories, and nesting females may have trouble competing with resident male Broad-tailed and Calliope hummingbirds and with migrant Rufous Hummingbirds for nectar. The male Broadtails—who are more aggressive and dominate the females—stake out choice feeding territories where the best nectar-producing flowers are clustered. This restricts the females to more widely scattered flowers where nectaring is poorer. Also, stormy weather frequently occurs in the mountains during the summer. When the days are cold and wet, the female has to closely brood her nestlings, and she has fewer opportunities to forage for nectar or insects. Female Broad-tailed Hummingbirds in Colorado have been known to abandon their nestlings in late July and early August, after the flower supply dwindles and the migrant hummingbirds arrive.[281]

Our smaller hummingbirds weigh only about three or four grams apiece. There are two dire, interrelated consequences attendant on such a minuscule body size in a homeotherm. Hummingbirds have extremely high metabolic rates but only limited storage capacities for fuel in their crops and livers. During a cold night, when energy demands for body temperature regulation are high, hummingbirds face the danger of running out of fuel before the arrival of sunrise and the start of a new nectaring day. This problem is compounded if the hummingbird does not begin the night with its storage capacity filled to the brim.

If a female Broad-tailed Hummingbird has had limited foraging time, she may go into cold torpidity at night while incubating eggs or brooding nestlings. Of course, the eggs or young must be tolerant of low temperatures if this is to be a viable practice. Temperatures as low as 6.5 to 13.7° C have been recorded in Broad-tailed Hummingbird nests containing eggs or young when the females were presumably in cold torpor. It is significant that the females did not enter cold torpidity at sunset but maintained high body temperatures until late at night. Then, between 1:00 A.M. and 2:45 A.M., they allowed their body temperatures to drop and regulated them at a level below normal until the approach of sunrise at 5:00 A.M.[282] By not entering cold torpidity until late at night, when her energy reserves are low but not dangerously so, the female exposes her eggs and nestlings to cold temperatures for only a few hours; and she is rendered helpless by torpidity for a minimum amount of time.

Rufous Hummingbird
Selasphorus rufus

The breeding range of the Rufous Hummingbird extends from the Pacific Northwest of the United States northward through Canada into southeastern Alaska. The Rufous Hummingbird is a common, often abundant migrant through the Great Basin on its southward passage in the summer and early fall. In the spring it moves northward mainly through the foothills and lowlands west of the Sierra Nevada; in its southward passage it migrates on a wide front through the mountains of the West. The southward migration is under way by late June or early July.

The male Rufous is our most aggressive and pugnacious hummingbird. During migration many males and females defend feeding territories. The males often dominate the females and obtain the choicer, denser clumps of nectar-producing flowers for their territories. Feeding competition occurs where hummingbirds are numerous, since nectar is not a bountiful crop. Each flower produces only a small amount of nectar, and energy must be spent hovering and cruising from flower to flower while harvesting nectar. If many of the flowers visited have little or no nectar because of prior visits by nectaring hummingbirds, nectaring could be a nonprofit business. Feeding territories help provide a more reliable food supply for hummingbirds.

An interesting time budget study was made during the summer in the Chisos Mountains of Texas. On the average, territorial male Rufous Hummingbirds devoted more of their time during the day to defending feeding territories (6.7 percent) than to actually feeding (5.5 percent). The remainder of their time was spent perching. Some of the hummingbirds—mainly females and immatures—were nonterritorial. These birds tended to start feeding earlier in the day, trespassing on feeding territories before their holders had started to tap the nectar for their own use.[283]

Other Trochilids

The Magnificent Hummingbird, *Eugenes fulgens*, is of accidental occurrence in the Great Basin. There are several records for Utah, including one for Salt Lake City.[27] Prior to 1982 this bird was called Rivoli's Hummingbird.

Anna's Hummingbird, *Calypte anna*, is of accidental occurrence along the western edge of the Great Basin. Its breeding range lies west of the Sierra Nevada and extends northward to southern Oregon and southward to Baja California. The Anna's Hummingbird has been recorded from the Carson River Basin in western Nevada.[75] There are also sight records from Verdi, Nevada, in late November. During August 1983 at least three different Anna's Hummingbirds began visiting the feeders that Jack and Ella Knoll maintain in northwest Reno. One individual, believed to be a young male, was still present as of December 4. The November-December weather was cold and several snowstorms struck Reno—including a blizzard on December 3. There is a postbreeding, up-mountain movement of Anna's Hummingbirds in California, which may carry a few wanderers into the western edge of Nevada.

Costa's Hummingbird, *Calypte costae*, is of accidental occurrence in the Great Basin. Its breeding range includes the Mohave Desert, due south of the Basin. There is a winter record at a winterized feeder in Salt Lake City.[56]

15
Woodpeckers

WOODPECKERS are members of the Family Picidae of the Order Piciformes. This family has an almost worldwide distribution. The name woodpecker commemorates the close associations these birds have with trees—they peck at wood as they excavate nest cavities, extract insects, and drum. Woodpeckers are designed to stand upright on vertical surfaces such as tree trunks; they seldom use horizontal perches. Their stiff-pointed tail feathers enable them to use their tails, along with their legs, to form tripods for bracing themselves against vertical surfaces. Most woodpeckers have zygodactylous feet: the inner and outer toes point backward, and the two middle toes point forward. There are a few species of three-toed woodpeckers—lacking inner toes. Most of the Great Basin species of woodpeckers show some sexual dimorphism, with males possessing red markings on their heads or faces. Woodpeckers have harsh voices. They often communicate by drumming on resonant surfaces.

The woodpecking activities of these birds are poorly understood. When excavating in wood or drumming on wood, they vigorously use their necks, heads, and bills as adzes or drumsticks. Yet woodpeckers do not stun themselves or develop brain concussions. Necks and bills are strong and skulls are thick-walled. The brain is suspended and protected by membranous meninges and tightly packed within the cranial cavity of the skull. Obviously, woodpecker brain tissues are tolerant of a lifetime of vigorous jarring. Exactly how the suspension, packing, and tolerance are accomplished would appear to be fruitful subjects for research.

Woodpeckers have peculiarly designed hyoid apparatuses and tongues capable of being extended far beyond their bills. The hyoid apparatus is a skeletal element of the throat region. A projection of the hyoid extends forward and is embedded in the tongue. The hyoid possesses two long flexible horns. These horns loop backward into the neck. Then they curve upward, just beneath the skin, over the back and top of the skull, and down the forehead to end in front of the eyes or in the external nostrils. When a woodpecker wishes to spear a wood-boring larva with its tongue, accelerator (geniohyoid) muscles associated with the two horns contract. This jerks the far ends of the horns back up the forehead and straightens out the loops in the neck. These movements drive the tongue forward, beyond the tip of the bill, to impale the prey. Tongue and prey are retracted into the mouth by the elastic recoil of the horns. Hence, the hyoid and tongue together form a double-shafted spear.

There is a question about how woodpeckers locate larval wood-boring beetles that live out of sight beneath the bark or in the solid wood of trees. Do they rap and then feel with their bill tips for vibrations, which vary according to whether underlying tunnels are present or not? Or do they hear the larval beetles munching on wood or moving about?

FAMILY PICIDAE

Lewis' Woodpecker
Melanerpes lewis

The breeding range of Lewis' Woodpecker lies in western North America in southern Canada and the United States. It is an uncommon breeding species in the Great Basin and is occasionally present during the winter. The Lewis' is opportunistic in its nesting and overwintering occurrence—its occurrence in a given locality may vary from year to year, depending on the abundance of insects during the nesting season and the abundance of acorns during the winter.[284] Prior to 1976 its scientific name was *Asyndesmus lewis.*

Most of our woodpeckers make their living by scaling bark or chiseling into dead or living trees after insect larvae. The flickers have deviated from this pattern in that their food niche involves foraging on the ground for ants. Lewis' Woodpecker is another nonconformist—it feeds on flying or moving insects and shifts to acorns and fruit seasonally or when insects are difficult to obtain.

The Lewis' Woodpecker hawks insects. At first inspection, flycatching appears to be an unlikely trade for woodpeckers—who are structurally adapted for clinging in a vertical position to trees, bracing themselves while in this position, and chiseling into wood. A Lewis' Woodpecker often initiates its flycatching from a scanning perch, a vantage point from which the bird can survey its surroundings for flying insects. Depending on the nature of the prey, the scanning perch may be high on top of a tree or utility pole or close to the ground on a low post. While scanning for insects, the bird turns or cocks its head in various fixed positions as it surveys its surroundings.

When prey is sighted the woodpecker flies out, follows the maneuvers of the insect, and captures it in its mouth. In flight the Lewis' Woodpecker continually flaps its wings, so that its flight is strong and direct, not undulatory as it is in so many woodpeckers. After capturing an insect, the woodpecker usually returns to its starting point on the scanning perch. If it had to climb to reach the flying insect, it may glide back, spiraling downward or descending in a series of downward-arching swoops. Sometimes a Lewis' will capture flying insects as do swallows, by flying to and fro, not returning to a perch between catches. It also captures insects from bushes or off the ground. Using a low perch, the woodpecker scans the surrounding vegetation and ground for moving insects. It may glean insects from limbs or trunks of trees; it may even flake off small bits of bark in search of prey.

The Lewis' Woodpecker also feeds on fruit, both wild and cultivated. In the fall it may store acorns or cultivated nuts such as almonds for use during the winter. In many places acorns form the winter food staple. Each woodpecker harvests, shells, and caches a supply of acorns for its own use—the caches are defended against raids by other birds. Natural crevices such as cracks in trees or utility poles are used as storage places.

Lewis' Woodpeckers nest in a variety of habitats in the valleys, foothills, and mountains of the Great Basin. Because of their flycatching, they do not frequent closed forests or dense woodlands. They need trees, but trees interspersed with open areas.

The territorial "song" of the male Lewis' Woodpecker consists of a harsh churr, repeated three to eight times in rapid succession, and of some weak drumming. The drumming is a simple roll, followed by three or four single taps. Churring and drumming occur together.[284]

The male Lewis' has two territorial displays—wing-out and circle flight. Both of these displays are used in courtship and territorial defense. Further, wing-out and chatter-calls may be used as a precopulatory display, circle flight as a postcopulatory display. In wing-out the male extends his wings and lowers his head while fluffing out the feathers of his throat and upper breast—this exposes the pink feathers of the belly and flanks. In circle flight he circles the nest tree in a glide, wings extended and high, with the pink feathers of his belly and flanks exposed. The flight usually terminates at the nest hole, where the male churrs.[284]

Yellow-bellied Sapsucker
Sphyrapicus varius

The range of the Yellow-bellied Sapsucker extends from southeastern Alaska down and across much of Canada and the United States. This sapsucker is a common permanent resident in the Great Basin. In the mountains it nests mainly in aspen groves or in aspens mixed with conifers. In the valleys it nests in stands of cottonwoods or other deciduous trees in riparian situations.

Sapsuckers drill small holes in trees and drink the sap which collects in these wells—hence the name sapsucker. In addition to feeding on the sap, these birds eat the delicate plant tissues lying under the outer bark—the cambium tissue and the bast or inner bark. They feed on fruit seasonally. Sapsuckers also feed on insects, which they capture with a number of techniques: by flycatching, by gleaning from trunks and branches of trees, and by ambushing insects attracted to the sap wells. Sapsuckers do not chisel into trees to remove wood-boring insects in the classical woodpecker manner. Woodpeckers adapted to feed on wood-boring insects have long, extensible tongues, with backward-directed spines at the tip. But the sapsucker tongue is short, barely extensible, and has a fringe of stiff hairs instead of spines.

The holes drilled by sapsuckers penetrate the bark and sometimes even the cambium layer. Some holes have a circular outline, others are more elliptical—being broader than they are high. On small tree trunks the wells are often aligned to form rings or partial rings around the trunk. The rings often spiral around the trunk, so that the individual holes are aligned not only in horizontal circles but also in vertical columns. On larger trees the wells are grouped into vertical columns, with no horizontal alignment, and are located on one side of the trunk. The columns often contain compound holes, formed when several wells are merged vertically.

In our quest for sweets we tap maple trees, particularly the sugar maple, and concentrate their sap by boiling it into syrup and sugar. Sapsuckers are much more omnivorous in their use of sap than we are. As far back as 1911, McAtee was able to compile a list of 258 species of trees, shrubs, and vines, belonging to 45 native families of plants, which by then were known to be attacked by sapsuckers in North America. In addition, he listed 38 introduced species known to be drilled by sapsuckers.[285] Sapsucker punctures can damage or kill trees by removing so much of the growing tissue that vigor is impaired, sap movement is interfered with, and entry portals for disease-causing organisms such as fungi are provided. The punctures also blemish wood to the extent that its value as lumber decreases. In McAtee's words, "These defects consist of distortion of the grain, formation of knotty growths and cavities in the wood, extensive staining, fat streaks, resin deposits, and other blemishes. All of them result from injuries to the cambium, their variety being due to differences in the healing. Besides blemishes, ornamental effects are sometimes produced during the healing of sapsucker wounds, such as small round stains, curly grain, and a form of bird's-eye."[285]

During the winter in temperate regions such as the Great Basin, the trees are dormant. The availability, abundance, and nutritional composition and value of sap and plant tissues vary seasonally. A detailed study made in northern Michigan of the seasonal changes in the diet of the Yellow-bellied Sapsucker showed that, in the early spring, the sapsuckers drilled to obtain bast and some sap from conifers. As spring progressed, the birds drilled to obtain the copious sap from a variety of angiosperms such as maples, elms, oaks, and hackberries. By midsummer sapsuckers were concentrating on birches for sap—even after their leaves are out, birches still produce sap with a sugar content of up to 20 percent or so. In fall the sapsuckers obtained bast from maples. During the winter they scaled bark from trees to obtain insects, fed on frozen fruit, and collected a little sap on warm days.[286]

A diversity of animals are attracted to the sap wells of the Yellowbelly. Among the mammals, the red squirrel is a frequent sap-stealing visitor. Insects such as flies, wasps, hornets, butterflies, moths, and ants are also attracted to the wells; when discovered by the sapsucker, these insect visitors are eaten. Sapsuckers are adept at flycatching insects as well. They sally forth from a scanning perch or alternate between two scanning perches when flycatching.

Other birds visit or hang around the sap wells to feed on the sap and insects there. Warblers and hummingbirds are especially prone to do this. In addition, the Pine Siskin, American Goldfinch, White-crowned Sparrow, Ruby-crowned Kinglet, Downy Woodpecker, and Hairy Woodpecker have been seen at sapsucker wells.

Upon arrival at the nesting ground from migration or the wintering ground, the female sapsucker goes to her nest site of the previous year. If she does not have a previous site to return to, she selects a tree with old nest holes. Here, in her temporary territory, she spends her time feeding, preening, and signal-drumming. The drumming of the sapsucker has been described as being "of about three to five seconds' duration and starts with a steady roll for one or two seconds which is followed by a series of loud taps at irregular intervals for two to four seconds."[287] Eventually a male will hear her and respond—sometimes he will be her mate of the previous year, returning to his past nesting site. When the male approaches the female, the two may engage in mutual displays. One sapsucker may eventually fly off in "rapid moth-like flight," followed by the other. For a time before copulation occurs, the drumming and mutual displays will continue between the male and female. During the displays the two birds face each other and engage in bobbing movements.[288] Sometimes other sapsuckers are attracted to the site and participate in the displays.

Williamson's Sapsucker
Sphyrapicus thyroideus

The breeding range of Williamson's Sapsucker lies in western Canada and western United States. This is an uncommon breeding species in some of the higher mountain ranges of the Great Basin. It is present in the western edge of the Basin in the Warner Mountains, Carson Range, and Sweetwater Range. Then, there is an apparent hiatus in its distribution eastward into central Nevada, where it has not yet been detected in the Pine Forest, Santa Rosa, Desatoya, or Toiyabe ranges. But it reappears again in eastern Nevada and Utah. In eastern Nevada it is known to occur in the Ruby Mountains, Spruce Mountain, Pequop Mountains, and the Snake Range.[289] It has nested along the South Fork of the Humboldt River at Twin Bridges, Elko County.[211] Williamson's Sapsucker is a permanent resident in the western edge of the Great Basin, although there may be some movement to lower altitudes during the winter. In the eastern part of the Basin the sapsuckers are evidently migratory, with some casual overwintering there.

The Williamson's is a bird of the high montane forests, typically frequenting the Canadian and Hudsonian life zones at altitudes of from about eight thousand feet upward. In the Sierra Nevada this sapsucker is closely associated with lodgepole pines.

Little is known about the natural history of this quiet, retiring bird. The male and female are so unlike in appearance that for many years they were considered to be separate species. Then, in 1874, Henry Henshaw found them nesting and discovered their secret. Like its close relative, the Yellow-bellied Sapsucker, this bird drills wells in trees and drinks sap and eats plant tissues. Its wells have been found in the lodgepole pine, red and white firs, ponderosa pine, Jeffrey pine, sugar pine, mountain hemlock, and quaking aspen.

The Williamson's Sapsucker feeds on insects as well as on sap and plant tissues. Merrill has observed that it is "especially partial to young pines," has a "noticeable habit" of working its way "down as well as up a trunk," and when feeding "will often work up and down a favorite tree repeatedly."[290]

Downy Woodpecker
Picoides pubescens

The breeding range of the Downy Woodpecker extends from southeastern Alaska and northern Canada to southern California and Florida. The Downy is a rare to common permanent resident in the Great Basin. It is considered to be common in Utah.[27] In the western part of the Basin it is uncommon, and in some localities it is rare or missing altogether. It is also scarce in the northwestern corner of the Basin in Oregon.[103] It is missing from the southwestern mountainous rim of the Basin—along the eastern slope of the Sierra Nevada from Lake Tahoe southward.[291] But it is present in adjacent Nevada along the lower reaches of the Carson and Walker rivers. The uncommonness of this woodpecker over much of the Great Basin is evidently not of recent origin. Writing of the 1860s, Ridgway stated, "We found this bird to be unaccountably rare in all portions of the country, even where its larger cousin, *P. harrisi* [Hairy Woodpecker], abounded; indeed, it was seen at only two localities along the entire route."[25] Prior to 1976 the Downy Woodpecker's scientific name was *Dendrocopos pubescens*.

The Downy Woodpeckers of the Great Basin nest primarily in montane aspen groves and in riparian cottonwood and willow stands in the valleys. During the fall and winter the males and females remain apart, and individuals of both sexes may defend feeding territories as well as the area immediately around their roost holes. Studies have shown that there is a tendency for females and males to occupy different feeding niches during the winter; when they meet while foraging, hostile encounters may occur. The females forage mainly on the trunks and larger branches of medium-size trees. The males forage in a greater variety of sites, including small trees, but concentrate most on small branches. During the cold time of the year, Downies are forced to concentrate more on subsurface insects and less on surface prey. They find more subsurface prey on dead trees and limbs than on live ones. During the warm part of the year, they tend to feed more on surface prey in live trees.[292]

The courtship behavior of Downy Woodpeckers gets under way between mid and late winter. Drumming serves to bring the female and male together. Both sexes drum, using regular drum trees which have limbs with good resonance. The territorial disputes between males involve bill-waving dances. As courtship progresses, the pair eventually seek a nest site. While searching for a nest site, a Downy Woodpecker will drum or tap to summon its mate to look at a potential location. Once a site has been selected, both sexes work at excavating the nest hole. Once the eggs are laid the birds take turns incubating by day, but the male does the night incubating and roosts in the nest hole.[293]

Hairy Woodpecker
Picoides villosus

The Hairy Woodpecker is a resident of the wooded and forested areas of North America from central Alaska southward; it is also found in the mountains of Central America. The Hairy is a common permanent resident in the montane forests and riparian valley woodlands in the Great Basin. It may be missing from the northwestern corner of the Great Basin in Oregon.[103] Prior to 1976 the scientific name of the Hairy Woodpecker was *Dendrocopos villosus*.

The Hairy Woodpecker's feeding techniques are the classical ones we commonly associate with woodpeckers. It forages mainly up the trunks of trees but may go out on the larger limbs as well. It hitches itself up the trunk in a series of short hops, bracing itself with its tail, often spiraling around the trunk on its way to the top of the tree.

Some of the Hairy's prey is located on the surface of the trunk or in the bark. This prey is obtained by pecking, by peering and pecking, or by scaling or flaking off bark. Females may be more prone to seek superficially located prey than are males. Also, the larvae of wood borers may be excavated from deep within the wood. Sometimes it takes fifteen minutes or more to chisel into the tunnel of a wood borer, before the larva can be removed by the highly extensible tongue of the woodpecker. Males may be more prone to excavate than females.[294]

Sometimes, when moving along a limb, a Hairy Woodpecker will deliver a quick rap to the wood here and there. This percussion act may be involved in locating prey, either by startling it so that it moves or by furnishing sound clues to the whereabouts of wood borers' tunnels.[294] Experiments testing the hearing range of the Hairy Woodpecker indicate that it is similar to that of other kinds of tested birds—it is good enough that the bird may hear some of the sounds its insect prey makes when feeding or moving around.

Drumming with their bills on a resonating or resounding surface is a method of signaling widely employed by Hairy Woodpeckers. Even the females engage in considerable drumming. The drumming of the Hairy consists of rolls, each being a series of about a dozen or so raps; the rolls may be repeated up to ten times or so a minute. Hairy Woodpeckers drum under a variety of circumstances, and observers believe they have been able to detect some of the functions of drumming. Territorial drumming is mainly done by males on the nesting territory, and it may stimulate drumming duets with neighboring males. Females carry out territorial drumming in the fall when establishing winter territories. Drumming occurs in other contexts beyond that of territory. The members of a pair call one another by drumming when separated. They use drumming as location signals when apart. They drum as precopulatory behavior. The members of a pair carry on stimulatory drumming duets during the long, up to three months or so, engagement period prior to nesting. They drum when an avian intruder approaches their nest or roost hole.[295] And sometimes they apparently drum just for the sheer enjoyment of drumming.

White-headed Woodpecker
Picoides albolarvatus

The White-headed Woodpecker occupies a restricted range in western North America. It is present in the western edge of the Great Basin in the Carson Range. It is also present along the mountainous western rim of the Basin on the eastern slope of the Sierra Nevada. Prior to 1976 the scientific name of the White-headed Woodpecker was *Dendrocopos albolarvatus*.

The dietary habits of the Whitehead are quite unusual for woodpeckers. At certain times of the year the birds may subsist largely on the seeds of pines, especially the seeds of ponderosa pines, with which they are often closely associated. A woodpecker harvesting seeds will alight on a pine cone, cling to the side or bottom of the cone, and chisel it open to expose the seeds.[296] In northern California, the White-headed Woodpecker has been known to have a serious impact on the seed production of the sugar pine. Small parties of four to six woodpeckers have been seen harvesting seeds from green cones—a woodpecker would alight on a cone and slash a deep vertical cut down its length, exposing the seeds.[297] As a mark of this close association, the plumage of White-headed Woodpeckers is often smeared with pine pitch.

In addition to feeding on pine seeds, White-headed Woodpeckers also feed on animal food. When foraging on the trunks and branches of pine trees, they scale bark by prying it loose with their bills or flake it away with glancing blows instead of with straight-ahead blows. They may probe in crevices in the bark with their tongues.

Three-toed Woodpecker
Picoides tridactylus

The Three-toed Woodpecker is a permanent resident of the taiga—the great circumboreal coniferous forests of Eurasia and North America. In North America its range extends across the forested regions of Alaska and Canada, with extensions southward into the United States. In the West one of these southward extensions crosses the eastern part of the Great Basin. The Three-toed Woodpecker is a rare resident of the mountainous eastern rim of the Great Basin in the Wasatch Range. It is also a rare resident of the Snake Range in eastern Nevada. Prior to 1982 the common name of this woodpecker was the Northern Three-toed Woodpecker.

In North America the Threetoe is closely associated with spruce trees. Because of its narrow interest in spruces and their bark insects, this woodpecker can be regarded as a feeding-niche specialist, whereas the Black-backed Woodpecker—associated with pines as well as with other kinds of conifers such as firs and spruces—can be regarded as a feeding-niche generalist. No doubt, the Three-toed Woodpecker is not found in the western mountainous rim of the Great Basin because of the lack of spruce trees in the Sierra Nevada.[298]

Both the Three-toed and Black-backed woodpeckers are attracted to forested areas where there is an unusually good supply of dead and dying trees. Therefore, they may show up in concentrated numbers in places where wood-boring insects have greatly multiplied following disasters which have killed many trees—after forest fires, windstorms, and insect epidemics. Other woodpeckers besides these two species also aggregate at these places. The question is still unanswered as to how woodpeckers so readily find these areas of dead and dying trees. But there must be nomadic movement after the nesting season, on the part of at least some woodpeckers, to account for their rapid buildup at a major, newly developed food source.

On past occasions there have been winter invasions of Three-toed and Black-backed woodpeckers into northeastern United States from adjacent Canada. These irruptions are generally attributed to spillovers from unusually high population buildups in Canada.

Black-backed Woodpecker
Picoides arcticus

The Black-backed Woodpecker is a resident of the boreral coniferous forests of North America. It ranges from central Alaska across Canada and, in the West, down through the Cascade Mountains into the Sierra Nevada. It is a rare resident in the western edge of the Great Basin, where it has been recorded from Eagle Lake, Lake Tahoe, and the Carson Range. Prior to 1982 the common name of this woodpecker was the Black-backed Three-toed Woodpecker.

As its former name indicated, this woodpecker possesses only three toes on each foot instead of the usual four of a woodpecker; the hallux or big toe is missing from each foot. Woodpeckers possessing three toes per foot are highly specialized for digging out wood-boring insects—studies indicate that 75 percent or so of their food may consist of the larvae of these insects. The loss of the hallux from each foot apparently enables the woodpecker to deliver a more powerful blow when excavating wood-boring larvae without impairing its climbing ability. The hallux is a backward-directed toe, and toes pointing backward are of little importance in climbing vertical surfaces—they are mainly of importance when the bird is perching. The woodpecker develops most of the momentum of the blow with its body, not with its head and neck. It can develop maximum momentum by rotating its heels outward prior to delivering the blow, then rotating them inward as the blow is delivered. There is reason to believe that, if the hallux was present, it would probably impede the free rotation of the heel when the bird strikes a blow.[299]

Northern Flicker
Colaptes auratus

The breeding range of the Northern Flicker blankets North America, from the northern tree line in Alaska southward into Mexico. The Northern Flicker is a common permanent resident in the Great Basin, where it is the most abundant and widespread species of woodpecker present. Its status in the Basin has apparently changed little since the 1860s, when Ridgway wrote, "Being the most abundant and generally distributed of the Woodpeckers, this species was found in all wooded localities; and though it appeared to be rather partial to the deciduous trees of the lower valleys, it was far from rare among the pines of the mountains, excepting in the denser portions of the forest."[25]

During the past ten years, flickers have received rough and inconsistent treatment at the hands of the Committee on Classification and Nomenclature of the American Ornithologists' Union. In the thirty-second supplement to the checklist of North American birds, which appeared in 1973, the committee created an enlarged species called the Common Flicker, *Colaptes auratus*. This was done by lumping three previously recognized species—the Yellow-shafted, Red-shafted, and Gilded flickers. These three forms were now considered to be subspecies of a single species, since they hybridized readily when they met in nature. The new name for this enlarged species was obviously a poor choice by the committee, since the word common when used as a modifier in an English name is often virtually meaningless. Repenting, the committee made still another flicker name change in 1982. The Common Flicker then became the Northern Flicker.

The Northern Flicker present in the Great Basin is the Red-shafted Flicker. However, the Yellow-shafted Flicker shows up occasionally during migration and in the winter, and hybrids between Red-shafted and Yellow-shafted flickers have been found in the Basin. The occurrence of this type of hybrid here is not a recent development. In January of 1868, Ridgway collected a Red-shafted x Yellow-shafted hybrid in Washoe Valley.[25]

The Northern Flicker is our most unique woodpecker, since its ecological niche includes the ground as well as trees. This flicker does much of its

Young Northern Flicker

foraging on the ground, even away from the presence of trees, out in sagebrush shrublands. It visits the ground in search of ant nests, and up to 45 percent or so of its food consists of ants.

Although the flicker is visually prominent in the Great Basin the year around, it also becomes aurally prominent in the spring. Then, the loud drumming of male and female flickers is widely heard, along with their loud, long, rolling location calls of *keck-keck-keck-keck*.

Upon returning to the breeding grounds in the spring, male flickers seek out their territories and nesting sites of previous years. They announce their presence with long series of *keck-keck-keck* calls, with drumming playing a minor role. Females also return to their previous nesting territories, and some remating of last year's pairs results. Other flickers are attracted to a calling male, and sometimes more than one female will be attracted to compete for the male. Flickers have an aggressive social display—females direct this display at rival females, males at rival males. The rivals face each other on the branch of a tree, clinging to a tree trunk, or even on the ground. The head is tilted back, with the bill pointing upward. The wings are open and spread, the tail turned sideways and twisted, so that the bright undersides of the wings and tail are displayed. The head is bobbed, so the head, neck, and bill are in constant motion. Chasing may occur, around tree trunks and through branches.

The Northern Flicker tends to reuse nesting cavities from year to year, but it often loses them to other birds and must excavate a cavity. Cottonwoods, aspens, and willows are often used.

Other Picids

The breeding range of the Red-headed Woodpecker, *Melanerpes erythrocephalus*, lies east of the Great Basin, but there are records of occurrence for this woodpecker along the eastern edge of the Basin. In June of 1869, Ridgway observed one near Salt Lake City.[25] A specimen was obtained at the Bear River National Wildlife Refuge in 1941.[56] The Redhead is a rare nesting species in the Uinta Basin—just off the eastern edge of the Great Basin.[27]

The breeding range of the Acorn Woodpecker, *Melanerpes formicivorus*, lies to the west and to the south of the Great Basin. This is a breeding species in the western edge of the Great Basin, in the area between Susanville and Janesville, California.[92]

The Red-breasted Sapsucker, *Sphyrapicus ruber*, resides along the mountainous western rim of the Great Basin—in the Carson Range and in the Sierra Nevada. Prior to 1982 the sapsuckers with red breasts were classified merely as subspecies of the Yellow-bellied Sapsucker.

The Pileated Woodpecker, *Dryocopus pileatus*, is a peripheral member of the Great Basin avifauna. Its breeding range loops around the Basin, extending southward down the mountainous eastern and western rims. A specimen was collected in 1905 at a site near Eagle Lake, California.[300]

Western Grebe. *John Running.*

American White Pelicans at Pyramid Lake. *John Running.*

American White Pelicans at Bear River Refuge, Utah. *Stephen Trimble.*

Cattle Egret at feeding station along head of cow. *Robert Alves.*

Great Blue Heron. *John Running*.

White-faced Ibises feeding at Bear River Refuge, Utah. *Stephen Trimble.*

Virginia Rail. *Robert Alves.*

Sandhill Cranes dancing at Malheur Refuge, Oregon. *Stephen Trimble.*

Canada Goose preening. *John Running.*

Female Mallard. *John Running.*

Male Mallard. *John Running.*

Pair of Cinnamon Teal, Malheur Refuge, Oregon. *Stephen Trimble.*

Male American Wigeon with American Coot in foreground at Virginia Lake in Reno, Nevada. *Robert Alves.*

Male Ruddy Duck. *Tony Diebold.*

Distraction display by parent Killdeer. *Robert Alves.*

American Avocet, Bear River Refuge, Utah. *Stephen Trimble.*

Black-necked Stilt. *John Running.*

Long-billed Curlew. *Robert Alves.*

Marbled Godwit. *Robert Alves.*

Common Snipe on perch prior to taking off on a winnowing flight.
Robert Alves.

Migratory phalaropes in July at Great Salt Lake near Antelope Island.
Stephen Trimble.

California Gulls at Mono Lake. *John Running.*

Turkey Vultures at roost basking in the morning sun. *Tony Diebold.*

Turkey Vulture basking. *Tony Diebold.*

Turkey Vulture gliding with primaries highly slotted at wing tips. *Stephen Trimble.*

Young Northern Harrier. *John Running.*

Nestling Swainson's Hawk. *Stephen Trimble.*

Red-tailed Hawk, Hart Mountain, Oregon. *Stephen Trimble.*

Prairie Falcons at aerie. Note the black axillary region visible on the flying falcon. *Tony Diebold.*

Golden Eagle near Gerlach, Nevada. *Tony Diebold.*

Golden Eagle. *John Running.*

Two Golden Eagle nests on transmission tower, Buena Vista Valley, Nevada. Upper nest fledged two young in 1980. *David Worley.*

Young Prairie Falcon. *Stephen Trimble.*

Great Horned Owl. *Stephen Trimble.*

Great Horned Owl nest with three young near Gerlach, Nevada. *Tony Diebold.*

Young Great Horned Owls. *Tony Diebold.*

Burrowing Owl family at their burrow near Gerlach, Nevada. *Tony Diebold.*

Male California quail. *Tony Diebold.*

Male Sage Grouse strutting. *Tony Diebold.*

Male Sage Grouse strutting before three females at the mating center, Bedell Flats, Nevada. *Fred Ryser.*

Chukars near Gerlach, Nevada. *Tony Diebold.*

Mourning Doves. *Tony Diebold.*

Broad-tailed Hummingbird nectaring. *Stephen Trimble.*

Female Black-chinned Hummingbird incubating eggs at Hardscrabble Canyon, Pyramid Lake, Nevada. *Keith Giezentanner.*

Say's Phoebe. *Stephen Trimble.*

Horned Lark. *Stephen Trimble.*

Violet-green Swallow. *Stephen Trimble.*

Cliff Swallow nests on artificial nesting structure at Bear River Refuge. *Fred Ryser.*

Barn Swallow. *Stephen Trimble.*

Steller's Jay. *Stephen Trimble.*

Scrub Jay. *Stephen Trimble.*

Pinyon Jay. *Stephen Trimble.*

Male Brewer's Blackbird. *Tony Diebold.*

Female Northern Flicker. *Tony Diebold.*

Black-billed Magpie. *Stephen Trimble.*

Common Raven nest and young. *Tony Diebold.*

Loggerhead Shrike with nesting material. *Tony Diebold*.

American Robin nest with eggs and newly hatched nestling. *Stephen Trimble*.

Male Yellow-headed Blackbird at Stillwater National Wildlife Management Area, Nevada. *Stephen Trimble.*

Warbling Vireo on nest in aspen. *Stephen Trimble.*

16

Perching Birds

THE ORDER Passeriformes, by far the largest order of birds, contains the passerine or perching birds, including all the so-called songbirds. Passeriforms have feet that are adapted for perching on branches and similar objects. Each foot bears four toes, all arising at the same level. Of these four toes, three are directed forward, and one is directed backward. The hind toe is usually the best developed of the four and is not reversible.

When the bird is perched, its four toes curl around the perch. The muscles which flex or curl the toes are located higher up on the leg; they exert their pull by means of long tendons which run through grooves and sheaths and insert on the underside of the toe bones. A passerine bends its legs in order to settle down on a perch. This bending increases the tension exerted by the tendons, curling the toes around the perch. The ends of the tendons lying underneath the toe bones bear many tiny projections on their underside, and the opposing walls of the tendon sheaths bear riblike projections. The weight of the perching bird on its toes forces the tendon projections to mesh with the opposing sheath ribs—this locks the tendons, and consequently the curled toes, in place. Thus, even during sleep the bird remains on its perch, since its toes cannot uncurl.

FAMILY TYRANNIDAE

Tyrannidae is a New World family of tyrant flycatchers. These flycatchers obtain most of their food by intercepting flying insects. They await prey while perched in a characteristically upright posture on prominent scanning perches, then sally forth to snap passing insects out of the air. Their behavior toward other birds is often tyrannical—aggressively attacking and driving them away. Most tyrannids are plainly colored in browns, grays, olive greens, and yellows. Their heads are relatively large, and their bills are broad, flat, and strongly hooked at the tip. Prominent rictal bristles surround the base of their bills.

Olive-sided Flycatcher
Contopus borealis

The breeding range of the Olive-sided Flycatcher lies in the conifer forests of North America from northern Alaska across Canada, with extensions into the United States. The Oliveside is a migrant in the valleys and an uncommon summer resident in the mountain ranges of the Great Basin. Prior to 1982 its scientific name was *Nuttallornis borealis*.

The breeding habitat of the Olive-sided Flycatcher lies in coniferous forests, particularly fir or spruce forests. This flycatcher is most commonly found in the coolness of higher altitudes and in the more open or broken sites in a forest. The scanning perches of this species are often near or at the very top of towering trees—it flycatches in the highest air spaces of all our flycatchers. The diet of the Oliveside consists almost entirely of flying insects, especially bees and wasps. Nonflying arthropods are often included in the diets of most flycatchers, but forms such as caterpillars, spiders, and millipedes have not been found in this bird's diet.

The song perches of the males are very high ones, and during the breeding season singing occurs early in the morning and in the evening. There are three syllables in the song—the first syllable is short, sharp, and not very loud; the next two are loud and strongly accented. The song has been variously described as, for example, *whip-whée-péeoo* and *pi-pée-pa*.

Little is known about the displays of the Olive-sided Flycatcher. However, several interesting flight displays have been briefly witnessed. In one, two flycatchers engaged in a mutual pendulum display while bill snapping. The two flew back and forth in an arc parallel to each other, three or four times, then perched separately. The performance was repeated several more times. The arc of the pendulum was about thirty to forty feet across and had a depth of six to eight feet; the display took place about forty-five to fifty-five feet above the ground. Unfortunately, the sex of each bird was unknown.[301]

Western Wood-Pewee
Contopus sordidulus

The breeding range of the Western Wood-Pewee lies in western North America, extending from southern Alaska through Mexico, and in Central America. This pewee is a common migrant and an uncommon to common breeding species in the Great Basin. Prior to 1982 its common name was unhyphenated.

Among Great Basin flycatchers, the Western Wood-Pewee is a habitat generalist. In describing the distribution of this pewee in Nevada, Linsdale went so far as to say that it "occurs wherever there are trees, especially in the northern part of the state."[91] This is an exaggeration if interpreted to mean that trees in Nevada invariably contain pewees. But there is a ring of truth about the statement if it is interpreted to mean that this pewee is not tied to any particular type of tree. The Western inhabits wooded and forested areas in the valleys and mountains of the Great Basin, regardless of whether the trees present are coniferous or deciduous ones. However, it does prefer broken areas where the trees are not densely packed and where some dead trees or trees with dead limbs are present.

The Western Wood-Pewee does its flycatching well above the ground in airspaces twenty to seventy-five feet up. Its scanning perches are vantage points on projecting stubs on dead trees, on dead branches, or in widely branching trees where the foliage is thin. While scanning for flying insects, the pewee sits in an upright position and moves its head to scrutinize the air sectors around it, much like *Empidonax* flycatchers do. When prey is sighted, the pewee makes a dashing sally out from its scanning perch, snaps up the insect with a clearly audible click of its mandibles, and usually returns to the same perch. Upon alighting it quivers its tail and wings.

The male's territorial song, sung after sunset and again before sunrise, consists of two phrases repeated over and over again: the first is a hoarse *pee-ee*, followed by the same phrase delivered in a softer tone and with a downward inflection.[84] Ridgway, aware of the Western's vesperal song, commented, "It seems, however, to be more crepuscular than the eastern species, for while it remains quiet most of the day, no sooner does the sun set than it begins to utter its weird, lisping notes, which increase in loudness and frequency as the evening shades deepen. At Sacramento we frequently heard these notes about our camp at all times of the night."[25]

The Western Wood-Pewee saddles its nests on horizontal limbs or sometimes places them in forks or even crotches; the nests are usually positioned well above ground. Ridgway collected nine pewee nests, eight with eggs, for the United States National Museum while in California, Nevada, and Utah in the 1860s. Later, innocently commenting on the nests he had seen and collected, he wrote, "The nest of this species, as is well known, differs very remarkably from that of *C. virens* [Eastern Wood-Pewee], being almost invariably placed in the crotch between nearly upright forks, like that of certain *Empidonaces*, as *E. minimus* and *E. obscurus*, instead of being saddled upon a horizontal branch."[25]

Another early ornithologist—the doughty Charles Emil Bendire, captain and brevet major, U.S.A. (retired)—took offense at Ridgway's conclusions regarding the nesting habits of the Western Wood-Pewee. Bendire mounted an attack on Ridgway with all the courage and skill available to an old Indian fighter, a fighter who had been decorated for bravery and who is reported to have entered Cochise's camp during hostilities to parley with the great Chiricahua chief. Charging ahead, Bendire wrote:

> My observations regarding the position of the nest of the Western Wood Pewee are radically different from the above, and all that I have seen, some twenty in number, were saddled directly on limbs, or at points where branches forked, and never in crotches; and the seventeen specimens now before me were all similarly placed. Among these is one collected by Mr. Ridgway himself, No. 15200, United States National Museum collection, collector's No. 1282, from Parley's Park [Utah], June 25, 1869, which is catalogued in the above-mentioned report as "Nest in crotch of a dead aspen along stream," but which shows distinctly that it was saddled *on* a horizontal fork and not *in* an upright crotch. If the Western Wood Pewee places its nest occasionally in a crotch, which I do not deny, it is exceptional and not the rule, and the many records I have of its nesting from Texas, Arizona, Nevada, Utah, Colorado, California, and Oregon confirm my assertions fully, and show conclusively that this species does not differ in this respect from the [Eastern] Wood Pewee.[272]

Willow Flycatcher
Empidonax traillii

The breeding range of the Willow Flycatcher covers southwestern Canada and much of the United States, except for the Southeast. This is a common breeding bird in the Great Basin. The Willow Flycatcher was formerly included in a species called the Traill's Flycatcher. In 1973 the Traill's Flycatcher was split into two species—the Willow Flycatcher and the more northerly Alder Flycatcher, *E. alnorum*.

The Willow is one of five species of Great Basin flycatchers belonging to the genus *Empidonax*. The members of this genus look so much alike that they usually cannot be identified with certainty in the field by plumage alone. One often ends up identifying them merely as *Empidonax* flycatchers or empidonaces, without trying to label them as to species. They are all small, with a light-colored ring around each eye and two white bars on each wing. Their backs are darker-colored than their underparts.

Empidonax flycatchers can best be identified as to species during the nesting season. Then, they have different habitat preferences, habits, and territorial songs. During migration identifications carried beyond the genus level are iffy, since the males are usually not singing then, and migrants will linger and feed in all types of habitat. Each species has various call notes, and familiarity with these call notes may be helpful in arriving at an identification. But it is the song of the male on his territory which will be of most help to the birder. Remember, singing will be at its peak in the early morning and evening.

The nesting habitat of the Willow Flycatcher effectively isolates it from the company of other empidonaces during the breeding season. As its English name denotes, this flycatcher lives among willows, often nesting in thickets along streams and along the margins of ponds and lakes. Here, in patches of willows, alders, or shrubs, all the basic necessities of life are present—nesting sites, scanning perches and flycatching airspaces, song perches, and loafing and roosting sites. The Willow Flycatcher nests low in a willow or shrub or clump of ferns. It flycatches around and in breaks in the foliage canopy of the willow thickets. Because of the low profile of most of the plants in its environment, its scanning perches and flycatching airspaces are positioned very close to the ground. The nest of the Willow Flycatcher at times is characterized by long pieces of grass projecting out in various directions or dangling from the bottom. The usual song of the male is a low, rough *fitz-bew*.

Hammond's Flycatcher
Empidonax hammondii

The breeding range of the Hammond's Flycatcher lies in western North America, extending from central Alaska southward to the mountains of central California in the west and northern New Mexico in the east. The Hammond's Flycatcher is a migrant in the Great Basin. It is also an uncommon breeding species in the mountainous eastern and western rims of the Basin and in some of the mountain ranges within the Basin. Breeding specimens have been collected from the Carson Range in the west and the Snake Range and Raft River Mountains in the east.[289]

The nesting habitat of the Hammond's lies in dense, shady, rather humid montane forests. In the Sierra Nevada this flycatcher frequents lodgepole pine forests on up into the red fir forests of the Canadian life zone; it rarely nests in deciduous groves such as those of aspens. In Utah and Colorado it nests in both aspen and coniferous forests.

The territorial song of the male Hammond's is delivered from high song perches. There are various descriptions of it in the ornithological literature. Following field studies, Johnson described the three syllables of the song as a snappy *whee-suk* or *whee-sup*, with the last part of the syllable at either a higher or a lower pitch; a rough, burred *bzrrp*, falling off in pitch at the end; and a rough, burred *bzeep*, rising in pitch at the end. The three syllables are put together in "couplets of two or in various sequences of three" by the singing males.[302]

During the breeding season activities of the Hammond's Flycatcher are carried on well above the ground, at heights of twenty to one hundred feet. Here nesting, foraging, singing, loafing, and roosting all occur. The flycatching airspaces are shaded and cramped, lying "within the branch-work of old trees or between rather close-standing younger trees."[26] The foraging heights of this flycatcher were studied in montane aspen-conifer habitat in southern Colorado. There the bird foraged principally in a zone twenty to forty feet high, at the level of the middle canopy of taller conifers and aspens.[303]

Dusky Flycatcher
Empidonax oberholseri

The breeding range of the Dusky Flycatcher lies in western Canada and the United States. This is a common and widespread migrant and breeding species in the Great Basin. It is much more widely distributed here as a breeding species than is the Hammond's Flycatcher.

The montane breeding habitat of the Dusky Flycatcher may be chaparral—low, thick, scrub vegetation containing scattered coniferous or deciduous trees; or it may be open, sunlit conifer or aspen forests with a bushy or shrubby understory; or it may be around meadows or clearings in dense forests. The Dusky usually nests in a shrub or small tree within four to seven feet or so of the ground. Most of its flycatching takes place within thirty feet of the ground.

The song of the male Dusky Flycatcher is composed of three syllables. Johnson has described these as follows: a *prllit* or *prit* of intermediate pitch, a *brrltt* or *prrddrt* or *prrlrt* of low pitch, and a *peet* or *preet* or *seep* or *zeet* of high pitch. The song may be a three-part sequence composed of one of each of the three syllables, or it may be a four-part sequence where one of the syllables is repeated. Sometimes one of the syllables may be left out, and by the end of the nesting season the song sequence may consist of only one or two syllables separated by long pauses.[302]

Gray Flycatcher
Empidonax wrightii

The breeding range of the Gray Flycatcher lies in arid western United States. The Gray is a common and widespread migrant and breeding species in the Great Basin.

The pinyon-juniper woodlands form the principal breeding habitat of the Gray Flycatcher in the Great Basin. In the western end of the Basin, it also nests in big sagebrush shrublands—where the sagebrush is luxuriant and tall and sometimes almost "with the stature of small trees."[103] Here, some nesting may also take place in open, dry yellow pine or mountain mahogany areas where sagebrush grows in among the trees.

Because of the low profile of the shrubs and pinyon and juniper trees in its breeding habitat, most of the activities of the Gray Flycatcher are played out close to the ground. The scanning perches are on top of shrubs or trees, and the flycatching airspaces are close to the ground. The flycatcher will often capture insects on the ground or on low plants. Nests are usually built within six feet of the ground. The song of the male consists of two syllables, which can be sung separately or put together in various combinations. These two syllables have been variously described as a vigorous *chi-wip* and a fainter but higher-pitched *cheep*[84] or as a vigorous *chulup* and a softer *keep*.[92]

Western Flycatcher
Empidonax difficilis

The breeding range of the Western Flycatcher extends from southeastern Alaska to Baja California and western Texas. The Western Flycatcher is an uncommon to common migrant and breeding species in the Great Basin. It has been recorded as a breeding species in the Carson Range, Toiyabe Mountains, Ruby Mountains, Spruce Mountain, Deep Creek Range, Raft River Mountains, and others.[289]

The breeding habitat of the Western is usually near running water in well-shaded montane forests or canyons. This flycatcher's activities are usually carried out within forty to fifty feet of the ground in the forest canopy. However, although its scanning perches and song perches may be as far as forty feet above the ground, it may on occasions nest on the ground.

The song of the male Western Flycatcher involves three syllables. These three syllables have been variously described: as a *ps-séet ptsick* and, after a short pause, *sst*[84] and as a whistled *pu-eet*, then a snappy *pit-tu*, and then a high-pitched *seet*.[92] The order of the syllables may be varied.

Say's Phoebe
Sayornis saya

The breeding range of the Say's Phoebe extends from central Alaska to central Mexico in western North America. The Say's is an uncommon to common summer resident and breeding bird in the Great Basin. Although most of our phoebes are migratory, an occasional winter bird is seen here. The Paiute name for the Say's Phoebe was To-que'-oh.[25]

Say's Phoebes inhabit open areas both in the valleys and foothills and where canyon mouths open onto the desert. They prefer sparsely treed localities, where the shrubs are low and scattered and the ground shows through. Because phoebes capture prey close to the ground, sometimes on the ground, they can forage effectively only in low, sparse vegetation. Their scanning perches are located on weed tips, low bushes, fence posts and wires, rocks, and any other available low vantage point. Phoebes also do some scanning while hovering within a few feet of the ground. Of all our flycatchers, the feeding niche of the Say's Phoebe is positioned closest to the ground.

These phoebes are among the earliest of migrants to return in the spring, sometimes arriving in late February or early March. This is especially early for flycatchers, who typically constitute the tag end of migration. Flycatching is highly rewarding only on warm days in the insect-filled air of late spring and summer—winter comes and goes all spring long in the Great Basin. Say's Phoebes are almost entirely insectivorous, and their early spring passage may be possible because their foraging niche is positioned in a warmer microclimate than that of other flycatchers. When the sun shines in the spring, sites on or close to the ground warm up to well above the general ambient temperature, and insect activity is more heated on and close to the ground.

Say's Phoebes build their nests in sites where there is solid overhead cover. In the wild, phoebes nest on recessed rocky ledges, on ledges in caves, beneath overhanging rocks on cliffsides, in holes in banks, and under overhanging banks. Phoebes are not shy around human company. As the West was settled, they built their nests in barns and sheds, on porches, under bridges, inside abandoned buildings, and in the entrances of old mine shafts.

Phoebes were quick to form attachments with settlers in the Great Basin. In describing the Say's here in the 1860s, Ridgway wrote:

> It was even noticed at several stage-stations in the midst of the Humboldt and Carson Deserts, where no water occurred except in the ar-

tificial wells. About the larger settlements it was found to be more numerous, and at Unionville, in the West Humboldt Mountains, had, with *Sialia arctica* [Mountain Bluebird] and *Salpinctes obsoletus* [Rock Wren], taken possession of the abandoned houses in the upper portion of the town. At this place we observed a nest which was attached to the under side of the eave of a large stone building, being apparently built upon the base of a deserted nest of the Cliff Swallow.[25]

Ash-throated Flycatcher
Myiarchus cinerascens

The breeding range of the Ash-throated Flycatcher lies in western United States and Mexico. This bird is an occasional to common summer resident in the Great Basin.

The Ash-throated Flycatcher's generic name is derived from the Greek: *myia* for fly and *archon* for ruler. Although *Myiarchus* flycatchers are not quite as bellicose as kingbirds, they surely deserve better than the title of ruler of flies.

The nesting behavior of North American members of the genus *Myiarchus* is unusual for flycatchers, since they nest in cavities in trees or in substitute cavities such as old drainpipes and artificial bird boxes. Ash-throated Flycatchers nest in old woodpecker holes and in natural cavities in junipers and cottonwoods in the Great Basin. Sometimes an Ashthroat will live up to its family name of tyrant flycatcher and not wait until a woodpecker has had the use of a newly excavated nesting cavity—it will dispossess small woodpeckers like the Downy from freshly excavated nesting cavities by claim jumping.

The Ash-throated Flycatcher does not inhabit densely forested or wooded areas; in the Great Basin it prefers situations such as pinyon-juniper woodlands or riparian cottonwood groves. Scanning perches are employed for launching sallies on passing insects. After an insect is captured, the flycatcher may make a spread-tail turn and alight on a new scanning perch—it does not show the strong tendency of most flycatchers to return repeatedly to the same perch between sallies. Sometimes flycatching is supplemented by gleaning. The flycatcher will hover in the foliage as it gleans insects. This may be how it captures many of the hemipterans—true bugs—and caterpillars in its diet. Hymenopterons—bees and wasps—usually form the major item in its diet.

Flycatching is a skilled profession, accomplished on the wing against prey which are not always passive targets: the insect frequently becomes aware of a fast-closing bird and takes evasive action. The question has arisen of whether young birds acquire proficiency in the art and science of flycatching strictly by trial-and-error learning, or whether some parental teaching is involved. Irene Wheelock witnessed an Ash-throated Flycatcher drilling her young just four days out of the nest. Appearing with a small butterfly in her bill where her young were perched in a low tree, the female fluttered close in front of one of the young and dropped the insect. It "flew lamely down just in front of an eager baby," who darted down after it, alighting on a still lower perch to eat it. Despite her fluttering, scolding, and coaxing, the young bird did not resume a higher perch again until after finishing its meal. Then, off and on during the day, the female repeated, with variations, the lesson of flycatching.[304]

Western Kingbird
Tyrannus verticalis

The breeding range of the Western Kingbird lies in western North America, extending from southern Canada to Mexico. The Western is a common migrant and breeding species in the Great Basin.

The Western Kingbird nests in valleys, preferring streamsides where stands of cottonwoods and other deciduous trees are present. Frequently, it is abundant on ranches and along the edges of towns and cities; in towns it may nest in residential districts. Although it flycatches in open areas and over shrublands and grasslands, it requires trees or utility poles upon which to build its nest. The status of the Western Kingbird in the Great Basin has apparently not changed much with settlement. When describing its status in the 1860s, Ridgway wrote, "Generally distributed throughout all fertile districts of the west, this species was extremely abundant in favorable localities."[25]

Ridgway was probably one of the first to note that the Western Kingbird was more aggressive than the Eastern Kingbird. In describing the Western, he wrote:

> In its habits, this Kingbird is remarkably similar to the eastern species, *T. carolinensis*, and their nest and eggs cannot be distinguished; but it is of an even more vivacious and quarrelsome disposition, continually indulging in aërial combats, sometimes to such an extent that half a dozen or more may be seen pitching into each other promiscuously, but apparently more from playful than pugnacious motives. They are also of a very sympathetic disposition, for when a nest is disturbed, the owners soon bring around them, by their cries, all the others in the neighborhood; but no sooner do they assemble than they begin their playful con-

> tests, and fill the air with their twitterings. Their notes are all weaker and less rattling than those of the eastern species, partaking more of the character of a tremulous, though rather shrill, twitter. . . . We know of no other bird so easily tamed, or which so thoroughly *enjoys* the society and protection of human beings, when once domesticated, as this species.[25]

At one time the common name for this kingbird was the Arkansas Flycatcher or Arkansas Kingbird. Its present name is more appropriate. However, during the twentieth century, particularly during the latter half, the Western Kingbird has been extending its breeding range and migratory wanderings eastward. Some breeding has been recorded from Illinois, southern Michigan, northwestern Ohio, and southern Ontario; and the Western Kingbird is now a "sparse but regular migrant in fall on [the Atlantic] coast."[305]

The Western Kingbird makes its living by flycatching—feeding heavily on hymenopterons (bees and wasps), coleopterans (beetles), and orthopterans (grasshoppers and crickets). Kingbirds use scanning perches and flycatch in wide-open areas. After capturing a high-flying insect, a Western Kingbird may lazily glide back down to its perch.

Kingbirds have voracious appetites. During the United States Geological Exploration of the Fortieth Parallel, the survey team had a succession of three Western Kingbirds as pets while in the Great Basin. The first of these pets was called Chippy. Robert Ridgway, the expedition's zoologist, was intrigued by Chippy's almost insatiable appetite. After much camp speculation, a count was made of the number of grasshoppers Chippy consumed in a day. Each member of the expedition kept track of the number of grasshoppers he fed Chippy. In one day's time, Chippy consumed 120 grasshoppers.[25]

Eastern Kingbird
Tyrannus tyrannus

The breeding range of the Eastern Kingbird extends from central Canada southward over much of the United States—except for some regions along the Pacific Coast and in the Southwest. The distribution of the Eastern in the Great Basin is quite complex. In some localities it is a rather common breeding species; in others it is of rare or casual occurrence or missing altogether. It is most abundant in the northern and eastern ends of the Basin—it is listed as a common breeding species on the official checklists of the Bear River National Wildlife Refuge in the northeastern corner and the Malheur

National Wildlife Refuge in the northwestern corner. In general, the Eastern Kingbird decreases in commonness both southward and westward in the Great Basin.

Kingbirds are flycatching hunters of insects. They earned the lordly title of kingbird not by their demeanor in the company of such lowly subjects as insects but, rather, by their tyrannical behavior toward other birds. Kingbirds do not tolerate the presence of other birds, especially large birds. When a vulture, hawk, raven, or other such intruder wanders into a kingbird's airspace, it is soon under attack. The kingbird launches into the air, climbs above the intruder with shrill cries, and mounts a fierce attack on the defenseless back of its victim. It may even alight on the intruder's back and loosen feathers in the fury of its attack. Sometimes its mate will join in the attack.

The breeding range of the Eastern Kingbird overlaps that of the Western Kingbird; they live in sympatry over about 60 percent of the breeding range of the Western.[306] Usually, when two closely related species of similar habits are sympatric, one species may have some competitive advantage over the other and be more prosperous. In the Great Basin the Western Kingbird is much more abundant and widespread than the Eastern Kingbird. It may well be that the Western is better adapted to the hot, arid climate. Also, it has been noted in Colorado that, although the two kingbirds may sometimes nest in the same area without too much conflict, the Western Kingbirds "are more aggressive and often chase the other away."[307]

Other Tyrannids

The breeding range of the Black Phoebe, *Sayornis nigricans*, lies mainly south of the Great Basin. The Black Phoebe is of rare occurrence in the Basin. There are records for Salt Lake City in the eastern end and from the Lahontan Valley, Lake Tahoe, and Carson River Basin in the western end.

The breeding range of the Eastern Phoebe, *Sayornis phoebe*, lies to the east and to the north of the Great Basin. This phoebe is of casual occurrence in the eastern part of the Great Basin, where there are sight records from Fish Springs and Salt Lake City.[27]

The Vermilion Flycatcher, *Pyrocephalus rubinus*, is of accidental occurrence in the Great Basin. Its breeding range lies to the south of the Basin. In mid May of 1981, a male and female Vermilion appeared on Valley Road in Reno and were seen there for several weeks.

The breeding range of the Cassin's Kingbird, *Tyrannus vociferans*, curls around the eastern and southern sides of the Great Basin. This kingbird is an occasional summer resident in the eastern part of the Basin as far north as

Salt Lake City.[27] It is of accidental occurrence in the western part of the Basin. There is a September record for Minden, Nevada; one kingbird was found dead along a highway, with a live one nearby.[75]

The Scissor-tailed Flycatcher, *Tyrannus forficatus*, breeds southeast of the Great Basin. This flycatcher is of accidental occurrence in the Basin. There are several sight records for Utah.[27] The Scissortail is on the accidental list at Malheur National Wildlife Refuge.

FAMILY ALAUDIDAE

The lark family, Alaudidae, is an Old World one. Only one species, the Horned Lark, reached the New World. Larks are small birds, plainly colored in cryptically streaked browns and grays. Their bills are pointed and slightly downcurved. They have long, straight, sharp hind claws. Larks inhabit open country where the vegetation is low and often sparse; they feed, nest, loaf, and sleep on the ground. Many larks are fine songsters, and some are renowned for their flight songs. They are highly gregarious and are found in flocks when not breeding.

Horned Lark
Eremophila alpestris

The breeding range of the Horned Lark extends over much of the Americas, Eurasia, and northern Africa. The Horned Lark is a widespread and abundant permanent resident in the Great Basin. Migrants and winter visitants are present in season.

The Horned Lark is the only native North American member of the Family Alaudidae—meadowlarks and other "larks" are not true larks. Horned Larks inhabit areas where the vegetation is low or widely scattered with some bare ground showing. They are not found in forested or wooded areas, tall grasslands, or dense shrublands. In the Great Basin the Horned Lark is most abundant in the arid valleys but may occur in suitable habitat on mountain plateaus or in montane fields.

Ridgway captured the very essence of the Horned Lark in the Great Basin in his account of the 1860s. Since larks in the Basin are today as they were then, his words are worth repeating:

> Few birds are more widely distributed than this one; and if the sagebrush deserves the title of "everlasting," from its abundance and uniform distribution, it would be as proper to designate this species as "omnipresent," so far as the more open portions of the western country

> are concerned. No locality is too barren for it, but, on the contrary, it seems to fancy best the most dry and desert tracts, where it is often the only bird to be seen over miles of country, except an occasional Dove (*Zenædura carolinensis*), or a solitary Raven, seen at wide intervals. Neither does altitude appear to affect its distribution, except so far as the character of the ground is modified, since we saw them in July and August on the very summit of the Ruby Mountains, at an altitude of about 11,000 feet, the ground being pebbly, with a stunted and scattered growth of bushes.[25]

Although the male Horned Lark sings from lowly, earthbound perches, he also sings in flight, as does his much more famous relative, the Skylark of Europe. The flight song of the Skylark has inspired two major poets, Shelley and Wordsworth, to capture it in moving verse. The flight song of the Horned Lark is unlikely to inspire even a minor poet: it has been described as "thin and unmusical" and "suggesting the syllables tsip, tsip, tsée-di-di."[84] The male Horned Lark commences a song flight by suddenly rising with rapid wingbeats, climbing in irregular circles or almost vertically if flying into a strong wind, to a height of several hundred feet. High in the sky, he flaps and soars in circles or curves, singing again and again. Often the song barely reaches a human listener below. After a prolonged performance, the lark folds his wings and plunges earthward, only to open his wings at the last moment and sail along over the ground before alighting.

There are many subspecies of Horned Larks in North America. In several regions observers have heard two types of song during song flights. A recitative song or rapid, monotonous series of *pit-wit, wee-pit, pit-wee, wee-pit* is sung for several seconds while the lark is flapping at the beginning of the flight song. Occasionally the recitative song is delivered at other times during a song flight, but the rest of the composition consists mainly of bouts of intermittent song. The intermittent song of *pit-wit, pit-wit, pittle wittle, little, little, leeeeee* is sung for several seconds while the lark soars. Longer silent periods interrupt the bouts of intermittent song as the lark climbs upward at the end of the glide.[308]

The female Horned Lark builds her nest in a depression in the ground. She may excavate her nest hollow so that it is sheltered by, or even partly under, a tuft of grass or plants or a cow chip. The most unusual feature of the nest site is the "pavement" of small pebbles or flakes of cattle dung or pellets of mud that some females build to one side of the nest. Sometimes pavement materials are piled around the rim of the nest. As yet there is no satisfactory explanation of this custom.

The open, arid areas inhabited by Horned Larks expose them to heat, cold, and water stresses. They do not have any special ability to hold down their evaporative water losses during the hot desert day. At an ambient temperature of 25° C, their daytime evaporative water losses occur at an average rate of about 0.6 percent of their body weight per hour, about average for a twenty-six-gram bird. By the time the ambient temperature has reached 45° C, the evaporative water loss rate is up to almost 4 percent of their body weight per hour.[309] The larks' water losses are greater than their gain of metabolic water from their food, and they need drinking water or water of succulence to remain in water balance. If surface water is available, they will drink it. However, it is believed that their shift from a diet of almost entirely seeds during the winter to one heavily larded with insects in the spring and summer allows them to obtain enough water of succulence to remain in water balance. In one experiment, Horned Larks could not survive indefinitely on a diet of commercial birdseed without drinking water, even at a mild ambient temperature (23° C) and high relative humidities (40 to 60 percent). However, some larks got by for as long as sixteen to thirty-one days before they were severely water-stressed.[309]

Horned Larks have a degree of physiological adaptation for life in hot environments. Their standard metabolic rate is about 25 percent lower than is typical for a twenty-six-gram bird, and their thermoneutral zone is high at 35° C. Horned Larks regulate their body temperature at about 41.5° C, but at night their temperature drops 2 to 2.5° C, lowering their heat and water losses. Under heat stress the body temperature may rise to as high as 45° C, allowing for heat storage and for more favorable heat exchanges with their environment; and this delays the onset of heavy expenditures of water for body temperature regulation.

Horned Larks also practice behavioral temperature regulation. During the heat of a summer day, they remain relatively inactive in the shade cast by rocks, shrubs, or other vegetation. While foraging they fly from shade to shade and avoid sunlit, superheated ground. This helps hold the heat burden down even for these ground foragers. On a hot day when the wind is blowing, larks often perch at the top of shrubs where they are fully exposed to the wind. On cold, windy days, if the sun is out, larks will sun themselves on the lee side of rocks or shrubs. During stormy weather they seek the shelter of thicker clumps of vegetation. On cold nights Horned Larks dig roost holes in the ground with their bills and sleep in these holes with their backs about level with the ground—thus only a minimal amount of body surface is exposed. This minimizes radiative and convective heat losses and possibly conductive heat losses as well. The roost holes are often located out of the wind on the lee side of vegetation. In soft earth, larks may dig new roost holes each evening; but in hard earth, where digging is difficult, they may use their roost holes on subsequent nights.[309]

FAMILY HIRUNDINIDAE

Hirundinidae is a family of swallows of nearly worldwide distribution. Hirundinids are graceful aerial artists who spend more time airborne than any other group of passerines. Their legs and feet are poorly developed, and they walk with difficulty on the ground—which they seldom visit. Although their triangular bills are short and flat, their mouths can be opened widely, and the surrounding rictal bristles enhance the insect-netting abilities of the mouth. Swallows cut, turn, swoop, climb, and often skim over vegetation or water as they fly down insects.

These birds are highly gregarious and not only nest in colonies but also feed in groups—often in mixed-species groups. Swallows are darker-colored above than below. Often their dark backs of blue, purple, or green possess a metallic sheen or luster. The call notes of hirundinids are twittering or squeaking ones. During the fall, huge concentrations of swallows build up along their migratory pathways, particularly around marshes, where they feed by day and roost in the emergent vegetation by night. At this time, utility wires and barbwire fences may be lined with thousands of swallows of various species—sunning, loafing, and preening between feeding bouts.

Purple Martin
Progne subis

The breeding range of the Purple Martin extends from southern Canada to northern Mexico, mainly in regions east of the Rocky Mountains or west of the Cascades and Sierra Nevada. The Purple Martin is an uncommon breeding species in the eastern part of the Great Basin. In Utah it is reported to be a "sparse and localized breeding species in the mountains throughout the state."[56] There are no modern breeding records for the Purple Martin in the western part of the Basin, although it has supposedly nested on the California side of Lake Tahoe in recent years. Past records of breeding on the western edge of the Great Basin include Eagle Lake in 1905 and possibly Carson City in the 1860s.

The Purple Martin was apparently more common in the Great Basin in the late 1800s and early 1900s than it is today. Purple Martins have not been recorded in west central Nevada since Ridgway's observations of martins at

Carson City and Virginia City in the 1860s.[25] In comparing the past and present occurrences in Utah of the Purple Martin it has been said that "Henshaw . . . reported its occurrence throughout Utah in large colonies, both in towns and cities as at Salt Lake City, Salt Lake County, where it was breeding in bird boxes. . . . We are not aware of any such large colonies in the state at the present time, although it is found consistently in mountainous areas especially where there are aspen forests and ponds or lakes over which the birds feed."[56]

Tree Swallow
Tachycineta bicolor

The breeding range of the Tree Swallow spans the North American continent from coast to coast and extends from northern Alaska, through Canada, down into northern and central United States. The Tree Swallow is a common breeding species and a common, often abundant migrant in the Great Basin. Prior to 1982 its scientific name was *Iridoprocne bicolor.*

The nesting of these swallows is restricted to localities where tree cavities are available. In the Great Basin there is keen competition for tree cavities, since the supply is limited—conflicts frequently arise over possession of a nesting cavity. In the valleys the situation is more critical than in the mountains, since so much of the valley land is shrubland, where both trees and woodpecker excavators are in short supply.

Tree Swallows seem to have a decided preference for nesting in the vicinity of water or wet meadows over which they can forage. In the 1860s Ridgway found this swallow in greatest abundance "among the cottonwoods of the Lower Truckee, near Pyramid Lake, in May," where "every knot-hole or other cavity among the trees seemed to have been taken possession of by a pair."[25] Today, Tree Swallows are probably more abundant nesters in the mountains than in the valleys of the Great Basin. Subsequent to the 1860s many Great Basin valleys have been invaded by the European Starling, an exotic pest. The starling is a cavity nester and an aggressive, pugnacious bird—it readily displaces native cavity-nesting birds. The European Starling has undoubtedly played a role in depressing the breeding populations of Tree Swallows and other native birds in lowland sites in the Basin since the 1950s.

The Tree Swallow will nest near human habitations, even in artificial bird boxes. Ridgway noted that, following settlement in the Great Basin, these swallows were quick "to have taken advantage of the abode of man in localities where there are no trees to accommodate them. Such was conspicuously the case at Carson City, where they were quite numerous, and built their nests under the eaves, behind the weather-boarding, or about the porches of dwellings or other buildings, and were quite familiar."[25]

Although Tree Swallows consume more plant material than do other swallows, most of their diet consists of flying insects. These birds forage on the wing, coursing back and forth in true swallow fashion. They alternate flapping flight and sailing or gliding—they have a tendency to sail more than do Violet-green Swallows. Tree Swallows share a trait with some other swallows of coming out of a glide and, with several quick wingbeats, climbing steeply upward. They momentarily pause at the top of the upward climb; then, just before stalling, they sail back downward for a way before entering flapping flight again.

Tree Swallows are not true colonial nesters, even though they are social birds and forage together away from their nests. Bonds of real affection are evidently formed between mates or even between the members of a foraging flock. Ridgway noted displays of affection when he was collecting museum specimens. He wrote, "The specimens in the collection were shot on the wing; and when one was brought down the rest would exhibit great concern, circling about the victim, and uttering a plaintive twitter, as their suffering companion lay fluttering on the ground."[25]

Violet-green Swallow
Tachycineta thalassina

The breeding range of the Violet-green Swallow encompasses much of western North America, from northern Alaska south to central Mexico. This is a common migrant and breeding species in the Great Basin.

During the Great Basin summer, Violet-green Swallows nest in both the valleys and the mountains. Because of their swift, powerful flight and aerial foraging skills, these swallows can find food in practically any locality in the Basin during good weather. But they are limited to nesting at sites where suitable cavities are available. Two general kinds of nesting cavities are employed—crevices in cliffsides and canyon walls and old woodpecker holes or natural cavities in trees.

Early on, in the 1860s, Ridgway concluded that the Violet-green Swallow was a "strictly saxicoline" or rock-inhabiting species in the Great Basin and was to be found "only where precipitous rocks, affording suitable fissures, occurred."[25] Either Violet-greens have added the option of nesting in tree cavities since Ridgway's day, or his field observations were not extensive enough to detect this habit.

Ridgway first encountered the Violet-green Swallow nesting in "cliffs of calcareous tufa" on Anaho Island in Pyramid Lake. Subsequently, he found it nesting in the "limestone walls of the eastern cañons of the Ruby Mountains." Although possibly in error about the pervasiveness of its habits, Ridgway did make some sound ecological generalizations about the composition of the saxicoline nesting community in which the Violet-green can be found. He noted that this swallow commonly nested on the same cliffsides and canyon walls as the White-throated Swift and Cliff Swallow and that, at sites where suitable caverns were present on the cliffsides and canyon walls, the Barn Swallow was also a member of the saxicoline nesting community.

In the Great Basin, the Violet-green Swallow also nests in old woodpecker holes and natural cavities in deciduous and coniferous trees, particularly in the mountains. However, competition for these cavities is keen—there are just not enough Great Basin woodpeckers at work to meet the demands for housing on the part of cavity-nesting birds. Because of competition with other birds and since tree cavities are usually not highly bunched, there are fewer opportunities for Violet-greens to nest colonially in tree cavities than on cliffsides and canyon walls, and much solitary nesting takes place. Nevertheless, in some localities small groups of cavities are available in the same tree or in neighboring trees. The tree cavity–nesting community of which the Violet-green Swallow is an associate includes the Yellow-bellied Sapsucker, Hairy Woodpecker, Northern Flicker, Mountain Chickadee, White-breasted Nuthatch, House Wren, Mountain Bluebird, and Tree Swallow. Violet-green Swallows prefer to nest in open or broken forested areas or along the borders of clearings, meadows, and chaparral where the trees are not densely packed.

Violet-green Swallows spend much of each day on the wing, capturing flying insects. They are swift, strong, tireless fliers who alternate flapping flight with sailing or gliding. They hunt low, they hunt high, they hunt wherever flying insects are to be found. Unlike flycatchers—who return to a scanning perch between captures—swallows, like swifts, are continually on the wing while foraging. When not foraging they perch, often in the company of other swallows, to loaf, preen, and sun themselves.

Like many of their swallow kin, Violet-greens do not shun towns and even cities when seeking nest cavities or foraging. This swallow is an early spring migrant, often arriving in west central Nevada in early March along with the Tree Swallow. Violet-green Swallows suffer during spring storms and may experience some mortality because of their early return. However, they can successfully undergo cold torpidity and possibly may get by during cold spring nights by entering this state. Violet-green Swallows found on the ground in cold torpidity in southern California revived and flew away after the sun warmed them up in the morning.

Northern Rough-winged Swallow
Stelgidopteryx serripennis

The breeding range of the Northern Rough-winged Swallow extends from southwestern and south central Canada, down across the entire United States, and into South America. The Northern Rough-winged Swallow is a common migrant and breeding species in the Great Basin. Prior to 1982 its common name was the Rough-winged Swallow and its scientific name was *S. ruficollis*.

The Northern Roughwing nests in burrows in earthen banks, much like the Bank Swallow. Unlike the Bank Swallow, however, it does not excavate a completely new burrow but modifies the existing old burrow of a ground squirrel or other animal. And the Northern Roughwing has much more flexible breeding habits than the Bank Swallow. It nests in all types of crevices and holes in a variety of situations, including the foundations or masonry of human-built structures. I have even found this swallow nesting in unused drainpipes.

A number of factors make the Northern Rough-winged Swallow a more versatile nester than the Bank Swallow. Not only does it nest in a wider variety of burrows than the Bank, but it is often a solitary nester and fits into many sites which would not accommodate a Bank Swallow colony. It is not as closely tied to water as the Bank Swallow and is often well removed from water when nesting. There are many dry arroyos in the Great Basin which provide bankside nesting sites and airspaces for catching flying insects along their length. Sometimes Northern Rough-winged Swallows will nest in association with Bank Swallows, and sometimes several pairs of Northern Roughwings will nest in the proximity of each other.

This small swallow is saddled with the genus name of *Stelgidopteryx*. This name, derived from the Greek, means scraper-winged: *stelgis* is a scraper, *pteryx* a wing. The name refers to a characteristic of the outermost flight feather which confers roughness to the edge of a wing. The outer web or vane of this primary flight feather has been modified into a series of stiff, recurved hooks. The function of this roughness is unknown. Many years ago, the great ornithologist Elliott Coues speculated that the "hooks assist the birds in crawling into their holes, and in clinging to vertical or hanging surfaces."[310] More recently, it has been suggested that the roughness of the wing edge may produce a high-pitched sound when the bird is in flight.

Bank Swallow
Riparia riparia

The Bank Swallow is widely distributed in North America, Eurasia, and northwestern Africa. In North America its breeding range extends from northern Alaska, through Canada, and over all but the southern end of the United States. The Bank Swallow is an uncommon to common migrant and breeding species in the Great Basin. It is a common summer resident in Utah,[27] but it is the least abundant of the breeding swallows in the western end of the Great Basin, where nesting colonies are widely scattered.

Both the common and the scientific names of this swallow—*riparia* is derived from the Latin *ripa*, meaning the bank of a stream—refer to the bird's way of life. The Bank Swallow nests colonially but in individual burrows, excavated in the vertical sides of stream and lake banks, sand or gravel pits, bluffs, and road cuts. The nesting burrows are usually positioned as high up on the side of the bank as possible. Much of what we know about Bank Swallows is due to exhaustive field studies by Dayton and Lillian Stoner, a husband-and-wife team. Their field studies were carried out in the Lake Okoboji region of Iowa and the Oneida Lake region of New York.

The Bank Swallow excavates its nesting burrow by digging with both bill and feet; the male and female alternate in digging. In starting a new burrow, the digger clings to the side of the bank, frequently using its tail as a brace, and pecks at the bank with rapid side-to-side movements of its head. A concavity is formed. Then, as the pecking produces somewhat of a shelf for the bird to stand on, it starts digging with its long, sharp claws as well as with its beak. Once the actual burrowing progresses inward, the swallow ejects dirt by vigorously kicking its feet and rapidly shuffling its wings. In easily dug banks, the burrow will progress inward at a rate of three to four inches a day. The burrow is formed with an arched ceiling and a flat floor. Although burrows may range in length from fifteen to fifty inches, the average depth may be around twenty-eight inches.[311] The entrance to the burrow is about one and a half to two and a quarter inches wide. The claws of the swallow may become rounded and blunt due to abrasion while digging.

Because earthen burrows are readily dug, Bank Swallows may lose young to weasels, minks, and skunks in areas where the bank allows these predators access to the burrows and an opportunity to dig. American Kestrels may also prey on Bank Swallows. Although Banks are seldom captured in flight, the young swallows are captured by kestrels who perch at the edge of the colony and wait for the nestlings to come out to the burrow entrance. When about fourteen to sixteen days old, young swallows spend much time there. A kestrel will charge the entrance and frequently nail a young swallow before it can retreat inward. Kestrels may fly from burrow to burrow, pausing to reach into each burrow with a foot, in an attempt to surprise a nestling near the entrance.[312] Then, the European Starling and the House Sparrow, two exotic pest birds, may enter a colony and appropriate nesting burrows from the swallows.

The Bank Swallow feeds almost entirely on flying insects, which it captures by flying rapidly and low over water or low vegetation such as meadows and marshes. It is usually seen in flapping flight. It seldom flies high or sails upward and then coasts downward as do other swallows.

Cliff Swallow
Hirundo pyrrhonota

The breeding range of the Cliff Swallow extends from central Alaska and Canada to central Mexico; this bird does not breed in southeastern United States. The Cliff Swallow is a common, often abundant migrant and breeding species in the Great Basin. Prior to 1982 its scientific name was *Petrochelidon pyrrhonota.*

Not only is the Cliff the most abundant and widespread swallow in the Great Basin today, but it was so even in Ridgway's day. In describing the status of the Cliff Swallow in the 1860s, Ridgway wrote:

> It was also noticed along every portion of our route across the Great Basin, especially in the vicinity of rivers or lakes, or at the settlements, whether large or small. The species may be considered the most abundant one of the family throughout the West. . . . In localities most remote from settlements it of course built its nest only on the face of overhanging cliffs, but if near a settlement, any large building, as a barn or church, was almost sure to be selected; in either case, vast numbers congregating together and fixing their peculiar gourd-shaped nests side by side or upon each other, the same as in the east.[25]

The Cliff Swallow is a colonial nester. A site suitable for occupancy by a nesting colony must meet certain basic requirements. The actual nesting site must possess a vertical surface, protected from above by an overhang or ledge or eave, for attachment of the mud nests—mud adheres best to surfaces such as rough rock, rough cement, or weathered board. There must be water with a wet mud supply in the vicinity. And there must be adequate airspaces for productive foraging for flying insects over meadows, marshes,

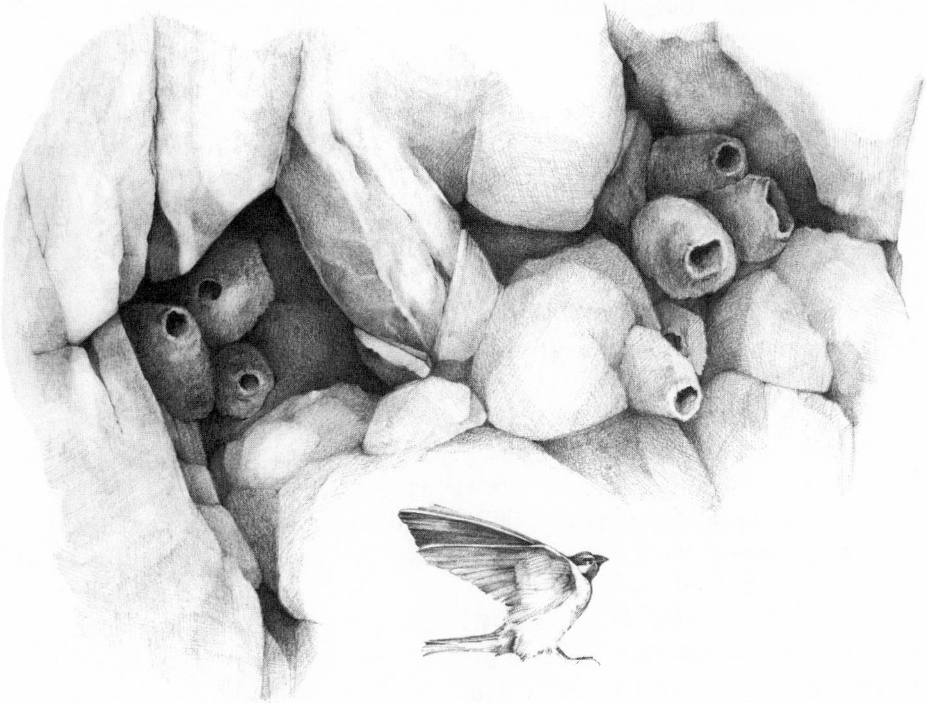

Cliff Swallow nesting colony

grasslands, or water. However, Cliff Swallows may fly several miles out to forage.

Cliff Swallows often simply reuse old nests that are in good condition or repair old ones that are in poor condition. At a new building site the initial nests are built at the junction of the vertical surface and the overhang; a nest positioned just beneath the overhang usually lacks a mud roof. Subsequent nests are built below the initial nests on the vertical surface and, if there is room, horizontally outward under the overhang—the surfaces of the first nests are used as vertical or horizontal attachments for subsequent nests. The male apparently initiates nest building by carrying pellets of mud in his beak to the nest site before he is fully pair-bonded with a female. But he soon has a female who joins in. While one bird is out gathering mud, the other remains at the nest. About one week is required to build a nest, which contains an estimated nine hundred to twelve hundred mud pellets.[313]

In the flat, marshy world of the Bear River National Wildlife Refuge, Cliff Swallows build their nests on the underside of the observation platforms that have been erected as vantage points for refuge visitors. A wooden tower with sides designed as attachment surfaces for Cliff Swallow nests has been erected near the refuge headquarters. The swallows have responded by forming a nesting colony on this structure.

The nest-building activities of Cliff Swallows go much beyond decorating Great Basin cliffsides and buildings with temporary housing for themselves. The nests of most birds are ephemeral structures which are wide open to the elements. The gourd-shaped nests of the Cliff are enclosed, however, and when built under a protective overhang have some permanence. So in their nest building Cliff Swallows join those other master builders, the woodpeckers and magpies, in providing nesting and roosting shelter to an array of wildlife. When you visit the nesting colonies at Bear River Refuge, you will probably see some House Sparrows nesting in swallow nests. House Sparrows regularly nest in Cliff Swallow colonies, and other species such as the Say's Phoebe, Plain Titmouse, and Eastern Bluebird have been reported occasionally nesting there. The swallow nests also afford roosting shelter to a variety of flying forms, including bats. In the Great Basin and elsewhere in the West, flocks of Rosy Finches regularly find winter shelter in the nests of Cliff Swallows. The Canyon Wren has also been reported roosting in Cliff Swallow nests.

Barn Swallow
Hirundo rustica

The Barn Swallow, like the Bank Swallow, is widely distributed in North America, Eurasia, and northwestern Africa. In North America its breeding range extends from northern Alaska to central Mexico. The Barn Swallow is a common transient in the Great Basin, and it is a common breeding bird in the valleys and at lower elevations.

Barn Swallows have greatly increased their distribution and numbers in the Great Basin since settlement has occurred. Since they require a solid roof over their nesting sites, prior to settlement they had to rely on caves and caverns in rocks to furnish overhead cover. In the 1860s in the Great Basin, Ridgway found them "inhabiting the same localities as the Cliff Swallows," but they were "everywhere much less numerous." He found them nesting in caves among the tufa domes at Pyramid Lake and in caverns in the limestone cliffs in the Ruby Mountains. In 1869 he found a Barn Swallow's nest in a stable at Parley's Park, Utah.[25] Then, as more stables, barns, sheds, and bridges were built, the swallows had more nesting sites.

These swallows nest in the vicinity of water, where they can obtain wet mud for their nest building. The number of pairs nesting at one site varies from one to many, depending on how many can be accommodated beneath the overhead shelter with some spacing between nests. Carrying mud back to the site in their beaks, the swallows build their nests on vertical surfaces or on a horizontal beam or rafter; the mud is mixed with straw or grass. If the nest is built on a vertical surface, the top rim of the nest is semicircular in shape. If the nest is built on a horizontal surface, the rim is more circular. Barn Swallows will often repair and reuse old nests.

Barn Swallows are very graceful fliers. While foraging for flying insects, they often barely skim the surface of water or the tops of vegetation. Swallows are flocking birds, and often a number of swallows of various species may forage together. The largest gatherings of swallows occur in late summer prior to migration. Hundreds, often thousands, of swallows of various species may assemble in the evening to roost together in a marsh. By day they may be seen strung along telephone wires or fence wires, preening and loafing and periodically launching into foraging flights. These vast congregations often vanish so abruptly on their fall migration that in earlier years it was commonly believed that the swallows simply disappeared under the marsh waters to hibernate in the bottom muds. Today we know that swallows do not lead a froglike winter but fly south to bask in warmer climes.

FAMILY CORVIDAE

Corvidae is a nearly worldwide family—of jays, crows, magpies, ravens, and their allies. Some of the corvids are the largest in size of all the passerine birds. In Great Basin bird society the corvids form the first family of the land. Seen and heard throughout the land, from valleys to mountains and from shrublands through woodlands to forests, corvids are conspicuously present. Corvids are a birder's delight. They are easy to find and follow in the field. Medium to large in size, they are bold, noisy, and often present in flocks.

Corvids rank among the most intelligent of birds, and many facets of their reproductive, feeding, food-storage, flocking, and general behavior are fascinating. They are not great beauties—with black, gray, white, and subdued blue colors prevailing. The sexes are colored alike. Although they possess a varied repertoire of notes and frequently sound off, their voices are harsh or raucous. They do not contribute to the more melodious music of nature.

Corvids are omnivorous in appetite, with stout bills which allow for variety in food. They have a taste for the eggs and young of other birds—this may be why they are often chased by small nesting birds. Most corvids are permanent residents in a region and do not migrate, although some movement does take place as the birds search for food, particularly during years of profound crop failure. Some migration may occur in regions where snow is heavy and winters are severe. The external nares or nostrils of most corvids are covered with forward-projecting feathers or bristles. Some ornithologists have considered these nostril muffs to be an adaptation to help hold down evaporatory water and heat losses in the cold. But experiments by Bruce Wunder and Joseph Trebella in which these feathers were removed from cold-stressed American Crows indicated that the feathers made relatively little difference in reducing evaporatory losses.[314]

Steller's Jay
Cyanocitta stelleri

The Steller's Jay ranges over much of western North America on into Central America. It is present in both the eastern and the western ends of the Great Basin but is apparently missing from much of the interior. Ridgway was the first to detect this seeming hiatus in the distribution of the Steller's in the Great Basin. Reporting on his fieldwork of 1867 to 1869, he wrote, "We found this jay only among the pines on the Sierra Nevada, since it did not, like the Nutcracker (*Picicorvus*), occur on the higher ranges of the Great Basin, though it was represented on the eastern side by the *C. macrolopha*—neither the latter nor the subject of these remarks occurring at any point intermediate between the Sierra and the Wahsatch, along the line of our route."[25]

We now know that the Great Basin is not as devoid of Steller's Jays as Ridgway thought it to be. Steller's Jays do occur at some localities between the Sierra Nevada and Wasatch Mountains. In the western part of the Basin, Steller's Jays are present in ranges along the western edge of Nevada, from the Carson Range to as far south as the White Mountains. They are also present in the Desatoya Mountains, well over a hundred miles into the interior. In the eastern part of the Basin, Steller's Jays occur deep in the interior, in ranges along the Utah-Nevada border such as the Deep Creek and Snake ranges. They also occur in northeastern Nevada in the Cherry Creek and Pequop mountains.[315]

More extensive searches must be made for Steller's Jays in the interior of the Great Basin. Apparent gaps in distribution can reflect gaps in knowledge as well as gaps in nature. However, it is interesting that Steller's Jays apparently have such a feeble presence in that part of the Great Basin where montane boreal plant zones of closed forests are poorly developed or missing, and where sagebrush-grass vegetation may extend from the valleys up to almost ten thousand feet.[1]

The Steller's is a creature of montane forests, especially coniferous ones. A simple way to become acquainted with this bird is to visit a picnic area or campground in a montane forest. Steller's Jays often abound at sites where food scraps are readily available—campground jays become unafraid of people and visit picnic tables in quest of food. However, being reared on handouts can have serious consequences for young jays. Richard Boyer told the story of what happened in Sequoia National Park after most of the campgrounds were closed in late September:

> The birds, having lived from campers' scraps all summer, were suddenly thrown on their own forage efforts, and the young birds, not knowing how to gather their natural food, suffered from near starvation. Groups of 25 to 40 birds would gather around the remaining habitations in the closed areas, and in order to keep more of them from starving, rations of rolled barley were thrown to them to supplement their diets, and until they accustomed themselves to natural food items. It was not unusual for the birds to be picked up by hand, such was their condition. Many starved and were found dead, and almost . . . all were this year's young birds.[316]

Steller's Jays are very noisy birds, except when close to their own nests or when searching out the nests of others to loot. They are quick to discover intruders, and their raucous alarm notes alert all birds and mammals within earshot to the threat of danger—human hunters detest this jay because of its big mouth. The jays delight in mobbing owls, and their harsh, nasal *wah* has several functions, including instigating a predator mobbing; several quick *shook* calls constitute a hawk alarm.[317] In the 1860s Ridgway collected a series of Steller's at Carson City, Nevada, during which one wing-wounded jay, hopping from limb to limb, uttered "a very perfect imitation of the squealing note of the Red-tailed Hawk (*Buteo borealis*), apparently for the purpose of preventing pursuit."[25]

These jays are omnivorous feeders. The bulk of their food is of plant origin—including acorns and nuts, pine seeds, and fruit. Being good corvids, they do have a taste for insects, eggs, fresh meat, and carrion. When food is present in surplus, they will store it in crevices in trees and even in the ground. They not only obtain fresh animal food by sneakily robbing the nests of other birds, but on occasions they go after moving meat in the bold fashion of raptors. The literature reports instances of Steller's Jays pouncing on a young quail and carrying it off, capturing a flying Pygmy Nuthatch in midair, and swooping down on a platform feeder to seize a Gray-headed Junco.

Steller's Jays remain paired the year around. With the approach of the breeding season varied courtship behavior is enacted, including courtship circling, during which the male closely circles around the female. The nest is usually constructed in a small conifer. Both sexes participate in building a bulky twig nest, using mud as a binder and fine plant materials for the lining. Apparently the female incubates the eggs, but both parents feed the young.

Steller's are usually permanent residents within a region. However, there are fall and winter movements which result in jays appearing at lower elevations. Ridgway interpreted these downward movements as the result of harsh montane winters. He wrote, "Except when driven to the lower ravines and foot-hills by the unusual continuance of cold weather or by violent snow-storms upon the mountains, it was not observed to descend to below the coniferous woods, though it was common in the lower edge of this forest-belt."[25] In addition to being responses to montane weather conditions, some of the fall and winter movements may represent the dispersal of young jays, away from their natal grounds, following the breakup of family groups.

Scrub Jay
Aphelocoma coerulescens

The range of the Scrub Jay includes parts of western United States, Florida, and Mexico. This bird is common over much of the Great Basin, except for the northwestern corner, where it is rare and local in occurrence. The following statement by Ridgway indicates that its status has not changed with settlement of the Great Basin: "This very interesting bird we found to be the most generally-distributed species of the family, since it occurred on nearly every range where there was water in the main cañons, or extensive woods of nut-pine and cedar on the slopes."[25]

Its English name indicates that this jay frequents areas where the vegetation profile at best is only of medium height. The Scrub Jay is a permanent resident in thickets and woodlands on foothill ridges and slopes, in lower canyons, and in ecotones or transitional areas between closed forests and woodlands or shrublands. Because of its habitat requirements, it overlaps more with the woodland-inhabiting Pinyon Jay than with the forest-inhabiting Steller's Jay.

As we have just seen, the English name of Scrub Jay is derived from the bird's habitat preference. Another rationale for coining names is to refer to the bird's call notes with an echoic name—such as Killdeer. Among Great Basin Indians, the Paiute name for the Scrub Jay was We'-ahk. In describing the call notes of the Scrub Ridgway wrote, "That most frequently uttered is a shrill screech, sounding like *we'-ahk, we'-ahk*, whence the name bestowed upon it by the Paiute Indians."[25]

The Scrub Jay feeds on a wide variety of animal and plant food. Pinyon pine seeds and acorns, when available, are important food items. And the Scrub Jay has the corvid family taste for the eggs and young of other birds. The nineteenth-century ornithologist Charles Emil Bendire reported that the Scrub Jay not only eats the eggs from other birds' nests but may then rip open the nests with its bill. Bendire accused the Scrub of being "especially destructive to the nests of the Black-chinned and Anna's Hummingbirds and the Ground Tit" or Wrentit. A number of observers have accused these jays of being highly destructive to domestic hens' eggs and chicks. Bendire charged that "as soon as they hear a hen cackle after laying, three or four of these birds go to the spot at once."[272]

The Scrub Jays' appetite for nuts and fruit has been troublesome for growers of almonds, English walnuts, cherries, and prunes. The birds' crop depredations, plus their reputed fondness for the eggs and young of game birds, have triggered a lot of jay shooting by growers and sportsmen. During the years jays were unprotected by law, California sportsmen organized

Scrub Jay

hunts, sometimes as contests between two teams, to thin out the jays. Steller's Jays as well as Scrub Jays were zapped on these hunts, for a jay is a jay is a jay.

Scrub Jays have been observed landing on mule deer to feed on ectoparasites, a behavior reminiscent of that of the Black-billed Magpie, another corvid. Jays alight on the backs of deer to search for ticks and flies. One jay was seen repeatedly landing on the legs of a deer and clinging there while capturing prey—probably flies. This is apparently not unusual behavior, because deer are unconcerned while being used as a feeding ground by jays.

Much bobbing characterizes Scrub Jays: they bob up and down when perched, when calling, and upon alighting. The postlanding bobbing has been described as follows: "Just after alighting a jay will often execute a deep bow involving the entire body, and this may be repeated a number of times and in different directions. The purpose of this bowing is not clear to us."[318] On the ground and when progressing through the limbs of a tree, a Scrub Jay hops. In flight it often alternates sailing or gliding with flapping.

Scrub Jays are the most elusive of jays. Although small parties of jays may be encountered in the field, especially family groups following nesting, adult Scrub Jays come mostly in pairs. In some parts of their range these jays maintain territory the year around. When foraging together for food or nesting materials, one member of a pair will often remain on sentinel duty, while the other forages in exposed places on or close to the ground.

Pinyon Jay
Gymnorhinus cyanocephalus

The Pinyon Jay inhabits much of western United States. It is a common permanent resident in the Great Basin. Prior to 1973 its scientific name was *G. cyanocephala*, and prior to 1982 its common name was the Piñon Jay. The Pinyon is one of the most fascinating members of the Great Basin avifauna, as it was in Ridgway's day. Ridgway wrote:

> This extraordinary bird was found to inhabit exclusively the nut-pine and cedar woods on the mountain ranges of the Interior, of which it was the most characteristic species. It was eminently gregarious, even breeding in colonies, and in the winter congregating in immense flocks, which sometimes consisted of thousands of individuals, all uttering their querulous notes as they swept to and fro over the hills, in their restless migrations.[25]

The Pinyon Jay thrives in regions where pinyon pines, *Pinus monophylla* or *Pinus edulis*, occur, but the jay is not totally tied to pinyon pines. Especially in the northern part of its range, it lives in the complete absence of pinyon pines. Here it replaces pinyon seeds with the seeds of other pines and with other items such as juniper berries. Over much of the Great Basin, Pinyon Jay economy is based mainly on the seeds of the single-leaf pinyon pine, *Pinus monophylla*. However, in the northwestern and north central parts of the Basin, roughly north of a line through Pyramid Lake and the length of the Humboldt River, there are no pinyon pines to provide seeds for the jay. In the eastern end of the Basin, *Pinus monophylla* is replaced by *Pinus edulis*, the two-leaf or Colorado pinyon pine.

The association between Pinyon Jays and pines is not a casual one, with the jays feeding on pine seeds when they are available—it is an association of deep-seated benefit to both bird and tree. Over the ages Pinyon Jays and trees have coevolved to the point where each possesses adaptations of value to the other—these being structural and physiological ones on the part of the trees and behavioral ones on the part of the birds. It is easy to understand how jays benefit by eating pine seeds, but it is not all that easy to comprehend how pines benefit by having jays eat their seeds, since on the surface this appears to be counterproductive. Let us first consider the least obvious aspect of the association between jays and trees, that is, how the trees benefit by having jays feed on their seeds.

Pinyon Jays not only harvest pinyon pine seeds for consumption in the late summer and fall, but they transport most of the harvested seeds to communal caching areas and store them in the ground for use during the winter and spring. On those occasional years when a bumper crop of pine cones is produced, a flock of jays will store a tremendous number of seeds during the long harvest period. Many of these caches will never be reclaimed by the birds. Some of the unreclaimed seeds will germinate to produce new pinyon pine trees. Consequently, pinyon pine trees employ Pinyon Jays to do their seed dispersal and planting. The seed dispersal service rendered by jays is a very effective one, since the birds transport the seeds away from the seed-producing trees. By caching the seeds the birds are inadvertently planting them in the soil, something difficult for a seed to accomplish on its own but essential for successful propagation. Further, the jays differentiate between seeds and harvest and cache only sound, healthy seeds. Then, by placing a number of seeds in each cache, they insure that competition will ensue between germinating seeds for a position in the soil. To make the system work, the trees must produce more than enough seeds to support the needs of the jays, and the jays must cache more than enough seeds to meet their needs. Pinyon Jays are abetted as pinyon pine planters by

Pinyon Jay

some of the other corvids, especially by Clark's Nutcrackers, and by some rodents, who also harvest and cache pine seeds.

For services rendered, pinyon pines provide Pinyon Jays with cover from the weather and predators, with perching and nesting sites, and above all with an excellent food staple. Not only are pinyon pine nuts very large and thin-shelled, as pine seeds go, but they are both highly nutritious and of high caloric value. *Pinus monophylla* produces seeds that are approximately 10 percent protein, 23 percent fat, and 54 percent carbohydrate.[319]

As we have seen, Pinyon Jays harvest and cache only sound pine seeds. Experimental studies have shown that they use visual, tactile, and auditory cues in distinguishing between good and bad seeds. Pinyon seeds with uniformly dark brown shells are usually sound. Those with tan-colored or medium brown shells are usually empty—containing only the shriveled remnants of an aborted embryo. When harvesting, the jay extracts only the dark brown seeds from the cone. Then, by briefly holding the seed in its beak, the jay can bill-weigh the seed to detect the occasional dark brown seed which contains an aborted embryo. A further test of the soundness of the seed is done by bill clicking. While the seed is in its bill, the jay rapidly opens and closes its mandibles on the seed. A bad seed has a more hollow sound to it during bill clicking. Thus, using a battery of quick and simple tests, the jay harvests and caches only sound pinyon seeds, which is of advantage to both bird and tree.[320]

During their coevolution with jays and nutcrackers, corvid testers and cachers of sound pinyon seeds, the pinyon pines have evolved the large, thin-shelled, nutritious seeds which attract these corvids. They have also evolved characteristics which abet the detection of sound seeds by the corvid planters. Vander Wall and Balda have shown that cones grow on a pinyon pine tree so that they are "oriented outward and upward which increases illumination of the seeds and tends to hold them in the cone." Then a pinyon pine "displays the seeds in its cone more readily than wind-dispersed pines by opening the seed-bearing cone scales to a uniform angle." Finally, "seeds are retained for a long period of time in the cone because they are held in deep depressions on the cone scales by small flanges."[320]

The external nares or nostrils of the Pinyon Jay are unfeathered and exposed. This unusual feature in corvids gave rise to the Pinyon's genus name of *Gymnorhinus*, from the Greek meaning naked-nosed. As we have already seen, experiments with American Crows indicated that nasal feather tufts probably do not reduce water and heat losses in corvids during respiration.[314] Consequently, possession of a "naked nose" is probably not a liability to the Pinyon Jay during the cold winters and hot, dry summers in the Great Basin. A naked-nosed approach may be an asset to the jay early in the season when removing seeds from sticky, green, unopened pinyon pine cones. There would be no nasal feather tufts to sop up resin and become smeared with pitch.

Sociality is more highly manifested in the Pinyon Jay than in any other Great Basin corvid. The adults apparently remain pair-bonded the year around, although they live in loosely organized flocks during the winter. A flock occupies a home range area and does not migrate, despite Ridgway's poetic passage describing the birds congregating in immense flocks in the winter, "all uttering their querulous notes as they swept to and fro over the hills, in their restless migrations." Ridgway's description well fits the foraging movements of a flock, however. When the pine seed crop is poor, the flock may move nomadically in search of food but does not perform a rhythmical, set migration. At such times irruptions may occur, and Pinyon Jays often appear in places where they are not normally found. But, on the whole, the home range of a flock is fairly stable. For at least the past thirty years one flock has had a home range around the foot of Geiger Grade, the road leading up to Virginia City in the Virginia Range, Nevada.

Courtship between members of an established pair and the establishment of pair bonds between younger Pinyon Jays take place during the winter and early spring. As is the case in practically all temperate zone birds which have been studied, the onset of reproduction is controlled by the photoperiod or day length, which is daily increasing during the winter and spring. Under the influence of the photoperiod, the gonads undergo recrudescence—they increase in size and eventually start producing mature sex cells and higher levels of sex hormones. Ligon believes that "piñon pine seeds, when available, interact synergistically with vernal photoperiod to accelerate gonadal development."[321] His belief is based on experimental studies on testicular development in male Pinyon Jays and on observations that, in the spring following a major pinyon pine seed crop, the jays nest earlier than usual.

Courtship behavior is more difficult to study when it occurs in flocking birds than it is in species where the males are spread out and isolated on territories prior to the commencement of courtship and the formation of pairing bonds. We are indebted for much of our knowledge about courtship behavior in Pinyon Jays to field studies carried out by Balda and Bateman near Flagstaff, Arizona.[322] Pinyon Jays are early nesters. Ridgway discovered this fact too late in the season of 1868 to study nesting Pinyons. He wrote, "The breeding-season of this bird is remarkably early; for on the 21st of April, before we had thought of looking for their nests, full-grown young were flying about in a cedar and piñon grove near Carson City. In this grove we found the abandoned nests, perhaps a hundred or more in number, and also one containing young nearly ready to fly; but we were too late for the eggs."[25]

Since birds cannot produce fertile eggs on demand but must first experience complex physiological and psychological changes, early breeding must be preceded by much earlier courtship. Balda and Bateman discovered that silent courtship feeding was already under way within their study flock by mid November. This involved both adult jays and gray-colored, first-year birds; it probably functioned to strengthen and stimulate existing pair bonds between adults and helped initiate pair bonding between young jays. By mid December Balda and Bateman noticed that "not only were seeds passed silently, but females were seen to chase the males and beg fairly loudly for food; occasionally males begged from females," and during the begging "the partly outstretched wings were fluttered vigorously." By January or early February "courtship parties or chases" were observed, sometimes "involving as many as 12 individuals and including both adult and first year gray birds. These groups flew rapidly through and over the trees, performing sharp turns and steep dives. A very low *cluck* was often given by the birds performing these flights. . . . Pairs also began to move as far as 1000 ft from the feeding flock where they simply fed together, fed each other, or the male presented twigs and grasses to the female."[322]

Pinyon Jays nest in a loose colony, not locating more than one nest in a single tree; the flock will often use the same nesting area for a number of years. Nest building is generally under way by the end of winter. Usually only adults nest, although some observers have reported first-year females nesting. The male may select the actual nest site and attempt to entice his mate to the site by carrying nesting materials to it and feeding her if she follows. Both sexes work at nest building. Nests are usually built in juniper or pine trees. Ridgway's hundred or more Carson City nests in 1868 "were all saddled upon the horizontal branches, at a height of eight or ten feet from the ground."[25] During the time the colony is nest building, the jays still roost together at night and feed, at times during the day, as a flock. The nests, bulky yet compactly built, have a deep cup. They are not only well insulated but are often located on the south side of the trees.

Only the females develop a brood patch; they incubate the eggs for a period of about seventeen days. While females are incubating, males form foraging flocks and feed and roost together. On occasions during the day, males will leave the foraging flocks to feed their mates. Because of the coldness prevailing during their early nesting season, females must provide eggs with almost continual incubation to prevent cold damage. Before full-time incubation gets under way, eggs laid early may freeze during cold weather.

Some females start incubating with the laying of the third egg, and the eggs of a single clutch may hatch over a two- or three-day period.[323]

Upon hatching a nestling weighs in at about six grams and is blind and naked. By the end of the first week of life, its eyes are beginning to open. By the end of the second week it is quite well feathered, especially on its upper side. During cold weather, nestlings are almost continually brooded by their mother. Although the males are still in foraging flocks, they regularly leave these flocks to bring food back to the nest to feed their mates and young by regurgitation. The nestlings receive some pinyon pine seeds, but their main food staple appears to be insects. When the nestlings become better feathered, the adult females are able to forage for food during the warmer times of the day. During the final four or five days of nest life, the young jays are fed communally. Adults will feed nestlings other than their own, and as many as seven adults have been seen around some of the nests.[322]

By the end of the third week of life the young have fledged and are leaving the nests, but they are poor fliers and do not venture far. A number of families will merge to form a nursery group of about twenty-five to fifty birds. The group of fledglings will remain quietly perched in trees and shrubs, guarded by some of the adults, while the rest of the adults go foraging for food. When the foraging adults return with food, the young will flutter their wings and beg loudly. The adults feed the young communally until they are about six weeks old and have become quite adept at flying and foraging. After this, the young are fed only by their own parents. By midsummer even the parents refuse to feed their own young and drive them back when they attempt to beg.

All during the nesting season, most yearling Pinyon Jays are in flocks by themselves. These birds do not attempt to breed, nor do they play any role in the communal feeding of young jays.

By late summer the harvesting and caching of pinyon pine seeds by jays are under way. The pinyon cones are still green, so the cones must be hammered open by the adult jays before the seeds can be removed. The seeds are transported in the jay's esophagus to a caching area, which is often located in the vicinity of the nesting area on the home range of the flock. Caching sites are often located under the south side of trees. Under a tree the ground is somewhat shielded from snowfall, and the snow on the south side is the first to melt. The ground litter is moved aside with beak and feet, and about fifteen or so seeds are thrust into the ground. Then the site is covered over again with litter.[322] The large fall-winter flock has reformed during this time, as the various nesting groups and the yearling flock have reunited.

Pinyon Jays forage both on and off the ground. When ground-foraging, jays walk along and probe with their bills. Often the flock will progress in a rolling, leapfrog fashion—with the rear of the flock flying up and dropping beyond the front edge of the flock. An early description of this characteristic foraging maneuver was written by Henshaw: "A large flock of these birds were seen near Silver City, N. Mex., October, busily engaged on the ground feeding upon grass seeds. Those in the rear kept flying up and alighting in the front rank, the whole flock thus keeping in continual motion."[324] In another early account, Bendire pointed out that flock leapfrogging in these jays was much like that practiced by blackbirds: "A great deal of time is spent on the ground, where they move along in compact bodies while feeding, much in the manner of Blackbirds, the rearmost birds rising from time to time, flying over the flock and alighting again in front of the main body."[272]

Another bird which employs flock leapfrogging when foraging on the ground is the European Starling. This exotic species was introduced into eastern North America from Europe around the turn of the century. As the starling multiplied and spread over North America it eventually entered the range of the Pinyon Jay. When this happened in Colorado, Beidleman and Enderson discovered the two foraging together in 1963. They wrote, "The Starlings and Piñon Jays on the ground moved together in the 'rolling' pattern, the back part of the mixed flock rising from the short grass and dropping back to the ground in front of the flock."[325] In other places where the two have met, such as in Arizona and the Great Basin, they have sometimes associated together.

Mixed-species foraging flocks are not of uncommon occurrence in nature. Throughout the bird world, foraging flocks composed of more than one species can be found. In fact, some birds even form mixed foraging associations with mammals. Although the European Starling has much in common with the Pinyon Jay—in the way that it walks, forages, flies, and maneuvers—it is not the only flock associate of the Pinyon. Balda and his coworkers have identified four other flock associates of the jay in Arizona: the Hairy Woodpecker, Downy Woodpecker, Northern Flicker, and Clark's Nutcracker. When foraging a Pinyon flock spreads out over the ground and in the trees, and a wide variety of feeding niches are available to the flock associates within the confines of the flock. The flock associates can give their undivided attention to feeding, since they are protected from danger by the efficient sentinel system employed by the Pinyon Jays. Upon detection of potential danger the sentinels will sound a "rhythmic *krawk-krawk-krawk*" alarm. Then the associates can promptly take cover, or they can join the mobbing party of jays that will shortly descend upon the unlucky owl or hawk which has been detected.[326]

Clark's Nutcracker
Nucifraga columbiana

The Clark's Nutcracker inhabits the high mountains of western North America from southern Canada to Baja California. It is a permanent resident in the montane coniferous forests of the Great Basin. It was common in the Great Basin in Ridgway's day, and it is common today. The Clark's Nutcracker ranges over the entire height of a mountain, dropping down into pinyon woodlands to harvest pine nuts and foraging up beyond the upper tree line into the alpine zone on higher peaks.

There is a single species of nutcracker widely distributed over Eurasia. The presence of a new species of nutcracker in North America was first detected during the Lewis and Clark Expedition. When Captain William Clark discovered this nutcracker near the Lemhi River in Idaho, he thought he had discovered a new species of woodpecker. Some of the behavior and the undulatory flight of the nutcracker are certainly reminiscent of woodpeckers. Ridgway wrote, "It acts like a Woodpecker, screams like a Woodpecker, and looks so much like one that the best ornithologists are apt to be misled, by the first glimpse of it, into believing it an undescribed species of the Woodpecker family."[25]

Others have seen some behavioral resemblances to crows in nutcrackers, and an often used common name in the past for the nutcracker was that of Clark's Crow. The nutcracker was originally described and named in 1811 by Alexander Wilson from specimens collected on the Lewis and Clark Expedition. Wilson recognized its resemblance to the crow by naming it *Corvis Columbianus*. There was a period in the late 1800s and early 1900s when some were spelling Clark as Clarke—some of the older books therefore refer to Clark's Nutcracker as the Clarke's Nutcracker or Clarke's Crow. The Washo name for this bird was Pah'-bup, and the Shoshone name was Toh'-o-kōtz.[25]

Nutcracker economy is based mainly on pine seeds. The dependence of the Clark's on pine seeds was dramatically demonstrated by a study carried out in western Montana. There, over the course of a year, the stomach contents of 426 nutcrackers were analyzed. The study showed that conifer seeds made up 83 percent of the food ingested, that pine seeds were present in 98 percent of the 426 stomachs, and that the most commonly taken seed when available was that of the ponderosa pine.[327] Nutcrackers harvest seeds from a number of different kinds of conifers. Depending on which species of pines are locally available, nutcrackers have been reported feeding on the seeds or nuts of the single-leaf pinyon pine, two-leaf pinyon pine, ponderosa pine, Jeffrey pine, whitebark pine, limber pine, foxtail pine, and bristlecone pine and also on the Douglas fir.

Nutcrackers not only harvest pine seeds for immediate use but store them for future use. When extracting seeds from large cones, nutcrackers may stand on cones or cling beneath them. When the birds extract seeds from small cones, such as from pinyon pines, the cones are usually detached from the tree before the seeds are removed. Nutcrackers twist the cones off, sometimes by first pecking to loosen them. Then, the detached cones are carried in the bill to a seed-extracting perch on some horizontal tree limb. Here, one foot is used to hold down the cone, while the bill is employed to peck and pry the pine cone scales apart to expose the seeds. Sometimes prying is not necessary, if the birds select open cones to begin with. When harvesting limber pine seeds in the Sierra Nevada, nutcrackers selected open cones.[328] In contrast, when harvesting pinyon seeds in Arizona, nutcrackers selected closed cones over open ones.[320]

As in any nut crop, all pine seeds are not equally edible. Some are diseased or aborted. Edible pinyon seeds have a uniformly dark brown coat, whereas the aborted seeds have a tan to medium brown coat. Limber pine seeds do not show this color difference. But, while the Clark's Nutcracker holds the seed in its bill, the mandibles are "opened and closed rapidly, rattling the seed."[320] This rattle test helps the bird determine the condition of the seed within the seed coat. A nutcracker may also briefly hold a seed in its bill, as if bill-weighing the seed. Possibly the bird tests for seed density by doing this.

If a harvested seed is to be eaten, the seed coat is cracked open between the two mandibles, or the seed is held down with one foot and hammered open with the bill. If the seed is to be cached, it is retained by the bird until a quantity of seeds have been harvested. Nutcrackers do not have a crop, the pouchlike enlargement or outgrowth of the esophagus, to use as a shopping bag—as do many birds. Instead they have a sublingual pouch in the floor of the mouth. This pouch has the capacity to hold as many as ninety-five pinyon pine seeds. After filling its sublingual pouch with edible seeds, the nutcracker flies from the harvest area to one of the communal caching areas. Nutcrackers may fly there individually, in pairs, or in small groups.

The communal caching areas are located in exposed places where sun and wind work most effectively on snow. Caching areas must be located where deep snow does not remain much beyond late winter. Exposed, steep, south-facing slopes or ridges make good caching grounds, for here

Clark's Nutcracker

the sun and wind can melt or sweep the snow away quicker than on level ground or on sheltered slopes. If conifers are present, their foliage will further shelter the ground from snow.

Within the communal caching grounds, small groups of seeds are buried together, about an inch underground, and covered by soil or litter. During harvesttime in the San Francisco Mountains of Arizona, a single bird may store an estimated 22,000 to 33,000 pinyon pine seeds.[320] In the Sierra Nevada, Tomback discovered that caches averaged 4.4 whitebark pine seeds per cache. Since a single nutcracker stores up to 32,000 whitebark pine seeds, over 7,000 separate caches are required to house each bird's seeds. Storage of 22,000 to 33,000 pine seeds per bird means that each nutcracker is storing two to three times more than enough food to support itself from winter to midsummer.[329]

Over the years, observers have been intrigued by the ease with which nutcrackers reclaim their food caches even when the ground is covered by snow. The question arises whether this is done by random searches or whether some element of memory is involved. Because of the high percentage of success achieved by nutcrackers searching for food caches, most observers have concluded that some sort of memory is involved. In a study carried out on the eastern slopes of the Sierra Nevada, Tomback found an overall prod hole success rate of 72 percent in the spring, compared to a much lower summer success rate of 30 percent. Of course, rodent looting being in operation longer would more severely depress summer retrieval rates than spring ones. Tomback observed that, upon landing on the ground in the cache area, a nutcracker would select a site and either probe into the ground with its bill or dig a hole with sideswiping movements of its bill. Usually, after locating a cache, a nutcracker would shell the seeds on the spot and leave the seed coats behind—thus providing a fairly reliable method of distinguishing between successful and unsuccessful prod holes.[329]

Pine seeds constitute the food staple of Clark's Nutcrackers over most of the year. During the warm months of the year, arthropods, especially insects, are sought. True to their corvid heritage, nutcrackers will feed on the eggs and young of other birds and will attack and kill small vertebrates.

Because of their almost total reliance on pine seeds for winter food, a widespread failure of pine trees to produce seeds has a profound effect on resident nutcrackers. They must depart from their home forests. During these late summer irruptions, nutcrackers may wander far down out of the mountains. They may remain in lowland areas until the following spring or summer.

Nutcracker movements in the northeastern end of the Great Basin were studied during the summer and fall of 1977, 1978, and 1979. Over two thousand emigrating birds were seen during this time, mostly flying southward in small, loose flocks of about ten birds each. The resident nutcrackers did not seem to be involved, since they continued to forage locally and cache pine seeds. The emigrants appeared to be coming from farther north—they apparently overwinter in the pinyon-juniper woodlands of the Great Basin, and some apparently remain there long enough to breed the following spring. During the summer of 1978 a northward movement of nutcrackers was observed, probably the return movement of the 1977 emigrants.[330]

Clark's Nutcrackers breed early, often when the mountains are still under deep snow. Because of the difficult and hazardous travel conditions prevailing over the rugged terrain in early spring, few human observers have been around to witness courtship and nesting in this species. Mewaldt made a number of observations on courtship in western Montana. He described what appeared to be paired nutcrackers in pursuit flights—usually with one bird carrying a dead twig in its beak. He also witnessed courtship feeding. Once paired, nutcrackers probably remain together for their remaining years. Mewaldt stated that the territory around the nest he most thoroughly studied "contained at least 2.1 acres and was actively defended by the male against trespass by other nutcrackers. He used pursuit, body contact, and voice in territorial defense."[331]

Nests are built in conifers—often in conifers located on cold, steep slopes. The prime requisite for a nesting site appears to be not warmth but rather shelter from the wicked spring winds which buffet the mountainsides. Early on in human experience with the Clark's Nutcracker, Bendire noted this. He commented:

> The majority of sites chosen offered little concealment, but in every case especial care was observed in selecting one affording thorough protection from the assaults of the fierce March winds which prevail in this mountain region. The nests examined by me near Camp Harney, Oregon, were all found in sheltered situations on side hills where they were well protected from heavy winds, and the horizontal limbs selected for building sites were usually strong and bushy, with numerous small twigs among which the nests could be securely built.[272]

In describing the nesting of nutcrackers in the eastern part of the Sierra Nevada, Dixon wrote:

> All of the nests were in juniper trees on steep slopes at the 8000-foot level and contrary to our expectations were located in the coldest spots, where the snow stayed on the ground the longest. It is quite likely that

> these locations are the freest from the wind which blows so hard at these elevations, and I feel certain the juniper trees are used because of their sturdy build and ability to withstand the wind action. All nest locations seemed to have been selected with protection from the wind in mind, as the nests were either on top of a large limb, or, if supported by a small branch, were surrounded by heavy limbs that gave protection.[332]

Although the female Clark's Nutcracker may do more of the work than the male, both participate in nest building. The nest platform is constructed of dry twigs. The outer sides of the nest are also of twigs, sometimes bound together by strips of bark. The inner bowl may consist of bark fiber or wood pulp, lined with dry grasses and fine strips of inner bark. In Montana, Mewaldt found a half-inch-thick layer of mineral soil under the grass lining in the bottom of the bowl.[331]

The deep, thick-walled bowl of the nutcracker nest aids in combating coldness by limiting heat transfer between the inside and the outside of the nest. Also, both the male and the female nutcracker develop a brood patch. By sharing duties, the male and female nutcrackers can virtually provide round-the-clock incubation to the eggs and brooding to the nestlings. So, as long as the nest doesn't blow away, the eggs and nestlings remain fairly snug during the cold, windy montane spring.

Black-billed Magpie
Pica pica

The Black-billed Magpie is widely distributed over Europe and Asia as well as northern and western North America. It is a common and widespread permanent resident in the valleys and foothills of the Great Basin. To the Paiute Indians this magpie was known as Que'-tou-gih, gih; to the Washo it was known as Tah'-tut.[25]

The presence of the Black-billed Magpie in North America first became known to science on September 16, 1804. The Lewis and Clark Expedition was in South Dakota on that day when a hunter "killed a bird of the *Corvus* genus and the order of the *pica*, about the size of a jack-daw with a remarkable long tale [*sic*]."[333] Members of the expedition soon became well acquainted with the Blackbill—they found it to be a bold, voracious camp visitor. Magpies would enter their tents, pilfer meat from their dishes, and snatch morsels of meat from under the noses of hunters at work dressing game.

Soon, another expedition to the West was to bear witness to the fact that the magpie was not only bold but ferociously so. Zebulon Pike's journal entry for December 1, 1806, reads as follows:

> The storm still continuing with violence, we remained encamped; the snow by night one foot deep; our horses being obliged to scrape it away, to obtain their miserable pittance, and to increase their misfortunes, the poor animals were attacked by the magpies, who attracted by the scent of their sore backs, alighted on them, and in defiance of their wincing and kicking, picked many places quite raw; the difficulty of procuring food rendered these birds so bold as to light on our mens arms and eat meat out of their hands.[334]

Today, the Black-billed Magpie is one of the most frequently encountered birds in the shrublands and woodlands of the Great Basin. There is little possibility of overlooking its long-tailed black-and-whiteness or of not hearing its harsh loudness. Even if unseen and unheard, its presence is still marked, wherever one's gaze may wander, by countless huge, domed stick nests, in various stages of aging and disarray. The magpie is most abundant in the vicinity of ranches—particularly in thickets or groves along streams. But it can often be found foraging for food far from ranches or the cover afforded by thickets and groves. It forages along Interstate 80 in Nevada, where low shrubs afford only feeble cover and where utility poles are not always present to offer shade and protected high perches.

There is reason to suspect that the Black-billed Magpie has increased in abundance in the eastern end of the Great Basin since settlement. Following extensive field studies during the 1860s, Ridgway made the following comment about the magpie's abundance: "In western Nevada, from the Sierras eastward to the West Humboldt Mountains, it was one of the most abundant species, but on the opposite side of the Great Basin its entire absence from many favorable localities was noted as the most striking peculiarity of the fauna."[25] Times have changed, and Ridgway's comment is no longer appropriate. Today, the magpie is abundant eastward from the West Humboldt Mountains. Since the magpie thrives around ranches, it may well be that this increase in occurrence followed the establishment of ranches.

The annual cycle of life begins early in the year for the Blackbill. Courtship commences during the winter—although influenced by weather conditions, it is usually under way by late January. Observing magpies in the field in Wyoming, Erpino noted that the earliest annual sexual behavior involved pursuit flights. These flights, he believed, may have been involved in the formation of pairing bonds between males and females.[335] Most observers believe that magpies mate for life. The members of a pair usually are

Black-billed Magpies

not as intimately associated during the nonbreeding season as they are during the breeding season—it is possible that chases involving the two sexes merely serve to rekindle the flame. There is evidence that, once a female reaches the fertile stage of her reproductive cycle, her mate remains almost continually at her side—possibly to ward off strange males.

As the days become milder, nest building commences. Nest building is usually under way by March, and magpies will finish nesting before the heat of June arrives. Both sexes participate in nest building. A pair may spread the construction out over a span of six weeks or so, and a number of pairs often nest in close proximity to each other. The pairs tend to favor certain sites. Although a nest may be repaired and used for several seasons, a new nest is usually built each year. Hence, several abandoned nests are often located close by an active nest.

Magpie nests are generally located within twenty-five feet of the ground. The oval-shaped nest generally bears a domed roof of sticks. The roof may be lacking on some nests, especially those located in thorny shrubs. Usually one but occasionally two side entrances lead into the nest chamber. The floor of the nest chamber is a mud bowl, lined with rootlets, grasses, hair, and other fine materials.

Erpino recognized four sequential phases to nest building: anchor, superstructure, mud bowl, and lining. Initially, an anchor of mud was put in place to form the base of the nest. Then, sticks and twigs were stuck into the mud anchor to initiate the building of the superstructure. Unlike the bottom and roof, the sides of the superstructure were sparsely constructed during this phase. After a mud bowl was built upon the anchor, the bowl was lined. Finally, the sides were filled out.[336]

Once built, magpie nests are very durable structures. They may persist for years, allowing other animals to profit from the architectural skills of magpies. A diversity of birds nest in old magpie nests or use dilapidated magpie nests as nesting platforms, and a variety of birds and mammals occupy magpie nests as sleeping chambers or as refuges during storms. The first recorded observation of this nature for the Great Basin was made when Ridgway found a Gadwall nest near Pyramid Lake in 1868—he described it as a "nest of down, placed on top of a dilapidated nest of a Magpie, *in a willow-tree, about 8 feet from the ground.*"[25] The Gadwall is not the only duck indebted to the magpie. Mallards sometimes nest in old magpie nests. In recent years along the Carson River near Dayton, Wood Ducks have nested in old magpie nests. Here suitable tree cavities are scarce, whereas magpie nests are abundant.

Of all the magpie's neighbors, owls seem to make the most frequent use of abandoned magpie nests. Great Horned and Long-eared owls frequently

Black-billed Magpie nest

nest on top of broken-down magpie nests. Western Screech-Owls will nest in abandoned magpie nests. Among hawks, American Kestrels will nest in old magpie nests, and Sharp-shinned and Swainson's hawks will nest on top of dilapidated magpie nests. Other birds known to have used abandoned magpie nests for nesting platforms include the Black-crowned Night-Heron and the Mourning Dove. A considerable array of small birds—including blackbirds, bluebirds, robins, and warblers—has been known to seek shelter inside of abandoned magpie nests during severe storms.[337]

Mammals as well as birds appreciate abandoned magpie nests. Erpino found an adult female and three young raccoons occupying one.[336] Gilman reported the use of one by four young house cats, and Warren reported a gray fox using an old nest for daytime sleeping.[338]

A fascinating aspect of Blackbill behavior involves this bird's readiness to engage in protective nesting associations with humans and with raptors. We usually do not tolerate the company of predators such as snakes and raccoons and eliminate those which persist in visiting our habitations. The presence of dogs also discourages predators. So magpies may find some relief from nest predators by nesting close to human habitations. Although raptors vigorously protect their nest sites from approach by potential predators, strangely enough quite a few large raptors will allow smaller birds to nest close to them. In North America the Black-billed Magpie has been observed forming protective nesting associations with hawks—including the Red-tailed, Swainson's, and Ferruginous hawks. The magpie's relations with the Ferruginous Hawk are particularly close, and an active magpie nest may be located in the same tree as an active Ferruginous nest. Bowles and Decker have reported finding "a nest of a magpie built into the side of one of the hawks'" in eastern Washington, with each nest containing a full set of eggs.[339] During the Lewis and Clark Expedition, Captain Lewis observed, "The nests of the Bald Eagles, where the Magpies abound, are always accompanied by those of two or three of the latter, who are their inseparable attendants."[324] Also, the magpie has been reported forming protective nesting associations with Ospreys.

Like so many other birds, Black-billed Magpies engage in courtship feeding. Courtship feeding is of practical, as well as stimulatory, value to the magpie, since it continues even after the female has laid her eggs. The female incubates the eggs. All during the incubation period of eighteen days or so, the male regularly brings food to the nest and feeds his mate. Therefore, the eggs are not uncovered for lengthy periods of time.

Young magpies are naked and helpless at hatching. Both parents feed the nestlings. The fecal sacs voided by the nestlings are carried away by their parents, preventing nest fouling. Young magpies become fully feathered at an average age of about twenty-seven days.[336] They may remain in the nest tree for some time after fledging. After they leave the nest they can fly well, and they join the company of other magpies.

Black-billed Magpies associate together in loose flocks outside of the breeding season. At night a flock or several flocks will roost together in a sheltered place which will afford protection from the elements and from heavy radiation losses to the night sky. An unusual winter roost was once reported from Manitoba, Canada, where a small group of magpies used a cattle shed—they spent the nights sleeping on the warm backs of the cattle there. Magpies may also use their old nests for winter roosting.

The magpie is not a true creature of the sky. When aloft there is nothing flashy about this bird but its coloration—its short wings do not provide fast or strong flight, and its long tail is a hazard on windy days. The magpie is more a creature of the ground or trees. When on the ground it moves about by walking. If in a hurry it may hop or even take a wing-aided jump.

Black-billed Magpies can be seen feeding both along highways on roadkills and on the ground on a variety of food. They usually forage over open landscapes—characterized by low vegetation, with a sprinkling of thickets or trees for cover from aerial attacks by raptors. Although omnivorous in appetite, they much prefer animal to plant food; magpies are the most insectivorous of the Great Basin corvids. The greatest intake of vegetable food occurs during the winter, when animal food is difficult to come by.

The magpie's appetite for the eggs and young of other birds has been well publicized. Magpies are not well liked by Great Basin sportsmen, who on past occasions have sponsored magpie hunts. Magpies have been reported working over the nests of ducks in the Bear River marshes, and they have been seen removing eggs and young from Tree Swallow nests at Lake Tahoe.

Mammalogists have been troubled by magpies following along their small-mammal traplines and flying off with the snap traps which have caught mice or other mammals. Magpies have even been credited with possessing enough insight to search for traps around the bits of cotton placed by the trapper on the tips of shrubs to mark the sites. Years ago in Nevada and Utah, when coyotes were being systematically killed off, magpies attracted by the poisoned bait were also killed off in great numbers. The magpies' impact on short-circuiting coyote poisoning was severe enough that on occasions preliminary campaigns were instituted to deplete the magpie populations before going after the coyotes! Such a campaign occurred on the Pyramid Lake Reservation in the winter of 1921–22. A population of magpies estimated at over one thousand birds was reduced to less than a dozen by the following winter.[338]

There is a much more basic ecological relationship between magpies and coyotes than magpies having served as a buffer between coyotes and poisoned bait. Magpies are quick to find coyote kills and move in to feed after the coyotes have finished. Because of their quickness in dodging and taking off, they may even feed on a kill while a coyote is still present. Some observers have suggested that magpies have enough insight to keep an active eye on hunting coyotes in anticipation of an eventual kill. Feeding opportunities have also been suggested as constituting one of the fringe benefits magpies gain by forming protective nesting associations with receptive raptors. Leftover food is often available at raptor nests.

Not only will magpies feed on the abandoned kills of predators but, on occasions, they will cooperate to harass a large raptor and steal its prey. We witnessed such an event on the Lahontan Audubon Society's 1965 Christmas Bird Count. A group of eleven Black-billed Magpies had surrounded two adult Bald Eagles on the ice-covered surface of Little Washoe Lake. Some of the magpies encircling the eagles would simultaneously feint attacks on them, and when an opening developed in front of a magpie the bird would snatch at the meat. The eagles just could not protect themselves in all directions at once against the much smaller, quicker, and more agile magpies. Dixon observed an attack by three magpies on a Golden Eagle in Alaska. Two magpies swooped repeatedly at the eagle's head. The eagle released its prey, a ground squirrel, to strike back at the magpies. Then, the third magpie sneaked in, on the ground, and carried off the squirrel. After some fighting among themselves, the three magpies shared their stolen prey.[340]

Surely, the Black-billed Magpie is a bird of striking contrasts—not only of plumage but also of behavior. Certain of its behavioral traits—such as building durable, roofed nests—are of value to other animals, but still other traits are downright detrimental, at least in our human judgment. Perhaps the most humanly deplorable behavior of the magpie is its habit of feeding on the flesh of living animals, by pecking holes in their backs or by pecking out their eyes. We have already seen how in 1806 hungry magpies landed on Zebulon Pike's sore-backed horses and tore their flesh away. These happenings still occur. I arrived in Nevada in 1953. As soon as I had occasion to talk to ranchers about birds, I heard several bitter tales about magpie attacks on the flesh of living animals, usually animals who were sore-backed or wounded. Mammals who have been attacked include bison, mules, horses, cattle, sheep, and pigs.

The live-flesh-eating behavior of the Blackbill may have evolved out of an entirely different type of behavior. In the bird world certain birds visit the backs and heads of large mammals to feed on the ectoparasites present there. Probably the best-known practitioners of this specialized type of feeding behavior are the oxpeckers of Africa, who are said to obtain all their food from the hides of large wild mammals and domestic livestock. Oxpeckers have been accused of occasionally enlarging the wounds left by the removal of ticks. But, unlike magpies, oxpeckers do not visit large mammals solely to feed on living flesh. In livestock regions where cattle dips are used extensively to control ectoparasites, oxpeckers have decreased in numbers or disappeared entirely. Although their economy is not based on ectoparasites as is that of the oxpeckers, magpies do occasionally land on large mammals to feed on ticks and insects; the birds have been seen doing this on horses, bighorns, elk, and mule deer. Though the practice by itself is mutually beneficial to both magpies and mammals, it may have provided the initial opportunity for live flesh eating and been involved in the evolution of that behavior.

American Crow
Corvus brachyrhynchos

The breeding range of the American Crow extends from coast to coast in North America—over much of Canada and the United States. This bird is an uncommon to common summer resident and winter visitant at a limited number of localities in the Great Basin. Prior to 1982 its English name was the Common Crow.

Before settlement and the advent of agriculture and ranching, the American Crow was a rare bird indeed in the Great Basin. When describing the 1860s Ridgway wrote, "The Crow was so extremely rare as to be met with on but two occasions, when the number of individuals was limited to a very few."[25] The two occasions to which Ridgway referred took place at Truckee Meadows, today's Reno-Sparks area, and near the Humboldt marshes, both in the month of November.

Nowadays, crows are frequently encountered in the valleys in the western and central parts of the Great Basin—especially around farms and ranches, often up to hundreds at a time. They can often be seen and heard in cities and towns such as Reno, Carson City, Fallon, Lovelock, Winnemucca, Elko, and Wells. I have occasionally encountered them in the lower ends of wide mountain canyons, such as those in the Toiyabe Mountains.

My observations indicate that the breeding populations of crows in the western end of the Great Basin have been slowly increasing during the last fifteen years or so. Crows are continually showing up at new sites to nest, such as on the University of Nevada campus in Reno. In the eastern end of the Basin, crows do not seem to share this breeding prosperity. In Utah the American Crow is common only during the winter; it remains an uncommon and localized breeder.[27]

In recent years the American Crow has become conspicuous in the towns and cities of western Nevada. I have been impressed by the obvious process of urbanization that crows are experiencing in Reno, Carson City, Lovelock, and Winnemucca. During the spring and summer of 1981, the entire area in and around the cemetery in western Winnemucca was alive with crows. During the winter months of 1983, a large group of crows frequented a residential district in Reno just south of the Truckee River on California Avenue. Apparently, the urbanization of American Crows in the West is not unique to the Great Basin. Tony Angell, in his recent book on corvids, observed that "in the Pacific Northwest there is little doubt that crows have adapted well to urban life, with many cities now supporting populations of these birds, along with those of starlings and pigeons."[341] It is interesting that a relative of the crow, the Black-billed Magpie, has also been undergoing a recent bit of urbanization in the Great Basin. In northwest Reno, magpies have been penetrating two or three blocks into residential districts during their daily activities.

The crow bears one of the most recognizable bird names in all of North America. It has also been one of the most highly persecuted birds here. It has been declared guilty of being as black of heart as it is of plumage—of being a despoiler of corn and grain and a ravager of the eggs and young of other birds, especially waterfowl.

The farmer and the crow have long been enemies; the scarecrow stands as a symbol of this enmity. In his classic *The Crow and Its Relation to Man*, E. R. Kalmbach summarized the results of a study of 2,118 stomachs of adult and nestling crows, which were collected in forty states, in the District of Columbia, and in several Canadian provinces. The results showed that 28 percent of the yearly food of the crow was animal food and 72 percent was vegetable food. The principal animal food was insects, including deleterious ones such as May beetles and their larvae and grasshoppers; some birds and bird eggs were also taken. Over half of the vegetable food was corn, but most of this was waste corn gleaned during the cold months of the year. However, Kalmbach concluded that the offenses of the crow outnumbered its good deeds and that local control measures were justifiable.[342]

The bitter hatred between the crow and the sportsman has apparently had a peculiar origin: it has arisen mainly from the profit-making greed of sporting magazines and ammunition manufacturers. Ellsworth Lumley pointed out that prior to the early 1930s the crow was "looked upon as a nuisance but not as a menace," that "most men would not waste shells on him," and that "all sportsmen and sporting organizations were agreed that the shortage of waterfowl resulted from drought, drainage, and the advancement of civilization." Then, on December 29, 1932, the magazine *Field and Stream* issued a form letter which stated, "If you have been following the reports from Canada you know that practically everyone competent to judge is convinced that the greatest destroyer of North American wildfowl is the crow. Canadian authorities are emphatic in their opinion that our duck shortage of the present day is due more to crows than to droughts and drainage put together."[343]

Beginning in 1933, soon after the *Field and Stream* letter appeared, the sporting magazines started featuring article after article vilifying the crow—the evil scoundrel responsible for game shortages—and giving advice on crow shooting. Lumley believed that the sporting magazines and ammunition companies needed a target to stimulate slumping ammunition sales and the sales of ammunition advertisements—resulting from the deteriorating waterfowl hunting of the time. Getting a year-round war declared against crows by sportsmen, who quickly organized crow hunts and crow-killing contests, caused sales to boom. Not only were ammunition and advertisements sold, but also crow decoys, blinds, calls, and guns. Sportsmen are still being conditioned to regard the crow as a major enemy. But they should remember that crows and waterfowl have lived together in good health since time immemorial. Game shortages have resulted both from the destructive impact civilization has had on wildlife habitat and from ever increasing hunting pressure.

The American Crow has the general reputation of being a wily, wary bird. Many observers have commented on how very difficult it is to closely approach crows if you are carrying a gun. Yet, the crows in the West do not appear to be as wary as are those in the Midwest and East. Early on, Ridgway commented on this from the stage station near the Humboldt marshes. He wrote, "Three individuals only were found there, and these walked unconcernedly about the door-yard with the familiarity of tame pigeons, merely hopping to one side when approached too closely."[25] Ridgway was surprised to learn, when he questioned the station-keeper, that these three crows were not pets but wild birds. Several authors have commented on the tameness of crows in the Pacific Northwest. Of course, crows in the West have not been as severely persecuted as they have been in the Midwest and East. Our tolerance of them may have had something to do with their invasion of the cities and towns in the Great Basin and Pacific Northwest.

Crows are wary enough to post sentinels in situations where they are vulnerable to close approach by predators. A recent paper told how crows always posted sentinels on vantage points before feeding in a sanitary landfill in Virginia. Common Ravens and European Starlings would join the crows to feed in the landfill. The ravens would never enter the trench until the sentinel crows were in position and some crows were feeding. With the approach of potential danger the sentinel crows would intensify their cawing, and all three species—crow, raven, and starling—would fly away. The crows and ravens at the landfill would also become more alert upon hearing the alarm notes of Blue Jays in the woods adjacent to the landfill.[344]

Common Raven
Corvus corax

The Common Raven is widely distributed over the northern hemisphere in Eurasia, Africa, North America, and Central America. In the United States it is mainly confined to the West. It is a common and widespread permanent resident in the Great Basin. To the Paiute Indians the raven was Ah'-dah, to the Washo it was Kah'-gehk, and to the Shoshones it was Hih.[25]

The Common Raven is the most highly visible bird in the Great Basin. Regardless of the land, whether it be arid or lush, desert or ranch, valley or mountain, hot or cold, the raven will be there. And, as it is today, so it was in the past. When describing the 1860s Ridgway wrote, "This large bird is one of the most characteristic species of the Great Basin, over which it appears to be universally distributed, no desert-tract being so extensive or sterile that a solitary Raven may not be seen any day, although in such regions it is most usually observed winging its way silently, or with an occasional hoarse croak, from the mountains on one side the desert to the range opposite. It is also plentiful in the most fertile sections."[25] If anything, the Common Raven's Great Basin realm may now be more far-flung than it was in Ridgway's day, when the bird was forced to wing its way across the desert from mountain range to mountain range. Today these desert shrublands are spanned in many places by telephone and power lines and by occasional transmission towers. Now the raven can leave the desert skies and find elevated perches and nesting sites on utility poles and transmission towers.

The raven quickly moves out onto shrublands as a breeding bird when human construction projects provide elevated nesting sites. For example, a transmission line was constructed between Valmy and the Idaho border near Jackpot, Nevada, in 1979. In early June of 1980, during the very first breeding season after the line was in place, I flew along its entire length in a heli-

Common Raven

copter. Since we were studying raptors, I did not have much time to devote to ravens; but I did note at least thirty active raven nests on the brand-new transmission towers, some with young ravens still in the nest. Only one active Red-tailed Hawk nest and two large empty nests, which appeared to be those of Golden Eagles, were on the new towers. It may well be that the Common Raven pioneers in the nesting invasion of shrublands, by trying out transmission towers in advance of any significant movement by hawks and eagles.

Humanity has never been indifferent to the presence of ravens. Europeans have long feared the raven as a bird of evil omens. In his poem *The Raven*, Edgar Allan Poe addressed the bird as a "thing of evil . . . prophet still, if bird or devil." Yet to the Indians of the Pacific Northwest the raven was a great being—on occasions sly, tricky, or greedy but a great benefactor since it was involved in the creation of the world.[345] Although reference is made to the raven in Judaism and Christianity, it plays a much lesser role there than in Indian religions. Genesis tells us that, following the forty-day flood, the raven was the first scout Noah sent forth from the ark to see if the waters had started to recede. Since the raven never bothered to return to the ark, Noah then turned to more reliable observers, the doves, to check on the recession of the waters.

I confess to a certain fondness for ravens. We have shared the pain of being afield together in the numbing cold of the subarctic Alaskan winter and in the searing summer heat of North American deserts. We have wandered together through the quiet beauty of remote mountain and desert country. To me, ravens are a hardy race—an intelligent, alert, lively presence—which can ameliorate the harshness of climate and countryside.

It is difficult for me to comprehend the contempt and revulsion with which wildlife biologists often regard the Common Raven. I was quite surprised by the reception given to my account of how pioneering ravens had immediately moved out onto the shrublands to nest on newly constructed transmission towers, apparently breaking the trail for raptors to follow. Some wildlife biologists regarded the entire sequence of events as a major environmental disaster: providing shrubland-nesting opportunities for ravens was too great a price to pay for providing shrubland-nesting opportunities for eagles and hawks. I am reasonably sure that most of the ill will directed toward the raven by wildlife biologists is not generated by superstitious fear of the bird as the bearer of evil omens. Rather, it is directed toward the raven's reputedly keen appetite for the eggs and young of other species of birds. But then, according to the Gospel of Saint Luke, Jesus said, "Consider the ravens: for they neither sow nor reap; which neither have storehouse nor barn; and God feedeth them."

Common Ravens are omnivores, utilizing a wide variety of animal food supplemented by some plant food. They function both as predator and scavenger, killing some live game and utilizing available carrion. They possess the corvid family appetite for the eggs and young of other birds, and their impact on ducks and other nesting water birds can be heavy. Ravens possess a catholic palate and feed upon insects and other invertebrates along with all kinds of vertebrates. Among the mammals, voles, ground squirrels, hares, and rabbits are often exploited by nesting ravens.

Overall, carrion plays a major role in the economy of ravens. Ravens are frequently encountered, flying or perched, along the highways of the Great Basin. A few even nest here on roadside utility poles. High-speed highways promote mortality among wildlife and provide a table richly set with carrion. The valley shrublands of the Great Basin further encourage scavenging as a way of life, since carrion is highly visible in the sparse, low vegetation, and the Common Raven hunts by sight. Speaking of ravens locating carrion by sight, I recall being told by a condor watcher that California Condors often appeared to rely on ravens to locate carrion for them—a condor becomes aware of the presence of carrion by noting the descent of ravens to the ground to feed. Whether this is common practice or not, the California Condor and the Common Raven are often seen feeding on the same carrion, sometimes in the company of the Turkey Vulture.

Common Ravens interact with a variety of birds and mammals. Some of the most interesting interactions involve ravens and wolves. Ravens have been suspected of following wolves or wolf tracks in the snow with the expectation of eventually finding a wolf kill to feed upon. Researchers working in Minnesota with wolves outfitted with radio transmitters attempted to evoke additional responses from these animals by uttering imitation wolf howls. In some instances the human howls either attracted ravens to the caller or persuaded ravens to alter their flight in order to investigate the calls.[346] In the past some of the raven's ecological relationships with the wolf have not been happy ones. When wolves were being poisoned with strychnine, ravens often met their death through secondary poisoning by feeding on wolf carcasses, or they were directly poisoned by feeding on the toxic bait.

There is evidence that the Common Raven thrived in the presence of herds of wild ungulates. Here carcasses are of common occurrence due to disease, accidents, and predators. During northern winters, ravens have been seen accompanying herds of caribou and musk-oxen. During the glory days of the bison, ravens apparently accompanied the herds and flourished. The late, great ornithologist Margaret Morse Nice believed that the almost complete disappearance of ravens from Oklahoma and Kansas was due to extermination of the bison coupled with unintentional poisoning of ravens by cattlemen during their campaigns against the wolf.

Common Ravens are thought to mate for life. Members of a pair tend to remain together the year around. Even during the winter months ravens usually come in twos in the Great Basin. Often two ravens will be perched side by side on a utility pole or be flying together. Even when groups of ravens are encountered, the spacing often suggests that there are pairs within the group.

Courtship has its onset early in the year. By late winter, members of a pair are showing closer interest in each other. During pursuit flights, members of a pair chase each other. Often the aerial displays involve spectacular maneuvers or acrobatics—dives, tumbles, somersaults, and side or back flips. The male and female may soar together, wing to wing, high in the sky.

Soon nest building is under way. Most raven nests in the Great Basin are built on cliffsides, often on the same line of cliffs or in the same canyons with nesting Golden Eagles and nesting Prairie Falcons. In the absence of cliffs in valley shrublands, nests are often built on utility poles or transmission towers. Nests are sometimes built in trees or on such structures as windmills and old buildings.

Ravens, eagles, and hawks have a tendency to play musical chairs with old nests. This situation, where one species will use a nest one year and a different species will use it the next year, is referred to as interspecific nest appropriations. In the Great Basin, interspecific nest appropriations involving the Common Raven, Red-tailed Hawk, and Golden Eagle appear to occur quite frequently. An interesting case involving the Swainson's Hawk and the Common Raven occurred along Upper Maggie Creek in northeastern Nevada during the breeding seasons of 1979 and 1980. Here, in 1979, a pair of Swainson's Hawks succeeded in fledging three young in a low tree nest despite problems with ravens—in advance of egg laying, David Worley saw the two adult hawks near the nest site diving at ravens. In 1980 the ravens appropriated the nest long before the Swainson's returned from migration and were raising young when the hawks arrived. Despite some aerial encounters, the ravens were able to hold on to the nest.

Some observers have reported that both parents incubate eggs. Others have reported that only the female incubates, while the male either is on sentinel duty near the nest or is out foraging for food for the incubating female. Incubation commences with the first egg laid, and the incubation period has a duration of about three weeks. The parents will defend the nest site against intrusion by raptors or by humans. A most unusual nest defense was reported by Stewart James in Oregon. Both parents silently departed when James' group approached the cliffside nest for the first time. However, when the men were climbing back down from inspecting the nest, the ravens returned. One raven picked up rocks in its beak and flipped them down at the departing men.

Young ravens are poorly developed at the time of hatching—being blind, naked, and helpless. They are brooded by the female and fed by both parents by regurgitation. Slow in developing, they do not fledge until they are about six weeks old.

Common Ravens may congregate at night at winter roosts. Stiehl has described the happenings at a large roost located on the western edge of Malheur Lake. This roost, located in a dense growth of bulrushes and sheltered by a low ridge of ground, has been in use for a number of years. Stiehl's count of 836 ravens at the roost on January 4, 1977, is the highest count ever for a raven roost. Interestingly, the ravens did not fly directly to the roost in the evening. They first went to preroost sites on a dry lake or in short vegetation within six-tenths of a mile of the roost. Here they would feed, fly together, or engage in apparent food chases. Just before sunset, they would fly low and directly to the roost. The openness of the preroost sites and the conspicuous activity of the ravens there may have functioned to advertise the presence of a nearby roost to passing ravens.[11]

Large numbers of Common Ravens may congregate to fly and soar together—they carry out aerial maneuvers and apparently play. The largest such concentration I have seen in the Great Basin was one I encountered on a hot July day. I was southwest of Winnemucca in Dun Glen Canyon in the East Range. High in the air, over Auld Lang Syne Peak, about two hundred and fifty ravens were milling around, soaring and maneuvering in the sky. Above a ridge to the north, in line with Dun Glen Peak, another hundred or so ravens were likewise engaged.

I encountered my most interesting conclave of ravens on a November day in the Toiyabe Mountains, just east of Austin. Along a two-mile stretch of U.S. 50, small parties of ravens were seen, here and there, mainly over ridges and Austin Summit, engaging in various aerobatics. In all, there were probably close to one hundred and fifty ravens present. Parties of fifteen to twenty ravens each were milling around in the sky. Interactions between ravens involved one raven playfully diving down or flaring up to another raven, while the party drifted along as a loose group. The "target" raven would often roll over on one side or on its back to meet its "attacker," but no physical contact was ever made. This air show had all the earmarks of high-spirited play. The most complex interactions involved two, sometimes three, ravens engaged in pursuits and mock dogfights. Usually the pursuits continued long enough that the participants moved quite a distance from the group with which they were formerly associated. The pursuits entailed soaring, gliding, and flapping flight. The lead raven was frequently buzzed from below or from above by the raven tailing it. Often the lead raven would roll over on one side or flip over on its back to meet the near approach of the tailing raven.

Other Corvids

The breeding range of the Gray Jay, *Perisoreus canadensis*, brushes the eastern, northern, and western edges of the Great Basin. The Gray Jay is not a full-fledged member of the Great Basin avifauna, since it does not range far into the interior of the Basin. In the eastern part of the Basin, it is a resident on high mountains in central Utah.[56] In the western part it is a resident in the Warner Mountains.[289]

The breeding range of the Blue Jay, *Cyanocitta cristata*, lies in Canada and the United States to the east of the Great Basin. The Blue Jay is of accidental occurrence in the Basin. A specimen was collected on the outskirts of Salt Lake City.[27] In the western part of the Great Basin, this jay is listed as being of accidental occurrence at Malheur National Wildlife Refuge;[61] there is a sight record from Carson Valley, Nevada; and there is a record from Incline Village at Lake Tahoe, Nevada.[347]

FAMILY PARIDAE

The Family Paridae contains the titmice—the best known of which are the chickadees. Formerly, the Verdin and Bushtit were included in this family. However, in the new sixth edition of the AOU checklist, each is placed in a family by itself. Parids are distributed over much of the world but are best represented in the northern hemisphere. They are very small birds with soft, thick plumage, usually colored in grays and browns but sometimes boldly marked with black or white. The sexes are alike. Their bills and legs are short but strong. Practically all parids are cavity nesters. They are highly arboreal and gregarious; outside of the nesting season they travel through the trees in small, mixed-species foraging flocks. They are very active, even restless, but will tolerate a close approach by humans. Parids utter lisping and chattering call notes, and their whistled songs are often quite pleasant to the ear.

Black-capped Chickadee
Parus atricapillus

The breeding range of the Black-capped Chickadee extends across North America from central Alaska and central Canada southward into central United States. In the western end of the Great Basin, this chickadee is of accidental or occasional occurrence. But in the eastern part, at localities such as the Snake Range in eastern Nevada and in western and central Utah, it is an uncommon to fairly common permanent resident.

During the fall and winter, Blackcaps may wander about, sometimes as members of small, mixed-species foraging flocks. Chickadees nesting in mountains in aspen groves or along streams may perform altitudinal migrations and spend the winter at lowland sites. Some chickadees reside in the valleys, nesting in riparian woodlands.

The Black-capped Chickadee is a cavity nester, but it is not heavily obligated to woodpeckers—it prefers to excavate its own nesting cavity. For this it needs a dead stub with a firm outer shell and rotten inner wood. Chickadees are difficult to attract to artificial nest boxes; but it can be done if the nest box is filled with sawdust, for the chickadees to clean out and dump.[348]

This bird's English name is imitative of its call notes of *chickadee-dee-dee.* In the Great Basin you must listen to as well as look at your chickadees when in the field. The much more widespread and abundant Mountain Chickadee, which looks much like the Black-capped Chickadee, is here as well. The call notes of the Mountain Chickadee are *chick-a-zee-zee, zee.* The spring song of the male Blackcap is a "slow, two-toned 'fee-beee,' the 'beee' at a lower pitch," compared to a "three-toned 'fee-bee-bee,' the 'bees' at a lower pitch," for the male Mountain Chickadee.[348]

Mountain Chickadee
Parus gambeli

The breeding range of the Mountain Chickadee lies in the interior of western North America. This bird is a common permanent resident in mountain ranges throughout the Great Basin.

Mountain Chickadees nest in the montane conifer and aspen forests and in the denser pinyon-juniper woodlands of the Great Basin. They are most abundant at higher elevations in the Canadian life zone. During the fall and winter Mountain Chickadees appear at lowland sites. In field studies conducted by Dixon and Gilbert in northern Utah, eighteen out of nineteen chickadees collected below 5,900 feet in the fall were immatures, according to skull ossification criteria. Most, perhaps all, of the adult chickadees studied appeared to remain in the vicinity of the montane sites they occupied during their first breeding season. Dixon and Gilbert concluded that Mountain Chickadees are partial altitudinal migrants: some of the immatures migrate down into the lowlands, while the adults are sedentary and do not wander far from their first-year nesting territories. As a measure of their sedentariness, the adults studied "did not visit points more than about 0.8 mile apart in an area of discontinuous woodland." Finally, Dixon and Gilbert speculated that the chickadees frequently encountered above the upper tree line in late summer are probably immature birds.[349]

Mountain Chickadees are cavity nesters. However, unlike the Black-capped Chickadee, they usually do not excavate their own cavities but occupy natural cavities or woodpecker-excavated cavities in trees, stumps, or snags.

Mountain Chickadees feed mainly on insects; they do most of their foraging within fifty feet or so above the ground. They glean insects and their eggs and larvae off the ends of living branches and off the dead twigs in the outer foliage of conifers and aspens. While foraging they may circle around a tree, progressing from branch to branch.

Outside of the breeding season, Mountain Chickadees forage in small parties or in mixed-species flocks. A social dominance hierarchy, called a peck-right system, emerges from intraflock clashes between chickadees. In a peck-right system, a bird can automatically supplant any other bird from a feeding position who ranks below it in the dominance hierarchy. In turn, it can be automatically supplanted from a feeding position by any bird who holds a higher position within that hierarchy. Obviously, the individual aggressiveness of a bird will play a role in determining what rank it gains in the hierarchy. But a study by Dixon in northern Utah led him to conclude that seniority in the area frequented by the flock also influenced social position. The senior resident in the area occupied the alpha or number 1 position in the hierarchy. When two rival flocks of Mountain Chickadees clash at a food site, the outcome is dependent on which flock's foraging range the site is on. The home flock is dominant over the intruding flock.[350]

Mountain Chickadees often forage outside of the breeding season in flocks with other species of birds. Panik followed foraging flocks on ninety-eight occasions through the pinyon-juniper woodlands of the Virginia and Pine Nut ranges in northwestern Nevada during the months of July to April. Forty-five of these flocks were mixed-species foraging flocks; fifty-three were single-species flocks. Mountain Chickadees were present in fourteen of the mixed-species flocks. They were flock leaders in four of the fourteen flocks and coleaders with Bushtits three times. The largest mixed-species foraging flock in which Panik found Mountains present was led by Bushtits and contained forty-eight birds of seven different species. In addition to the Bushtits and Mountain Chickadees, there were Dark-eyed Juncos, Red-breasted Nuthatches, Plain Titmice, Rufous-sided Towhees, and a Golden-crowned Kinglet in the flock. As the birds foraged, the flock progressed through the woodland at a rate of eighteen hundred feet per hour. In addition to the six species mentioned as flock associates of the Mountain Chickadee, on one occasion Panik saw a Bewick's Wren as a flock associate.[351]

At the northwestern edge of the Great Basin at Eagle Lake, California, Manolis studied the foraging relationships of Mountain Chickadees and Pygmy Nuthatches in a Jeffrey pine–western juniper woodland. Chickadee flocks were observed foraging alone on six occasions and in mixed flocks with Pygmy Nuthatches on twenty-one occasions. Other species observed as flock associates with Mountain Chickadees and Pygmy Nuthatches were the Bushtit, Golden-crowned Kinglet, Brown Creeper, and White-breasted Nuthatch.[352]

There is a minimum of competition for feeding sites and food between the various species in mixed-species foraging flocks, and clashes between them are seldom seen. The participating species differ somewhat in their foraging techniques and niches. Ornithologists have puzzled over what, if anything, birds achieve by participating in mixed-species foraging flocks beyond the company of other birds. One possible gain is the service of additional eyes and ears on the alert for predators without any additional competition for food.

Plain Titmouse
Parus inornatus

The breeding range of the Plain Titmouse lies mainly in western United States, extending from Oregon southward into Baja California and to western Texas. This titmouse is a common permanent resident in the pinyon-juniper and juniper woodlands of the Great Basin. Elsewhere on its geographic range, where oak woodlands are extensive, this titmouse is partial to oak.

Thus, there is a strong bias on the part of the Plain Titmouse toward residing in evergreen woodlands, since junipers, pinyon pines, and oaks retain their foliage during the winter and provide dense canopy cover the year around. Field studies in California by Dixon revealed that the very cautious titmouse exploits the dense evergreen canopies the year around for predator protection. A titmouse is seldom seen outside of the foliage canopy, except when it is perched so high that it has an unobstructed view of its surroundings. It will forage on the ground only in areas where the understory and ground cover are thin, and even then it will carry food from the ground to a perch with better viewing before eating. When foraging in the canopy, a titmouse may move from the spot where it captured prey to a perch with a better view, before consuming the prey. While foraging it is constantly on its guard and frequently looks around to survey its surroundings.[353]

Titmice frequently participate in mixed-species foraging flocks. They were members of twenty-two of the forty-five mixed-species flocks Panik encountered in pinyon-juniper woodlands in northwestern Nevada.[351] They were never seen as flock leaders, but the constant vigilance they exercise when foraging would enhance the security of the flock. While foraging titmice glean insects from the bark of the small branches and twiggery which support the canopy foliage. They also take insect galls, fruit, seeds, and other vegetable matter when available.

Plain Titmice are sedentary like Mountain Chickadees. They are cavity nesters, using natural cavities and old, abandoned woodpecker cavities. In Dixon's California study, titmice tended to form permanent pair bonds. The paired male and female remained on their territory the year around. Both defended the territory. Most of the territorial defense was of a vocal nature and consisted of "scolds, calls or song, often in combination with wing vibrations." Dixon heard two basic types of song from the males, both given in a clear whistle and repeated three to five times. There was a *peter peter peter*, with the first syllable of each phrase being higher in pitch, as well as a less common *sweet sweet sweet*, each phrase rising in inflection, sometimes given so slowly that each note sounded as if it were composed of two syllables.[353]

FAMILY AEGITHALIDAE

In the fifth edition of the AOU checklist, two species of bushtits were recognized—the Black-eared Bushtit and the Common Bushtit. Both species were in the Family Paridae. Then in 1973, in the thirty-second supplement to the checklist, the Black-eared was proclaimed to be merely a black-eared morph of the Common Bushtit, and the two were combined into one species with the English name of Bushtit. In 1982, the Bushtit was transferred out of the Family Paridae into the Family Aegithalidae, a family of long-tailed tits.

Bushtit
Psaltriparus minimus

The breeding range of the Bushtit lies in western North America. The Bushtit is an uncommon to common permanent resident in the Great Basin.

This 5.5-gram mite is among our very smallest birds. The Bushtit is very gregarious—outside of the nesting season it is always found in flocks. It frequents a variety of habitats in the Great Basin, including riparian woodlands and thickets in valleys, pinyon-juniper woodlands, areas of very tall big sagebrush, the lower ends of mountain canyons, and mid-altitude mountain mahogany thickets and brushlands.

Bushtits remain in cover while foraging, much as do titmice, and like titmice they glean insects from the surfaces of leaves and twiggery within the foliage. While foraging they assume a number of different feeding postures, including hanging upside down like chickadees. They remain in constant motion, shifting through the trees. A foraging flock is loosely organized; its members maintain auditory contact with each other by continually uttering soft, twittering location notes of *Tsit, tsit; tsit; tsit.* When traveling rapidly and not tarrying to glean, the flock members remain in auditory contact by loudly uttering one to five location notes, followed by a long, shrill, quavering note—such as *Tsit, tsit, tsit, sre-e-e-e; tsit, sre-e-e-e.* A lackadaisical Bushtit left behind by the flock soon becomes agitated: it flies to the highest available perch and loudly utters *Tsit', Tsit', sre-e-e-e'* notes. Any flock member who hears these distress cries will respond with answering cries to guide the lost bird back to the flock. When winged danger is near, such as a Sharp-shinned Hawk, the entire flock joins in a shrill chorus of quavering *Sre-e-e-e-e-e* calls until the danger is past.[354] The birds remain motionless. The *Sre-e-e-e-e-e* cries have physical characteristics which make it extremely difficult for a listener to trace them back to their source.

Bushtits are often present in mixed-species foraging flocks. Panik found them in eighteen of the forty-five mixed-species foraging flocks he encountered. They were flock leaders in thirteen of these flocks and coleaders along with Mountain Chickadees in three other flocks. The mixed-species foraging flocks led by Bushtits moved at rates of nine hundred to eighteen hundred feet per hour.[351]

Master builders, Bushtits construct an elongated, gourd-shaped, pendent nest. Either the top of the nest is completely roofed over, with the entrance located in the neck of the structure, or the nest opening is at one side of the top and roofed over with a hood. The first step in nest building involves constructing a rim or circle of nest materials, tied together with spiderwebs and located within the fork of a twig or between two parallel twigs. Once the rim is in place, the birds proceed using one of two techniques. Usually they build a small bag, one inch or so deep, beneath the rim. Then, working inside this small bag, a Bushtit will stretch and shake it, progressively deepening it. As the bag stretches and deepens it becomes thinner. The birds thicken and fill in any thin or loose places which develop during stretching. Eventually, the top is roofed over. In the second, less frequently used technique, a long, loose, thin bag is constructed below the rim. Then, all the thin and loose places in this long bag are filled in and thickened, and the nest is roofed over.[355]

As we have seen, the Bushtit is a very tiny bird at 5.5 grams. As a consequence of its smallness, it pays a greater price for its homeothermism. In a laboratory study, Bushtits had to consume 80 percent of their body weight per day in mealworms to maintain themselves at an ambient temperature of 20° C. Birds caged in pairs or groups spent part of their time perched together, and they huddled together at night in a tight mass. Huddling is of survival value to Bushtits, since the nocturnal metabolic rate or nightly energy cost of a paired Bushtit is only 79 percent of that of a single Bushtit.[356] Huddling in the cold during the night is not just a laboratory artifact. Observations of wild, free-living Bushtits have shown that on mild nights they roost perched no closer than about five centimeters to each other, but on cold nights they perch so that they are closely packed together.[357]

FAMILY SITTIDAE

The nuthatches of the Family Sittidae are mainly northern birds—best represented in Eurasia, with but four North American species. They are mostly nonmigratory in habit. Nuthatches are small, stocky, short-necked birds with thin, straight, pointed bills. The sexes are alike or nearly so: they are colored gray to blue above, with the top of the head sometimes being brown or black; they are whitish or brownish below. A dark eye line is usually present. Sittids are highly arboreal—they nest in cavities and forage on the trunks and branches of trees. Some of their foraging movements are quite unique. They may hunt insects and spiders by climbing headfirst down tree trunks or by moving along the undersides of tree limbs while hanging upside down. Nuthatches cling to bark with the sharp claws of their long, strong toes. Their very short, truncated tails cannot be positioned to function as braces—as in woodpeckers. Even when descending a tree trunk headfirst they do not step along, a foot at a time, but move with short, jerky hops. Sittids are essentially solitary birds, although some form foraging flocks or join mixed-species foraging flocks.

Red-breasted Nuthatch
Sitta canadensis

The breeding range of the Red-breasted Nuthatch lies mainly in the boreal coniferous forests of extreme southeastern Alaska, southern Canada, and northern United States. Its breeding range extends far into the United States along the coast, in the mountains of the West, and in the Appalachian Mountains in the East. Red-breasted Nuthatches are erratic migrants; during years when the pine seed crop is poor they may migrate south in unusually high numbers. The Redbreast is an uncommon to common permanent resident in the mountains of the Great Basin, but there is some altitudinal migration into the lowlands during the fall and winter. Latitudinal migration also occurs, and some years widespread southward movements can be detected at valley sites in the Basin in the late summer and early fall.

This nuthatch frequents the montane aspen and conifer forests. It is a cavity nester and usually excavates its own nesting cavity. The cavity is excavated in the decayed wood of a dead stub or tree or in a dead limb on a living tree. Often nuthatches will start excavating at several different sites before one is found that is suitable.

The Red-breasted Nuthatch almost invariably surrounds the entrance into its nest cavity with the sticky pitch of fir, spruce, or pine trees. It does so even when nesting in a hardwood tree or in a nest box. The pitch is carried in the tip of the bill, a globule at a time, and smeared onto the bark. A pair may continue to add pitch to the nest entrance all during the nesting season.

The female incubates the eggs, but both parents feed the young. Nuthatches cannot alight at the entrance of the cavity without getting stuck in the pitch, so they have somewhat of a problem in entering the nest—they either fly directly into the hole or hover next to it and then dive in. The pitch around the entrance is certainly sticky and strong enough to trap a nuthatch. There is a report in the literature of a nuthatch getting stuck and being held till her death in her entrance pitch.

Not much light has been cast on the rhyme or reason for surrounding the cavity entrance with pitch. Some have speculated that this keeps ants out of the nest, which it probably does, if that is a problem. Others believe that since the pitch is as tenacious as birdlime it may discourage visits by small avian nest predators, or by that brood parasite the Brown-headed Cowbird, or by small mammalian nesting cavity competitors such as deer mice. It has even been suggested that the pitch may be a nonfunctional evolutionary carry-over from the days when ancestral Red-breasted Nuthatches used natural cavities and modified or caulked the entrances into their cavities.

The food staples of this species are apparently insects, spiders, and conifer seeds. The insect and spider foraging beats lie chiefly along trunks and main branches—high up in tall conifers. Red-breasted Nuthatches usually forage in the top one-third of a tree. Their high foraging stations quite effectively segregate them from White-breasted Nuthatches, who may be foraging closer to the ground on the trunks and larger branches of the same trees. Still a third species, the Pygmy Nuthatch, may be foraging high up in the same trees but mostly out in the terminal twiggery and clusters of pine needles, rather than on the inner branches and trunks. Thus, three closely related species of nuthatches may occur together in the Great Basin but be segregated by feeding niches.

Red-breasted Nuthatches move more rapidly than White-breasted Nuthatches when foraging. Sometimes they may flycatch insects. During the winter they rely heavily on conifer seeds for sustenance. They pry the pine cone scales apart with their bills to remove the seeds. While feeding nuthatches do not hold their food down with their feet but wedge food items into cracks and pick them apart with their bills. Sometimes nuthatches store food in crevices.

Redbreasts are not gregarious, although they may group together during migration. Mated adults often remain together during the winter. Outside of the breeding season nuthatches are sometimes members of mixed-species foraging flocks. Panik found them in seven of the forty-five mixed-species foraging flocks he encountered in pinyon-juniper woodlands. They were leaders in two of these flocks. Five of the seven flocks each contained two nuthatches, one flock contained one nuthatch, and one flock contained five.[351] The high incidence of two nuthatches per flock suggests winter pairing. When the birds are foraging or moving apart, their common call note serves as a location call. This call note has been likened to a blast from a tiny tin trumpet; it resembles a loud, nasal *yna* or *kng* sounded either singly or repeated.

White-breasted Nuthatch
Sitta carolinensis

The breeding range of the White-breasted Nuthatch extends from southern Canada into Mexico. Although this nuthatch occurs from coast to coast in the United States, it is absent from extensive regions, most conspicuously from the Great Plains region. The Whitebreast is a rare to uncommon permanent resident in Great Basin mountain ranges. It occurs in many mountain ranges here but is not yet known to be a breeding species in three ranges along the northern edge of the Basin: the Pine Forest and Santa Rosa mountains of Nevada and the Raft River Mountains of Utah.[289]

This nuthatch is mainly present in montane forests, but it may be encountered in pinyon-juniper woodlands and in riparian woodlands. Much of its foraging occurs on the deeply creviced bark on the trunks and larger branches of trees. Its foraging beat lies partly in an upside-down world, which it views while moving headfirst down a tree trunk or upside down on the underside of a tree limb. This down-trunk foraging is unique to nuthatches. The normal way to pursue a foraging beat on a tree trunk is to move up it, headfirst, as do woodpeckers and creepers.

Birds are better designed to assume a head-up position on a tree trunk than a head-down position. In the head-up position the tail can be braced against the tree trunk and, along with the two legs, it can form a tripod to support the bird. Also, in this position more toes and claws can be brought into play in clinging to the bark, since most birds possess a foot with three toes directed forward and only one directed backward. In moving headfirst down the tree trunk, the nuthatch cannot use its tail as a brace, and only one claw on each foot is in a position to be anchored to the bark.

Curiosity compels us to question how the nuthatch maintains its position in its upside-down world. Lacking support from the tail, its body is positioned over a bipod—the two legs. However, the tarsus or foot region of a bird is long, and the nuthatch braces its two long tarsi flat against the bark for support. Further, the bird may keep its two feet spread, with one foot positioned up the trunk from the other. The foot highest up on the trunk is then rotated, so that the three forward-directed toes are then anchored in the bark. The nuthatch thus appears to hug the tree as it creeps along.

White-breasted Nuthatches feed on a variety of larval and adult insects. They also take acorns, nuts, and other plant food, especially in the winter. The name nuthatch means nut pecker and is probably a corruption of nut hacker, in reference to the bird's habit of wedging a hard food item into a crevice and hammering it apart. This nuthatch is known to store food in crevices or behind loose pieces of bark. It occasionally conceals a food-storage site by wedging bits of bark or lichen into the crevice on top of the food.

The White-breasted Nuthatch is not very gregarious; it is usually encountered foraging by itself or with its mate. Sometimes one becomes a temporary member of a mixed-species foraging flock. It has been seen in foraging flocks of Mountain Chickadees and Pygmy Nuthatches in a Jeffrey pine–western juniper area near Eagle Lake.[352] However, Panik did not de-

White-breasted Nuthatch moving down-trunk

tect it in any of the forty-five mixed-species foraging flocks he encountered in pinyon-juniper woodlands in northwestern Nevada.[351] The pair bonds in this species are more or less permanent ones. When foraging or traveling together in the fall and winter, the male and female are often out of sight of each other. They maintain vocal contact by frequently sounding their *hit-tuck* notes, which function as location notes.[358]

Whitebreasts are cavity nesters. They usually nest in natural cavities in living or dead trees, although sometimes they use an old woodpecker cavity. Infrequently they may excavate their own nesting cavity. During the nesting season male and female nuthatches engage in a most unusual type of bill-sweeping behavior. While holding some object in its bill, usually an insect, the nuthatch sweeps its bill back and forth on the bark around the entrance to the nest cavity. Sometimes bark located in the vicinity of the entrance is swept as well. At times the inside of the nest cavity is swept. Although the nuthatch usually bill sweeps with an insect, it occasionally will sweep with a bit of fur, feather, or plant material or even with its bill alone.

Kilham has noted that the insects used in bill sweeping may be mainly blister beetles—beetles producing a "copious, oily, vesicant fluid." He believes that nuthatches sometimes compete for natural cavities with tree squirrels and that "bill-sweeping may serve to deter or deflect squirrels by spreading repellant or other substances present in the bodies of the crushed insects."[359] The White-breasted Nuthatch sometimes uses natural cavities which are large enough to accommodate nesting squirrels. The near intrusion of a tree squirrel may also precipitate a mothlike distraction display by one of the parent nuthatches. The nuthatch will display in front of the squirrel by widely spreading its wings and tail, pointing its bill upward, and slowly swaying back and forth.[360]

Pygmy Nuthatch
Sitta pygmaea

The Pygmy Nuthatch has a discontinuous distribution in western North America, extending from British Columbia to Mexico. It does not have much of a presence in the mountains of the Great Basin. Within the interior it is an uncommon permanent resident in the Snake Range of eastern Nevada. Along the western edge of the Basin it is a permanent resident in a few localities such as the White Mountains, the Carson Range, and the Lake Tahoe and Eagle Lake regions. The Pygmy is of common occurrence in the Sierra Nevada. In Utah, it is a common permanent resident in the eastern and southern parts of the state.[27]

The distribution of the Pygmy Nuthatch over western North America coincides fairly well with the distribution of the ponderosa pine. This nuthatch may be found inhabiting other types of conifer forests or utilizing other conifers in its foraging, but it is most frequently associated with open forests of ponderosa pines. These nuthatches are cavity nesters and commonly excavate their own cavities. Sometimes they use natural cavities or abandoned woodpecker holes.

Pygmy Nuthatches feed mainly on insects, with some pine seed utilization; their foraging beat lies high above the ground in the crowns of ponderosa pines and other conifers. They spend much of their time foraging out among the smaller branches, twigs, and pine needles, gleaning insects. During the course of Manolis' field study in a Jeffrey pine–western juniper area at Eagle Lake, they modified their feeding stations somewhat between summer and winter. In the winter they spent only about half of their time foraging among the smaller branches and needle clusters; the rest of the time they foraged on the larger branches and trunks.[352] Sometimes a nuthatch will flycatch insects or flutter in front of the foliage while gleaning insects.

Pygmy Nuthatches are gregarious outside of the nesting season. Not only do they form flocks by themselves, but they also participate in mixed-species foraging flocks. During the winter on the Jeffrey pine–western juniper study area at Eagle Lake, single Pygmy Nuthatches were observed on eleven occasions; on twenty-two occasions they were in flocks by themselves; and on twenty-one occasions they were in mixed-species foraging flocks, usually with Mountain Chickadees.[352] In describing the location notes of these nuthatches Hoffmann wrote, "They call to one another incessantly with a high staccato *tĭ-dĭ*, *tĭ-dĭ*, *tĭ-dĭ*, which becomes a rapid series of high cheeping notes when a number are together."[84]

As a further sign of their gregariousness, Pygmy Nuthatches have been known to practice communal roosting. Al Knorr observed about one hundred fifty of these nuthatches roosting in holes in a broken ponderosa pine trunk in Colorado in October, with at least one hundred birds sharing one of the cavities. Communal roosting, in association with huddling, would be of survival value by holding down the heat losses of these tiny birds during cold nights. But this does not always hold true. At the bottom of another roosting cavity in Colorado, the mummified remains of nine Pygmy Nuthatches were discovered. At still another roosting cavity the bodies of thirteen dead nuthatches were found. In both instances the roosting cavities were deep and cylindrical and had but one entrance hole. It may well be that, in cavities such as this, communal roosting sometimes leads to the suffocation of some of the nuthatches.[361]

As is the way of nuthatches, Pygmies sometimes move downward on a tree trunk. A study of their down-trunk locomotion revealed that they progressed by what appeared to be "a series of sidling hops or drops." When moving downward, this nuthatch's body rarely, if ever, points straight down the tree trunk—rather, it points off to one side of the trunk in a transverse position. Each time the nuthatch pauses in its downward movement, the up-trunk foot is turned sideways. The nuthatch may turn its body and head so they point down the trunk while at rest, but the upper foot remains turned sideways and clinging to the bark. The down-trunk leg and foot do not function in clinging; rather, they brace the bird and support its weight.[362]

FAMILY CERTHIIDAE

Certhiidae is a northern hemisphere family of creepers. One species is present in North America, and six species inhabit the Old World. Creepers live in wooded regions. They are slender little birds with pointed, often downcurved bills and long, stiff, pointed tail feathers. Their legs are short and slender, and their toes are slender with long, sharp claws. Certhiids are brownish and streaked above and whitish below; the sexes are alike. They are solitary arboreal birds. Creepers forage for insects and spiders on the trunks and limbs of trees. They spiral their way upward on tree trunks, using their tails as braces while they climb.

Brown Creeper
Certhia americana

The breeding range of the Brown Creeper lies in the boreal and montane coniferous forests of North America, extending from southeastern Alaska into the United States and through western United States to Central America. This creeper is an uncommon to common permanent resident in the mountainous eastern and western rims and in some of the mountain ranges of the Great Basin. As yet it has not been detected as a breeding species in some of the interior mountain ranges, including the Pine Forest, Santa Rosa, Desatoya, Toiyabe, and Shoshone mountains.[289] Nor is it known as a breeding species in the northwestern corner of the Great Basin in Oregon.[103] Prior to 1982 the scientific name of the Brown Creeper was *C. familiaris*, and it was regarded as being subspecies of the Treecreeper, *C. familiaris*, of Eurasia.

The Brown Creeper frequents montane forests where the trees are set fairly close together. It is more often first heard as a high, thin *screep* or *seee* or a wiry *scree*, then seen as a tiny bit of brown creeping up a tree trunk. Its preferred nesting site lies behind a piece of loosened bark still partially attached to a dead or living tree, often within five to fifteen feet of the ground. Here, between the bark and the trunk, the female builds her hammock- or crescent-shaped nest. The nest is fairly easy to find once you know where to search—the parents use a direct approach to the nest and are not at all cautious about revealing its presence.

Brown Creepers feed mainly on insects. Their foraging beats lie along the trunks and main branches of trees. In covering its beat a creeper starts at the base of the tree and follows a spiral course upward, encircling the trunk several times on its way to the top. It may forage out on the underside of some of the larger limbs. Once a creeper is high up in the tree, it breaks away and flies down to the base of a nearby tree to repeat the performance. While climbing it hitches itself upward, using its stiff pointed tail feathers as a brace against the tree trunk. Its long, sharp-pointed, and downcurved beak is used to pick insects out of crevices in the bark.

Outside of the breeding season, creepers are solitary birds. Sometimes one will briefly fall in with a mixed-species foraging flock of chickadees, nuthatches, and kinglets before going its own way. Creepers perform altitudinal migrations and appear at foothill and valley sites during the winter. Communal roosting has been observed. There is also an observation in the literature on huddling—two creepers were seen snuggling close together in the corner of a porch on winter nights.

FAMILY TROGLODYTIDAE

The wren family, Troglodytidae, is a New World family. Only one species has spread into the Old World. Wrens are small, somewhat chunky birds, plainly colored in browns and brownish grays. They are usually barred, streaked, mottled, or spotted. Their wings are short and rounded, and they often carry their tails cocked upright. Troglodytids frequent a wide variety of habitats—they forage close to the ground in tangles of branches, underbrush, or brush piles. Wrens are inquisitive, restless birds. Their calls are harsh chattering ones; they scold much; but their songs are of a melodious and bubbling nature. In some species both sexes sing, and in some species song may be heard the year around. Wrens are solitary birds, except when paired for nesting.

Rock Wren
Salpinctes obsoletus

The breeding range of the Rock Wren extends from southern British Columbia through western United States to Central America. This wren is a common summer resident in the Great Basin. Although the species is migratory, an occasional wren may be seen here during the winter.

Wherever rocky areas occur—in valleys, foothills, or mountains—Rock Wrens will be there. It is our most abundant and widespread wren, as it was in the 1860s. Writing of those bygone days, Ridgway observed:

> The Rock Wren is by far the most common and generally distributed species of the family in the Western Region, since the prevailing character of that country is so well suited to its habits. It was first met with near the summit of Donner Lake Pass of the Sierra Nevada, but this was on the eastern slope, and in a district where the pine forests were interrupted by considerable tracts of open country, of a more or less rocky nature. Eastward of this point, as far as we journeyed, it was found in suitable localities on all the desert ranges. Its favorite resorts are piles of rocks, where it may be observed hopping in and out among the recesses or interstices between the boulders, or perched upon the summit of a stone, usually uttering its simple, guttural notes. It is not strictly rupicoline [rock-inhabiting], however, for along the eastern base of the Sierra Nevada, where the pine forest reaches to the very base of the mountains, it was common in cleared tracts where there was much rubbish of old stumps, prostrate logs, and piles of brush, seeming as much at home there as among the rocks. At that place the males were occasionally observed to fly up to a naked branch of some dead tree, and remain there while they sang their simple trill. This species also freely accepts of the accommodations and protection afforded by man, for in many towns, notably those among the mountains, it nests about the old buildings and inside the entrance to mining-shafts, displaying as much familiarity and confidence as the little House Wren, or Bewick's Wren.[25]

As Ridgway noted, Rock Wrens are not totally rupicoline. They may be found away from rocks, at locales where log and brush piles abound or where earth banks have been riddled by the diggings and burrows of rodents. These alternative sites possess one attribute in common with the natural rock world of the wren—they are rich in crevices, interstices, passageways, recesses, and nooks and crannies of diverse shapes and sizes. Confined spaces provide a vital dimension to the Rock's world. The wren does some foraging on the exposed surfaces of rocks, and its song perches, loafing sites, preening sites, and sunning sites are here. But much of its foraging is done on the inner surfaces of crevices and interstices, within or between rocks. Then, too, it nests and seeks protection from the elements and danger in these enclosed spaces.

The Rock Wren is built for a shallow world, where the ceilings are not far removed from the floors. With flattened head and body, it creeps mouse-like, on short legs and small feet, into narrow horizontal fissures to forage. Its foraging reach within a narrow fissure can be extended into tight corners by probing with its long, slender, slightly decurved bill. This wren feeds mainly on insects, supplemented by spiders and other arthropods. An analysis of the stomach contents of eighty-eight Rock Wrens collected throughout Utah showed that the birds had fed extensively on injurious insects—such as grasshoppers, crickets, ants, and others of their ilk.[363]

The shallow nest of the Rock Wren is built in a tightly confined space in, between, or under boulders. A peculiar building trait of this species still puzzles ornithologists. The wrens usually pave the nest cavity with small stones, one-half to two inches long. The nest is then built on this stone foundation, and there is a stone walk leading from the nest to the entrance of the cavity. Sometimes the stone walk is eight to ten inches long. In situations where there is no room for a stone walk, the wrens still build a stone foundation for the nest or mix stones with the nesting material. Some Rock Wrens pile stones so high at the entrance to the cavity that a barrier is formed, over which the wrens must squeeze when entering and leaving the cavity. As yet, there is no reasonable explanation of what this stonework is all about. However, the stone walk can help human observers find Rock Wren nests that are tucked away out of sight.

Canyon Wren
Catherpes mexicanus

The range of the Canyon Wren extends from British Columbia, southward through western United States, to southern Mexico. This bird is a rare to uncommon permanent resident in the Great Basin. Prior to 1982 its English name was Cañon Wren.

The Canyon Wren inhabits steep-walled canyons, extensive, rugged areas of cliffsides and rock walls, and even the vicinity of human habitations. On the surface of the rock faces and within the crevices and interstices, the Canyon forages for food and finds nesting sites. Ridgway's description of the Canyon Wren in the Great Basin in the 1860s merits repetition:

> We found it everywhere more rare than the other species, and apparently confined to the more secluded portions of the mountains, where it frequented rocky gorges and the interior of caves more often than the piles of loose rocks on the open slopes. It was generally observed to be rather shy, and prone to elude pursuit by retreating to the deeper recesses of the rocks, now and then slyly peeping from some crevice but an instant, and

> then very unexpectedly reappearing at some distant place. While thus engaged, or while hopping about, examining each crevice for a spider or other insect, it utters a simple ringing note, which sounds somewhat like *dink*, uttered in a metallic tone; while now and then he pauses to pour forth his piercing song, which is of such volume as to fill the surrounding cañons with its reverberations. In many of its movements it greatly resembles the common Rock Wren, particularly in its manner of bowing and swinging oddly from side to side, when its attention becomes attracted by the presence of an intruder. It was frequently seen to cling to the roof or sides of a cave with the facility of a Creeper, and on one occasion to fly perpendicularly up the face of a cliff for a considerable height.[25]

This wren ranks among our foremost musicians, and its wild, haunting song is a thing of beauty. Hoffmann described it as "a cascade of sweet liquid notes, like the spray of a waterfall in sunshine. The opening notes are single staccato notes followed by long-drawn double notes, *tsee-i, tsee-i*, slower and descending in pitch, ending with still lower *tóo-ee tóo-ee tóo-ee.*"[84]

Bewick's Wren
Thryomanes bewickii

The range of the Bewick's Wren extends discontinuously from southwestern British Columbia to southern Mexico. The Bewick's is missing from most of the northwestern, north central, and eastern parts of the United States. This wren has a limited presence in the Great Basin. It is an uncommon to common permanent resident in northeastern California, in northwestern Nevada, in the White Mountains, and along the eastern base of the southern Sierra Nevada. In west central Nevada its breeding range extends at least as far east as the Fallon-Stillwater region, and there are winter records for Unionville in the West Humboldt Mountains. To my knowledge there are no occurrence records for north central or northeastern Nevada. The Bewick's Wren occurs in the eastern part of the Great Basin in Utah but is not considered to be as common there as in southern and eastern Utah.[56]

The Bewick's is now apparently more abundant and widespread in the western end of the Great Basin than it was in the 1860s. Ridgway was familiar with this species, having frequently observed it during his stay in Sacramento. However, during his years in the field in the Great Basin he did not collect or positively identify a single Bewick's Wren. He had one quick look at a wren at Glendale, along the Truckee River, in November of

1867 which may have been either a Bewick's or a House Wren. Today, Bewick's Wrens are quite common at the various localities visited by Ridgway.

This wren inhabits shrubby areas where the shrubs abut on riparian thickets or woodlands or intermingle with pinyon-juniper woodlands. It often frequents the vicinity of human habitations and is found in towns and cities.

The Bewick's Wren makes its living as a gleaner of insects. Its foraging stations embrace the leaves and branches of shrubs, the ground, and the lower branches and trunks of trees. It picks insects off the bark and leaves, sometimes probing for or pecking at insects in crevices in the bark. On the ground a wren may flip aside litter with its bill, but it does not scratch with its feet as do typical ground foragers.[364] A wren is continually on the move while foraging, seldom slowing down or stopping at one spot. It progresses by short hops from branch to branch or clings to the bark and creeps up a tree trunk by short hops. It seldom flies while foraging, except when changing from one tree or shrub to another. Sometimes a wren will hang upside down while gleaning on the underside of vegetation.

Outside of the breeding season, Bewick's Wrens may occasionally join mixed-species foraging flocks. In four of the forty-five mixed-species flocks Panik encountered in pinyon-juniper woodlands in northwestern Nevada, a single Bewick's was present. The principal flock associates of the wren were Bushtits, Mountain Chickadees, Dark-eyed Juncos, and Plain Titmice.[351]

The nest of this wren is usually located in cavities, crannies, or holes close to the ground or even in the ground. Crevices in buildings and rock walls and even bird boxes are used. In Sacramento in 1867, Ridgway found Bewick's Wrens frequenting the outhouses of the city, along with Barn Swallows and Black Phoebes, in their nesting.[25] Sometimes a nest is constructed at a site not completely enclosed by wood, rock, or dirt—such as in a brush heap or in a tree. Then, the wrens may arch over the top of the nest, building a domed structure with a side entrance. Sometimes a cast-off snake skin may be incorporated into the nest structure. Both of these traits are reminiscent of those of the House Wren.

The male wren is a capable musician. He sings from conspicuous, elevated song perches in the spring—short bursts of song may even be heard on mild winter days. A number of observers have commented on the general resemblance between the songs of Bewick's Wrens and Song Sparrows. During a field study in western Washington, Song Sparrows were seen defending their territories against trespass by Bewick's Wrens, although the wrens were not as aggressive in return toward the sparrows.[365]

Although differences exist between populations of Bewick's Wrens in regard to song, there are generally three recognizable parts to a song.

Hoffmann observed that "when the song is analyzed it is found generally to consist of three distinct parts, a high quick opening of two or more notes, then lower notes rather *burry* in quality, and in closing a very delicate fine trill. There are endless variations, but the three distinct parts can almost always be distinguished in the full song; occasionally the first or last are left off."[84]

Bewick's Wrens may have territorial conflicts not only with Song Sparrows but also with House Wrens. Bewick's Wrens defend territory the year around.[364] During a California study, Bewick's Wrens moved in the fall into habitat left vacant by migrating House Wrens. Although the Bewick's were nest building in these areas when the House Wrens returned in the spring, they were displaced by the House Wrens and moved into adjacent areas to nest.[366]

House Wren
Troglodytes aedon

The breeding range of the House Wren extends from southern Canada, down over most of the United States, to northern Baja California. This wren is an uncommon to common summer resident in the Great Basin. Although the species is migratory, an occasional House Wren may be found here during mild winters.

The House Wren can be found as a breeding species wherever trees in sufficient quantities stand, in both the mountains and the lowlands of the Great Basin. Ridgway's words about this wren in the Great Basin in the 1860s also ring true today:

> The range of this Wren is apparently co-extensive with the distribution of the timber, or governed strictly by the presence or absence of trees, without special regard to their kind. Its vertical range, like that of the Robin, Louisiana Tanager, and many other species, was consequently very considerable, it being equally abundant among the cotton-woods of the river-valleys and the aspen copses of the higher cañons of the mountains. Indeed it is the only strictly arboreal species of this family which resides in summer in the Middle Province, and there much more rarely seeks the society of man or the protection of his presence than the Rock or Bewick's Wrens.[25]

The House is a cavity nester, usually nesting in natural cavities or abandoned woodpecker holes located within ten feet or so of the ground. It may nest behind loose bark—Ridgway found this to be the common situation in the 1860s. He wrote, "Very numerous nests of this species were found, their situation being various, although most of them were similar in this respect; the prevailing character being that of a large mass of rubbish filled in behind the loosened bark of the trunk of a tree, usually only a few feet from the ground, the entrance a natural crevice or a woodpecker's hole; it was always warmly lined with feathers, and very frequently possessed the ornamental addition of a cast-off snake-skin."[25] Today's House Wrens are just as inclined as their ancestors of Ridgway's day to stuff their nesting cavities to the hilt with rubbish. They may still add cast-off snake skins to their nests.

The House Wren is an extremely aggressive nesting community member. It is quite intolerant of neighboring nesting birds, especially other cavity nesters, but will harass noncavity nesters as well. Despite its tiny size, a House Wren is often able to evict other birds from their nesting cavities or harass them until they leave. It may go so far as to puncture eggs, toss eggs out of the nest, or kill nestlings. One of the few cavity nesters who has been able to outmuscle the House Wren is an introduced pest—the burly, pugnacious House Sparrow. In some localities in eastern United States, as the fortunes of the House Sparrow have waxed those of the House Wren have waned. Nest destruction by the House Wren has often been condemned by human observers. In the 1920s a bitter controversy sprung up in the ornithological literature, as writers debated the merit and character, or lack thereof, of the House Wren.

Another House Wren debate has questioned whether the western subspecies will associate as readily with people as will the eastern subspecies. Early on, as we have seen, Ridgway commented on the infrequent association of wrens and humans in the Great Basin. Today, I too mostly encounter my House Wrens living away from human habitations in the Basin. But I am not sure what motivates this infrequent association—is it more the reluctance of House Wrens to live next to settlement or our reluctance to build and hang nest boxes for wrens?

The House Wren feeds almost entirely on insects and other arthropods. Its foraging station is located close to the ground, usually within four feet, and often in dense underbrush. This wren is often heard before it is seen. It has "an extensive, rippling, laughing song which reminds one strongly of a musical waterfall or purling brook."[367] The Chippewa Indians had a descriptive name for the House Wren of "O-du-ná-mis-sug-ud-da-we'-shi, meaning a big noise for its size!"[368]

Winter Wren
Troglodytes troglodytes

The breeding range of the Winter Wren lies in the boreal forests of Eurasia and northern North America. Its breeding range barely extends into the

United States, except in the Appalachian Mountains in the East and in the humid coastal forests and on the western slopes of the Sierra Nevada in the West.

The Winter Wren is a rare to uncommon winter visitant in the Great Basin. However, it is reported as being of rare occurrence during all four seasons of the year at Malheur National Wildlife Refuge.[61] It has been reported nesting in Juab County, Utah, in the past, but this may have been an error in identification.[56] Most surprisingly, it was reported as being a resident in "The Birds of Carson River Basin."[75] However, I know of no other claim or documentation for its resident status in western Nevada or adjacent California.

Marsh Wren
Cistothorus palustris

The breeding range of the Marsh Wren includes marshy areas scattered from southern Canada, southward over the United States, into Mexico. The Marsh Wren is a common summer resident in the Great Basin. It is an occasional to rather common overwintering species at some Great Basin marshes. The number of wrens overwintering in an area is influenced by the severity of the winters and by whether the waters mainly remain open or not. Prior to 1982 the English name of this species was Long-billed Marsh Wren, and prior to 1976 its scientific name was *Telmatodytes palustris*.

As its name suggests, the Marsh Wren is found where tall, emergent vegetation—particularly cattails, bulrushes, and tules—dominate in marshy areas. It is a bird well named, since within the marsh world it finds all the necessities of life, including food, song perches, courting centers, nest sites, cover, roosts, and loafing sites.

The nest of the Marsh Wren is lashed to stems of emergent vegetation, one to three feet above the water. The seven-by-five-inch nest is a woven, oblong, domed structure with an opening on one side. The outer shell is woven of coarse plant material, and the lining is of fine plant material and feathers.

A peculiar behavioral trait is associated with nest building. Not only is a brood nest built, in which eggs are laid and young are raised, but during the nesting season the male builds a number of unlined nests on his territory. These nests have been variously named by ornithologists as multiple nests, incomplete nests, cock nests, dummy nests, and male nests. The male Marsh Wren is not unique in building multiple nests; this practice occurs in certain other species, including House Wrens. Male House Wrens may go around filling up a number of potential nest cavities with trash in advance of the breeding season.

Most of the interest and debate about multiple nest building has focused on the probable function of this behavior. Does it somehow allow the male to release pent-up sexual energy? Does it provide roosting shelter? Does it provide the male with a pleasurable activity? Does it supply a lot of fake nests to confuse and distract potential predators? Or what?

Field studies by Verner suggest that multiple nests function in courtship. The multiple nests are all built in a small area on the male's territory. This area then serves as a courting center. The male does most of his singing here. When a potential mate visits his territory, he displays to her, then flies to his courting center. Sometimes she follows him there. At the courting center she may examine the multiple nests, entering them one after another, while escorted by the male. If she pair-bonds with him, she will either line one of the multiple nests to use as the brood nest, or she will start a new nest—which will be built mainly by the male—for this purpose.[369]

A male who has been unable to obtain a mate may end up with several mating centers on his territory, each center containing about one to five nests. If a male is successful in pair bonding, he will start building a new courting center on his territory and advertising for another mate, about the time his first mate starts egg laying. Male Marsh Wrens engage in polygyny whenever possible.[369]

A detailed study of the reproductive success of eighty male Marsh Wrens was conducted in the state of Washington. Of these males, thirteen were unable to obtain mates; forty-seven attracted one female apiece; and twenty had two females apiece. The average number of courting nests built and completed by the males correlated with their mating success. Bachelor males averaged 17.4 nests apiece, monogamous males 22.1 nests, and bigamous males 24.9 nests.[370]

Experimental studies have revealed that Marsh Wrens have a killer instinct when in the presence of the eggs or nestlings of other birds.[371] Not only do they puncture the eggs and kill the nestlings of other species of small marsh-nesting birds, but they attack the eggs and nestlings of other Marsh Wrens. Red-winged Blackbirds and Yellow-headed Blackbirds have low nesting success when they nest in the vicinity of Marsh Wrens. Often the nesting colonies of these two blackbirds are spatially separated from Marsh Wren breeding areas within a marsh. Blackbirds chase Marsh Wrens when they encounter them.

The most puzzling aspect of the Marsh Wrens' killer instinct is their readiness to attack the eggs and nestlings of others of their kind. Not only do strange males, females, and juveniles do this, but within a family of wrens this will take place. A male will puncture the eggs and kill the nestlings of his mate when she leaves them unguarded. Within a polygynous family, one female will attack the nest of another harem member when she leaves it unguarded. When a nest is attacked, not only are eggs punctured or broken and nestlings killed or removed, but some of the nest lining is usually removed by the attacking wren.[372]

If a female is to successfully raise a family, she must be vigilant while incubating the eggs and raising the young by herself. Since she cannot leave her nest unguarded, she must restrict her food foraging to a small area around her nest. She unhesitantly attacks other wrens, including her mate, who venture near her nest and drives them away. She lines her nest with very soft, shock-absorbing material. Therefore, it is difficult for an attacking wren to puncture eggs sitting on this bouncy lining.[372]

FAMILY CINCLIDAE

The dippers form the Family Cinclidae, a four-species, Old World–New World family. Dippers are small chunky birds, plainly colored in browns, grays, or black. Their wings are short, rounded, and concave. Their tails are short, and their legs and toes are long and strong. Cinclids, our only truly aquatic passerines, are highly adapted for an aquatic life. They live in and along clear, swift, rocky mountain streams. Their plumage is soft and filmy with a thick undercoat of down. Their preen glands are extraordinarily large.

American Dipper
Cinclus mexicanus

The American Dipper ranges from north central Alaska, through western Canada and western United States, to Central America. It is a common permanent resident in and along many of the streams of cold, swift water in the Great Basin. During the winter it may perform altitudinal migrations downstream to stretches of open water. Prior to 1982 its English name was the Dipper.

The American Dipper, to some the Water Ouzel, inhabits cool, clear, swift-moving streams; it makes its living along the banks of streams and underwater. The dipper frequents streams rich in aquatic insects, where there are small waterfalls or cascades and where some rocks project, here and there, above the surface of the water. Dipper streams in the Great Basin are restricted almost entirely to mountains and foothills—particularly mountain canyons and slopes.

Dippers are among our most fascinating and delightful birds. Their name is descriptive of their constant dipping or bobbing: the entire body moves up and down, as the dipper repeatedly crouches down on bent legs and then straightens up again, at rates of up to sixty dips a minute. Dipping begins early in life. It has been observed in seventeen-day-old nestlings after they were handled. It is apparently more frequent when the dippers have been disturbed.

The foraging behavior of the American Dipper is varied. It may forage by swimming on the surface of the water, by plunging its head underwater while standing on a rock, by moving along the shore, or even by flycatching insects. But its unique technique of diving underwater and foraging while progressing along the bottom of the stream has excited the most interest. Disagreement has sprung up around the question of how a dipper manages to propel itself along the bottom of the stream. There have been two schools of thought about how this action is accomplished—represented by those who believe it is merely a matter of leg and foot action and by those who believe that wing strokes primarily keep a dipper submerged and propel it along. These questions have been difficult to resolve, since all the action occurs under turbulent water, and the legs and wings may be in motion simultaneously.

Motion picture photography employed under laboratory conditions eventually showed that a dipper must always use its wings to remain submerged and that both horizontal and vertical progression underwater are principally attained by strokes of the partially opened wings. The legs may assist somewhat in propelling the dipper along the bottom of the stream, when they are used alternately in a running fashion. But there is no evidence that the legs can do it all by themselves. When a dipper briefly loses contact with the bottom while moving forward, the legs are folded or are allowed to trail behind, and the wings provide all of the propulsion. In underwater wing action, the front edge of the wing is depressed when the wing is moving backward. Depression of the front edges of the wings during the backward stroke drives the dipper downward as well as forward. Thus, the bird can stay submerged as well as move forward.[373]

In moving underwater a dipper usually travels against the current. Various speculations have centered around why a dipper is so faithful to an upstream orientation. Maybe the dipper is then better prepared to seize food items being carried by the current, or it could be that aquatic insects and their larvae are more numerous on the lee sides of rocks in swift water. Some have suggested that somehow moving against the current helps the bird remain submerged. However, it seems more plausible that the bird benefits from moving against the current because the water pressure plasters down its plumage and doesn't ruffle it. The skin of the dipper must remain dry and shielded from contact with the water, if the bird is not to become chilled in the cold water in which it forages.

Sometimes a dipper resurfaces from a dive and, without stopping at the

American Dipper

surface of the water, flies directly into the air. If it does not stop at the surface first, or if it has been swimming on the surface, a dipper takes off with its first few wing strokes striking against the water and with a backward kick against the water with both feet.[373]

The American Dipper is well adapted for diving and remaining submerged under cold water. Each nostril opening is located behind a large scale partly covered with feathers, which helps prevent water from entering the nostrils. Dippers are very well insulated. This is reflected by their very broad range of thermoneutrality (11.5 to 34° C); by a very low lower critical temperature for a fifty-gram bird (11.5° C); and by the low energy cost of regulating their body temperature under cold stress of only 0.33 calories per gram of body weight per hour for each decrease in ambient temperature of 1° C below the lower critical temperature of 11.5° C.[9] The plumage of the dipper is very full and compact. Goodge counted an average of 4,200 (ranging from 3,700 to 6,300) contour feathers—all the feathers visible on the surface—on the body of an American Dipper, based on a sample size of five dippers; this is high for a fifty-gram passerine, since the average value is typically less than 3,000.[9] The dipper also has an unusually dense supply of down feathers under the contour feathers. Finally, the dipper has a large uropygial gland at the base of its tail, from which to extract oil for preening and waterproofing the contour feathers.

Observers have considered the question of how long dippers remain submerged during a dive, and dive durations of ten to thirty seconds have been timed. Estimates in the literature of up to two minutes seem unduly long, although no one has attempted to measure the bird's ultimate capacity to remain submerged. While underwater a dipper experiences hypoxia or oxygen insufficiency, since it cannot breathe and must get by on the oxygen it obtained just before diving. Various diving birds studied by comparative physiologists have been found to possess in general several physiological adaptations, which allow them to contend with the problem of hypoxia. Murrish has found some of these adaptations in dippers. He discovered that dippers experience "an immediate, initial bradycardia upon submersion followed by a gradual further decline in heart-rate. They show an immediate tachycardia upon surfacing."[374]

The bradycardia or slowing down of the heart rate at the beginning of the dive results in a greatly lessened circulation of blood. If bradycardia is accompanied by constriction of peripheral blood vessels, as it typically is in divers, the arterial flow of blood to the muscles and other noncritical organs is greatly reduced. The blood, which is the oxygen transport system, is then supplied mainly to the heart wall, eyes, and central nervous system—all extremely sensitive to oxygen deprivation. The tachycardia or speeding up of the heart rate at the termination of the dive quickly resupplies the musculature and noncritical organs with blood and oxygen. Murrish was able to induce the bradycardia merely by dropping water on the nostrils of the dipper, without actually forcing the bird to submerge its head beneath the water. The diversion of oxygen-carrying blood from the skeletal muscles during diving is not of serious consequence, even though these muscles are hard at work while the dipper is submerged. Muscle tissue contains stored oxygen; then, too, for a while muscles can obtain energy for contraction from anaerobic chemical reactions—or reactions not requiring molecular oxygen.

Murrish also looked at the oxygen-carrying capacity of dipper blood. It was not unusually high for a bird, although the blood contained slightly more hemoglobin than that of some nondiving birds. The oxygen capacity of dipper blood was only about 19 volume percent.[374] That of the nondiving humans is slightly higher at 20 volume percent, and some mammalian divers like seals and whales have oxygen capacities of between 30 and 40 volume percent.[375] So the blood in the dipper is not carrying an unusually high concentration of oxygen at the start of a dive.

The dives of the dipper are very shallow (from six inches to several feet) and of very short duration (ten to thirty seconds) when compared to the dives of seals and whales and even other diving birds like loons. Because of its short, shallow dives, the dipper does not face any threat of the bends or caisson disease—caisson disease can result only after prolonged dives at depths below about twenty-two yards.

Since Murrish's experiments involved restrained dippers, it would be informative to obtain some data by telemetry on free-diving dippers. This would reduce the possibility of experimental artifacts. Also, calculations of the total volume of the lungs, the posterior air sacs, and the blood of the dipper are needed to ascertain how much molecular oxygen is available for circulation during diving. Further, a determination on free-living dippers of the average time span between as well as within dives would be interesting.

The location and architecture of the American Dipper's nest are somewhat unique. The nest is often positioned behind a small waterfall or in the spray of cascading water, and the mosses which comprise the main structural component of the nest remain alive and green. Sometimes the dipper will forgo the sound and spray of falling water and build above a smooth-flowing stream on the underbeam of a bridge. This they have been doing at a bridge over the Truckee River at Verdi, Nevada. The dipper builds a large, domed, Dutch oven–shaped nest. The entrance hole grows progressively larger along with the nestlings. Often nests are repaired and used in subsequent years.

The American Dipper is more beautiful of voice than of dress. This drab bird is a rare musician among birds—performing during most of the year, in storm and calm, under clouds and sun. Hoffmann wrote, "The Dipper's song is strong and sweet, made up of a great variety of trills and flute-like passages, delivered with great spirit and brilliance."[84] The spirit and brilliance of this montane musician truly touched John Muir. Muir wrote:

> During the golden days of Indian summer, after most of the snow has been melted, and the mountain streams have become feeble,—a succession of silent pools, linked together by shallow, transparent currents and strips of silvery lacework,—then the song of the Ouzel is at its lowest ebb. But as soon as the winter clouds have bloomed, and the mountain treasuries are once more replenished with snow, the voices of the streams and ouzels increase in strength and richness until the flood season of early summer. Then the torrents chant their noblest anthems, and then is the flood-time of our songster's melody. As for weather, dark days and sun days are the same to him. The voices of most songbirds, however joyous, suffer a long winter eclipse; but the Ouzel sings on through all the seasons and every kind of storm. Indeed no storm can be more violent than those of the waterfalls in the midst of which he delights to dwell. However dark and boisterous the weather, snowing, blowing, or cloudy, all the same he sings, and with never a note of sadness.[376]

The above passage from the work of John Muir was taken from his poetically charming essay "The Water-Ouzel," which appeared as a chapter in his book *The Mountains of California*. In this essay Muir captured the true wild essence and spirit of the American Dipper. Paperback editions of *The Mountains of California* are available in bookstores, and this delightful essay on the ouzel merits reading in its entirety.

FAMILY MUSCICAPIDAE

The muscicapids form a large family of mainly Old World species. In the sixth edition of the AOU checklist, several groups of birds who were up to now given familial rank lose this rank and are included as merely subfamilies of the Family Muscicapidae. The Family Sylviidae, containing our kinglets and gnatcatchers, now becomes the Subfamily Sylviinae. The Family Turdidae, containing our thrushes, now becomes the Subfamily Turdinae.

Golden-crowned Kinglet
Regulus satrapa

The breeding range of the Golden-crowned Kinglet lies in the boreal forests of southern Alaska and Canada, with extensions southward into the western and eastern parts of the United States. The Goldencrown is an uncommon permanent resident in both the western and the eastern ends of the Great Basin. There may be a hiatus in its distribution here—records are lacking for Nevada mountains in the central part of the Great Basin. Altitudinal migrations may occur; some kinglets appear at lowland locales, including pinyon-juniper woodlands, during the winter.

The Golden-crowned Kinglet is a summer resident in closed montane forests, where there are densely needled boughs in which to nest and forage. Its foraging station centers around the outer twiggery and needle tufts of densely foliaged conifers, especially firs and spruces. Foraging kinglets often engage in aerial gleaning—they flutter or hover in front of sprays of foliage to pick off small insects and spiders. They also flycatch passing insects. Kinglets are very active birds, almost constantly on the move and continually opening and closing their wings even when perched.

The thin call notes of this kinglet are often confused with those of the Brown Creeper, but the *zee, zee, zee* of the kinglet is broken into syllables, whereas the creeper's long *scree* is not. The faint, high-pitched song of the male kinglet has been described as *tse; tse, tse, tse-tse-tse-tse, tse, tse*. Hoffmann said it "sounds as if the bird were mounting a tiny stairway of lisping notes and then tumbling down at the end."[84]

Golden-crowned Kinglets are often present in small foraging flocks outside of the nesting season. Sometimes they join mixed-species foraging flocks. They were present in seven of the forty-five mixed-species foraging flocks Panik encountered in pinyon-juniper woodlands in northwestern Nevada.[351]

Ruby-crowned Kinglet
Regulus calendula

The breeding range of the Ruby-crowned Kinglet lies in the taiga or boreal coniferous forests of Alaska and Canada and penetrates deeply into western United States. This is a common breeding species in montane forests in the Great Basin; it is much more abundant and widespread than the Golden-crowned Kinglet. Altitudinal migrations occur, and kinglets are common in lowland localities during the winter.

This kinglet is more of a solitary bird than the Golden-crowned Kinglet. Single Rubycrowns are usually encountered outside of the breeding season and may occasionally be seen in mixed-species foraging flocks. Like the Golden-crowned Kinglet, this species has a jerky, undulating flight and continually opens and closes its wings even while perched.

The two species of kinglets differ in a number of other ways. The Rubycrown is not tied as closely to dense conifer forests and may even be found in open, sunlit areas in the forests. Its foraging station is located closer to the ground, and it exploits the shrub layer much more than does the Goldencrown. It is a more active and rapid forager than the Goldencrown, as well as a more accomplished songster.

The diet of the Ruby-crowned Kinglet is mainly one of insects and other small invertebrates. In foraging the Rubycrown concentrates on the outer foliage of trees and shrubs. Constantly on the move from perch to perch, it twists and turns and even jumps up and flutters its wings to reach gleaning sites. Sometimes one gleans by hovering in front of clusters of pine needles. A passing insect or one dislodged by the movement of a kinglet will be snapped out of the air, flycatcherlike. The wells of sapsuckers will be visited by Ruby-crowned Kinglets, as well as by Golden-crowned Kinglets, in search of sap and insects.

The male Rubycrown is held in high regard as a songster. Hoffmann wrote, "In March, and occasionally earlier the wintering males begin to sing. The song varies greatly in different individuals but at its best is extremely musical and astonishingly loud for so tiny a performer. It begins with two or three high notes which are uttered at increasing speed, then a soft *kew, kew, kew*, followed by a phrase of three syllables *teé-di-di*, which is repeated two or three times."[84]

The nests and nest sites of the two species are much alike. A kinglet nest is a hanging cup, suspended in dense foliage, beneath the tip of the horizontal limb of a conifer. The deep, thick-walled cup is composed mainly of mosses, grasses, and fine bark, bound together and lashed to conifer twigs with spiderwebs. Often the nest is lined with feathers. Nest locations vary from positions very low to very high in conifers.

Blue-gray Gnatcatcher
Polioptila caerulea

The breeding range of the Blue-gray Gnatcatcher extends across the United States and southward to Central America. This bird is an uncommon summer resident, decreasing in both numbers and occurrence northward, in the Great Basin.

In the Great Basin, the Blue-gray nests principally in pinyon-juniper woodlands. It also frequents stands of willows and cottonwoods along streams and desert shrubs adjacent to treed areas. The nest of this tiny bird is a thing of beauty: a small cup-shaped structure, built of plant down, tied together with the silk of insects and spiders, decorated with lichens, and saddled on a horizontal limb or in a fork well above the ground. Sometimes the gnatcatcher will dismantle an old nest and reuse the material in building a new one. If disturbed while nest building, it may dismantle the partially or even completely built structure and use the material to build a new nest nearby.

The name gnatcatcher implies that this six-gram mite is a hunter of small game. And so it is, feeding mainly on insects and other small arthropods. Much of the gnatcatcher's foraging involves gleaning insects and their eggs and larvae from the open branchwork and foliage of small trees and shrubs. While foraging it is continually on the move, pausing but briefly at a perch to scan the vegetation and glean before moving on to a new perch. Aerial gleaning is also employed, as the bird hovers in front of the foliage. While gleaning a gnatcatcher may tumble off its perch after dislodged insects or chase flying insects—when hawking insects it does not engage in typical flycatching by waiting motionless on a scanning perch and returning to a perch after each capture. Sometimes a gnatcatcher will capture insects on the ground or on ground-level vegetation. Then, it operates from low perches or even hovers above the ground to locate insects. If large prey such as a grasshopper or butterfly is captured, it is broken up by being battered against a perch before being eaten.[377]

Western Bluebird
Sialia mexicana

The breeding range of the Western Bluebird lies in western North America, extending from southern Canada into Mexico. The Western is a rare to common summer resident in the Great Basin. There is apparently somewhat of a hiatus in the distribution of this species in north central and northeastern Nevada; at least, there is a paucity of summer records from these parts of the state. Ridgway noted a large gap in the distribution of the Western Bluebird in the 1860s. He commented, "Yet we lost sight of this species entirely after we left the eastern water-shed of the Sierra Nevada, and never saw nor heard of it in the Wahsatch or Uintah Mountains, notwithstanding the latter country appeared equally adapted to the requirements of the species. The last individuals seen, as we journeyed eastward, were a few families of young birds, with their parents, in the wooded valley of the Truckee River, near the Big Bend."[25] Western Bluebirds are occasionally seen in the Great Basin during the winter.

Western Bluebirds frequent lightly wooded areas in the valleys, pinyon-juniper woodlands, and open montane forests at lower altitudes. They are cavity nesters, using abandoned woodpecker holes or natural cavities. They will readily come to bird boxes and have been known to nest in the gourd-shaped mud nests of Cliff Swallows.

During the warmer months bluebirds feed mainly on insects, but they take berries and other fruit in season. During the fall, foraging flocks may wander to higher altitudes in search of food. The foraging station for capturing insects is on or close to the ground, in areas where the vegetation is low and often interrupted by bare or rocky patches. Scattered low trees and bushes and tall weed stalks such as mullein must be present to serve as scanning perches. If the scanning perches are inadequate, bluebirds may locate insects by hovering low over the ground. They have even been seen briefly soaring, and alternating soaring with hovering, when hunting insects in the face of strong updrafts.

The Western has a fondness for fruit such as elderberries, juniper berries, and especially mistletoe berries—from the family of evergreen plants that parasitize a wide variety of trees and shrubs in the deserts and mountains. The edible part of the mistletoe berry is the soft, mucilaginous pulp surrounding the inner seed. After feeding on mistletoe berries Western Bluebirds, along with their relatives the robins and solitaires, disseminate the seeds. The seeds are not affected by their passage through the alimentary tract of the bird, and when voided they still possess enough of the mucilaginous coating to stick to the branches of new host trees. It has been suggested that the Western Bluebird's eating habits may beneficially affect its reproductive success, since in spreading mistletoe seeds the bluebird is creating future nesting sites for itself: the parasitic mistletoe promotes the death and eventually decay of the tree branches and the formation of new cavities in trees. In some regions, such as the Sierra Nevada, Western Bluebirds move to higher altitudes in the fall and early winter to feed on mistletoe berries. Mistletoe can also be used as cover—small flocks of Western Bluebirds have been seen roosting at night in the center of clumps of mistletoe.

Mountain Bluebird
Sialia currucoides

The breeding range of the Mountain Bluebird extends from central Alaska through western Canada and western United States. This bluebird is a common summer resident in mountain ranges in the Great Basin. It is a common migrant and occasional wintering bird in Great Basin valleys.

Mountain Bluebirds most commonly nest at moderate to high altitudes—from seven thousand to twelve thousand feet. They nest near their foraging areas, which lie in montane meadows, in rock fields, in logged or burnt-over tracts, or up in the Hudsonian life zone, where the trees are stunted and widely spaced. Ridgway's description of the nesting distribution of the Mountain Bluebird in the 1860s still applies:

> Its favorite haunts are the higher portions of the desert ranges of the Great Basin, where there is little water, and no timber other than the usual scant groves of stunted cedars, piñon, or mountain mahogany. In these elevated regions it is abundant during the summer, and even remains in winter, except when violent storms or severe cold drive it to the more clement valleys, where it may be seen, either singly or in considerable but scattered flocks, whenever a snow-storm prevails on the mountains.[25]

Over the years some species of birds have been able to better adapt to the encroachment of human civilization than others. Ridgway noted how in the 1860s the Mountain was already availing itself of "the superior accommodations and protection afforded by civilized man." He wrote, "In June, the 'Mountain Blue-bird' was observed to be common in Virginia City, Nevada, where it nested in the manner of the Eastern species, in suitable places about buildings in the town, the old mills and abandoned shafts of the mines being its favorite haunts, which it shared with the House Finch (*Carpodacus frontalis*) and the Rock Wren (*Salpinctes obsoletus*)." Ridgway also noted the spread of nesting Mountain Bluebirds down into the valleys, writing, "For although it is naturally a bird of the high mountains, we noticed that at Salt Lake City they were quite numerous, although, were the locality unreclaimed from its primitive state, they would not have been found there except during their vertical migrations, influenced by changes in the climate. Even on Antelope Island, in the Great Salt Lake, a few pairs were seen about the buildings of the ranche."[25]

Changes come and go. According to Ridgway, Mountain Bluebirds overwintered in west central Nevada in numbers in the 1860s. He noted that winter storms drove them out of the mountains into the valleys around Carson City, then stated that the "visits of this species to the lower valleys are only occasional, however, for as soon as a storm in the upper regions subsides, they return to their own haunts."[25] Nowadays, Mountain Bluebirds are seen only occasionally here during the winter; most migrate to overwinter.

Then, according to some Utah observers, the spread of nesting Mountain Bluebirds into the valleys has not persisted. They have commented that this bluebird was "formerly a common summer resident in valleys and mountains throughout the state. Now, less common and mostly confined to mountains and high valleys as a breeding species."[56] One wonders what motivated the retreat of nesting bluebirds from the valleys. Has our increasing reliance on chemical pesticides been involved? Has our mania for cleaning up the landscape and eliminating nesting cavities by trimming off dead limbs and cutting down dead trees and snags been involved? Has the spread of competing cavity nesters such as starlings into the valleys been involved?

Mountain Bluebirds are probably the most highly insectivorous of all the members of the thrush family, although they do add berries and other fruit to their diet seasonally. Their feeding station is located on or close to the ground. They hunt over wide-open country, where the vegetation is short or the ground bare and where there are scattered trees, shrubs, and even boulders which can be used as scanning perches. In areas where scanning perches are not available, bluebirds use a hovering technique. They hover ten to twenty feet above the ground on rapidly beating wings, with feet dangling and tails spread. When prey is sighted they flutter down to make the capture. Some flycatching of passing insects also takes place.

Townsend's Solitaire
Myadestes townsendi

The breeding range of the Townsend's Solitaire extends from central Alaska, southward through western Canada and western United States, into Mexico. This bird is an uncommon to common permanent resident in the Great Basin. It nests in the mountains and often appears in the foothills and valleys during the winter.

The Townsend's nests in open coniferous forests on mountain slopes or in mountain canyons at moderate to high altitudes, on or near the ground in sheltered, well-drained sites. The nest is frequently located in a depression or pocket beneath an overhang of cliff, bank, tree trunk, stump, or fallen-tree roots. The bulky nest of plant materials frequently features a long apron of nesting matter hanging below its front.

The male solitaire possesses one of the finest voices to be found in our mountains. His territorial-courtship song has been described as "a rather prolonged, warblelike series of rapid notes, each note on a different pitch than the last. The notes are clear, sweet, and loud, and follow each other almost as rapidly as those of the winter wren."[378] Song perches are often dead stubs in the crowns of tall trees, one hundred feet or more above the ground. Sometimes the male will leave his perch and spiral upward in a song flight—he will sing off and on, as he climbs in a series of spirals to as high as five hundred feet or so above the ground. The song flight may terminate at a song perch to which the male plunges, on set wings, in a series of steep, downward zigzags. Or he may pull out of his plunge when near the ground and spiral upward again in another song flight.

During the warmer part of the year, insects constitute the food staple of the solitaire. Solitaires flycatch from scanning perches on outer, exposed branches of trees. They sit quietly and fly out to intercept passing insects. At times they ground forage by operating from scanning posts located close to the ground. They may flutter up against a tree trunk to snap up some wandering spider or insect.

The winter behavior of Townsend's Solitaires is fascinating. Solitaires feed mainly on berries and other fruit then. They must often leave their breeding grounds, or at least perform altitudinal migrations, to obtain sufficient berries to carry them through the winter. Not only are their food-directed movements of interest, but on the wintering grounds territories are established and chasing, displaying, singing, and fighting occur.

The winter behavior of Townsend's Solitaires was studied along the western edge of the Great Basin near Eagle Lake by Lederer. Solitaires overwinter here, feeding on the berries of the western juniper. When the solitaires arrive in the fall there is much chasing, fighting, and singing as winter territories are established. After they are established much of the territorial defense is passive: solitaires announce possession by perching near the tops of trees, calling some but seldom singing. A solitaire only defends its territory and supply of juniper berries from other solitaires. American Robins also overwinter here and feed on juniper berries. When a robin approaches a perched territorial solitaire, the solitaire flies away, yielding to the robin. When the ground is free of snow, solitaires forage for berries under the juniper trees; when snow blankets the ground, solitaires feed in the junipers. Much of their tree foraging is accomplished by flying around the outside of a juniper and hovering while picking berries from the outer tips of the branches—in effect, a type of aerial gleaning.[379]

During the course of a winter study near Flagstaff, Arizona, nine territorial Townsend's were sexed, and all were males. In addition to territorial males, nonterritorial solitaires were present. These nonterritorial solitaires employed sneaky behavior to trespass on territories and feed on juniper berries. Five of the nonterritorial birds were examined, and three were adult females. Here, the territorial males defended their territories against trespass not only by other solitaires but sometimes by other species, including juncos and bluebirds. Their food staple was the berries of the one-seed juniper, but they also fed on mistletoe berries.[380] In some regions of the West, mistletoe berries are their winter food staple.

Veery
Catharus fuscescens

The breeding range of the Veery lies mainly to the east and north of the Great Basin. The Veery is an uncommon summer resident in northern Utah.[27] It is a rare summer resident in northeastern Nevada, and it is listed as being of accidental occurrence at Malheur National Wildlife Refuge.[61] Prior to 1973 the scientific name of the Veery was *Hylocichla fuscescens*.

Since the Veery still frequents willow thickets along valley streams, Ridgway's 1869 observations are still relevant:

> It was extremely abundant along the Provo River, especially just above the *debouché* of that stream through its picturesque cañon between two lofty snow-clad peaks of the main range of the Wahsatch Mountains; and it was also seen in the valleys of the Bear and Weber Rivers, farther northward. In all these localities it frequented the dense willow-thickets in the immediate vicinity of the rivers, where it was extremely difficult to discover, and next to impossible to secure specimens after they were shot.
>
> We never tired of listening to the thrilling songs of these birds, for they were truly inspiring through their exceeding sweetness and beautiful expression. The modulation of their notes was somewhat similar to that expressed by the syllables *ta-weél-ah*, *ta-weél-ah*, *twil'-ah*, *twil'-ah*, the latter portion subdued in tone, thus seeming like an echo of the first. In the valley of the Provo it was not unusual to hear a dozen or more of these exquisite songsters uniting in their rivalry, the most favorable time being the afternoon and evening.[25]

The foraging station and foraging techniques of the Veery are quite like those of other members of the genus *Catharus*. It forages for insects, spiders, and other invertebrates mainly on the ground, occasionally venturing into the lower parts of shrubs. When berries and other fruit are in season it expands its feeding station above the ground. While foraging on the ground the Veery hops along, turning over dead leaves with its bill to expose the ground.

Swainson's Thrush
Catharus ustulatus

The breeding range of the Swainson's Thrush extends from central Alaska, across much of Canada, with some extensions into the United States. This thrush is an uncommon to common summer resident in some of the Great

Basin mountain ranges. It is an occasional to common migrant at lowland localities. Prior to 1973 the scientific name of the Swainson's was *Hylocichla ustulata*.

William Swainson was a nineteenth-century English naturalist whose memory is also commemorated by the Swainson's Hawk and the Swainson's Warbler. He is also remembered as a leading proponent of a weird quinary system of taxonomy derived from numerology. Five was considered to be a divine number. God not only created flowers whose parts, such as petals and sepals, came in fives; but, supposedly, even species, genera, and families were created in quincunxes or groups of fives.

The Swainson's Thrush is not as widely distributed in the mountains of the Great Basin as is the Hermit Thrush. These two thrushes are sometimes found together. Although both may nest in stands of aspens or conifers, the Swainson's is more closely tied to streamsides, with a strong preference for dense thickets of willows.

For capturing insects and other invertebrates, the Swainson's and Hermit thrushes both have ground-level feeding stations, which they expand above ground seasonally when berries and other fruit are available. Both forage like robins—running, suddenly stopping, and scrutinizing the ground from an upright posture. They differ in that the Swainson's Thrush does not nervously twitch its wings or raise and lower its tail like the Hermit Thrush.

Although the Swainson's Thrush is a fine songster, he is not nearly the equal of the Hermit and Wood thrushes:

> The song is set in character and each individual thrush begins his song on about the same key—not changing from song to song as does the Hermit. The first syllables of any individual's song are always of the same pitch, and full, clear, and deep; the remainder are more wiry, ascending, and sometimes the last one goes up so high in pitch as to become almost a squeal: *wheer*, *wheer*, *wheer*, *whee-ia*, *whee-ia*, *whee-ia*, or *quer*, *quer*, *quer*, *quee-ia*, *quee-ia*, *quee-ia*. The call note oftenest heard is a soft liquid whistle, *what* or *whoit*, sounding much like the drip of water into a barrel. An imitation of this note by the observer will often bring a thrush into close range.[318]

Hermit Thrush
Catharus guttatus

The breeding range of the Hermit Thrush extends from central Alaska and central Canada into the United States. This thrush is a common summer resident in Great Basin montane forests. It is a common migrant and rare winter visitant in the valleys here. Prior to 1973 its scientific name was *Hylocichla guttata*.

Its name refers to the Hermit's shy and retiring habits. This thrush lives in the solitude of the mountains away from human settlement in the Great Basin. It frequents middle to high altitudes, nesting in mountain mahogany thickets, aspen groves, and dense to moderately open coniferous forests—often near streams.

Most critics of bird music recognize the Hermit Thrush as the finest songster in North America, among the very greatest bird voices in the world. In describing the song Hoffmann wrote:

> It includes three or four passages, separated by considerable intervals and at higher or lower pitch, but each opens with a flute-like note that is held a moment. It is this opening note that gives the performance the effect of a chant of sacred music. None of the other fine performers among the mountain birds, the Fox Sparrow, the Green-tailed Towhee, or the Solitaire, has the same spiritual quality of tone, or gives the effect of religious ecstasy.[84]

In a more detailed description of the song, Saunders wrote:

> The song of the hermit thrush is a long-continued one, made up of rather long phrases of 5 to 12 notes each, with rather long pauses. All the notes are sweet, clear, and musical, like the tone of a bell, purer than the notes of the wood thrush, but perhaps less rich in quality. The notes in each phrase are not all connected. The first note is longest and lowest in pitch, and the final notes are likely to be grouped in twos or threes, the pitch of each group usually descending. Each phrase is similar to the others in form but on a different pitch, as if the bird sang the same theme over and over, each time in a different key. If one listens carefully for each note, however, two different phrases are rarely exact duplicates in form, but slightly varied, a likeness to certain symphonies of some of the great composers . . . the average individual has a range of about an octave.[378]

In paying homage to our greatest songster, it seems fitting to conclude with the words John Burroughs wrote over one hundred years ago:

> Ever since I entered the woods, even while listening to the lesser songsters, or contemplating the silent forms about me, a strain has reached my ears from out the depths of the forest that to me is the finest sound in nature,—the song of the hermit thrush. I often hear him thus a long

> way off, sometimes over a quarter of a mile away, when only the stronger and more perfect parts of his music reach me; and through the general chorus of wrens and warblers I detect this sound rising pure and serene, as if a spirit from some remote height were slowly chanting a divine accompaniment. This song appeals to the sentiment of the beautiful in me, and suggests a serene religious beatitude as no other sound in nature does. It is perhaps more of an evening than a morning hymn, though I hear it at all hours of the day. It is very simple, and I can hardly tell the secret of its charm. "O spheral, spheral!" he seems to say; "O holy, holy! O clear away, clear away! O clear up, clear up!" interspersed with the finest trills and most delicate preludes. It is not a proud, gorgeous strain, like the tanager's or the grosbeak's; suggests no passion or emotion,—nothing personal,—but seems to be the voice of that calm, sweet solemnity one attains to in his best moments. It realizes a peace and a deep, solemn joy that only the finest souls may know.[381]

The voice of the Hermit Thrush may well be, as Burroughs believed, the "finest sound in nature." Since the Hermit performs in the late spring and early summer in many of our Great Basin mountains, it is a sound well within your hearing. Listen for it along forest streams at middle and upper elevations—preferably near dawn and dusk. Although its song perches are usually in trees well above the ground, the Hermit Thrush will often be encountered on or close to the ground. In the West this thrush nests mainly in small conifers or shrubs close to the ground. In the East it nests mainly on the ground.

During spring and summer the foraging station of this thrush is on the ground, where insects, spiders, and other small invertebrates are to be found. When ground foraging a thrush does not visit wide-open areas; it is seldom far from cover. While foraging it moves much like a robin—hopping a short distance, stopping suddenly, and examining the ground from an upright posture. If dead leaves are present, a thrush may lift leaf after leaf in its bill and toss them aside to expose the ground. After berries and other fruit become available, the Hermit Thrush will supplement its diet with these items—then its foraging station expands above ground. Sometimes migrants or winter stragglers will be found in towns and cities in the Great Basin, feeding on the fruit of cultivated trees and shrubs. Note the Hermit Thrush's nervous mannerism: it twitches its wings every few seconds, often accompanying this with a slow downward movement of its tail.

American Robin
Turdus migratorius

The breeding range of the American Robin covers much of North America, extending from just south of the arctic tundra to southern Mexico. This bird is an abundant and widespread summer resident in the valleys and mountains of the Great Basin. Although robins perform both altitudinal and latitudinal migrations, they are common winter residents in the Basin, but in lesser numbers than during the summer. Prior to 1973 this bird's English name was merely the Robin.

The American Robin is our true national bird, by fact if not by law. It actively seeks our company, forages in plain view on our lawns, and nests in our shade trees. It is the first bird, and often the only bird, a child learns to call by name. The American Robin is our most visible and beloved songbird.

Before settlement of the Great Basin, the robin frequented the forests, particularly aspen groves, and streamside vegetation in the mountains; and it frequented the riparian woodlands and thickets in the valleys. Of the 1860s Ridgway wrote, "From the Sierra Nevada eastward, however, it was continually met with in all wooded localities, the aspen groves of the higher cañons being its favorite resort during the summer, while in winter it descended to the lower valleys, and passed the season among the willows or cotton-woods and attendant shrubbery along the streams."[25] After settlement of the Basin, the robin became much more widely distributed; it enlarged its habitat to include our lawns, gardens, orchards, pastures, and irrigated fields.

The robin's spring and summer foraging station is centered on the ground, where the vegetation is short and the soil organically rich and moist and soft. Here earthworms and insects thrive. In the fall and winter the feeding station lies above the ground, where berries and other fruit are to be found on shrubs, vines, and trees.

Probably the most frequently viewed hunting scene in nature is the sight of an American Robin stalking earthworms on a lawn. The robin hunts by making short runs, stopping abruptly, and, while leaning forward, cocking its head from side to side to eye the ground directly in front. When an earthworm is sighted, the robin jumps ahead and strikes at the worm with its bill. If the bird succeeds in grasping the earthworm, it shifts its weight back on both feet and attempts to tug the worm out of its burrow. If the first pounce is unsuccessful, the robin may change position with a hop and strike again. Earthworm hunting usually occurs early or late in the day, when the worms are at the entrance to their burrows or even partly out of their burrows. Hunting may also occur during midday if rain or lawn watering has driven the worms out of their burrows.

A debate of long standing among bird-watchers has been waged over the questions of how robins locate earthworms and what the head cocking is all about. Is it to better see or better hear earthworms? The consensus now is that earthworms are seen and not heard. Studies have shown that the sounds made by earthworms burrowing in the soil are of very low intensity, yet experimentally produced high-intensity background noise did not interfere with the robin's ability to locate earthworms. During other experiments robins did not peck at artificially produced earthworm holes; but, when pieces of dead earthworms were placed just inside the artificially produced holes, they found and removed the pieces.[382]

With the return of spring, male American Robins select and defend territories. A male tends to return to the same general locality each year to do so. Then, the females return to the neighborhood of their previous nestings to form a pairing bond with a male. A territory may or may not include sufficient feeding ground to support a family of robins. Often the territories are too small to do so; then the robins in each locality will have communal feeding grounds where they all can forage.

Males defend their territories against trespass by other robins by song, threats, and fighting. The females participate actively by vigorously defending the area around the nest site. When an intruder appears, the resident robin dashes at it with feathers erect while chirping and snapping its bill. A chase may ensue among the trees or in the air. When fighting erupts, robins strike at each other with bills and wings. They may fight on the ground, flutter up into the air fighting, or carry the fight into a nearby tree.

The territorial song of the male American Robin is one of the most commonly heard voices of spring. Saunders has written:

> The song of the robin is long-continued; made up of phrases with short pauses between them. These phrases are repeated, alternated, or otherwise arranged in groups of two to five, with longer pauses between the groups. Each phrase is composed of one to four notes, but most commonly two or three. The notes are frequently joined by liquid consonant sounds like r or l. . . . The time of the song is regularly rhythmical, the phrases and pauses being of even length. Ordinarily the robin sings at a rate of two phrases per second. In the very early morning they often sing faster and more continuously, the phrases not being broken up into groups. Then the rate is about two and a half phrases per second. Individual robins differ from each other in the phrases they use and the order in which they sing them.[378]

The phrases of the song rise and fall in pitch. Words such as "Cheer up! cheeri-ly, cheer-i-ly, cheer up!" have been fitted to the song. Sometimes, when his mate is nearby, the male will sing as if for her ears alone. He then sings with his bill closed, and his song is a whisper.[383]

The courtship and pair-bonding activities of the robin are simple ones: there are no intricate and highly stereotyped performances. The song of the territorial male serves as advertising, and, once a female is attracted to a male, courtship and pair formation are accomplished without structured displays. In company, the female and male seek out a nesting site on the male's territory; often a number of trees are closely inspected before a nest site is selected. The female usually builds the nest with little or no help from her mate. Since robin nests are frequently built in conspicuous sites close to human habitations, the mud nest of the robin has become the most widely recognized avian architectural accomplishment in all of North America. The nest is a bowl-shaped structure; coarse grasses and twigs are welded into a mud bowl, which is lined with fine blades of grass. The female carries the mud to the nest site in and on her bill, drops the pellets inside the developing bowl, and kneads them into place with her body.

American Robins are gregarious birds outside of the nesting season and are usually found in small flocks. Even during the nesting season robins may gather at communal roosts to spend the night. Observations made in eastern United States revealed that the adult males may leave their females and nests in the evenings and gather at neighborhood roosts to sleep. After the first broods of young were well able to fly, they joined their fathers at these roosts. Then, at the end of the nesting season, the adult females and second broods of young began using the roosts. A population of robins may have several roosts, and the number of robins present at any one roost may vary from night to night—depending on the population density, the number of birds at a roost may vary from a few on up into the thousands. When the robins are settling down for the night, they may contend for roosting sites. At these times, bill snapping is used as a threat display. Often other species of birds will join the robins at their roosts.

Winter roosts are more prevalent than summer roosts. Some winter roosts, located on favorable wintering grounds, are spectacular—several hundred thousand robins may be present at a single roost. The life span of a roost varies. It may be used but for a season, or it may be reused year after year. One winter a roost was located on the University of Nevada campus in Reno. Several thousand robins would gather nightly to roost in the trees bordering the campus quadrangle. During the early morning flocks and individual robins would fan out to forage, returning at dusk to roost. Of course, the mess formed on campus sidewalks by their droppings was not appreciated. The University of Nevada at Reno has never had a reputation for hospitality as far as bird visitors are concerned.

Other Muscicapids

The breeding range of the Gray-cheeked Thrush, *Catharus minimus*, lies far to the north of the Great Basin in North America and Siberia. This thrush is of accidental occurrence in the Basin. There are several undocumented sight records of its occurrence in north central Utah.[27] Prior to 1973 its scientific name was *Hylocichla minima*.

The Wood Thrush, *Hylocichla mustelina*, breeds in eastern United States and the southeastern edge of Canada. The Wood Thrush is of accidental occurrence in the eastern part of the Great Basin. There is a specimen record for Salt Lake City and a sight record for Utah County.[27]

The breeding range of the Varied Thrush, *Ixoreus naevius*, extends from north central Alaska, through the Pacific Northwest, southward to northern California and northern Idaho. The Varied Thrush is an occasional to uncommon transient and winter visitant in the Great Basin. It is much more regular and abundant in its occurrence along the western edge than elsewhere here.

FAMILY MIMIDAE

The mimids constitute a strictly New World family of mockingbirds, thrashers, and catbirds—ranging from southern Canada deep into South America. Mimids are rather trim, medium-size birds, with long tails and legs. Their wings are relatively short and rounded. They are mainly brown, gray, or slate-colored. Their lighter-colored underparts are often spotted or streaked. The sexes are colored alike. Most mimids feed on the ground or close to it, in and around thickets or brushy growth. Song is highly developed in this family. As the family name implies, there are mimics in the family.

Gray Catbird
Dumetella carolinensis

The breeding range of the Gray Catbird extends from southern Canada over much of the United States. The Great Basin marks the west central edge of its breeding range in the United States. This species is said to be an uncommon summer resident in Utah.[27] The Gray Catbird rapidly decreases in abundance westward within the Great Basin. There are a few sight records and specimens from northeastern Nevada and a nesting record by Al Knorr from the Toiyabe Range in central Nevada. It is of rare occurrence in the spring and summer at Malheur National Wildlife Refuge.[61] The Gray Catbird may be extending its range westward in the Basin. Prior to 1973 the English name of this species was the Catbird.

Its name was derived from the catlike mews which this bird frequently utters. However, not all its vocalizations resemble cat calls, for the catbird comes from a family of gifted musicians. Like its relative the Northern Mockingbird, the catbird is capable of fine song and is an excellent mimic. It may sing at night. The genus name of *Dumetella*, derived from the Latin diminutive meaning little shrub, refers to the bird's fondness for dense shrubbery or thickets.

Northern Mockingbird
Mimus polyglottos

The range of the Northern Mockingbird extends from southeastern Oregon and Newfoundland, southward over the United States, to southern Mexico. This bird is a rare to uncommon resident in the Great Basin. It is not a regular migratory species, but some individuals wander in the fall and winter, sometimes northward. In the Reno area the Northern Mockingbird may be more numerous during the winter than during the summer. In Utah mockingbirds decrease in number during the winter; they are considered to be uncommon in the summer but rare in the winter.[27] Prior to 1982 the English name of this species was the Mockingbird.

Both the English and the scientific names of this bird refer to its vocal abilities. It is truly a mocking bird, since it can readily imitate or mock any sound it hears. Its scientific name was derived from the Greek meaning many-tongued mimic, and it does speak in tongues and sings the songs of many species of birds. Forbush has written, "He equals and even excels the whole feathered choir. He improves upon most of the notes that he reproduces, adding also to his varied repertoire the crowing of chanticleer, the cackling of the hen, the barking of the house dog, the squeaking of the unoiled wheelbarrow, the postman's whistle, the plaints of young chickens and turkeys and those of young wild birds, not neglecting to mimic his own offspring. He even imitates man's musical inventions."[384] Hoffmann wrote:

> Individual Mockingbirds differ greatly in their repertoire. Many sing chiefly their own songs, made up of high clear notes repeated rapidly a number of times; these are clearer and more rapidly repeated than the

> phrases of the California Thrasher. Occasionally a bird adds to his own song remarkable imitations of the sounds he hears about him, chiefly the notes of other birds. These imitations are generally harsh sounds, like the notes of California Woodpeckers, Shrikes, Kingbirds or the Ash-throated Flycatchers, or rapidly repeated phrases like those of the Plain Tit or Cactus Wren. Besides the song, the commonest notes of the Mockingbird are a loud smack or kissing sound, and a harsh, grating *chair.*[84]

Another unusual aspect to the singing of the mockingbird is its tendency to sing at night, especially on moonlit nights. Nocturnal song among birds is a rarity. It is of interest that two of the greatest voices in the bird world are heard by night—that of the nightingale in the Old World and the mockingbird in the New World.

Insects and fruit figure heavily in the diet of the Northern Mockingbird. Much of the insect food is captured on the ground. A peculiar type of behavior called wing flashing often occurs when a mockingbird is foraging on the ground. Some ornithologists believe that wing flashing startles an insect into revealing its presence, either through the sudden appearance of the white wing patches or by casting a shadow over the insect.

A mockingbird wing flashes by lifting both wings away from the sides of its body and simultaneously spreading them open through a series of jerky motions or hitches. Usually one to three, sometimes up to five, hitches occur in a wing spreading. As the wings are hitched open the white wing patches appear to flash. The wings can be spread in a more vertical or in a more horizontal direction.[385]

Field studies by Hailman showed that wing flashing is strongly correlated with the bird's ground-foraging movements. Three types of movements can be seen during ground foraging: looking down at the ground, presumably to locate prey; running or hopping, presumably to close in on prey; and striking, presumably to capture prey. During his two years of field studies Hailman discovered that, 96 percent and 99 percent of the time, wing flashing was followed by one of the three foraging movements. He then concluded that wing flashing played a role in ground foraging, possibly to flush insects.[385]

The ontogeny or development of wing flashing in young mockingbirds has been studied using hand-raised birds. Both sexes showed wing flashing when they were nine to thirteen days old. The young hand-raised birds displayed this behavior when they were excited or uneasy due to exposure to strange objects or unexpected movements or noises.[386]

Wing flashing has been seen in contexts other than those involving ground foraging and uneasy, hand-reared young. Consequently, other explanations of this behavior have been offered. Some believe that it is hostile or threat behavior. It has been seen when mockingbirds were mobbing an owl, confronting a cat, and attacking a snake.[387] It has been seen during territorial boundary disputes between mockingbirds. Observers have noted that birds flashing white in their wings when flying, such as Evening Grosbeaks and Red-bellied Woodpeckers, were attacked more often and more vigorously by mockingbirds at a feeding station than birds without white in their wings. By now, it is obvious that mockingbirds engage in wing flashing in a variety of situations and that debates over the function of this behavior will continue.

Northern Mockingbirds are highly territorial and maintain winter feeding territories as well as breeding territories. The breeding territory is defended by the male, with little or no help from the female. The winter territory encloses the food supply; both sexes defend this territory. Sometimes a pair will remain together on a single winter territory, sometimes they separate and maintain individual territories. The winter territory may be in a different area, or it may be part of the breeding territory. Winter territory is often defended more vigorously than breeding territory, yet trespass commonly occurs.[388]

The mockingbird prospers with human settlement and thrives where fruit trees and ornamental fruit-bearing trees and shrubs, such as pyracantha and hawthorn, are planted. Although mockingbirds are basically nonmigratory, individuals do wander during the fall and winter. Thus, new breeding populations can become established as we create new habitat for them. The expansion of mockingbirds in California has been very noticeable since the early 1900s, and even some northward expansion has occurred. The expansion of breeding mockingbirds into west central Nevada has apparently been of recent origin. Although mockingbirds were regularly seen during the winter in the 1950s and 1960s in Reno, the first indication of nesting I am aware of occurred in the early 1970s.

Sage Thrasher
Oreoscoptes montanus

The breeding range of the Sage Thrasher extends from south central British Columbia, through the interior of western United States, to Texas and Oklahoma. The Sage is a common summer resident in the shrublands of the Great Basin. It is only rarely seen here in the winter.

Its English name is highly appropriate, since this thrasher is almost always found in shrublands dominated by big sagebrush. Therefore it occurs mainly in the higher valleys and foothills and on mesas. Its scientific name is as inappropriate as its English name is appropriate. *Oreoscoptes* is derived from the Greek meaning mimic of the mountains. Although this thrasher belongs to the Family Mimidae, it is not a mimic like the mockingbird and catbird. Nor is it a montane bird, *montanus* being derived from the Latin meaning of mountains.

The earliest account in the ornithological literature of the breeding behavior of the Sage Thrasher is based on observations Ridgway made in the spring of 1868 in west central Nevada. He wrote:

> At Carson City, very favorable opportunity was afforded for observing the habits of this interesting species during the breeding-season. The males began singing about the 24th of March, or immediately after their arrival, but their notes were then subdued, while their manners were reserved in the extreme. They soon became numerous in the sagebrush around the outskirts of the city, and were often seen perched upon the summit of a bush, turning the head from side to side in a watchful manner, even while singing; when approached, disappearing by diving into the bush, and, after a long circuitous flight near the ground, reappearing some distance in the rear of the pursuer. This peculiar, concealed flight we found to be a constant habit of the species. As the pairing-season approached, with the advance of spring, the songs of the males became greatly improved, both in strength and quality; their manners also became changed, for they had lost their former shyness. About the 10th of April, the males were engaged in eager rivalry, each vying with the other as he sang his sweetest notes, his wings being at intervals raised vertically so as to almost touch over the back, and quivering with the ecstasy that agitated the singer. The first eggs were laid about the 20th of April, the nests having been commenced a week or more earlier; and by this time the males had become perfectly silent, their main occupation being that of sentinel on guard for the approach of an intruder.[25]

Early on, Ridgway described how male Sage Thrashers raise their wings vertically over their backs at intervals while singing from a song perch. Of late, the wing raising of singing Sage Thrashers has been envisioned as being homologous to the wing flashing in their relatives the mockingbirds, although thrashers raise their wings smoothly, not in hitches.[389] Male Sage Thrashers perform song flights over their territories. They fly in an undulatory fashion, low over the sagebrush, in zigzag or circular paths, singing as they go. At the termination of a flight, they land with their wings upraised and flutter them briefly. As they continue to sing from a song perch, they repeatedly raise and lower their wings.

Ridgway noted wing trembling during the song flights he witnessed in the spring of 1868 at Carson City. The entry in his field notebook dated April 9 reads:

> The Sage Thrasher is now one of the most common birds in this vicinity. To-day a great many were noticed among the brush-heaps in the city cemetery. Its manners during the pairing season are peculiar. The males, as they flew before us, were observed to keep up a peculiar tremor or fluttering of the wings, warbling as they flew, and upon alighting (generally upon the fence or a bush), raised the wings over the back, with elbows together, quivering with joy as they sang.[310]

The nest of the thrasher is built either in the branches of a shrub or on the ground under a shrub—usually sagebrush. Sage Thrashers use larger sagebrushes to nest in than to nest under. A field study carried out on forty-seven Sage Thrasher nests in southeastern Idaho showed that both ground and shrub nests were always placed to maintain an almost identical depth of foliage over the nest. Very little variability existed in the average distance from a ground nest to the top of the crown of the shrub above it or in the average distance from a shrub nest to the top of the crown of the shrub it was in.[390]

Why would Sage Thrashers be so careful to maintain a constant thickness of shrub cover above their nests? An obvious answer is that a cover depth of twenty-three inches or so functions most effectively in heat economy during nesting—that this cover depth is optimal for providing the nest site and its occupants with maximum shielding from the sun and from the night sky. Since the height of the nest varied considerably, from right on the ground to well up in a shrub, protection from ground predators must not be involved in nest placement.

Although the Sage Thrasher has a fondness for fruit in season, both wild and cultivated, its main staple is insects. Its foraging station for insects is mainly on the ground, between and beneath the shrubs, with limited visits to the foliage of the shrubs. This thrasher does some highly beneficial feeding in the Great Basin, since it is one of the very few birds that will feed heavily on the Mormon cricket, a very injurious pest, and its eggs.

Other Mimids

The Brown Thrasher, *Toxostoma rufum*, an eastern species, is of occasional occurrence in the eastern part of the Great Basin.[27] It is of accidental occurrence in the western end, where it has been reported from Malheur National Wildlife Refuge.[61]

The Bendire's Thrasher, *Toxostoma bendirei*, is a resident of the hot deserts to the south of the Great Basin. It is of accidental occurrence in the Basin. It has been recorded from Vernon and Faust and from the east shore of Utah Lake in Utah.[27]

FAMILY MOTACILLIDAE

The pipits and wagtails of the Family Motacillidae have an almost worldwide distribution. Motacillids are slender birds with thin, pointed bills and rather long, slim legs. Their long tails are often edged with white or yellow. They frequent open, treeless country and carry out most of their activities on the ground—here they walk or run but never hop. Pipits were named for their twittering call notes, wagtails for their incessant tail wagging. Motacillids frequently call when in flight, and some have flight songs. They are gregarious outside of the nesting season.

Water Pipit
Anthus spinoletta

The breeding range of the Water Pipit lies on the arctic tundra of North America and Eurasia, extending southward on mountaintops. In western North America breeding occurs in the arctic-alpine life zone on mountains as far south as New Mexico. The Water Pipit is a summer resident and breeding species on some of the higher mountains in the Great Basin. It is a common migrant and an occasional overwintering bird at lowland sites here.

The Water Pipit usually nests on mountains above the upper tree line in the arctic-alpine life zone or, as it is sometimes called, the alpine zone. Depending on local conditions, the upper tree line in the Great Basin is usually at an altitude of ten thousand to eleven thousand feet. Since even four-wheel-drive roads and trails usually never get anywhere near this far up the mountainside, the alpine life zone can be studied only after a long, strenuous climb under what are hypoxic conditions for most weekend hikers. Consequently, it is the most poorly studied life zone in the Great Basin.

As for breeding records of the Water Pipit in the Great Basin, there are none for northeastern California.[93] The Water Pipit may be a breeding species at Steens Mountain in southeastern Oregon.[103] It is said to be "a summer resident in the high mountains throughout the state" of Utah,[27] including the Deep Creek Mountains close to the Utah-Nevada border.[391] Until recently there were no breeding records for the Water Pipit for the

Nevada part of the Great Basin—or for the entire state of Nevada. Then, concerted fieldwork conducted by Al Knorr led to the discovery of pipits in the mountains of Nevada. To date, Al has found Water Pipits nesting on four mountains in Nevada, three of which are in the Great Basin: Wheeler Peak in the Snake Range, Arc Dome in the Toiyabe Range, and Mount Rose in the Carson Range.

The nesting environment of the Water Pipit is characteristically arctic (northern) or alpine (mountain) tundra. The life form of tundra vegetation is very low, being measured in inches. It consists of short herbaceous plants, including grasses and sedges, of lichens and mosses, and of greatly stunted shrubs. In the high Uinta Mountains, off the northeastern edge of the Great Basin in Utah, the Water Pipit is also a subalpine breeding species in "either dry or wet montane meadows in the coniferous forests in the vicinity of water."[392]

When on territory male pipits engage in song flights. A male will mount straight up in the air, fifty to two hundred feet, singing a simple song which has been variously interpreted as *che-whée, che-whée* or as *tjwee, tjwee* or as *te-cheer! te-cheer!* During his descent to the ground, he raises and spreads his tail, dangles his legs, and opens his wings fully, or even flutters his wings, as he floats downward; and he sings at a faster tempo. Some singing also occurs from the ground.

Although pipits feed on seeds, the bulk of their diet consists of insects. Their foraging station is on the ground, in short vegetation or on bare surfaces, preferably in moist areas. While foraging a pipit walks or runs after insects.

When in the valleys during migration or during winter, Water Pipits are usually in flocks. The most unusual instance of play I have yet witnessed among birds involved Water Pipits. It was a sunny winter day, and there were ice floes in the Carson River near Dayton, Nevada. We were near a stretch of the river where the water was moving the ice floes swiftly along through some rapids. Water Pipits were landing on ice floes and riding them downriver through the rapids. After traversing the swift water, each party of pipits would leave their floes and fly back upriver to catch other floes coming down.

Other Motacillids

A specimen of the Olive Tree-Pipit, *Anthus hodgsoni*, was collected in the Great Basin ten miles south of Reno on May 16, 1967. This specimen is now in the collection of the United States National Museum. It was the first record for this Asiatic species for North America.[393] The AOU committee elected to use the English name of Olive Tree-Pipit for this exotic species in the sixth edition of the checklist, but the name Indian Tree Pipit has also been used in the past.

The breeding range of the Sprague's Pipit, *Anthus spragueii*, lies far to the northeast of the Great Basin in the prairie provinces of Canada and adjacent United States. This pipit is a "rare winter visitant and transient" in the eastern end of the Great Basin in Utah.[27]

FAMILY BOMBYCILLIDAE

Three species of waxwings form the Family Bombycillidae. Waxwings inhabit the northern coniferous and birch forests of North America and Eurasia. They are small birds. Their soft, filmy plumage is colored in blended fawn-browns and grays. A prominent head crest is present. Their bills are short, thick, slightly notched, and hooked. The secondaries of the wings of two of the species are often tipped with what appear to be drops of red wax—hence the name waxwings. Bombycillids are highly gregarious, arboreal birds—even associating together on the nesting grounds. They migrate nomadically and are not much more predictable in their choice of breeding grounds than in their choice of wintering grounds.

Bohemian Waxwing
Bombycilla garrulus

The breeding range of the Bohemian Waxwing lies in northern Eurasia and northwestern North America. In North America it extends from Alaska, through western Canada, into the northwestern corner of the United States. The Bohemian is a rare to fairly common winter and spring visitant to the Great Basin. It is an irregular visitant in that during some years the flocks are large and widely distributed, but during most years they are small and few in number. This species is apparently most consistent in its winter appearances in the northeastern part of the Great Basin in Utah. Here observers rate it as a "common winter visitant."[27]

Although Bohemian Waxwings show up regularly in the winter south of their breeding range, major invasions occur periodically. Then numerous flocks appear, some of them containing hundreds of birds. These flocks wander widely in search of berries and other fruit. When invasions occur they are not local in extent but may involve the entire western or eastern part of the United States. Invasions are believed to result when profound food shortages develop in Canada.

Foraging flocks of Bohemian Waxwings are often accompanied by other birds, especially American Robins. When feeding the robins are quite aggressive toward the waxwings—a robin will frequently fly up at a waxwing and supplant it at a feeding site. Bohemian Waxwings are very gregarious. When a feeding flock is disturbed and put to flight, the individuals will take off in various directions, sounding their rough *scree* call. But the birds will then circle around until they can regroup into a tight flock again. Foraging flocks of waxwings are often attracted to towns and cities where ornamental trees and shrubs such as mountain ash, hawthorn, cotoneaster, and pyracantha provide them with berries and other fruit. They will feed on juniper berries and visit orchards to feed on frozen apples.

Cedar Waxwing
Bombycilla cedrorum

The breeding range of the Cedar Waxwing extends from southeastern Alaska, across southern Canada, into northern United States. The Cedar Waxwing is an uncommon summer resident and a fairly common overwintering species in the Great Basin.

Cedar Waxwings nest in the valleys of the Basin. They nest in streamside woodlands and in parks and parklike grounds in cities and towns. I have found them nesting, off and on over the years, on the Reno campus of the University of Nevada. Some of their breeding localities are hot ones, such as along Smoke Creek on the very edge of the Smoke Creek Desert in northwestern Nevada.

These waxwings are highly social birds. They are gregarious outside of the nesting season, and several pairs may nest in the vicinity of each other. Their courtship antics are quite engaging ones. Not only does the male feed the female during courtship, but sometimes he will pick a berry and perch beside the female—they will then pass the berry back and forth a number of times before the male either eats it or drops it. When a male and female are perched side by side, hopping dances may occur. Each takes a single hop away from the other, then a single hop toward the other—alternately parting and rejoining. Each time they meet on their inward hop, they may bring their bills together in an action called billing. On one occasion Cedar Waxwings were seen using an apple blossom petal in their hopping dance. The petal was exchanged each time they rejoined on the inward hop. Acts of courtship may continue into the actual nesting period as connubial displays, and courtship feeding continues. When perched close together the male and female may engage in billing, or they may touch each other's breast feathers with their bills. At times they may chase each other in wide circles around the nest tree.

The male Cedar has the reputation of lacking a real territorial-courtship song. Very few observers have ever heard anything even suggestive of a song. However, I heard and watched Cedar Waxwings "singing" from trees just below Virgin Peak in the Virgin Mountains of southern Nevada. Of course, I have the reputation of having a tin ear when it comes to music!

The behavior of Cedar Waxwings outside of the breeding season is also interesting. Sometimes a number of waxwings will perch close together, all in a row. Then one will pick a berry or flycatch an insect and pass it to another waxwing. That berry or insect may travel up and down the line several times before one of the birds will eat it. When flycatching flying insects, both Cedar and Bohemian waxwings operate from perches high in trees, and they usually return to a perch between captures. A most unusual type of flycatching was once witnessed. During a swirling snowstorm, Cedar Waxwings were flying out from perches and snapping snowflakes out of the air. Was it to drink or to play or a combination of both?

FAMILY PTILOGONATIDAE

Ptilogonatidae is a New World family of silky flycatchers. Four species of ptilogonatids inhabit the dry, brushy country from southwestern United States to Panama. Silky flycatchers possess soft silky plumage colored in black, grays, or browns—sometimes marked with white or yellow. They bear prominent head crests and have short wings and legs, long slender tails, and short broad bills. The sexes are not alike. Silky flycatchers are active, shy birds that move around in small flocks. Although they flycatch insects, their principal food consists of berries. They are especially fond of mistletoe berries.

The range of the Phainopepla, *Phainopepla nitens*, is located south of the Great Basin. This species is of accidental occurrence in the Basin. There are records of its occurrence in Churchill and Storey counties in Nevada. It is listed as an accidental species on the official checklist of Malheur National Wildlife Refuge.[61]

FAMILY LANIIDAE

Laniidae is an Old World family of shrikes; its main center of diversity is in Africa. Only two of its sixty some species are found in North America. Laniids have soft plumage. Many are gray or brown above and white below. They are characterized by relatively large heads and by bills that are stout, strongly hooked, and notched. Outside of the nesting season they are solitary birds. Shrikes are unique among passerines in being highly specialized raptors or predators. They feed on large insects and small vertebrates, which they kill with their strong bills. They hunt by watching for prey from elevated, exposed perches. Sometimes they hover over their prey before dropping down on it. Shrikes impale captured prey on thorns or other sharp objects—the thorns function as meat hooks.

Northern Shrike
Lanius excubitor

The breeding range of the Northern Shrike is located in the boreal forests of Alaska, northern Canada, and Eurasia; and in North America this bird overwinters as far south as northern United States. The Northern is an uncommon winter visitant in the Great Basin. The numbers overwintering vary from year to year; many of the birds are juveniles.

Although large insects are taken in season, the food staple of this small raptor consists of small rodents and birds. The appearance of large numbers of shrikes in southern Canada and northern United States during the winter is believed to be due to a scarcity of small rodents, especially mice, in Alaska and northern Canada. Voles and lemmings have cyclic populations, which periodically build up to a high density and then crash. Studies in eastern United States suggest that the major invasions of Northern Shrikes occur after some of the northern populations of rodents have crashed. From 1900 to 1935 a major shrike invasion occurred about every four years.

The Northern Shrikes visiting the Great Basin to overwinter often appear to be tamer than our resident Loggerhead Shrikes. Dick Ankers demonstrated this to my satisfaction years ago. Try to approach the Northern and Loggerhead shrikes you encounter in the field, and note the approach distance at which each flushes or flies away. Could it be that the Northern Shrike's experience with settlers in the thinly populated North has been kinder than what the Loggerhead Shrike has experienced here?

Loggerhead Shrike
Lanius ludovicianus

The breeding range of the Loggerhead Shrike extends from southern Canada into Mexico. This shrike is a common summer resident in the Great Basin. During the fall some migration occurs, particularly in the northern part of the Basin, and fewer shrikes are present in the winter than in the summer.

Loggerhead Shrikes frequent open country in the valleys and foothills of the Great Basin. They are present in desert shrublands, in juniper or pinyon-juniper woodlands, in mountain mahogany stands, and often around the outskirts of ranches and towns. Thickly foliaged trees and shrubs afford nesting and roosting sites as well as hunting perches. Telephone wires and fences also provide hunting perches. Shrikes hunt where the tall vegetation is scattered and where there is much bare ground or ground covered with short vegetation.

Shrikes are raptors and prey on large insects and small vertebrates. Like the falconiform and strigiform raptors, shrikes have a strong, hooked bill. But, unlike these other raptors, shrikes lack talons and have weak feet. They kill and tear prey apart with their powerful bills but receive little or no help in killing, carrying, and holding prey from their feet. The upper mandible is strongly hooked, and the tomium or cutting edge of each side of the upper mandible is modified for shearing or snipping. Each tomium is notched near the tip of the mandible, and in back of each notch there is a tomial tooth. Shrikes kill vertebrates such as mice and small birds by biting them at the back of the head or on the neck—the tip of the lower mandible forces the prey against the tomial teeth of the upper mandible. Sometimes several bites are required to kill the prey. Sometimes a shrike will beat larger victims against the ground or a perch.

Possessing the weak feet characteristic of perching birds, shrikes have difficulty butchering large prey. Since they are unable to hold prey firmly with their feet, they have had to come up with an alternative method of immobilizing the carcass. This they do by impaling it on thorns or barbwire or broken, jagged twigs or by wedging it into a crotch in a tree or shrub. Thus, shrikes utilize a meat hook while eating and are, in effect, tool users.

Shrikes do not always proceed to promptly eat the prey they have impaled. They may return within a day or two to do so, or they may never do so. Except during cold, refrigerating weather, the impaled meat is not going to remain very desirable for longer than a few hours or days at the most. Because shrikes do not eat all the prey they impale, some ornithologists have interpreted impaling as constituting food storage. But usually the meat left impaled for any length of time is mainly the least digestible and desirable parts of the mouse or bird carcass or insect body.

One of the vernacular names for shrikes is that of butcher-birds; the family name is from *lanius*, the Latin for butcher. Their habit of impaling prey, using meat hooks while butchering, and assembling carcasses in places where good meat hooks are available certainly suggests butchers and butcher shops.

Shrikes still-hunt, utilizing perches from which they launch their attacks when prey is sighted. In his monograph on North American shrikes Alden Miller described two kinds of still-hunting: active hunting and pas-

Loggerhead Shrike

sive hunting. Shrikes employ active hunting mainly in the early morning and early evening hours; they use hunting perches located within six feet of the ground from which to scrutinize the area immediately adjacent. Sometimes they get down on the ground or in low shrubs and hop around while searching for prey. This behavior is called active hunting, since the shrike moves on to another perch if no prey is captured within the first minute or two. Passive hunting usually takes place during the day, after the early morning bouts of active hunting have quenched the bird's hunger. Now the shrike can be more selective and leisurely about capturing prey. The perches used in passive hunting are high ones—on the tops of trees or on telephone wires—which allow the bird to scan the ground to a distance of one hundred fifty feet or so of the perch. The shrike may remain perched for as long as ten to thirty minutes, even if no prey is captured, before moving on to another hunting perch.[394]

Loggerheads are not gregarious birds. After the family group breaks up at the end of the breeding season, or following migration in migratory shrikes, each bird establishes a fall and winter feeding territory. Miller's observations in California indicated that in nonmigratory populations one adult may retain the breeding territory as a fall and winter feeding territory, and the other adult of the pair seeks a new, unoccupied area for its feeding territory. Both breeding and feeding territories are defended by advertising—by perching on conspicuous, high perches. "This advertisement is aided by song and by the familiar series of from four to ten or twelve screeches of progressively diminishing intensity." When an intruding shrike violates territorial boundaries, the resident shrike gives a series of sharp *bzeek, bzeek, bzeek* calls, and if the intruder does not leave immediately a fight may ensue. Each territory possesses a headquarters, with good cover and high observation perches. The roosting site is located here and, on a breeding territory, the nesting site as well. When not hunting shrikes spend much of their time at their headquarters.[394]

When the time comes to establish pairing bonds, nonmigratory shrikes may sometimes reverse the normal order of things. Typically, in passerine birds, the male establishes a breeding territory and advertises there. The female is attracted to an advertising male, and pair bonding normally occurs on the male's territory. But in Miller's nonmigratory Loggerhead Shrikes there were indications that probably "the male usually moves about seeking the territory of the female" and that pair bonding then occurs on the territory of the female.[394] Shrikes often use the same areas year after year as breeding territories. This may mean that the females frequently retain the breeding territories as their fall and winter feeding territories. In migratory Loggerheads a normal order of events may prevail, since winter feeding territories and breeding territories are not interchangeable. Upon returning from migration in the spring, males are soon seen establishing and defending breeding territories.

FAMILY STURNIDAE

Sturnidae is an Old World family containing well over one hundred species of starlings and related birds. One species, the European Starling, was introduced into North America with great success. Sturnids are chunky, medium-size birds. Many have dark-colored plumage with a metallic sheen to it. Many possess short legs and tails. Their bills are stout and either straight or slightly downcurved. Sturnids are often highly gregarious—they may form large foraging flocks and immense roosting concentrations, and some are colonial nesters. On the ground they usually walk or run, and in the air their flight is direct and swift. Sturnids prefer open, broken country, but some species such as the European Starling live in close association with humans.

European Starling
Sturnus vulgaris

The European Starling—prior to 1982 its name was the Starling—is a native of Europe introduced into North America. The first successful introduction was apparently made late in the nineteenth century: Eugene Scheifflin released eighty starlings in Central Park, New York City, on March 6, 1890, and another forty on April 25, 1891.[395] At least one attempt was made to introduce starlings into western United States. In 1889 and 1892 the Portland Song Bird Club released thirty-five pairs in Portland, Oregon, but by 1901 to 1902 the birds had completely disappeared.[103] However, from the successful New York City introductions starlings slowly spread across the continent. By 1978, eighty-eight years after the first successful introduction, Kessel was able to report that European Starlings were nesting for the first time at Fairbanks, Alaska—5,250 miles across the continent from New York City as the crow flies or, in this case, as the starling flies.[396]

During its spread across the continent, the European Starling reached the Great Basin in the late 1930s; a specimen was collected near Salt Lake City on February 26, 1939.[56] The starling was first reported in western Nevada near Fallon in 1947. "The first nest of the starling in Utah was found on 25 May 1949 in an old woodpecker or flicker hole on the west side of Salt Lake City."[397] The first reported nesting in Nevada was at Jiggs, twenty-eight miles south of Elko, in 1956.[398] By now the European Starling is a common and widespread breeding bird in the Great Basin. It is found mainly in the valleys and foothills, especially around human habitations in towns and cities and on ranches and farms. Common the year around, the starling feeds and roosts in flocks outside of the nesting season.

The European Starling is an undesirable alien—to birds and humans alike. Native cavity-nesting birds have suffered the most from the inundation of North America by starlings, and their numbers have decreased drastically in many regions of the country. Starlings are persistent, tough, pugnacious competitors—able to physically evict nesting birds from cavities. Bluebirds, wrens, nuthatches, swallows, and flickers have been adversely affected by starlings. Starlings have even been known to patiently await the completion of new nesting cavities by flickers before evicting them.

Starlings are highly gregarious outside of the nesting season—they forage in large flocks and concentrate by the thousands, even by the hundreds of thousands, to roost. Their roosts plague urban as well as rural areas. Sections of cities have been blighted by the presence of large starling roosts, where the din and filth of hordes of birds fill the air and litter the sidewalks, buildings, and parked automobiles.

The principal agricultural damage caused by these birds occurs at cattle feedlots, where immense swarms congregate to feed on grain. Poisoning programs are sometimes used at feedlots to help control the number of starlings. And starlings can pose health problems. They are involved in outbreaks of a fungal disease of humans called histoplasmosis. The birds do not directly transmit the disease, but the microorganism which causes histoplasmosis flourishes in soils under starling roosts where the ground is richly covered with fecal material.

But the European Starling is not all bad. Well over half of its annual food consumption is animal material—mainly insects. Its feeding station is on the ground in open areas where the vegetation is short. When foraging the bird walks along rapidly, on a zigzag path, pausing to capture any prey it encounters. A large foraging flock may move across the ground in a rolling fashion much like a flock of foraging Pinyon Jays: birds from the rear are continually flying up, leapfrogging the flock, and landing at the front. This constant change in relative positions within a flock insures all birds equal foraging opportunities—no bird has to subsist on leftovers. Not all the starling's foraging is done on the ground, however. It may capture flying insects by flycatching from a perch or by hawking while in flight like a swallow. It will forage in trees for fruit and on cornstalks for corn.

European Starlings often nest close together; the nesting territory of a pair may extend no further out than ten to twenty inches around the entrance to the nesting cavity. Their courtship is neither complex nor colorful. Unpaired males sing by their cavities and flap and flutter their wings at the approach of other starlings. Once paired, males sing even more often. A male may carry green leaves into the nesting cavity. He may pick up green leaves and hold them while he sings and while his mate moves to and from the nesting cavity. Various posturings and movements may take place.[399]

The song of the starling is not overly musical. It has been said that "the series of squeaks, chatters, creaking rattles, chirps, and wheezy notes is far from pleasing to the ear; but these are often interspersed with long-drawn, cheerful whistles, which are almost humanlike and easily imitated."[400] Starlings not only sing in the spring, but some birds sing in the fall and even on winter roosts at night. Starlings are versatile mimics. They have been heard imitating the songs and call notes of many other birds, including robins, nuthatches, quail, pewees, Killdeers, flickers, crows, chickadees, kinglets, meadowlarks, and cowbirds. They may even imitate the barking of a dog or the mewing of a cat.

FAMILY VIREONIDAE

Vireonidae is a New World family containing about forty species of vireos. Although represented by species in all three Americas, this family shows its greatest diversity in the tropics. Vireonids are small, four- to six-inch, nondescript birds. They are plainly colored in olive greens and brownish grays above and are lighter below. Vireos were formerly called greenlets, in reference to the high frequency with which greens appear as plumage colors in this family. Even the modern name of vireo preserves this thought, since it is derived from the Latin *virere*, to be green.

Some vireos possess light-colored wing bars, eye rings, or eye stripes. Their plumage is never marked with streaks or spots. Vireonid bills are short, quite straight, slightly hooked at the tip, and with a small notch or tooth near the tip. Vireos are among our most persistent songsters, endlessly repeating their songs. They are arboreal, their movements are slow and deliberate, and they spend much foraging time gleaning insects from the twiggery and foliage of trees. Their cuplike nests are placed in the horizontal forks of shrubs and trees. Outside of the nesting season, vireos are solitary.

Solitary Vireo
Vireo solitarius

The breeding range of the Solitary Vireo lies in central and southern Canada and extends southward into western and eastern United States. The Solitary is an uncommon summer resident in the Great Basin.

The Solitary Vireo nests in varied habitats—in riparian woodlands, pinyon-juniper woodlands, mountain canyons, and montane forests, especially ponderosa pine ones. This vireo is highly insectivorous. It gleans insects from the twiggery and foliage in the lower and middle levels of trees. When gleaning the bird moves at a slow pace through the branchwork of a tree, perching and capturing insects in its stout, hooked, notched bill. Sometimes it will support itself on fluttering wings to reach an awkward gleaning site.

The English name of Solitary Vireo is not a particularly good one. This bird is certainly solitary and nongregarious, but then so are the rest of the vireos. Although Solitary Vireos do not band together to form flocks after the nesting season is at an end, single birds will occasionally join mixed-species foraging flocks. Panik found a single Solitary Vireo in two of his forty-five mixed-species foraging flocks in the pinyon-juniper woodlands of northwestern Nevada. On both occasions the flocks were encountered in September and the vireo's flock associates were a group of Bushtits and a single Plain Titmouse.[351]

The western Solitary Vireo is a pleasant songster. It has been described as a "slow but persistent singer; the syllables of its song are set off from one another by long rests. . . . Each note is clear cut and loud so that the song rings out, and may be heard for a considerable distance. Successive notes are variously inflected, some rising, others falling; at times a bird will give a regular alternation of rising and falling inflections. Hence the name 'question-and-answer bird' has been suggested" for this vireo.[318]

Warbling Vireo
Vireo gilvus

The breeding range of the Warbling Vireo extends across North America from Canada southward into Mexico. The Warbling Vireo is a common summer resident and migrant through the Great Basin. It reaches its peak abundance in the deciduous stands of vegetation along valley and mountain streams and in montane aspen groves.

Ridgway's description of the Warbling Vireo in the Great Basin in the 1860s still rings true:

> Few, if any, of the western birds are more extensively distributed or more abundant than this Greenlet, for it abides in all fertile localities. Altitude makes no difference with it, since it is equally common among the willows or cotton-woods of the lowest valleys and the aspens just below the timber-line—the only condition required being, seemingly,

the existence of deciduous trees or shrubbery. The food of this bird consists in summer chiefly of worms and other insects, but in the autumn it seems to subsist almost exclusively on the small bluish berries of a species of cornel (*Cornus pubescens*), which grows abundantly along the mountain streams.[25]

The Warbling Vireo is highly insectivorous. Its foraging station is high in the dense canopy of deciduous trees, although it will occasionally forage in conifers. When gleaning insects and spiders it moves very deliberately, and it often sings. Over a hundred years ago Coues described the situation well: "We seldom see them, indeed; they are oftener a voice than a visible presence—just a ripple of melody threading its way through the mazes of verdure, now almost absorbed in the sighing of foliage, now flowing released on its grateful mission."[310]

This vireo is a persistent songster and sings even during the midday heat of summer. Males sometimes sing even when sitting in the nest incubating eggs. The song has been described as "not broken up into short, exclamatory phrases like those of . . . the red-eyed, the solitary, and the yellow-throated, but [it] continues on in a long series of slow, quietly delivered musical notes increasing in force to the end. The pitch undulates gently to the final note, which is generally the highest and the most strongly accented. . . . The most suggestive rendering of the vireo's song, perhaps, is Wilson Flagg's (1890): 'Brig-a-dier, Brig-a-dier, Brigate,' which, pronounced slowly, brings out the rhythm admirably."[400]

Other Vireonids

The breeding range of the Gray Vireo, *Vireo vicinior*, lies in southwestern United States, including southern Nevada and southwestern Utah. This vireo is of occasional occurrence in the eastern and central parts of the Great Basin. It has been recorded from Sevier County in Utah[56] and from the Toiyabe Mountains in central Nevada.[401]

The Yellow-throated Vireo, *Vireo flavifrons*, breeds in eastern United States; this vireo is of accidental occurrence in the Great Basin. There is a June record from near Logan, Utah.[402]

The breeding range of the Hutton's Vireo, *Vireo huttoni*, lies along the West Coast and in extreme southwestern United States southward to Central America. The Hutton's is of accidental occurrence in the Great Basin. It is on the accidental list from Malheur National Wildlife Refuge.[61] It has been recorded from Unionville in the West Humboldt Mountains of Nevada by R. E. Wallace.

The Philadelphia Vireo, *Vireo philadelphicus*, breeds far to the east and north of the Great Basin in northern United States and Canada. There are several undocumented sight records for this vireo from Salt Lake City.[27]

The breeding range of the Red-eyed Vireo, *Vireo olivaceus*, extends across Canada and the United States, except for the Far West and Southwest. The Red-eyed Vireo is a rare to uncommon migrant in the eastern and northern parts of the Great Basin; it is of accidental occurrence elsewhere here. It may be a rare nester in the eastern edge of the Basin in Utah.[56] There is evidence that this species is extending its breeding range in the West.

FAMILY EMBERIZIDAE

In the sixth edition the AOU checklist recognizes for the first time the Family Emberizidae. This extremely large family was formed by reducing three former families to subfamilial rank, combining them with two subfamilies removed from the Family Fringillidae and with a subfamily from the Family Coerebidae. The checklist subfamilies now in the Family Emberizidae are:

1. Subfamily Parulinae—formerly Family Parulidae (wood-warblers).
2. Subfamily Coerebinae—formerly a subfamily in the Family Coerebidae (honeycreepers).
3. Subfamily Thraupinae—formerly Family Thraupidae (tanagers).
4. Subfamily Cardinalinae—formerly Subfamily Richmondeninae, then changed in 1973 to Cardinalinae (cardinals and allies), in the Family Fringillidae.
5. Subfamily Emberizinae—formerly Subfamily Emberizinae (sparrows and buntings) in the Family Fringillidae.
6. Subfamily Icterinae—formerly Family Icteridae (meadowlarks, blackbirds, and orioles).

Creation of the Family Emberizidae by the merging and dismemberment of some fifth edition families represents the most drastic act performed by the AOU Committee on Classification and Nomenclature in preparing the sixth edition of the checklist. It will take time for some of us to feel comfortable with this new grouping of formerly strange bedfellows. However, there is a bright side to the issue. Despite the scientific acumen of the committee, out in the real world wood-warblers will continue to flit through the trees as wood-warblers, sparrows will continue to embrace the ground and low vegetation as sparrows, blackbirds will continue to be blackbirds, and so on.

SUBFAMILY PARULINAE

The wood-warblers of the Subfamily Parulinae are New World birds. There are about one hundred and twenty species in the three Americas, with well over fifty breeding in North America. Wood-warblers are small, exquisite birds—often brightly marked with yellows, blues, browns, or reds. The males can be readily identified when in their bright nuptial plumages but are difficult to tell apart during the fall migration, when they are in their drab winter plumages.

Warblers have slender pointed bills, pointed wings, and rounded tails. They remain in constant motion, restlessly flitting through the foliage of trees and shrubs. A few species are terrestrial. The singing voices of the males are thin and high-pitched and often flawed by hissing and lisping tendencies. Wood-warblers are solitary birds outside of the nesting season and do not associate together in organized flocks. However, they migrate at night, often in great waves. Then, during the day, the trees are alive with feeding warblers of various kinds.

Orange-crowned Warbler
Vermivora celata

The breeding range of the Orange-crowned Warbler extends from central Alaska, across much of Canada, and southward through western United States to Baja California. This warbler is an uncommon to common summer resident in Great Basin mountains. It is occasionally present during the winter in west central Nevada in the valleys. Its nesting habitat lies mainly in deciduous vegetation along streams, in canyons, and in aspen groves.

Male wood-warblers are generally colorful birds, often exquisitely so, but the male Orangecrown is an exception to the rule. Even in nuptial plumage he is but a plain, nondescript bird. High expectations are aroused by the name of Orange-crowned Warbler, but these expectations do not materialize when the bird is viewed. The orange color is confined to the basal portion of the feathers and is barely visible unless the crown feathers are ruffled. The orange crown patch is even duller and more indistinct in the female, and some lack one altogether. The trivial part of the scientific name of this warbler is derived from the Latin *celatus*, meaning hidden—as is the crown patch. The plain attire is not even adorned by contrasting tail spots or wing bars or eye rings. The voice of the male matches his attire; the notes of his nuptial song have been characterized as being tinny.

The Orangecrown lives mainly on animal food. Its genus name is misleading, since *Vermivora* is derived from Latin words meaning worm eating. But this warbler does not feed on worms, although less than 10 percent of its food consists of vegetable matter. Its foraging station is centered in deciduous trees and shrubs. Here it gleans insects and spiders from foliage of moderate density, at heights of five to thirty-five feet or so above the ground. While gleaning it moves in a somewhat methodical fashion through the foliage, and it frequently appears sluggish for a warbler. At times it will reach a difficult foraging station by fluttering its wings and momentarily hovering. Orange-crowned Warblers are not gregarious, although occasionally one or several will be seen foraging with other species of warblers or with vireos, kinglets, chickadees, and other birds.

The call note of the Orangecrown has been variously interpreted as a sharp *chip* or *chit*. The song of the male has been described as consisting of "about 18–22 notes repeated rapidly. At about half-way, the pitch drops a small but noticeable amount, sometimes rising again just at the end. Amplitude falls off slightly near the end." The song notes have been interpreted as *si-si-si-si-si-si-si-si-si-si-si-si-si-si-si-si-si-si-si*.[403]

Nashville Warbler
Vermivora ruficapilla

The breeding range of the Nashville Warbler is discontinuous, being partly in western Canada and United States and partly in eastern Canada and United States. This warbler is an uncommon breeding species along the western edge of the Great Basin in the Warner Mountains, Lake Tahoe Basin, and Carson Range and at Twin Lakes in Mono County, California.[404] It is an uncommon migrant throughout the Great Basin.

The Nashville Warbler nests in fairly open forested areas with a good shrub understory. Along the western edge of the Great Basin its favorite breeding habitat lies in montane yellow pine forests. The warbler forages in the foliage of the trees and nests on the ground under the dense cover of bushes.

The foraging station of this warbler is centered in the middle levels of the trees where the foliage is quite open, about twenty to fifty feet or so above the ground. Its diet is a highly insectivorous one. The Nashville is one of the most active of the warblers when gleaning, moving continually, sometimes hovering on fluttering wings to reach a difficult gleaning site. When deciduous trees are present, it will forage in them in preference to conifers.

The male sings from song perches in trees and also engages in song flights. His song consists of two phrases—the first part is sung four to seven times; then the second part is sung one to seven times. The first part is sung more slowly and at a higher pitch than the second part, and the first phrase of the song changes from two syllables to one as the tempo speeds up. The effects of increasing the tempo have been described as going from *see-bit, see-bit, see-bit, see-bit, ti-ti-ti-ti* when delivered more slowly, to *see-it see-it see-it see-it, ti-ti-ti* when delivered more rapidly, to *seet seet seet seet ti-ti-ti-ti-ti* when delivered even more rapidly.[403]

Virginia's Warbler
Vermivora virginiae

The breeding range of the Virginia's Warbler lies in southwestern United States. The Virginia's is an uncommon to common breeding species in the central and eastern parts of the Great Basin. There are small, isolated breeding populations in the southwestern part of the Basin in the Wassuk Range and in the White Mountains.[404] There are no known breeding populations in the western end of the Basin north of the Wassuk Range.

The Virginia's Warbler frequents pinyon-juniper woodlands, mountain mahogany thickets, and brushy areas of scrub oaks in the foothills and at middle elevations in the mountains. It may also be found in brushy areas in the valleys along streams. This warbler is a shy and timid bird. Its foraging station is in dense underbrush or thickets or trees close to the ground. At times it will forage on the ground or fly out to flycatch a passing insect. In the spring the males sing as they forage. Their song has been described as a "bright and rapid, che-wé-che-wé-che-wé, ché-a-ché-a-ché, or zdl-zdl-zdl-zdl, zt-zt-zt-zt."[403] The call note is a sharp *chip*.

Yellow Warbler
Dendroica petechia

The breeding range of the Yellow Warbler covers most of North America, extending from the northern tree line in Alaska and Canada southward into northern South America. The Yellow Warbler is a common summer resident and migrant in the Great Basin.

The status of the Yellow Warbler in the Great Basin has apparently changed but little since Ridgway was here in the 1860s. He wrote, "This common and familiar little bird was met with everywhere, except during the winter season; and in all wooded localities, with the exception of the higher forests, which it gave up chiefly to *D. auduboni*, was the most abundant and generally distributed member of the family. . . . throughout the Interior it was equally plentiful in every locality producing a growth of willows or other shrubbery, being most multiplied in the river-valleys or lower cañons, and gradually decreasing in numbers toward the summits of the mountains."[25] With settlement of the Basin an additional habitat dimension developed. Now, Yellow Warblers are common summer residents in both urban and rural localities where there have been lush plantings of ornamental trees and shrubs.

Soon after arriving back on the nesting grounds from spring migration, male Yellow Warblers establish territories on which they sing and display. Actual fighting is most frequent early on—when the territorial boundaries are being established and defended. Males engage in interesting territorial displays, which they share with certain other species of the genus *Dendroica*. Circling occurs when a territorial male flies toward a rival male but turns aside along an arc or circle. A male may engage in moth flight, flying slowly but on rapidly beating wings. Gliding may ensue during territorial interactions, as the male sails along with spread tail and on spread, set wings. A prolonged aerial chase may result. Following long encounters, a male may extend his wings and direct a wing-out display at his rival.[405]

Male Yellow Warblers are versatile songsters and have several types of songs. The Fickens have directed attention to what they recognize as the two principal song types of this warbler: an accented ending song, with a strongly upslurred, emphasized ending, and an unaccented ending song, lacking an emphasized ending. According to the Fickens over 90 percent of the songs sung by the males before the females arrive on the breeding grounds are accented ending songs. Unaccented ending songs become more common during egg laying and incubation and predominate by the time of hatching. The two types are given in different contexts. After territorial encounters between males, the unaccented ending song and muted songs are sung more frequently. At times of extreme aggression either type, the accented or unaccented ending song, can be sung in a muted fashion—at low volume. The Fickens believe the accented ending song to be the typical territorial song and the song most attractive to the females.[405]

Field investigations of the two principal song types of the Yellow Warbler have also been carried out by Morse. His observations indicated that unaccented ending songs were directed at rival male Yellow Warblers. Unaccented ending songs were heard in two general contexts: either during territorial encounters between rival males or when a male was singing near boundaries between his territory and those of other Yellow Warblers. In both situations the singing male has highly aroused aggressive tendencies. He may be highly motivated to attack his rivals, or he may be in a more ambivalent mood—not knowing whether to stand his ground or flee.

In contrast, Morse observed that the accented ending song was sung when the male did not have highly aroused aggressive tendencies. This type of song was directed mainly at the female. It predominated before and during pair formation, during precopulatory and postcopulatory behavior, and when the male was singing near the nest while the female was incubating eggs. Male Yellow Warblers also sang accented ending songs along the boundaries between their territories and those of another species of warbler—the Chestnut-sided Warbler.[406]

More recently investigators have looked for variability within the song types of the Yellow—for variability within the singing of a single male and between males from the same or different geographic locations. In a Michigan study, 745 songs sung by forty-five different males were subjected to spectrographic analysis. Forty different song figures or notes were detected in these 745 songs, a song figure being "a sound which produces a single, complete, and distinct impression uninterrupted by silences greater than two centiseconds." Much greater variability was found in unaccented ending songs than in accented ending songs. In all, the forty song figures were arranged into sixty different song patterns of unaccented ending songs and into only four different song patterns of accented ending songs. The investigators concluded that, since accented ending song is directed at females, having only a few song patterns of this type would make it simpler for female Yellow Warblers to identify singing males as male Yellow Warblers. But, since unaccented ending song is directed mainly toward other males, having many different patterns would allow territorial males to recognize each other as individuals and allow singing neighbors to be distinguished from singing strangers. The presence of a singing stranger offers more of a threat to a territorial male than does the presence of neighbors already established on territories.[407]

Yellow Warblers feed almost entirely on insects, spiders, and other arthropods. They glean their prey from the foliage and twiggery of trees and shrubs and may do some hovering and flycatching while foraging. A Canadian study indicated that males and females may have different foraging stations during the summer, thus partitioning the food resources during the nesting season. Up until the time the young were fledged, females generally foraged closer to the ground than did males. Females tended to forage more in the shrubs and in the lower and inner sections of trees. From the time of their arrival on the breeding grounds until eggs were present in the nest, males tended to forage in the upper and outer sections of trees. By using a high, conspicuous foraging station and by persistently singing, male Yellow Warblers are defending their territories while they forage. After egg laying commenced, the males tended to forage closer to the ground. Yellow Warblers usually locate their nests in forks or crotches of shrubs close to the ground. Hence, the low foraging station of the female keeps her close to the nest level. And the male lowers his foraging beat to nest level when he has an incubating female or nestlings to feed. Once the young have fledged, both males and females tend to forage higher up than before. There are several other differences in foraging behavior between male and female Yellow Warblers: females may hover more than males, and the sexes may differ in the species of trees they favor while foraging.[408]

Yellow-rumped Warbler
Dendroica coronata

In 1973 the Committee on Classification and Nomenclature lumped the Myrtle Warbler, *D. coronata*, with the Audubon's Warbler, *D. auduboni*, to form an enlarged species to be known as the Yellow-rumped Warbler, *D. coronata*. Since the Myrtle and Audubon's warblers are recognizable entities, the names of Myrtle Warbler and Audubon's Warbler are available to be used at the subspecies level. The breeding ranges of the two differ. The breeding range of the Myrtle extends from Alaska across Canada, barely penetrating southward into midwestern and northeastern United States. That of the Audubon's extends from British Columbia southward through western United States into Mexico.

The Audubon's Warbler is a common and widespread breeding species in the Great Basin, reaching its greatest density at middle to higher elevations in the mountains. It is a common migrant and an occasional to fairly common winter resident in the valleys and at lower elevations in the mountains. However, the Myrtle Warbler is only an occasional to rare migrant and winter visitant in the Great Basin.

During the summer Audubon's Warblers reside in montane conifer forests and in aspen groves. Their foraging station is in conifers, at heights from a few feet above ground on up to the top of tall trees. Much of their gleaning occurs in the outer foliage of the crowns of conifers. These warblers have a characteristic way of changing foraging positions. They often do so by flying out beyond the crown of the tree and circling around the outside of the crown to a new position. While flying they will snap up flying insects. Audubon's Warblers will often flycatch passing insects by making sallies out from perches. It is their activities outside of the tree crowns that make them so conspicuous to human observers. Although highly insectivorous, these birds take fruit and seeds seasonally.

Audubon's Warblers are on the move during the fall. Some even venture higher up the mountainsides to forage for a time at and beyond the upper tree line. But, before the onset of winter, they leave the mountains on latitudinal and altitudinal migrations. Those which overwinter in the Great

Basin do so in the foothills and valleys, below the level of persistent snow.

The song of this warbler has been variously described as, for example, a *chwee-chwee-chwee-ah-chwee* to a *tsil-tsil-tsil-tsi-tsi-tsi-tsi*. The call note is a metallic *chip*.[403]

Black-throated Gray Warbler
Dendroica nigrescens

The breeding range of the Black-throated Gray Warbler extends from southern British Columbia, southward through western United States, to northern Baja California. This warbler is a common summer resident and migrant in the Great Basin.

The Black-throated Gray frequents much drier habitat than is typical for wood-warblers. It is found in pinyon-juniper or juniper woodlands in the foothills and at lower elevations in the mountains of the Great Basin. This warbler is highly insectivorous and gleans insects in the dense terminal foliage of pinyon and juniper trees—it often remains out of sight by foraging just inside the outer edge of the foliage. Its movements are quite deliberate for a warbler.

The song of the Black-throated Gray is simple but pleasant to the ears. Composed of four or more notes, it has been described as *swee, swee, ker-swee, sick* or as *wee-zy, wee-zy, wee-zy, wee-zy-weet*. The terminal syllable may rise or decline in pitch.[403]

American Redstart
Setophaga ruticilla

The breeding range of the American Redstart extends from southeastern Alaska, across Canada, and southward into the United States. The western edge of this warbler's breeding range in the United States crosses the eastern edge of the Great Basin. So the redstart is an uncommon summer resident in the northeastern part of the Great Basin in northern Utah. It is a rare transient elsewhere in the Basin.

American Redstarts nest in stands of deciduous trees, especially in riparian willows, in the valleys and lower canyons of Utah. This warbler has evidently decreased in number in northern Utah in recent years, since early naturalists found it to be quite common there.[56] However, over most of its breeding range this warbler is probably more abundant today than formerly—it prefers the second-growth, open deciduous woodlands of today to the mature, closed deciduous forests that formerly prevailed. Some observers believe that this warbler is one of the three most abundant warblers in North America.[403]

MacGillivray's Warbler
Oporornis tolmiei

The breeding range of the MacGillivray's Warbler lies in western North America, extending from southeastern Alaska through Canada and the United States. This warbler is a common summer resident and migrant in the Great Basin.

The preferred nesting habitat of the MacGillivray's is dense, moist shrubbery in foothills or at middle elevations in the mountains. These warblers are found in a variety of sites: in canyons near streams, in thickets on north-facing slopes, in thick secondary growth following a fire or logging, or in the dense undergrowth of an aspen grove.

The foraging station of the MacGillivray's Warbler is located within a few feet of the ground. Here the bird gleans insects from the foliage. Its close ties to low shrubbery are manifested even outside the nesting season. The MacGillivray's is a very shy bird and even when foraging is usually concealed in the foliage. Indeed, this warbler is usually seen high above the ground only when a male uses a tree for a singing perch. The song of the male consists of three or four longer, often double, notes followed by several shorter ones. The song has been described as a louder *zweedle, zweedle, zweedle, zidl, zidl* or a somewhat softer *tswee-it, tswee-it, tswee-it, tswee-it, wik, wik.*[403] The common call note is a *tchek.*

Common Yellowthroat
Geothlypis trichas

The Common Yellowthroat occurs from coast to coast in North America, with a breeding range extending from central Canada to southern Mexico. This yellowthroat is a common breeding warbler in the Great Basin.

The Common Yellowthroat nests in the emergent vegetation of marshes, sloughs, and irrigation ditches and in dense riparian vegetation. This warbler's present-day status in the Great Basin is much like its status in the 1860s, about which Ridgway wrote:

> In all bushy places contiguous to water, this little bird was invariably to be found; but it was confined to the valleys, being replaced among the mountains, even in the lower cañons, by the *G. macgillivrayi*. Clumps of wild-rose briers and the banks of the sloughs seemed to be its favorite resorts, and in such localities near Pyramid Lake it was one of the most abundant species in May, and all day long enlivened the vicinity of one of our camps by its pleasant song of *witch'ity, witch'ity, witch'ity*—often from several rival males at the same time.[25]

The yellowthroat is almost entirely insectivorous in its diet. Its foraging station is within six feet of the ground in dense marsh or riparian vegetation. Even the singing perches of the male are located in this low vegetation, although the male may fly a few feet above the vegetation on a song flight.

Wilson's Warbler
Wilsonia pusilla

The breeding range of the Wilson's Warbler lies in the boreal forests of Alaska and Canada, extending further southward into western than into eastern United States. The Wilson's Warbler is a common migrant and an uncommon summer resident in the Great Basin.

The prime breeding habitat of the Wilson's lies in willow-alder thickets along mountain streams and lakes. Much of this warbler's activity, including its foraging, takes place within twelve feet or so of the ground. Even outside the nesting season, it spends much of its time in shrubby vegetation, only occasionally foraging in trees.

The Wilson's Warbler is highly insectivorous, gleaning insects from the twigs and foliage in streamside thickets. Sometimes a warbler will flutter up to pick insects off the underside of leaves or flit around in the foliage flycatching insects. *Wilsonia* is a genus of flycatching warblers. The Wilson's Warbler and others of this genus have a characteristic flycatching bill—the bill is broader at its base than it is high, and the base is surrounded by rictal bristles.

Males sing mainly within the foliage of the shrubbery without resorting to high, exposed song perches. Saunders has described the male's song as being "mainly a series of rapid chatterlike notes, dropping downward in pitch toward the end. It is not especially musical in quality. The notes are short, staccato, and with marked explosive consonant sounds."[409]

Yellow-breasted Chat
Icteria virens

The breeding range of the Yellow-breasted Chat extends across most of the United States, barely penetrating into southwestern Canada in the north and extending into northern Mexico in the south. The chat is a common migrant and summer resident in the eastern part of the Great Basin and an uncommon one in the western part.

The nesting habitat of the chat lies in dense riparian thickets in the valleys or along the foot of mountain ranges. This warbler nests and forages in thick, often almost impenetrable tangles of bushes, weeds, and vines—usually within twelve feet of the ground. It is highly insectivorous but will feed on fruit in season. While foraging a chat will remain within the thickets and, consequently, is heard more often than seen—its movements are quite deliberate for a wood-warbler.

The song of the Yellow-breasted Chat is unique among birds and must be heard to be appreciated. Taverner claimed that the male "laughs dryly, gurgles derisively, whistles triumphantly, chatters provokingly, and chuckles complacently, all in one breath."[409] Saunders found that the song "is long-continued, and consists of a variety of notes and phrases delivered in an irregular, mixed order, with pauses between them. The phrases vary greatly in quality, consisting of whistles, harsh cackles, squawks, squeals, and various explosive noises, not always easy to describe. Some of these are single short notes, short series of notes, or long series, often retarded in time."[409] It is difficult to compose lyrics for a chat song. One attempt by Forbush went as follows: "C-r-r-r-r-r,—whrr,—that's it,—chee,—quack, cluck,—yit-yit-yit,—now hit it,—tr-r-r-r,—when,—caw, caw,—cut, cut,—tea-boy,—who, who,—mew, mew,—and so on."[410]

The song perches of the male chat are concealed in thickets or in the crowns of trees. At times a bird may sing as it flies from perch to perch, "flapping its wings up and down and pumping its tail, with its legs dangling, the line of flight being exceedingly jerky."[409] A male may fly straight up into the air for a hundred feet or so on a song flight. He mounts on fluttering wings, head raised, legs dangling, singing as he goes. Hovering momentarily, he will sink earthward, engaging in aerial antics on his way down.

The Yellow-breasted Chat has been called a polyglot, a clown, and a ventriloquist—and these he is. He has also been called a mimic, but this he is not: his cacophonous repertoire is his very own. He often sings by night, especially on moonlit nights. Ridgway, awakened by his nocturnal concerts in the Great Basin in the 1860s, wrote, "We were often awakened at midnight by its notes, when, but for the yelping of the prowling Coyotes (*Canis latrans*), the stillness would have been unbroken. It was also observed that they were particularly musical on bright moonlight nights."[25]

Other Wood-Warblers

The breeding range of the Hermit Warbler, *Dendroica occidentalis*, lies in the Pacific Coast states. This warbler is an uncommon breeding species in the mountainous western rim of the Great Basin in the Carson Range and in the Sierra Nevada. It is of accidental to occasional occurrence as a migrant in the western and central parts of the Basin. I am not aware of any records of occurrence for the eastern part.

The breeding range of the Grace's Warbler, *Dendroica graciae*, extends from southern Utah and Colorado southward into Central America. The Grace's is of rare occurrence in the eastern part of the Great Basin; it has been reported from Provo[27] and Sandy.[401] If you are curious about whose namesake this warbler is, she was Grace Coues, the sister of Elliott Coues, the great nineteenth-century ornithologist.

Northern Wood-Warblers

There are some wood-warblers whose breeding ranges lie, at least in part, north of the Great Basin. Members of these species may pass through the Great Basin during migration on a casual basis, although their migratory flows may be directed along the West Coast or to the east of the Basin.

The Tennessee Warbler, *Vermivora peregrina*, is a casual migrant in the Great Basin. It has been recorded several times from the eastern part in the Stansbury Mountains[27] and at Provo.[401] It is a rare spring and fall transient at Malheur National Wildlife Refuge.[61]

The Magnolia Warbler, *Dendroica magnolia*, is of occasional occurrence in the Great Basin. There are sight records from Salt Lake City and Bear River National Wildlife Refuge.[27] In the western part of the Basin, two Magnolia Warblers were captured and banded at Little Valley in the Carson Range.[411] This warbler is on the accidental list at Malheur Refuge.[61]

The Cape May Warbler, *Dendroica tigrina*, is on the accidental list at Malheur Refuge.[61]

The Townsend's Warbler, *Dendroica townsendi*, is an occasional to uncommon migrant in the Great Basin.

The Blackpoll Warbler, *Dendroica striata*, is of rare occurrence in the Great Basin. It has been recorded from the Bear River Refuge and at Provo in Utah[56] and from the Goshute Mountains in eastern Nevada.[412] It is on the accidental lists at the Ruby Lake[235] and Malheur[61] Refuges.

The Northern Waterthrush, *Seiurus noveboracensis*, is a rare to uncommon migrant through the eastern end of the Great Basin in Utah[27] and eastern Nevada.[235] It is also a rare transient at Malheur Refuge.[61]

The Connecticut Warbler, *Oporornis agilis*, is of accidental occurrence in the Great Basin. It has been reported from the eastern end of the Basin from the Stansbury Mountains and from the Salt Lake City area.[27]

Eastern Wood-Warblers

Wood-warblers with breeding grounds to the east of us in Canada and the United States occasionally show up as vagrants in the Great Basin during migration. The diversity of eastern species which has been recorded out here is impressive, although the numbers of individual birds seen have been low. But then the total flow of migrant and vagrant wood-warblers through the western states in no way compares with the passage of warblers through the midwestern and eastern states. Back there, during the height of migration, the trees may be alive with the restless movement of tiny, colorful birds. Out West, at best, we may witness a quiet show of migratory warblers, but never anything like the eastern extravaganza.

There is a spring record of the Blue-winged Warbler, *Vermivora pinus*, from Logan, Utah.[413]

There is a sight record for the Northern Parula, *Parula americana*, from Farmington Bay, Utah.[27] This warbler is on the accidental list from Malheur Refuge.[61]

There are several records for the Chestnut-sided Warbler, *Dendroica pensylvanica*, in the Great Basin: from Logan, Utah;[414] from Carlin, Nevada; and from Dyer, Nevada.[179] This warbler is on the accidental list from Malheur Refuge.[61]

The Black-throated Blue Warbler, *Dendroica caerulescens*, is a rare transient in the Great Basin. There are records from the Salt Lake City area, the Stansbury Mountains, and Millard County in Utah.[56] This warbler is a rare spring and fall transient at Malheur Refuge.[61] There is a record of this species from the West Humboldt Mountains in western Nevada—a winter one from Unionville by R. E. Wallace. This warbler has also been found north of the Great Basin during the winter; Burleigh collected a specimen in early January in northern Idaho.[415]

There are several sight records for the Black-throated Green Warbler, *Dendroica virens*, on the eastern edge of the Great Basin at Salt Lake City and Parowan.[27] This warbler is on the accidental list at Malheur Refuge.[61]

There is a sight record of the Blackburnian Warbler, *Dendroica fusca*, for Salt Lake County on the eastern edge of the Great Basin.[27]

There is a sight record of the Yellow-throated Warbler, *Dendroica dominica*, for the eastern rim of the Great Basin in Wasatch County.[27] There are several records for the west central part of the Basin in Nevada: one in the vicinity of Austin[416] and one at Duckwater.[412]

There is a handful of records for the Palm Warbler, *Dendroica palmarum*, from the eastern part of the Great Basin from Bear River and Fish Springs Refuges and from Salt Lake City.[56] There is a record from Tonopah in central Nevada.[402]

There is a record of the Bay-breasted Warbler, *Dendroica castanea*, from the southwest corner of the Great Basin at Dyer, Nevada.[417] This warbler is on the accidental list at Malheur Refuge.[61]

The Black-and-white Warbler, *Mniotilta varia*, is probably the most frequently encountered eastern wood-warbler in the Great Basin. In Utah, there are two winter records from Provo,[418] a winter specimen was collected at Salt Lake City, and there is a spring sight record from Centerville.[27] In eastern Nevada there is a summer sight record from Wells.[417] In western Nevada we collected a summer specimen at the Eugene Mountains in Pershing County; and we have a number of spring sight records from the Verdi, Reno, and Wadsworth areas. There are records during migration from northeastern California.[93] There are both spring and fall records from Malheur Refuge.[61]

A spring specimen of the Prothonotary Warbler, *Protonotaria citrea*, was collected in the Carson Range on the western edge of the Great Basin.[419]

There are a number of records of the Ovenbird, *Seiurus aurocapillus*, in the Great Basin. There are records from the eastern end at Salt Lake City and in the Stansbury Mountains.[27] In the western end this warbler has been reported from Fallon[86] and from Sutcliffe at Pyramid Lake.[412] It is reported as being of rare occurrence during all four seasons of the year at Malheur Refuge.[61]

The Kentucky Warbler, *Oporornis formosus*, has been reported from Dyer in the southwestern edge of the Great Basin.[179]

The Hooded Warbler, *Wilsonia citrina*, has been recorded at Dyer.[179]

SUBFAMILY THRAUPINAE

Well over two hundred species of tanagers constitute the Subfamily Thraupinae. The members of this subfamily are confined mainly to tropical and subtropical regions in the Americas, but four migratory species breed in North America. The name tanagers is the anglicized version of *tangaras*—the name conferred upon these birds by the Tupi Indians of Amazonia.[420] Tanagers are brightly colored in solid patches, often with sharply contrasting areas. Their bills are usually rather conical and notched or hooked. Rictal bristles are present. Tanagers are typically solitary, arboreal birds.

Western Tanager
Piranga ludoviciana

The breeding range of the Western Tanager lies in western United States and Canada. The Western Tanager is a common migrant in the valleys and a common summer resident in the mountains of the Great Basin.

The Western frequents open coniferous forests, aspen groves, and mountain mahogany stands on mountainsides. Its status apparently did not change following settlement. In describing its status in the 1860s in the Great Basin, Ridgway wrote that from the Sierra Nevada "eastward it was met with in every wooded locality, being much more frequently seen on the mountains than along the rivers of the lower valleys."[25]

Little in the way of tanager courtship has been observed, except that males sing persistently during much of the day. Hoffmann described this singing as follows:

> A Tanager is always deliberate and often sits for a long period on one perch singing short phrases at longish intervals. The song sounds much like a Robin's; it is made up of short phrases with rising and falling inflections *pir-ri pir-ri pee-wi pir-ri pee-wi*. It is hoarser than a Robin's, lower in pitch and rarely continued for more than four or five phrases; it lacks the joyous ringing quality of the Robin's.[84]

After pair bonding the female does all the nest building and incubating by herself. Even when searching for nesting materials she is seldom accompanied by her mate. However, once the eggs hatch the male helps feed the nestlings.

The Western Tanager is characterized by an unhurried approach to life. Regardless of its type of activity, it moves slowly and deliberately. It has even been described as being "apathetic in temperament."[318] During the summer its diet consists largely of insects which are gleaned in among the larger twigs and branches of trees. Sometimes this tanager will flycatch insects. In the late summer it may resort to eating wild fruit.

Miller and Stebbins have observed the migratory passage of Western Tanagers across the hot, arid Joshua Tree National Monument of southern California. Often the migrants traveled in small parties of three or four birds. They took advantage of the larger trees and shrubs along their route for resting and foraging. When on the move they made long flights between stops in trees. They were not noted drinking water. But then they fed on insects, a good source of water of succulence, and "juicy fruits and buds may be taken." Migrants were usually very fat and presumably in good shape.[13]

Other Tanagers

The breeding range of the Hepatic Tanager, *Piranga flava*, lies to the south of the Great Basin. This tanager is of accidental occurrence here. There is a June sight record for Eureka, Nevada.[421]

The Summer Tanager, *Piranga rubra*, breeds to the south and east of the Great Basin. Of accidental occurrence here, it has been recorded from Parowan and Eureka, Utah, in the eastern end of the Basin;[27] from Dyer, Nevada,[401] and Mono Lake, California,[93] in the southwestern end; and from Malheur Refuge in the northwestern corner.[61]

The breeding range of the Scarlet Tanager, *Piranga olivacea*, lies to the northeast and east of the Great Basin in southern Canada and in midwestern and eastern United States. The Scarlet is of accidental occurrence in the Basin. There is a sight record from Salt Lake City[27] and a photographically documented record from Genoa in the Carson Range of Nevada.[422] This species is on the accidental list at the Malheur Refuge.[61]

SUBFAMILY CARDINALINAE

The Subfamily Cardinalinae contains the cardinals and their allies, such as the grosbeaks and buntings. Prior to 1973 the name of this subfamily was Richmondeninae. The cardinalines number about thirty-five species spread over the three Americas, with their center of diversity in tropical America.[420] They are rather stout of bill and body, and the males are often vividly colored in blues or reds. Although cardinalines may feed on the ground, they are essentially arboreal birds. They are not truly gregarious, but in some species the sexes remain paired the year around.

Rose-breasted Grosbeak
Pheucticus ludovicianus

The breeding range of the Rose-breasted Grosbeak lies far to the north and to the east of the Great Basin in Canada and the United States. This grosbeak is of occasional occurrence here. There are records of this bird from throughout the Basin, especially in recent years: from Malheur Refuge in the northwestern corner to Dyer in the southwestern corner and from Genoa and Reno in the west, through Eureka, Ruby Lake, and Fish Springs, to Salt Lake City and Cedar City in the east.

Black-headed Grosbeak
Pheucticus melanocephalus

The breeding range of the Black-headed Grosbeak spans western North America from southern Canada to Mexico. This grosbeak is a common summer resident in the Great Basin. Stragglers are occasionally present during the winter. To the Paiute Indians this grosbeak was Uni-gu'-eet; to the Washo it was Look'-em.[25]

The Black-headed Grosbeak frequents dense streamside vegetation in the valleys and in the lower parts of mountain canyons. It may also be found in pinyon-juniper woodlands, mountain mahogany stands, and open coniferous forests. Its status has apparently not changed much with settlement of the Great Basin. In describing the 1860s Ridgway wrote, "This fine bird was quite abundant in the fertile valleys and lower cañons along the entire route, from Sacramento to the Wahsatch and Uintahs." He then went on to say that its range was exactly the same as the Lazuli Bunting's, with both species reaching their upper altitudinal limit in the middle portions of mountain canyons, about where the summer range of the Western Tanager commenced.[25]

The foraging station of this grosbeak is in the crown foliage of deciduous trees, but some foraging occurs in shrubs and on the ground. Grosbeaks feed on insects, spiders, buds, seeds, berries, and other fruit. Sometimes a grosbeak will flycatch passing insects.

Solitary male grosbeaks begin arriving back on the breeding grounds about a week before the appearance of females. Upon arrival the males begin singing from high, exposed perches. Then, as females arrive, courtship commences. The males sing and fly about in the presence of females; sometimes one or more males chase a female in a pursuit flight. Only one type of courtship display, the song flight, was detected by Weston during field studies in California. After singing loudly from a song perch, a male will enter the air and fly along a horizontal course above the female—while flying he sings almost continuously. These song flights continue long after courtship is at an end.[423]

The song of the Black-headed Grosbeak is somewhat reminiscent of that of the American Robin. Of this resemblance Grinnell and Storer wrote, "The grosbeak's song is much fuller and more varied, contains many little trills, and is given in more rapid time. Now and then it bursts forth fortissimo and after several rounds of burbling, winds up with a number of 'squeals,' the last one attenuated and dying out slowly."[318]

Paired grosbeaks forage together—while foraging the male aggressively defends his mate from other males. The nest is built by the female, but while she is gathering building material she is accompanied by her mate, who sings as she works on the nest. The nest is positioned in a tree or shrub, often within twelve feet of the ground. It is very loosely constructed, and the bottom may be so thin that the eggs show through. Both parents incubate alternately by day, the female alone by night. Both parents attend the young. Black-headed Grosbeaks are most unusual in that the male and female may sing while on the nest incubating eggs or brooding nestlings.[423]

Blue Grosbeak
Guiraca caerulea

The breeding range of the Blue Grosbeak extends across the United States from California to New Jersey and southward to Central America. The Blue Grosbeak is a common breeding bird just south of the Great Basin. It is less common northward, and only sporadic or low-density nesting occurs in the Basin. During his fieldwork in the Basin in the 1860s, Ridgway did not encounter a single Blue Grosbeak. I have witnessed nesting in west central Nevada just once: Jack Knoll and I photographed fledgling grosbeaks along Steamboat Creek in the Truckee Meadows when two pairs of grosbeaks nested there one summer. Nesting occasionally occurs in the Stillwater area[255] and in the Carson River Basin.[75] In Utah the Blue Grosbeak also becomes less common northward.[56] Surprisingly, the Salt Lake Bureau of Land Management District checklist classifies it as a common summer resident in the northwest quarter of Utah;[424] this classification may be overly optimistic. It also nests in small numbers to the north of the Great Basin in southern Idaho.[415]

The Blue Grosbeak frequents valleys or foothill canyons in the Great Basin. Its nesting habitat lies along streams, irrigation ditches, and sloughs in short, dense vegetation such as willow thickets and weed patches. The male often sings for lengthy spells from an exposed song perch above the dense vegetation. His song is not as rich as that of the Black-headed Grosbeak, and some listeners have noted a resemblance to the song of the House Finch. Saunders has characterized the song as being "a series of notes, rather irregularly alternated up and down in pitch, the quality musical but burred."[425]

The nest of this grosbeak is usually located in dense vegetation within ten feet of the ground. Some observers maintain that the female alone builds the nest and incubates the eggs. Others such as Wheelock have observed participation by the male.[304] In some localities Blue Grosbeaks have been known to incorporate cast-off snake skins into their nests. Both sexes tend the young. The Blue Grosbeak feeds on insects, fruit, and seeds. Its foraging station is centered around the ground and in low vegetation.

Lazuli Bunting
Passerina amoena

The Lazuli Bunting's breeding range lies in the central and northern parts of western United States and in the southwestern end of Canada. The Lazuli is a common summer resident in the Great Basin.

This bunting is at home in both valley and mountain habitats—it often frequents the dense thickets along streams in valleys or in mountain canyons in the Great Basin. Its status here has apparently changed but little since the 1860s, when Ridgway found it to be a "very common species in all the fertile valleys, as well as in the lower cañons of the mountains." Ridgway also discovered that the Black-headed Grosbeak was invariably a habitat associate of the Lazuli Bunting.[25]

Lazuli Buntings arrive back in the Great Basin around mid May. Migratory buntings have been seen on occasions moving and foraging along with other birds, such as Chipping Sparrows and even wood-warblers. The males arrive in advance of the females and set up territories. They sing from conspicuous song perches, often from high above the ground, and engage in song flights. The male is a dedicated singer and will sing through the heat of the day. Grinnell and Storer described the song as being "rather high pitched . . . certain syllables may be added or dropped, but the general theme remains the same, and is uttered over and over again at intervals of about 12 seconds. One of our transcriptions of the song is as follows: *see-see-see, sweert, sweert, sweert, zee, see, sweet, zeer, see-see.*"[318] Hoffmann believed the male's "song is best distinguished by its marked division into short phrases which vary distinctly in pitch, generally beginning high, falling to successively lower levels and then rising again. The following syllables may give an idea of its character but the variations are endless—*tsip tsip tsip zwee tsit tsit tsit tsit.*"[84]

The Lazuli has a closely related eastern counterpart—the Indigo Bunting. Where the ranges of the two overlap and the buntings live in sympatry, they readily hybridize. Young males in regions of sympatry tend to blend into their songs phrases from the songs of both species.[348]

Both the nest sites and foraging stations of Lazuli Buntings are low. The nests are usually located within four feet of the ground in low, densely growing mixtures of vegetation, and the foraging stations have a similar location. Insects and seeds form the food staples.

Other Buntings

The Indigo Bunting, *Passerina cyanea*, is an eastern species. There are a few records of its occurrence from the eastern part of the Great Basin in Utah and northeastern Nevada. There are indications that this species is extending its range westward. I have encountered singing males on several occasions in the Virgin Mountains of southern Nevada.

The Painted Bunting, *Passerina ciris*, has a breeding range located southwest of the Great Basin. This species is on the accidental list from Malheur National Wildlife Refuge.[61]

Dickcissel
Spiza americana

The Dickcissel nests in midwestern United States. It is of occasional occurrence in Utah along the eastern edge of the Great Basin.[27]

SUBFAMILY EMBERIZINAE

Sparrows, buntings, towhees, juncos, and longspurs constitute the Subfamily Emberizinae. There are about two hundred and seventy species in the Americas, Eurasia, and Africa. Emberizinae is mainly a New World subfamily, and its center of diversity is in the tropical and subtropical Americas. Emberizines are small, stout-billed seed eaters. They are plainly colored in browns, often with darker streaks or mottling. Emberizines inhabit grasslands, shrublands, and open woodlands. They mainly forage and nest on the ground.

Green-tailed Towhee
Pipilo chlorurus

The breeding range of the Green-tailed Towhee lies in western United States. This towhee is a common summer resident and a rare winter visitant in the Great Basin. Prior to 1976 its scientific name was *Chlorura chlorura*. To the Washo Indians this towhee was known as Pooe-tse'-tse.[25]

Green-tailed Towhees are present in the higher valleys and mountains of the Great Basin. They are mainly associated with dense shrubbery, often with sagebrush thickets on dry, sloping ground. They may be present in chaparral openings in forests. The status of this species in the Great Basin has apparently changed little since the 1860s, of which Ridgway wrote:

> This very interesting species was met with on all the higher ranges, from the Sierra Nevada to the Uintahs, particularly in the elevated parks and cañons, where it was one of the most characteristic birds. We never observed it at a lower altitude than the beginning of the cañons, or, as happened rarely, in ravines of the foot-hills, while, in the river-valleys, it appeared to be entirely wanting. . . . this species is a bird of the chaparral, living chiefly in the brushwood of the cañons and ravines;

> but it is also found among the rank herbage of those flowery slopes so characteristic of the higher portions of that mountainous region.[25]

The male towhee sings from the top of shrubs. His song has been described as being "buzzy" and as a "rapid wheezy sequence," such as *sŭp-sĕ-tew'-sĭ-sĕ* or *eet-ter-te-te-te-si-si-si-seur*. The call notes of both the male and the female are a catlike *mē-ū* or a prolonged *mee a-yew*. Between song bouts the male may give the cat call.[318]

The activities of the Greentail are ground-centered: it nests on or close to the ground, and its foraging station is on the ground under or close to shrubs or in shrub tangles close to the ground. Towhees often nest in sagebrush where there are obvious open spaces between the shrubs. When flushed from a nest in such a situation, a towhee will often drop to the ground without opening its wings. Then, with its tail elevated over its back like a chipmunk, it will run rapidly, at an even speed, across the ground away from the nest. Alden Miller suggested that this "rodent-run" behavior of the Green-tailed Towhee functions in nest defense. Since it is often impossible for a human observer to initially distinguish between a fleeing chipmunk and a towhee on a rodent-run, nest predators such as coyotes might be fooled as well—they may mistakenly classify the fleeing animal as a chipmunk instead of a bird and be lured away instead of staying put and searching for a nest.[425]

Double scratching or bilateral scratching is a characteristic type of foraging behavior employed by many of the ground-foraging species in the Subfamily Emberizinae. Both of our towhees, the Green-tailed and Rufous-sided, as well as many of our sparrows, such as the Vesper, Sage, Savannah, Fox, Song, Lincoln's, and White-crowned, practice double scratching.[426] Double scratching consists of rapid backward kicks, executed simultaneously with both feet, which rake the ground and expose food items. In executing a double scratch the bird first hops forward, then immediately hops backward. During the backward hop the claws of the two feet rake the ground, displacing litter and exposing food buried in or under the litter. A bird may double-scratch in the same spot repeatedly, sometimes excavating a slight hollow.[427]

Rufous-sided Towhee
Pipilo erythrophthalmus

The Rufous-sided Towhee's breeding range extends across southern Canada and through the United States into Central America. This bird is a common permanent resident in the Great Basin. However, some altitudinal and latitudinal migrations occur, and there are fewer towhees present here in the winter than in the summer.

The Rufous-sided frequents stiff-branched shrub thickets on foothill slopes, in the lower reaches of mountain canyons, and along streams in the higher valleys. Along the western edge of the Great Basin it favors manzanita and ceanothus chaparral. Along the eastern rim it favors oakbrush.[56] Most activities are confined to these dense shrub thickets. The towhee's foraging station is on the ground. With stout feet and legs and heavy curved claws, it is a vigorous practitioner of double-scratch foraging. Towhees are more frequently heard scratching away under the cloak of some thicket than they are seen moving or flying about in the open.

The nests of this species are built in, on, or close to the ground. The female apparently does all the nest building, incubating of eggs, and brooding of young. When flushed from the nest a female runs across the ground to cover. Like the Green-tailed Towhee, she may then be readily mistaken for a small mammal.[425] After the young are fledged, the family may remain together in the nesting habitat for the rest of the summer, although some towhees may wander higher up the mountainside after the breeding season.

Rufous-sided Towhees have a catlike mewing call. The males sing from song perches at the top of shrubs or small trees. Each male has several different songs, and the songs vary from male to male and region to region. A song usually consists of an introduction followed by a trill. After extensive field studies, Borror concluded that the songs of *montanus*, the subspecies present in the eastern part of the Great Basin, were characterized by an introduction of up to eight similar syllables, followed by a trill so rapid that it sounded buzzy. He believed that Roger Tory Peterson's description of *chup chup chup zeeeeeeeeee* could be fitted to these songs. Borror also discovered that the songs of *curtatus*, the subspecies present over the rest of the Basin, generally lacked an introduction, consisting merely of a buzzy trill. Peterson's description of *chweeeeee* could be fitted to these songs.[428]

Chipping Sparrow
Spizella passerina

The breeding range of the Chipping Sparrow extends from the boreal forests of Canada, southward through the United States and Mexico, into Central America. The Chipping Sparrow is a common summer resident and migrant in the Great Basin. According to Ridgway, the Shoshone name for this sparrow was So'-ho-quoy'-e-tse.[25]

The Chipping Sparrow is found in a wide variety of valley and montane habitats. It lives where trees are scattered or in parklike stands or are interspersed with clearings. The ground must be well exposed to sunlight, so the ground cover must be short or discontinuous. Ridgway described the presence of this sparrow in the Great Basin in the 1860s with the following words:

> In the Interior it was found in all wooded districts, but, contrary to the rule elsewhere, was less abundant among the cotton-woods of the river-valleys than in groves of cedars and mahoganies on the lower slopes of the mountains, of which it was eminently characteristic. Nowhere did we find it in greater abundance than among these woods on the eastern slope of the Ruby Mountains, for there it was the most numerous of all the birds in July and August, associating in large flocks during the latter month, evidently preparing for their departure southward, which commenced in September.[25]

The foraging station of the Chipping Sparrow is ground-centered, although some sparrows forage in the spring foliage for insects, spiders, and buds. Chipping Sparrows pick food items off the ground and are not known to practice double scratching. Consequently, the vegetation on their foraging grounds need not be short or discontinuous. These sparrows feed heavily on seeds, insects, and spiders. When available, grass seeds constitute their food staple—this is why they often occur as common lawn birds.

The song perches of the male are located in trees, often high above the ground. The song is an insectlike buzz or trill, all on one pitch, a few seconds in duration. The call note of both sexes is a faint *tseet.* The female does all the nest building, although her mate may accompany her while she is gathering building material. Nests are placed in shrubs or trees, usually within six feet or so of the ground. The nests of grasses, plant stalks, and rootlets are lined with fine grasses and hair; horsehair is preferred above other types of hair. The lining is often so well constructed that it will maintain its identity when separated from the nest cup.

Brewer's Sparrow
Spizella breweri

The breeding range of the Brewer's Sparrow lies in western United States and western Canada. This species is a common summer resident and migrant in the Great Basin.

This tiny, nondescript sparrow is probably the most characteristic bird of our vast sagebrush country: it accompanies big sagebrush throughout the high valleys, foothills, and mountains of the Great Basin. But despite its ubiquity virtually nothing is known about the natural history of the Brewer's Sparrow. There has been no Boswell to record its biography.

The foraging beat of the Brewer's Sparrow lies mainly on the ground, with limited foraging up in the shrubs. Insects, spiders, and seeds are its food staples. Although a ground forager this sparrow is not known to double-scratch for its food, but then the ground litter in its sagebrush habitat is scanty indeed. The tops of shrubs are employed as loafing, lookout, and song perches. The male's song has been described as being a "series of trills and runs."[84] There are variations in pitch, and the phraseology is difficult to describe. Flock trilling, with a number of birds singing in chorus, may occur.[429] The call note is a faint *tsip*. The Brewer's nest is built in a shrub and lined with hair when available. At the close of the nesting season flocks form, each containing the members of one to several families.

Vesper Sparrow
Pooecetes gramineus

The breeding range of the Vesper Sparrow spans the width of North America and extends from central Canada to central United States. The Vesper is a common summer resident and migrant in the Great Basin.

The Vesper Sparrow resides in the higher valleys and mountains of the Basin. It mainly frequents sagebrush-grass habitats, where the shrubs are low and the ground is only thinly covered with grass. In the 1860s Ridgway most frequently encountered it on "the open grassy slopes of the higher cañons."[25] In Utah Behle and others have noted it occasionally nesting away from sagebrush in pinyon-juniper areas and in subalpine and alpine meadows of short grass.[429] I first detected it nesting in subalpine meadows in the Carson Range when I inadvertently caught fledglings in my small-mammal snap-trap lines.

This sparrow is closely bonded to the earth—even the song perches of the male are seldom higher than the top of a shrub. Its foraging station is on the ground, where it feeds on a variety of insects and spiders and on plant food, especially seeds. While foraging birds may employ the double scratch. The nest is usually sunk into a small depression in the ground under the cover of plants. When alarmed by the threat of danger, a Vesper Sparrow hunkers down motionless or seeks the cover of nearby shrubs. When flushed from the nest, some parents have been known to practice injury feigning by dragging a wing or a leg.[429]

The Vesper does not have much of a reputation for water bathing. However, dusting is a very common type of behavior in both young and old birds. Sparrows resort to regular sites where the earth is friable and fine particles of dust can be worked into their plumage. The birds soon fashion body-size hollows in the ground as they loosen the dust with their legs and work it into their plumage with body and wing movements. Dusting is a practice not to be confounded with water bathing. Wetting the plumage is believed to facilitate the spreading of oil from the preen gland through the feathers during preening; it also functions to clean feathers and skin and to dissipate body heat during hot weather. Dusting accomplishes none of these functions—possibly it may help the bird get rid of ectoparasites.[21] Vesper Sparrows are such vigorous dusters that when put into sudden flight a dusting bird may resemble a flying shower of dust.

The name Vesper Sparrow refers to the twilight song of the male. Of course, males sing all day long, but their song may be more inspirational to the human ear at the end of the day, when many other birds become silent. The song of the male opens with "two sweet notes"[84] which are "followed by 2 or 3 short higher trills, ending in a rapid melody."[348] The scientific name of the species is redundant—*Pooecetes* is from the Greek meaning grass dweller, and *gramineus* is from the Latin meaning grass-loving.[267]

Lark Sparrow
Chondestes grammacus

The breeding range of the Lark Sparrow extends from southern Canada, through the United States west of the Appalachians, into northern Mexico. The Lark Sparrow is a common summer resident and migrant in the Great Basin.

The available evidence indicates that the Lark Sparrow was originally a western species which extended its breeding range eastward following settlement. Ridgway noted this eastward spread in his 1877 report on the ornithology of the fortieth parallel:

> Though essentially a western species, it is not restricted to that portion of the country which extends from the Rocky Mountains westward, as is most often the case with the birds peculiar to the western division of the continent, but it also inhabits nearly every portion of the Mississippi Valley, where it is no less numerous than in the most favored portions farther west. Indeed, this species seems to be gradually extending its range to the eastward, probably in consequence of the general and widespread denudation of the forests, the country thus undergoing a physical change favorable to the habits of the species, having already become a regular summer resident in many sections of the country north of the Ohio.[25]

In his 1929 monograph, *Birds of Massachusetts and Other New England States*, Forbush classified the Lark Sparrow as a rare visitor to New England. He then cited Wheaton as the authority that the Lark Sparrow first appeared in Ohio in 1861, Butler as the authority that it first appeared in Ontario in 1861, and Eaton as the authority that the first known nesting in New York occurred in 1911.[410] In 1980 Peterson noted that the Lark Sparrow had experienced nesting reverses in the East, in that it formerly bred in West Virginia, western Pennsylvania, western Virginia, and western Maryland.[305]

The Lark Sparrow frequents open shrublands in the valleys and foothills of the Great Basin. Its foraging station is on the ground, where it feeds mainly on seeds and insects—particularly grasshoppers. The tops of shrubs are employed as lookout perches, and when disturbed on the ground a sparrow will fly up to such a perch.

The song perches of the male are on shrubs, on fence posts, and in small trees. The song is difficult to express in syllables since "there are certain 'words' or 'phrases' and the stringing together of these, in varying sequence, constitutes the song. The latter is therefore not a set utterance such as is given by so many birds. One recognizes the lark sparrow's song by this irregular combination of soft notes, trills, and buzzing or purring notes, by its varying intensity, and by its long continuance."[318] Ridgway, highly impressed by its song, wrote, "The principal characteristic of the Lark Sparrow is the excellence of its song, which far surpasses that of any other member of the family we have ever heard, while in sprightliness and continuity, qualities so often lacking in our finer singers, we do not know its equal in any bird."[25] However, Hoffmann may have been closer to the truth when he wrote, "The song just lacks considerable beauty, but it is never free and flowing; trills and sweet notes are interrupted by a rather unmelodious *churr.*"[84] The males sing not only all day long, even during the heat of the day, but often at night. They may sing on the wing between perches and sometimes climb into the air on song flights. Male Lark Sparrows are ardent suitors. According to Baepler, their "courtship involves much strutting, singing, chasing, and fighting."[429]

Black-throated Sparrow
Amphispiza bilineata

The Black-throated Sparrow's breeding range extends from the Great Basin, through southwestern United States, into northern Mexico. This bird is an uncommon to common summer resident in the Basin.

The Blackthroat inhabits the drier, hotter desert valleys of the Great Basin. It is found where the vegetation is sparse, but it is not closely tied to any particular associations of plants. It ranges from the alluvial plains, which slope downward from the bases of the mountains, all the way into the intervening valleys. It is even present in the lower, inner parts of the valleys, where saline or alkaline soils prevail and where low shrubs such as shadscale, rabbit brush, and greasewood are dominant. However, it does not frequent the bottoms of valleys occupied by playas or dry lake beds. Because it is truly an inhabitant of desert terrains, one of its vernacular names has been the Desert Sparrow.

Among the small, seed-eating birds of North America, the Black-throated Sparrow is probably the best adapted for desert life. Field observations and laboratory studies by Smyth and Bartholomew in southern California have revealed the general nature of this species' behavioral and physiological adaptations for desert life. When feeding mainly on dry seeds, Black-throated Sparrows regularly visited water holes to drink. But, when feeding mainly on insects or freshly sprouted grass or herbaceous material, the sparrows did not drink regularly and often lived far removed from surface water. Thus, in the winter, spring, and early summer in southern California, the sparrows exploited moist animal and plant food to remain in water balance.[430]

Laboratory studies revealed several physiological adaptations. In the absence of heat stress, sparrows survived on a diet of dry seeds and chick starter without drinking water—something few birds can do. And they did this without reducing their level of activity, which is very unusual. Most birds reduce their activity level when under water stress, which in turn reduces their pulmonary water losses. However, the kidneys of Black-throated Sparrows are superior to those of most other birds in their ability to concentrate electrolytes in the urine. Laboratory sparrows could gain water and maintain their body weight by drinking salt (NaCl) solutions of a concentration of up to 0.4 M strength. This means that Black-throated Sparrows could conceivably utilize brackish desert water as drinking water and get rid of the excess salt by concentrating it in their urine, providing there are no harmful chemicals in the brackish water. Finally, when under water stress sparrows were able to reduce their water losses from excrement—the water content of their excrement decreased from a normal average of 81 percent to an average of only 57 percent.[430]

Among the Indians of the Great Basin, the Paiute name for the Black-throat was Wut'-tu-ze-ze. Ridgway considered this to be an echoic name; he wrote:

> This species is remarkable for its peculiar song, which in pensive tone and sad expression harmonizes so perfectly with its desolate surroundings. It is from this song that the Indian name, *Wut'-tu-ze-ze*, is derived, for the notes are very nearly expressed by the syllables *wut'*, *wut'*, *zeeeeeè*,

repeated once or twice, the first two notes quick and distinct, the last one a prolonged, silvery trill. Frequently a singer reverses, at each alternate repetition of the song, the accent of the first and last portions, thus producing a very peculiar effect.[25]

Long after the Paiutes and Ridgway had listened to the song of Wut'-tu-ze-ze, Heckenlively carried out a modern study of its song in New Mexico. Instead of ears he used a tape recorder and sonagraph to prepare audiospectrograms from which the song's physical properties could be analyzed quantitatively. Heckenlively discovered that "Black-throated Sparrows are highly variable singers, both as individuals and as a population," and that typically a "song format includes a complex introduction followed by a buzz, trill, or both." The introduction contains a croak and note complexes, a croak being a "series of low-pitched fuzzy notes," a note complex being a "natural note group, distinguishable temporally, but containing dissimilar notes, unlike trills or buzzes." Both buzzes and trills are series of similar, one-note syllables. But the note in a buzz is repeated at a rate exceeding thirty times a second, usually with a broad frequency range, while in a trill the note is repeated at a rate less than thirty times a second, usually over a limited frequency range. The buzz is "unmusical" whereas the trill is "musical."[431]

Sage Sparrow
Amphispiza belli

The Sage Sparrow's breeding range lies in western North America, extending from the state of Washington and the Great Basin southward to Baja California. This species is an uncommon to common summer resident in the Basin, decreasing in abundance eastward. Sage Sparrows are migratory over the northern part of their range. Some overwintering occurs in the western part of the Basin. To the Paiutes the Sage Sparrow was Tok'-et-se-whah'.[25]

The Sage Sparrow is a denizen of the big sagebrush habitats of the Great Basin. It is curious that the Sage Sparrow is often missing from what appear to be suitable stretches of sagebrush here. Rich, who observed the same situation in Idaho, remarked that this sparrow "may have specific requirements that preclude it from large tracts of sagebrush which otherwise appear suitable."[432] In the 1860s Ridgway noted that "the distribution of this species seemed to be strictly governed by that of the sage-brush plants" but that "it is most partial to the moister valleys, where the growth is most thrifty." Ridgway further noted that "it was observed to be most numerous in the valleys" of the western part of the Great Basin, with "few being seen in the Salt Lake Valley."[25]

Field studies in southern Idaho[433] and in the northwestern part of the Great Basin[434] have shown that migratory male Sage Sparrows tend to return to the same area to establish territories in consecutive years. Curiously, yearling sparrows do not seem to manifest natal-site tenacity. After an intensive program of banding nestling Sage Sparrows was conducted in the northwestern part of the Basin, "careful searches of natal areas and their surroundings failed to reveal the return of any of the banded nestlings as yearlings."[434]

Another unusual aspect of reproduction in Sage Sparrows is that many birds arrive back on the breeding grounds in spring already pair-bonded.[433, 434] In most passerine birds the males first return and set up territories; then pair bonding occurs on the territories. Also, Sage Sparrows' territories can be extremely large for a passerine species. The average size of the territories of eight males in Rich's southern Idaho study was 10.94 ± 4.59 acres, the largest known for any species of North American sparrow.[433] Territorialism in Sage Sparrows shows other peculiarities. The boundaries between adjacent territories may overlap in places, and the size and shape of the territories may change daily during the course of the breeding season. Further, despite the overlapping, there are unoccupied spaces among the territories.

The principal territorial behavior of the male consists of singing—little or no chasing, fighting, and posturing occur. Bouts of singing alternate with bouts of foraging. After foraging on the ground, the male will fly to the top of a nearby shrub, engage in a song bout, then drop to the ground to forage again. Males do not have a favorite song perch; they sing from many different shrubs. However, they do have preferred areas within their territories in which they sing.[433]

The males of most species of songbirds have a repertoire of several distinct songs. However, a male Sage has only a single song type or pattern—once a young male acquires a song pattern, it varies little from song to song during his lifetime. Within a population of Sage Sparrows the song patterns of all the males may be alike or may be dissimilar, and groups of neighboring males may share a similar song pattern. There are often variations in song patterns between populations of Sage Sparrows, but these variations are not related to the distance between populations. The within-population and between-population variations do not fit any overall pattern with respect to habitats, population densities, or geography.[434] But, despite these variations in song patterns, all the patterns can be recognized by the human ear as Sage Sparrow song. This song has been characterized as being "a rather weak, high-pitched tinkling series of notes."[429] Hoffmann described it as "*tsit tsit, tsi you, tee a-tee*, the third note high and accentuated."[84]

The foraging station of the Sage Sparrow is on the ground. It gleans insects, spiders, seeds, and other food items from the ground between and beneath shrubs and from shrub branches within reach of the ground. A Sage Sparrow will run when alarmed. If closely pressed it may briefly fly up to a lookout perch in a shrub before dropping down out of sight again. Its nest is often placed in a shrub close to the ground. When flushed from a nest in a shrub, an incubating bird will drop to the ground and run off—with its tail held vertically. Sage Sparrows have a characteristic habit of continually jerking their tails up; they do this when perched in a shrub or even when on the ground.

Lark Bunting
Calamospiza melanocorys

The breeding range of the Lark Bunting lies east and north of the Great Basin in the prairie country of southern Canada and west central United States. The Lark Bunting is an uncommon transient in the eastern end of the Great Basin in Utah.[27] Nesting has occurred in Salt Lake County.[56] There are a few records of occurrence from farther west in the Great Basin—from White Pine, Humboldt, and Washoe counties in Nevada and from Sierra County in northeastern California.

Savannah Sparrow
Passerculus sandwichensis

The breeding range of the Savannah Sparrow covers most of North America from Alaska to Guatemala. This sparrow is a common summer resident and migrant in the Great Basin. Its English name, which is in reference to Savannah, Georgia, is a poor one. The Savannah Sparrow does not breed in southeastern United States, and it is at best but a transient or winter visitant at Savannah, Georgia.

The Savannah Sparrow frequents moist grassy areas in the valleys and lower canyons in the Great Basin. Therefore, it is usually restricted to the vicinity of streams, ponds, lakes, and irrigation systems—often where the waters and soils are very alkaline. It requires habitat where the ground cover of grasses, sedges, or alfalfa is dense but short.

The Savannah is another ground-centered sparrow. Its foraging station is on the ground, where it feeds on a variety of insects, spiders, seeds, and other items. When foraging a sparrow hops along gleaning food from the ground and vegetation, sometimes stopping to double-scratch. When closely approached by a human observer a sparrow will run rapidly through the grass with its head carried low, or it will fly for a very short distance before dropping down into the grass again. The Savannah flies in a jerky fashion along a zigzag course only a few inches above the ground. Its nest is usually placed in a depression in the ground. It may also loaf and preen on the ground by day and roost there by night. At times males sing from the ground.

Shortly after returning from migration, males establish territories and commence singing. Potter conducted a study of Savannah Sparrow territorialism in Michigan. Although there was only a limited amount of contact between adjacent territories, the average size for sixty-two territories was only 0.26 acre. All these territories were either partially or completely surrounded by unoccupied spaces. In four instances there was a little overlap between contiguous territories despite the available free spaces. Males sing while on their territories, usually from song perches on clumps of vegetation or from barbwire fences or from similar elevations. Some song perches are favored over others. Fights between neighboring males are infrequent—aerial chases are more common. As the resident male chases an intruder out of his territory, he may emit a buzzing call. During boundary disputes males may assume a face-to-face threatening posture, or they may walk together, side by side, along their common boundary.[435]

The Savannah has been referred to as a "buzzy-voiced" sparrow, with an insectlike song.[429] Hoffmann wrote that the song "consists of two or three preliminary chirps, followed by two long insect-like trills, the second pitched a little lower than the first, *tsip, tsip, tsip, tseeeeeeee tseer.*"[84]

Grasshopper Sparrow
Ammodramus savannarum

The Grasshopper Sparrow has, in places, a continent-wide breeding range which extends from southern Canada into southern United States. It also ranges from southern Mexico into South America. In the Great Basin the Grasshopper is a rare to uncommon summer resident with a spotty distribution. It has been reported from north central Utah,[56, 436] from northeastern and central Nevada,[91, 235] from western Nevada,[75, 437] and from northeastern California.[185]

The Grasshopper Sparrow is known from only a handful of valley sites in the Great Basin in grasslands where the ground is dry or well drained.

This sparrow requires a grassland habitat where the species are short or of medium height, growing in clumps or bushes instead of forming solidly covered sod. In optimal habitat not only do the grasses grow in clumps, but there are few or no shrubs present, and there is considerable bare ground. In eastern United States the Grasshopper Sparrow reaches its peak abundance in cultivated grasslands which are managed for hay crops—especially in fields of orchard grass, alfalfa, red clover, or bush clover, all of which grow in clumps.[438]

Before the settlement of North America, this sparrow possessed much less suitable habitat—it was restricted to natural openings in forested country and to natural grasslands. As the forests were cut and the land cleared, the Grasshopper Sparrow spread along with agriculture. But, today, over much of its vast breeding range this sparrow has only a spotty or local distribution, and much apparently suitable habitat is not occupied. Then, too, population densities fluctuate greatly from year to year, and sparrows may disappear from formerly inhabited localities and show up on new breeding grounds. Some regions, such as the Northeast and Florida, are now plagued with declining populations of Grasshopper Sparrows.[305] In some regions Grasshopper Sparrows are apparently being replaced by Savannah Sparrows, who can tolerate the invasion of grasslands by shrubs.[429]

Much of what we know about the Grasshopper Sparrow comes from field studies by Robert Smith. Grasshopper Sparrows are semicolonial nesters, nesting in small groups. The males arrive back on the breeding grounds from migration in advance of the females, about the time that the grasses have grown tall enough to conceal them. They immediately set up territories, where they display and sing their grasshopper song. The average size of twenty-two territories in Pennsylvania studied by Smith was slightly in excess of two acres. Males sing from an area near the periphery of their territories; the highest available song perches are used. A male may engage in song duels with rival males. He sings his grasshopper song from an erect posture, but while listening to the reply of a rival he engages in a threat display: he raises his wings and flutters one or both of them while crouched with his head lowered.[438]

The grasshopper song is so named because of its resemblance to the stridulations of long-horned grasshoppers, and from this song the sparrow derived its name. The grasshopper song has two variations sung by all males. Smith described these as *Tup zeeeeeeeeeeeee* and *Tip tup a zeeeeeeeeeee*; he concluded that the function of this song was to "proclaim and defend territory."[439]

A second type of song by the Grasshopper Sparrow has been called the sustained song by Smith. This song is more complex and musical than the grasshopper song; Smith described it as *Tip tup a zeeeeeeee zeedle zeee zeedle zeedle zeeeee.* Sometimes the grasshopper introduction is left out of the song. The sustained song is sung not only from perches but also in flight, by the male flying alone or pursuing a female. During a song flight the male flies up from the grass on quivering wings and, while in a "low, fluttering flight," sings the sustained song before dropping back into the grass again. Smith concluded that the main function of the sustained song was to attract a mate and then help maintain the pair bond with her. He also believed that when the grasshopper introduction was used it functioned as a threat toward other males.[439]

The third vocalization described by Smith was the trill, a "series of moderately loud notes on two tones," rapidly given from a perch or from the grass by a mated male near his nest. Smith described it as *Ti tu ti tu tiiiiiiii* and concluded that it functioned to help maintain the pairing bond and announce the approach of the male to the female and young at the nest.[439]

The female Grasshopper also has a vocalization, a trill which is weaker than the male's and lacks the descending pitch. Smith described the female's trill as *Ti ti iiiiiiiii*, delivered when she is hidden in the grass. The female may sing in answer to a male's sustained song or trill, or she may sing by herself. Smith concluded that the female's trill functioned to announce the sex of the bird and her availability as a potential mate to a territorial male. In species such as the Grasshopper Sparrow where the sexes look alike, territorial males can experience difficulty in readily identifying trespassing birds as rival males or potential mates. Mated females also use the trill to announce their location and to announce their approach to the nest to their mate and young.[439]

Fox Sparrow
Passerella iliaca

The breeding range of the Fox Sparrow lies mainly in western North America in Alaska, Canada, and the United States. However, it does extend across central Canada from the Pacific to the Atlantic. The Fox is an uncommon to common summer resident and migrant in the Great Basin. Stragglers may be present during the winter.

The Fox Sparrow inhabits densely twigged thickets along streams or montane meadows, in moist forest clearings or edges, or in springy areas in Great Basin mountains. Suitable thickets are formed by a variety of woody plants. Dense tangles of brush are sought which provide near impenetrable cover, leaf litter, and moist, cool ground-level haunts. In these densely twigged tangles the Fox Sparrow forages, nests, loafs, and seeks cover.

The Fox Sparrow is widely known for its shyness and its retiring habits—seldom does it venture out of its underbrush world during the nesting season. If danger threatens, it merely retreats deeper into the underbrush. When driven out of its thicket, a sparrow usually circles back home as soon as it can safely make the move. Fox Sparrows are more often heard than seen, as they loudly double-scratch in the leaf litter for their daily food. They forage for insects, seeds, and other plant food at their ground-level foraging station.

Males venture above the underbrush during the breeding season to sing from song perches in nearby small trees or from snags or other elevated sites. Fox Sparrows are among our better songsters. There is song variation between individuals and even within the singing of a single male. Hoffmann described the song as including "as one of the opening phrases a pair of loud sweet notes, *swee chew* or *wee chee* followed by trills and runs."[84]

Song Sparrow
Melospiza melodia

The breeding range of the Song Sparrow extends over much of North America from southern Alaska to central Mexico. The Song Sparrow is a common permanent resident in the Great Basin. Migrants and winter visitants also occur in the Basin.

Song Sparrows frequent brushy thickets adjacent to water in situations where wet ground is associated with low, dense vegetation. They are particularly fond of the borders of streams and ponds. As in Ridgway's day they are still common among the "tules or rushes fringing the sloughs and ponds near the larger bodies of water."[25] Their greatest presence is in the valleys and foothills, for they extend only up to moderate altitudes in the mountains.

That the Song is one of our best-known species has resulted mainly from eight years of research carried out in Ohio by the great ornithologist Margaret Morse Nice. Her studies still stand as perhaps the most comprehensive ones ever made by a single investigator on a single species of bird. Nice wrote her first major publication on Song Sparrows in German for the *Journal für Ornithologie*. The definitive report on her monumental studies was published in the Transactions of the Linnaean Society of New York in

1937 and 1943. Today, her reports are available as a two-volume set of paperbacks entitled *Studies in the Life History of the Song Sparrow.*[440]

Nice banded Song Sparrows so they were recognizable as individuals and followed the course of a number of generations on her study area at Columbus, Ohio. She discovered that a breeding population was always made up of two types of individuals—permanent residents and summer residents. About 50 percent of her males and 20 percent of her females were permanent residents; the remaining males and females migrated south for the winter. This condition, where only some of a population is migratory, is referred to as partial migration. Most of the birds in her breeding populations were consistently migratory or nonmigratory, but occasionally an individual's migratory status changed from year to year.

Song Sparrows are highly territorial, and over half of the year is spent on territories—nonmigratory birds even spend the winter on or close to their territories. Both migratory and nonmigratory young males frequently set up territories during their first fall of life. Some males occupy the same territories year after year, others may alter the boundaries slightly or occupy new territories. There is a strong tendency on the part of a female to return to the same territory or to its near vicinity to pair-bond each year. Territorial defense relies mainly on singing, but posturing, chasing, and fighting may occur. Males attempting to carve out territories in an occupied area may display in the presence of resident males by puffing out their plumage, holding a wing aloft and fluttering it, and singing softly. They may or may not be successful in physically resisting the attacks of resident males. Singing reaches its maximum intensity before the arrival of the females in the spring. It also occurs during the recrudescence of territorial behavior in the fall.[440]

Territorial males vigorously resist intrusions by other birds. This creates somewhat of a problem, since females must approach males on their territories in order to pair-bond with them and since Song Sparrows are homomorphic birds. The question of how unacquainted individuals of a homomorphic species, where the males and females look alike, announce their sex identity to each other is a fascinating one. Nice discovered that the male Song Sparrow does it by territorial singing and aggressive behavior, and "the female in the early breeding season announces her identity by her notes—either a high-pitched, nasal *eeeeeee* or a kind of chatter; she also indulges in various growling, grumbling expressions."[441]

Nice found male Song Sparrows to be unimaginative, chauvinistic suitors. Once a territorial male realized he had a female and a potential mate at hand, his sole courtship act was pouncing. He would "fly down suddenly, hit his mate and fly away with a triumphal song!" Males also pounced on neighboring females who were not in the close company of their mates. But Song Sparrow females were faithful mates, and they repulsed the advances of neighboring males until their mates came to their rescue. Females were not completely submissive to their mates, however; a male was subjected to some henpecking by his mate. "She often opens her bill at him, gives him small pecks or drives him to a moderate extent. She says *jee* to him, but he never says it to her."[441]

In species such as Song Sparrows where pair bonding occurs on the territory of the male, the females do the selecting of mates. The males have no choice but to accept or reject the females who visit their territories. Since the days of Charles Darwin observers have pondered the question of what criteria, if any, females use in selecting mates. Darwin believed that sexual selection was involved in at least some cases—that a female selects a mate on the basis of such overt and highly visible attributes as striking plumage coloration or adornment or musical ability or that these attributes were involved in competition between males for females. Darwin was trying to explain the evolution of sexual dimorphism in species where there were striking differences in plumage and song between females and males. He believed that, if sexual selection promoted nonrandom mating, the males could evolve in a direction different from that of the females with regard to plumage and song, even though the two sexes were being subjected to the same pressures of natural selection.

Margaret Nice looked for, but found no evidence of, sexual selection by female Song Sparrows. In describing her findings she wrote:

> In this study there has been an opportunity to watch for evidence of female choice each year among the twenty-five to seventy males all singing for mates. The males differ slightly in size, and notably in belligerency, in zeal of singing, in beauty of song (from a human standpoint), and in brightness or sootiness of plumage. I have no evidence that the female pays the slightest attention to the appearance, character, or singing ability of her mate, nor even to the number of legs he possesses. And it is not that her judgment is prejudiced by the attractions of a superior territory, for she is equally uncritical in this manner also. Old females try to come back to their former homes; otherwise their "choice" of mates appears to be perfectly haphazard.[440]

Although Nice did not find evidence for sexual selection by Song Sparrows, others have found evidence for it in some sexually dimorphic species such as Red-winged Blackbirds.[442]

The names we have for this species refer to its singing ability. Its common name does so, of course, and both parts of its scientific name do so.

Melospiza is from the Greek meaning song finch, and *melodia* is from the Greek for melody or melodious song.[267] To the Paiutes of the Great Basin the Song Sparrow was See'-hoot'-se-pah.[25]

The songs of the males are highly variable but always recognizable. An individual may have a dozen or more song patterns. When singing he may repeat a given pattern several times before switching to a different one. Then, too, the song patterns of different males vary; Saunders commented, "I have 885 records of the song, no two of them alike."[443] Although singing peaks in the spring, with a secondary peak in the fall, snatches of song may be heard at any time of the year. Males sing even during stormy weather and sometimes at night; females have also been known to sing. The song is "sweet and musical." Saunders described it as consisting of three parts, "strongly rhythmic introductory notes, a central trill, and a final series of rather irregular and indefinite notes."[443]

The female Song Sparrow performs most of the family tasks. She builds the nest, incubates the eggs, and broods the nestlings during their first five or six days of life. She also helps her mate feed the young and defend the territory. Young sparrows remain in the nest for about ten days. Then, upon leaving the nest, they are attended by their parents for a while. They gain their independence when they are about one month old.

Song Sparrows have been known to practice active anting. Anting is among the least understood types of avian behavior; to date, it has been observed worldwide in over two hundred species of passerine birds belonging to over thirty families.[21] Some birds engage in passive anting, during which they settle down on an anthill, spread their wings, and elevate their feathers so that the disturbed ants will swarm through their plumage. Other birds practice active anting—they pick up one or several ants in their bills and jab them into their plumage, especially under the wings and tail. The various species of ants that are employed in anting are ones which secrete formic acid or other repugnant fluids. Sometimes birds use substitute items when anting such as the flesh of lemons, orange peels, mothballs, live cigarette butts, onions, and smoke. Anting is believed to be related to feather care. Some ant secretions, such as formic acid, are insecticidal and may help rid the bird of lice and mites and other ectoparasites. Some ant secretions are essential oils and may help in oiling the feathers during the preening which usually follows the act of anting. Since anting occurs at a high frequency in the late summer when molting is under way, it has been suggested that it may help alleviate the skin irritation associated with feather growth.[444]

Lincoln's Sparrow
Melospiza lincolnii

The breeding range of the Lincoln's Sparrow lies in the boreal forests of North America. It extends from Alaska, across the width of Canada, barely into northeastern United States, but deeply into mountainous western United States. The Lincoln's is a common summer resident in Utah on the eastern edge of the Great Basin.[27] It is an uncommon summer resident in the Carson Range and a summer resident in the Warner Mountains on the western edge of the Great Basin. It is a rare to uncommon migrant within the Basin and is of rare occurrence during the winter. As yet there are no records of Lincoln's Sparrows as breeding birds in the interior mountain ranges of the Basin.[289] Field studies are needed to ascertain whether this is a real hiatus or merely a hiatus in our present knowledge of the breeding distribution of this species.

Lincoln's Sparrows frequent wet montane meadows where there are thickets of low willows or alders along the edges of the meadows or along streams. They are much shier and more retiring birds than their close relatives the Song Sparrows. Lincoln's Sparrows keep close to thickets; they forage, loaf, nest, and often sing there. Their foraging station is on the wet ground. Their diet consists of a variety of insects and seeds. Lincoln's Sparrows may engage in double scratching.

Males sing from thickets, from the ground, occasionally in flight, and from exposed song perches in nearby trees. Lincoln's Sparrows sing only during the nesting season. Their song has been described as "an extremely rapid gurgling utterance, remindful of Western House Wren: *zee zee zee ti ter-r-r-r-r-r.*"[318] Territorial defense mainly involves song, although some fighting between males may occur. Like male Song Sparrows, male Lincoln's Sparrows pounce on their mates during courtship. However, female Lincoln's Sparrows apparently elicit this pouncing behavior by special vocal and behavioral invitations.[443]

The Lincoln's Sparrow was named in memory not of Abraham but rather of Thomas Lincoln by Audubon. Thomas Lincoln accompanied Audubon to Labrador and collected a specimen of this new species for him to draw.

White-crowned Sparrow
Zonotrichia leucophrys

The breeding range of the White-crowned Sparrow covers much of Alaska and Canada and extends far southward into western United States. This sparrow is a common summer resident, migrant, and winter resident in the Great Basin. During the summer it resides in the subalpine meadows in most of the higher mountain ranges of the Basin and in its mountainous eastern and western rims. It has not yet been recorded as a breeding species in some of the interior mountain ranges such as the Desatoya Mountains and the Quinn Canyon Range.[289]

The Mountain White-crowned Sparrow, *Z. l. oriantha*, is the subspecies breeding in the Great Basin. The DeWolfes studied the nesting habitat requirements of Mountain Whitecrowns in the Sierra Nevada and at Lassen Peak in California. They observed that all the occupied habitat was characterized by patchiness and that five elements were intermingled in the patchiness. Most important, areas of bare ground were thoroughly intermingled with grass and dense shrubs or scrub conifers. Tall conifers occurred at the peripheries of the meadows, and fresh water was present. The sparrows showed a strong tendency toward nesting close to running water.[445]

In recent years, the White-crowned Sparrow has become the favorite subject of western ornithologists doing experimental research on passerine birds. Many facets of their morphology, physiology, ecology, and behavior have been explored. As our knowledge of an organism grows, that organism becomes increasingly valuable as a research subject. Past and present studies not only interact in a complementary and supplementary fashion, but past studies help formulate and direct new studies.

The White-crowned Sparrow is well qualified as a research subject. Its annual cycle of activities involves a number of interesting and complex physiological, ecological, and behavioral events. It is an amenable subject for human observation in both the field and the laboratory. As an abundant, widespread, and flocking ground bird, it can be readily netted or trapped in numbers and easily followed in the field. Interesting mixes of subspecies can be studied in western United States, where four subspecies with different migratory behaviors overwinter: *gambelii* is a long-range latitudinal migrant, breeding in Alaska and northwestern Canada; *pugetensis* is a short-range latitudinal migrant along the West Coast; *oriantha* is a short-range latitudinal and altitudinal migrant; and *nuttalli* is a nonmigratory form found along the coast of California. *Oriantha* is the breeding subspecies; *gambelii* is common in the wintertime; and the eastern subspecies *leucophrys* is of rare occurrence in the Great Basin.

As we have already noted, Mountain White-crowned Sparrows breed in suitable subalpine meadows. Subalpine meadows lie below the upper tree line on mountainsides; if the mountain has enough altitude to possess an upper tree line, the region above that constitutes the alpine zone. Subalpine meadows sometimes receive abnormally heavy snowfalls during the win-

White-crowned Sparrow

ter, and the onset of nesting may vary somewhat from year to year. Habitats on south-facing knolls or slopes are the first to emerge from the snow and be occupied by sparrows. Some years bare ground and grass may not be uncovered until late summer. During these snow-locked summers, female sparrows show some nesting flexibility. Many forsake their normal preference for locating their nests on the ground in dense dry grass and instead place them above the snow or wetness in willows or scrub pines. But overall the nesting density in an area may be reduced: some former breeding meadows may not be used at all, and the nesting density and productivity in others may decline below normal.[446]

Male Mountain Whitecrowns arrive on the breeding grounds in advance of the females. Upon arrival the males set up territories on snow-free areas on the meadows. Banding studies indicate that males tend to return to the same sites each year to establish territories. During spring, when the meadows are still covered with snow, males establish song perches in the tall conifers at the peripheries of the meadows and await the emergence of bare ground. In some instances pair bonds are established out on the peripheries before the pairs have snow-free ground to move onto.[447]

The female builds the nest and does all the incubating. In subalpine meadows the nights are often very cold and severe storms may strike. Yet thermal studies made at Tioga Pass at 9,800 feet in the Sierra Nevada revealed that, when females were sitting on eggs, the mean egg temperatures remained between 34 and 38° C even during cold nights. But egg temperatures fluctuated between 17.8 and 43° C during the early morning and late afternoon when the females were away from the nest foraging. The mean egg temperatures in ground nests were generally slightly higher and more stable than those in nests built above the ground. The females at Tioga Pass showed a preference for situating their nests on the northeast side of bushes. This apparently protected the nests from the prevailing morning winds and storms, which usually came out of the south, and reduced midday heat stress during sunny days.[448]

A bird nest is often more than an egg container. A well-built nest will help insulate the eggs and that part of the body of an incubating bird which fits into the nest cup from the outside environment, and the resistance to heat flow afforded by the nest may be important in the heat economy of birds incubating eggs in stressful thermal environments. The heat budget of incubating Mountain White-crowned Sparrows from a moderate elevation of about 6,200 feet on Hart Mountain, in the northwestern corner of the Great Basin, has been investigated by field and laboratory studies. The sparrows were nesting mainly on an open sagebrush flat dissected by small watercourses and riparian vegetation. On the basis of the studies, a female's expenditure of energy during a typical twenty-four-hour period was estimated "to average 15% lower in an incubating female than in a bird perching outside of the nest but exposed to the same microclimate." Unlike the high-altitude Tioga Pass Whitecrowns, the moderate-elevation Hart Mountain ones positioned their nests in the vegetation with respect to radiative heat flow, not forced convection. Their nests were placed so that the overhead cover occluded less of the eastern sky than of the western sky; the nests were exposed to the sun's rays 1.75 times as long during the cooler mornings than during the hotter afternoons.[449]

Considerable research has focused on the song behavior of White-crowned Sparrows. Particular attention has been given to the ontogeny of song—that is, how young males acquire their song and song dialects. During the breeding season, males sing from early morning to late evening and sometimes on into the night. Following the postnuptial molt after the breeding season, a recrudescence in singing may occur. Sporadic singing by both males and females occurs during the fall and winter. The females' song may be delivered at a lower sound intensity and faster tempo and may be less complete than the song of the males.[450] Singing by females normally ceases at the beginning of the breeding season.

Orejuela and Morton have described the song of the Mountain White-crowned Sparrow in the Sierra Nevada as consisting "of a whistled portion followed by a note complex and a trilled portion. Each of these three portions typically is made up of two phrases. In some cases the trilled portion is followed by a buzz. The song is delivered loudly and repeatedly from the same singing post. The duration of song is about 2 sec and the frequency range between 2 and 6 kHz."[451]

White-crowned Sparrow populations are well known for their song dialects. A song dialect results when the males of a local population share a song theme or themes or parts of song themes. A theme consists of "characteristic syllable types."[452] Song dialects therefore represent a "consistent difference in the predominant song type between one population and another of the same species."[453]

Orejuela and Morton studied Mountain White-crowned Sparrow song at fourteen breeding localities along the north-south axis of the Sierra Nevada. They detected ten song patterns and five song dialect areas. The Whitecrowns at Lassen National Park in northern California "sang a conglomerate of song patterns" and therefore did not constitute a dialect area.[451]

During a study of the song of eighteen populations of Mountain White-crowns in six western states, Baptista and King discerned seven dialects or song types. Five of their eighteen populations were located in the Great Basin. Song type 1 was sung at Incline Village in the Carson Range. This song type consists of "an opening whistle, followed by a buzz, two short trills, and then another buzz . . . Either or both of the two short trills may be replaced by single complex syllables or single simple syllables . . . the first buzz may be absent." Song type 2 was sung at Cottonwood Creek in the Wassuk Range of Nevada and in the Warner Mountains of northeastern California; this song type consists of "an opening whistle, followed by a buzz, two complex syllables, a buzz, and a trill." Song type 3 was sung at Hart Mountain and Steens Mountain in southeastern Oregon; this type consists of "an opening whistle, followed by one to five complex syllables (usually two), two buzzes, and then a long trill. . . . The terminal trill may be replaced by a buzz. . . . Sometimes the complex syllables may be replaced by a trill."[452]

The three descriptions by Baptista and King which I have quoted show that individual song variations and repertoires can exist within a dialect. Variations usually consist of "additions or omissions of terminal trills, or omissions of terminal buzzes when present. Rarely, an individual varies the number of complex syllables." When a song dialect contains several themes, individual males have been known to sing two of the local themes.[452] The whistled opening is the least variable part of the song and is the part which conveys the major cue that a White-crowned Sparrow song is in progress.[451]

The acquisition and development of song by young male Whitecrowns have been experimentally studied. Young males learn the song dialect of their natal population while still in the nest by listening to singing adult males. If isolated when very young and exposed to tape recordings, they can be taught to sing foreign song dialects. After a young male is about three months old, exposure to strange dialects has no effect on his song. If a young male is deafened before song is learned, his singing is abnormal; if deafened after he has learned a song, the song remains stable. Once a male has learned a song, he sings the same song year after year.

In nature, males and females learn the song dialects of their natal population early on in life and are not influenced by any dialects they subsequently hear. There had been long-standing speculation that song dialects could prevent gene flow between populations and promote inbreeding—inbreeding would occur if young males and females were attracted only to territories and mates in areas where they heard their parental dialect being sung. Then, a banding and recapture study by Baker and Mewaldt of two contiguous breeding populations of Whitecrowns with different song dialects indicated that gene flow was restricted by 80 percent and that inbreeding was occurring. Only 20 percent of the expected number of sparrows crossed the dialect border to mate with sparrows from the other population.[454]

Other Sparrows

The breeding range of the Cassin's Sparrow, *Aimophila cassinii*, lies to the southeast of the Great Basin, and this species is of accidental occurrence here. There are sight records from near Low and Parowan in Utah.[27] In 1891 a specimen was collected in Timpahute Valley, Nevada—just beyond the southern border of the Basin.

The breeding range of the Rufous-crowned Sparrow, *Aimophila ruficeps*, lies to the south of the Great Basin. This species is classified as being of rare occurrence on the Honey Lake Wildlife Area checklist.[455] There are other references to its possible occurrence as a straggler in the northwestern corner of the Basin.[93, 456]

The American Tree Sparrow, *Spizella arborea*, nests far to the north of the Great Basin in Alaska and northern Canada. This sparrow is a rare to uncommon winter visitant here. Prior to 1982 its English name was the Tree Sparrow.

The breeding range of the Clay-colored Sparrow, *Spizella pallida*, lies in the prairie regions of western Canada and northcentral United States. A specimen was collected in Tooele County, Utah, and there are sight records from Cedar City in the southeastern corner of the Great Basin.[27]

The Field Sparrow, *Spizella pusilla*, breeds in eastern United States. There are sight records of this sparrow from Salt Lake City.[27]

The breeding range of the Baird's Sparrow, *Ammodramus bairdii*, is located to the northeast of the Great Basin, in prairie country in south central Canada and north central United States. There are several sight records of this sparrow from northern Utah.[27]

The Le Conte's Sparrow, *Ammodramus leconteii*, has a breeding range which lies northeast of the Great Basin in south central Canada and north central United States. This sparrow is of accidental occurrence in the eastern part of the Basin, where it has been recorded from the Provo area.[27]

The breeding grounds of the Swamp Sparrow, *Melospiza georgiana*, lie far to the north and to the east of the Great Basin in Canada and the United States. Swamp Sparrows are uncommon winter visitants in the eastern part of the Basin in Utah.[27] A December specimen was collected at Ruby Lake.[91] Sightings have been claimed for the Reno area, Honey Lake,[455] and northeastern California.[93]

The breeding range of the White-throated Sparrow, *Zonotrichia albicollis*, lies far north and far east of the Great Basin in Canada and in the northeastern edge of the United States. White-throated Sparrows are rare to occasional stragglers and winter visitants at localities throughout the Basin.

The breeding range of the Golden-crowned Sparrow, *Zonotrichia atrica-*

pilla, extends from the northwestern coast of Alaska down through British Columbia. Overwintering occurs along the Pacific Coast. This sparrow is a rare to uncommon winter visitant in the Great Basin. It is considered to be a rare winter visitant in Utah.[27] It is an uncommon but regular migrant and winter visitant in the western end of the Basin.

The Harris' Sparrow, *Zonotrichia querula*, breeds in northwestern Canada. Overwintering occurs in south central United States. The Harris' is a rare to uncommon winter visitant in the Great Basin. Most winters a few individuals are detected around feeding stations maintained by bird-watchers in Reno.

Dark-eyed Junco
Junco hyemalis

The breeding range of the Dark-eyed Junco extends through the forests of Alaska and Canada into the United States. In the West it extends as far south as Baja California. Dark-eyed Juncos are common birds throughout the year in the Great Basin. They are summer residents in the mountains and migrants and winter visitants in the foothills and valleys. Depending on local conditions, some remain in mountain localities during the winter.

The fifth edition of the AOU checklist of North American Birds recognized seven species of juncos. The new sixth edition recognizes only two species: the Dark-eyed Junco and the Yellow-eyed Junco. In 1973 four of the fifth edition species were lumped to form an enlarged species called the Dark-eyed Junco. These were the White-winged, Oregon, Guadalupe, and Slate-colored juncos. Then in 1982 a fifth species, the Gray-headed Junco, was added to the mix. Consequently, five former species now form an enlarged species called the Dark-eyed Junco. The remaining two species were lumped to form an enlarged species called the Yellow-eyed Junco.

If you study the illustrations in bird guides, you will note that the five former species of juncos which were lumped to form the Dark-eyed Junco differ enough in their overt appearances that they can be readily identified in the field. Indeed, they differ much more than do some closely related species, such as the *Empidonax* flycatchers. The question then arises, why don't they constitute separate species in the eyes of the AOU Committee on Classification and Nomenclature? The answer is that the rationale of what constitutes a species differs in theory from what it is in practice. In practice, species are recognized in the field and laboratory by their appearances, by what geneticists call phenotypes. Taxonomists work mainly with a few of the structural phenotypes, rarely with physiological or biochemical or behavioral phenotypes. Thus, in practice, species are recognized on the basis of obvious overt differences in appearances. But in theory species are defined in terms of population genetics and reproductive isolation. Probably the most widely employed definition of what constitutes a species was offered by Ernst Mayr, who said that species "are groups of actually or potentially interbreeding populations, which are reproductively isolated from other such groups."[457]

In regions of sympatry, Dark-eyed Juncos readily hybridize. So, regardless of the fact that the genetically fixed phenotypic differences between these juncos are of greater magnitude than those separating many admitted species, they do not qualify as seven discrete species. They will not be recognized as separate species unless isolating mechanisms evolve which prevent hybridization and restrict gene flow between the different forms in areas of sympatry. What we require of evolution before we accept its products as new species is somewhat incomprehensible. Isolating mechanisms of the type called prezygotic—mechanisms which prevent pair bonding or copulation—are often very simple evolutionary accomplishments, such as minor differences in courtship or precopulatory behavior. Something as simple as the song dialects of White-crowned Sparrows, for example, functions as a prezygotic isolating mechanism.[454]

Several readily identifiable forms of the Dark-eyed Junco are members of the Great Basin avifauna. The Oregon Junco is the breeding form in the mountains on the western and northern edges of the Basin. In the northeastern corner, the Pink-sided Junco, formerly classified as a subspecies of the Oregon Junco, is the breeding form in the northern Wasatch Mountains.[27] The Gray-headed Junco is the breeding form in the mountains of the interior of the Basin and along its eastern edge. A zone of hybridization between Gray-headed and Pink-sided juncos has been identified in the northern part of the Wasatch Mountains and in the Raft River Mountains of northwestern Utah.[27] The Oregon Junco is a common migrant and common winter visitant; the Gray-headed Junco is a common migrant and rare winter visitant; and the Slate-colored Junco, an eastern bird, is a rare to uncommon winter visitant in the valleys and foothills of the Great Basin.

Dark-eyed Juncos nest in open montane conifer forests and aspen stands, often where meadows or streams interrupt the forests and create edges of low vegetation alongside of tall vegetation. They frequent sites where the ground bears grasses and low herbage but still receives midday shade in which they can forage. Often juncos are our most abundant montane breeding birds.

Juncos are predominantly ground-foraging and ground-nesting birds. Over much of the year seeds constitute their food staple, but insects are of importance when available. While foraging juncos hop along, not scratching for food but picking it up off the ground or from plants. When snow blankets the ground they obtain sustenance from the seed heads of plants projecting above the snow. They do desert the ground when danger threatens and fly up into the protection of trees.

Trees and shrubs afford song perches to territorial males. Their song has been described as "a quavering trill, metallic in quality, rapid in utterance, *eetle, eetle, eetle, eetle* . . . . continued for from one to three seconds, weakening in intensity toward the end; repeated at irregular intervals."[318] Thus, almost always, junco song consists of "a trill of similar, repeated syllables," all on the same pitch. There is little variation in song length, number of syllables, and duration of syllables. However, "the fine structure of the syllables themselves shows great individual variability." A male has several song types, but he may sing one type for quite a while before switching to another.[458]

The development of song by male Dark-eyed Juncos was studied by taking young birds from the nest and raising them in varying degrees of isolation. Each male developed several song types, which were longer with fewer syllables than those of wild males. Some abnormal songs developed when the captive males overheard the songs of other species of birds. Unlike wild males, the experimental males included more than one syllable type in a song. Unlike White-crowned Sparrows, there is apparently no critical age during which a male can develop song, at least for the males raised in captivity. Some of the experimental males developed new songs during their second year of life, modified other songs, and retained some songs unchanged. Thus, it appears as if the general pattern of song is inherited and not learned behavior in juncos, since the males raised out of hearing of wild, adult, singing males developed "a trill of normal, although more variable, over-all characteristics."[458]

One of the conclusions arising from the study of young males isolated in captivity was that song development "is intimately bound up with the so-called subsong, a long, rambling, and variable series of sounds, usually given at a lower intensity than the full song."[458] Subsong has been noted in a number of species of birds, characteristically in the early spring singing of yearling males and in older birds when they are motivated to sing by low but increasing concentrations of sex hormones at the beginning of a song season. Subsong is believed to provide "in some degree the raw material out of which, by practice and by elimination of unwanted extremes of frequency, the full song is 'crystallized.'"[21] Thorpe and Pilcher have enumerated the ways in which subsong differs from true or primary song: its fundamental frequency of pitch tends to be lower; its overall pitch range and that of its individual notes tend to be greater; it is quieter; and its phrases tend to be longer.[459]

Outside of the breeding season, Dark-eyed Juncos live in flocks. Sometimes when feeding they become part of a mixed-species foraging flock. In the shrublands of the Great Basin they are often associated with White-crowned Sparrows in foraging flocks. Panik studied ninety-eight foraging flocks in the pinyon-juniper woodlands of northwestern Nevada; he found Dark-eyed Juncos in sixteen of his forty-five mixed-species flocks, and in ten instances they were leaders. Their most common associates were the Plain Titmouse, Mountain Chickadee, Bushtit, and Red-breasted Nuthatch. Juncos often forage in flocks by themselves—seventeen of Panik's ninety-eight flocks were pure flocks of juncos. Flock size is usually small. In Panik's study, an average of 7.3 (range of 4 to 15) juncos were present in mixed-species flocks, and 6.9 (range of 3 to 16) individuals occurred in pure flocks of juncos.[351]

Sabine carried out revealing studies on winter flock behavior of Dark-eyed Juncos in both eastern and western United States. She discovered that winter flocks are not fortuitous associations of individual juncos. Banding studies showed that individuals tend to return to the same overwintering site each year. The juncos arrive irregularly over the course of several weeks. The early arrivals form flocks with stable membership; late arrivals are integrated into existing flocks. Each flock has its own foraging area and foraging circuit. The flock does not always move about as a single unit—groups of various sizes are continually leaving or rejoining it again, and single birds temporarily leave to move about, forage, rest, or loaf in solitude. Sabine evaluated this latter behavior as possibly being "an internal drive to be alone." She noted that a solitary junco "has been observed many times to eat more slowly and in a relaxed posture, sitting back on its 'heels,' but it straightens up, eats faster, and hops in the normal restless manner when companions arrive."[460]

Frequent hostile interactions between birds in a winter flock involving dominant-subordinate behavior can be observed. One junco will peck or run at another junco, who usually retreats or otherwise avoids the peck. Sabine showed that these hostile encounters reflect a type of social organization within the flock—its members are organized into a hierarchy with permanent dominant-subordinate relations between birds. A linear order of peck rights exists: "There is an alpha bird which can peck all others, a beta bird that can peck all but the alpha bird, and so on down to an omega bird that can peck no other." However, there are some "triangular relations such that A pecks B, B pecks C, and C pecks A." Sometimes a reverse peck occurs when a normally subordinate bird pecks a dominant one, who retreats. The peck is highly ritualized behavior; seldom is contact made or damage rendered. The peck may be a "mere gesture," a "stare," or the "head-raising posture."[461]

Sabine observed that an individual from one flock would occasionally visit another flock and did not meet with undue hostility as an intruder. Working with a captive flock of juncos, Balph discovered that when she introduced a foreign junco into the flock it was usually subordinate to the resident members. The foreign junco also had unusually intense hostile encounters, and "the most dramatic encounters between foreign juncos and residents typically occurred within minutes of the newcomers' introduction."[462] Balph has described hostile winter displays of juncos. In these displays contrastingly colored areas of the body are emphasized—the dark hood and white belly in the "head dance" and the black-and-white tail in the "tail up display."[463]

Longspurs

Longspurs occur in the Great Basin during migration and as winter visitants. They usually move about and forage with flocks of Horned Larks. Longspurs within Horned Lark flocks can often be heard before they are seen because of their distinctive call notes. Proven sites at which to look for all three of the species which occur in the Great Basin are the pastures and short-grass fields frequented by Horned Larks at Honey Lake, California.

The breeding range of the McCown's Longspur, *Calcarius mccownii*, lies on the short-grass plains of southern Alberta, Saskatchewan, and Manitoba southward into adjacent United States. Prior to 1973 this species' scientific name was *Rhynchophanes mccownii*. McCown's Longspurs have been recorded in the eastern part of the Great Basin in the Logan, Ogden, and Cedar City areas[27] and in the western part at Honey Lake.[455] This longspur may approach the northeastern corner of the Basin in its breeding. Burleigh stated that it was "apparently a rare summer resident in the extreme southeastern part" of Idaho.[415]

The Lapland Longspur, *Calcarius lapponicus*, has a circumpolar breeding range on the tundra of North America and Eurasia. The Lapland Longspur is a rare to uncommon migrant and winter visitant in the Great Basin; it has been recorded from a number of sites. In the 1860s Ridgway found it to be quite common in west central Nevada. He wrote, "During the more severe portion of the winter, individuals of this species were frequently detected among the large flocks of Horned Larks (*Eremophila alpestris*) around Carson City."[25]

The breeding range of the Chestnut-collared Longspur, *Calcarius ornatus*, lies on the short-grass prairies of southern Alberta and Manitoba southward into adjacent west central United States. This longspur is an occasional to rare winter visitant in the Great Basin. There is a specimen record from Tooele County[27] and a sight record from Antelope Island in Utah.[418] The Chestnut-collared is of rare occurrence at Honey Lake.[455]

Snow Bunting
Plectrophenax nivalis

The Snow Bunting has a circumpolar breeding range on the arctic tundra of North America and Eurasia—it breeds farther north than any other land bird. In North America the Snow Bunting winters in southern Canada and northern United States. It often associates with flocks of Horned Larks and longspurs. The Snow Bunting is an occasional to uncommon winter resident in the Great Basin, where it is more frequently encountered in northern Utah.[27] There are only two site records for Nevada: from Diamond Valley in northeastern Nevada[464] and from Sheldon National Wildlife Refuge.[465] This species has been recorded from Honey Lake in California[93] and from Malheur National Wildlife Refuge.[61]

SUBFAMILY ICTERINAE

The Subfamily Icterinae consists of ninety some species confined to the three Americas; its center of diversity is in the tropics. The North American species include blackbirds, grackles, orioles, cowbirds, meadowlarks, and the Bobolink. Icterines are small to medium-size birds. They are often black birds with a metallic surface sheen or with boldly contrasting areas of yellow, orange, red, chestnut, buff, or brown. Their wings are long and pointed, their tails often rounded, and their feet and legs stout. Icterine voices are often harsh ones, although some birds have clear whistles. They occur in a wide variety of habitats—including forests, woodlands, shrublands, grasslands, and marshlands. Many icterines are arboreal, some are terrestrial. On the ground they usually walk or run but don't hop. Many icterines are highly gregarious, and some are colonial nesters. Some species are polygynous, and the cowbirds are brood parasites.

Bobolink
Dolichonyx oryzivorus

The breeding range of the Bobolink spans most of northern United States and parts of southern Canada. The Bobolink is an occasional to uncommon summer resident in a few locales in the northern part of the Great Basin: it occurs in the northeastern part in Utah and northeastern Nevada and in the northwestern part in southeastern Oregon. It occurred in the past, and possibly can still be found, in the Eagleville area in northeastern California. Stragglers have been seen or collected at other Great Basin locales—such as the Toiyabe Mountains, Quinn River Crossing, and Dyer in Nevada and at Mono Lake, California.

The Bobolink's fortunes have ebbed and flowed over the years. With settlement of the eastern and midwestern parts of the country, the Bobolink prospered in agricultural regions, with heavy nesting occurring in long-grass hay fields. As mowing and raking machinery replaced hand-operated haying and as tractors replaced horses, nesting mortality increased, and breeding habitat decreased in quantity and quality. For many years Bobolinks were slaughtered in the rice fields of the South and for the market. The golden days of the Bobolink came to an end in midwestern and eastern United States in the early decades of the twentieth century.

Following the spread of irrigation and agriculture, the Bobolink extended its breeding range into the West. In describing the Great Basin in the 1860s Ridgway wrote, "The Bob-o-link seems to be spreading over all districts of the 'Far West' wherever the cultivation of the cereals has extended. We found it common in August in the wheat-fields at the Overland Ranche in Ruby Valley, and we were informed at Salt Lake City that it was a common species on the meadows of that section of the country in May, and again in the latter part of summer, when the grain ripened. We did not meet with it in summer, however, and doubt whether it breeds anywhere in the Interior south of the 40th parallel." In a footnote Ridgway went on to say that "according to Mr. Henshaw . . . the Bobolink apparently breeds at Provo, Utah, parent birds having been noticed feeding their young, July 25th."[25]

Red-winged Blackbird
Agelaius phoeniceus

The breeding range of the Red-winged Blackbird extends from northern North America southward across Canada and the United States into Central America. The Redwing is a common and abundant breeding species and migrant in the Great Basin. It is an occasional to common winter resident here. Its Paiute name was Pah-cool'-up-at'-su-que.[25]

The Red-winged Blackbird is the most familiar marshbird in the Great Basin. It nests in the valleys at wet meadows and in emergent vegetation in marshes, on the borders of lakes and ponds, and along irrigation ditches. In the mountains it nests at lower altitudes in meadows and in willow thickets along streams.

Red-winged Blackbirds are gregarious the year around, even nesting in colonies. Flocks of male blackbirds arrive on the breeding grounds in late winter in advance of the females, and soon thereafter instances of territorial behavior may be seen. In fact, during migration the reproductive urge is already stirring in some males, who show flashes of territorial behavior along the way. A male tends to return to the same or almost the same territory each year. Each year newcomers attempt to carve out territories within the colony, although they usually don't attempt to breed until their second year of life. Male Redwings are usually polygynous; harem sizes of two to nine females have been reported.

Much of our knowledge of Red-winged Blackbird behavior had its inception in pioneering field studies in Wisconsin by Nero. Males spend much time singing from perches on their territories while executing the song-spread display: in the full display the tail and wings are spread, the body feathers are erected, and the colorful shoulder epaulets are flared out. The song of the male is the familiar, often heard *conqueree* or *ko-klareeee* or *oak-a-lee* or *konk-la-ree*—an eagerly awaited sound of late winter or early spring. Redwings sing outside of the breeding season; bits of song may be heard the year around, wherever blackbirds have congregated to loaf or roost. The song is highly variable but always recognizable. Saunders wrote, "The song of the red-wing, well known to bird lovers as *conqueree*, is actually much more variable than this simple rendition. It generally consists of from 1 to 6 short notes, followed by a somewhat longer trill. The quality is pleasing, and the presence of prominent liquid and explosive consonant sounds give it a gurgling sound."[466]

Males execute song flights over their territories. They may do this when leaving their territories or when returning to them. In a song flight they fly on slow, deep wingbeats with erected epaulets. Then while gliding they utter a "long, rapid series of notes something like: 'tseeee···tch-tch-tch-tch···chee-chee-chee-chee···'"[467]

Another type of flight display has been described as taking place when territories are being established and females are returning. This is the fluttering flight, during which the male flies with shallow, rapid wingbeats.[468] Males engage in bill-up or bill-tilting displays during boundary encounters. When a territory is invaded, supplanting flights and occasionally actual fighting occur.

If a male wishes to join an established colony as a breeding bird, he first needs a territory. Sometimes the newcomer will challenge an established male in an attempt to usurp his territory. The interloper will enter an established territory and circle low around it, alternately flapping and gliding. The resident blackbird will promptly fly toward the intruder in an attempt to evict him. The interloper will take evasive action by retreating or dodging. If a determined challenge ensues, the birds circle about, climbing higher, and diving and pecking at each other, with the resident male endeavoring to stay above the intruder in the superior attack position. After they have drifted away from the disputed territory, one or the other will break away. However, the interloper may persistently return to the territory to challenge the resident male until the issue has been resolved. Sometimes the resident bird is eventually displaced.[469]

Flocks of female Red-winged Blackbirds start arriving back at the breeding colonies sometime after the males have returned. These females attract males, who approach them to display or pursue them when they fly. When a female enters a breeding colony, males perform song-spread displays. If a female alights on a male's territory, he will approach her closely to perform the song spread, and after he finishes singing he may continue to display. This after-song display may be "accompanied by a soft-whimpering 'ti-ti-ti-ti-. . .'"[467]

Male Redwings may engage in symbolic-nesting or nest-site demonstrations. This behavior often begins with the male crouched close to the female as he performs a song spread. Then he slowly flies away from her into the surrounding vegetation. Upon landing, "he commonly utters a low, harsh buzzing 'hahh...' or 'shhh,' the 'growl,' a call often given in threat."[467] Then he usually forces his way through the vegetation with his wings still elevated, often followed by the female, as he bows and pecks at nesting material. The display ordinarily ends here, but on several occasions Nero saw the male engaging in actual nest-building movements at an unfinished nest or in a clump of cattails on his territory.

Sexual chasing or pursuit flights are common behavior on the part of paired blackbirds, with the male chasing the female. Often the female flees her mate's territory during the chase. Usually the chase is terminated when the male breaks away. Sometimes the male overtakes the female, and he may hit her or catch her by her rump feathers. Sometimes other males join in chasing the female. During these chases male Redwings may sing and even attempt to perform a spread display while in flight.

One of the classical studies of territorial behavior was carried out on Red-winged Blackbirds by Nero and Emlen. This was a highly imaginative field study which involved ingenious experimental manipulations. Several nests containing eggs or nestlings were moved within the territory of the female's mate or from the territory of her mate into the territory of another male. These experiments showed that males "recognized sharp and stable territorial boundaries which could be defined within a few feet." Males defended these boundaries against "the intrusion of alien males, females or fledglings except when the latter were quiet," but they did not attempt to extend their territorial boundaries to defend their mates or nests moved into neighboring territories. "Males eventually accepted alien females which persistently invaded their territories to reach transported nests." The experiments also showed that females defended the vicinity of their nests against trespass by other females. Females "freely followed their nests as they were moved experimentally through the territory of the mate, but assumed a subordinate attitude and followed with difficulty when their nests were transported across territory boundaries."[470]

Both Red-winged and Yellow-headed blackbirds carry on continual warfare with the Marsh Wren, with whom they share the marshes. Marsh Wrens destroy eggs and kill young blackbirds and are a major mortality factor in blackbird reproduction. Female Redwings nesting at a distance from nesting Marsh Wrens have greater success than females nesting close to wrens: in a Canadian study, only 16 percent of the blackbirds nesting within six to ten meters of wren nests were successful, whereas 73 percent of the Redwing nests located between twenty-one and thirty-five meters from wren nests were successful.[471] Blackbirds are much larger than wrens, and both female and male Redwings chase Marsh Wrens when they encounter them. The territories of male Redwings are small, and the females in a harem usually nest close to each other, thereby gaining mutual protection against marauding wrens. Isolated females appear to have limited reproductive success. Marsh Wrens are also adversely affected by blackbird-wren interactions: wrens nesting close to Redwings experience a higher than normal failure rate.[472]

Tricolored Blackbird
Agelaius tricolor

The Tricolored Blackbird is principally a Californian. It ranges from the Klamath marshes of southern Oregon, southward through California—mainly west of the Sierra Nevada—into Baja California. The range of the Tricolored merely brushes the western edge of the Great Basin. Nesting by this blackbird has occurred in the Basin at Honey Lake.[117] There is a limited but regular migratory flow in the spring through Truckee Meadows.

Western Meadowlark
Sturnella neglecta

The breeding range of the Western Meadowlark extends from southwestern Canada, through western United States, into Mexico. The Western Meadowlark is a common and widespread permanent resident in the Great

Basin. During the winter meadowlarks are less numerous and widespread here than during the rest of the year. The Paiute name of this bird was Pah'-at-se'-tone.[25]

The Western Meadowlark is a grass-inhabiting species, living in stands of short grass, or in sagebrush-grass shrublands, or in meadows or alfalfa fields. It is a ground-inhabiting species and builds its nest, forages, and roosts there. Male meadowlarks even do some of their singing from the ground.

There are two species of meadowlarks in North America—the Eastern Meadowlark, *S. magna*, and the Western Meadowlark. To the eye these two look much alike, but to the ear they sound entirely different: the Western has a finer voice. The breeding ranges of the Western and Eastern meadowlarks overlap in midwestern United States and southern Ontario. With the clearing of the forests, the Western Meadowlark has extended its range eastward, through the Great Lake states, during the twentieth century.

Male meadowlarks are territorial during the breeding season, and they tend to return to the same nesting territories year after year. Males defend their territories against trespass by other males and transient females. In regions where their breeding ranges overlap, male Western and Eastern meadowlarks defend their territories against trespass by each other. Females do not engage in territorial defense.

The territorial behavior of these two species was studied by Lanyon in Wisconsin, where the two occur on a common breeding ground. When an intruder enters a territory, the resident male flies at it to chase it away. If the intruder does not retreat, the resident male will closely approach it and display in front of it. The resident male repeatedly tilts his bill up and engages in tail and wing flashing. Each time the tail and wings are rapidly opened and closed, the displaying male may utter his *chupp* call note. Sometimes the display involves fluffing out the body feathers and spreading the tail feathers. Occasionally the males fight with much bill jabbing, wing thrashing, and clawing.[473]

Singing and jump-flights are involved in boundary disputes between neighboring males. Rival males often sing close to each other along territorial boundaries—from song perches or from the ground. In executing a jump-flight a displaying territorial male springs into the air and flies a few yards before dropping to the ground again on vertically held wings. When flying the wings are fluttered high above the back, the legs dangle, and the tail is elevated. Then, the rival male will execute a jump-flight to land in the vicinity of the first male. Once the two birds are close together again, the displays continue.[473]

Females also tend to return to the same territories each year. If the previous year's territory is unoccupied by a male, the female goes to an adjacent, occupied territory to pair-bond. Consequently, a female is frequently reunited with the same male for several years in a row. But that is because she is attached to his territory, not to him. Polygyny is highly developed in both species of meadowlarks; the majority of males probably possess two mates. On rare occasions a male may even have three mates.[473]

Observations by Lanyon indicate that pair bonding is a very rapid, almost instantaneous process in meadowlarks which may be accomplished in a few hours' time. There is no sexual dimorphism in meadowlarks, and females apparently do not announce their sex by distinctive call notes. Lanyon concluded that a territorial male recognized an intruding bird as a female by her indifferent, passive behavior and her lack of aggressive posturing.[473]

As soon as the territorial male recognizes the intruder as a potential mate, he follows her about, singing off and on and displaying by expansion posturing—by fluffing out his body feathers and spreading his tail. The female entices the male into stimulating pursuit flight, often within minutes of entering his territory. She flies up and leads the male about, sometimes for several minutes and sometimes beyond the boundaries of his territory, before they return to the ground to loaf or display.[473]

Once the pair bond is formed, the birds remain close together until nesting begins. They feed, loaf, fly, and search for a nest site together. The female builds the nest and incubates the eggs by herself. The male may assist somewhat in feeding the nestlings, but he may be distracted from doing this if he is actively courting another female.[473]

The song of the male meadowlark has been well described by Saunders. In comparing the songs of the Western and Eastern meadowlarks, he wrote:

> The western meadowlark's song probably averages about the same in length, but contains more notes, and the notes are shorter and more rapidly repeated. The pitch is lower than that of the eastern species. Consonant sounds, both liquids like l and explosives like k or t, are much more frequent, occurring in practically every note of the song. Individual birds sing a great number of variations, and it is probable that the variation in this species is as great as in the eastern bird. Finally, the quality of the song is richer and fuller, resembling that of thrushes or the Baltimore Oriole.[466]

The male Western Meadowlark is said to be a discontinuous singer, in that there are pauses of up to eight seconds or so separating each bout of about two seconds of singing. Each male has five to twelve or so stereotyped song types; he will sing one type several times before changing to a different song.[474]

In addition to singing from song perches or from the ground, male meadowlarks engage in song flights. Chapman described this as follows:

> The flight song was uttered almost as frequently as the perch song. It was always preceded by a mellow, whistled *wheu*, repeated four or five times at increasingly shorter intervals, until it seemed to force the bird into the air to give freer utterance to a hurried, ecstatic, twittering, jumbled warble, as it mounted on fluttering wings to a height of twenty to forty feet, described an arc and sought a new perch.[34]

Western Meadowlarks occasionally sing at night. These nocturnal performances are usually short, consisting of only one or several bouts of song. It has been suggested that this song reflects occasional awakenings by males at night. Sometimes nocturnal song ends suddenly, after only two or three notes have been uttered. Another delightful aspect to meadowlark song is that there is a second concert season following a late summer cessation in song. This autumnal song occurs after the postnuptial feather molt.

Lanyon's observations indicated that primary (territorial) song was learned rather than inherited behavior and that the critical time for learning song was the two-month period following the third to fourth week of life. If deprived of this critical learning period by being isolated from its kind, the young meadowlark never acquired its species' song. Lanyon's observations indicated that the call notes were probably inherited behavior. But, if they were learned behavior, the learning took place during the nestling period.[473]

Meadowlarks feed primarily on insects and seeds. They pick their food off the ground, turn over clods of dirt, and drill into sandy or loose soil. In California meadowlarks have caused serious crop damage by drilling into the ground next to sprouting grain and pulling off or crushing the kernels of grain. Western Meadowlarks have performed notable service in the Great Basin by their predation on the Mormon cricket. LaRivers wrote, "This species is by far the ablest avian predator of the Mormon Cricket, for it specializes upon the eggs of the pest. Meadowlarks have been reported at various times as destroying entire, vast cricket egg-beds, and I have, on many occasions, seen them hard at work in such egg-beds, digging industriously for the palatable eggs, which are generally laid in clusters from a few to over fifty."[475]

Yellow-headed Blackbird
Xanthocephalus xanthocephalus

The breeding range of the Yellow-headed Blackbird lies in western United States and Canada. The Yellow-headed Blackbird is a common transient and breeding species in the Great Basin. A few overwinter here.

Yellowheads are colonial nesters. Their preferred nesting habitat is in tall, dense, emergent vegetation such as bulrushes, cattails, or phragmites, over standing water several feet in depth. Colonies of Red-winged Blackbirds often nest in the same marshes as Yellow-headed Blackbirds, but always at separate locations. Yellowheads are bigger and more aggressive than Redwings and preempt the best nesting habitats for themselves. Male Yellowheads sometimes return in the spring after male Redwings have already established territories in a marsh. In western Nevada they have been known to then supplant the Redwings at that marsh.[476] Males return in the spring in advance of females.

The territorial behavior of the male Yellowhead involves displays, songs, and fights. The song spread which accompanies singing is the most common display. Song-spread postures are accompanied by two major types of song. In the symmetrical song-spread posture, the wings are spread, the tail spread and lowered, and the head extended forward and upward; in the most extreme expression of this posture, the wings are held arched above the level of the back. In the asymmetrical song-spread posture, the wings are only slightly spread, with the right wing held out alone or at least further out than the left wing; and the head and neck are turned far to the left, so when the bird sings he sings over his left shoulder.[477]

Symmetrical song spread is accompanied by the male's accenting song, a song with "clearly defined syllables."[477] This song has a duration of one and a half to two seconds. It consists of "several liquid introductory notes followed by a short and highly variable trill which may even be completely missing."[468]

The asymmetrical song spread is accompanied by the buzzing song. The buzzing song is "much longer because the introductory notes are separated from the trill by a pause and because the trill is very prolonged." This song has a duration of up to four seconds.[468]

Some observers recognize a type of song between accenting song and buzzing song. This intermediate type accompanies the symmetrical song spread. It has a trill, and "in general, the longer the trill the more the song resembles the Buzzing Song and the more the bird turns its head to the left while singing."[468]

The two major song types are usually sung in different contexts. The accenting song is frequently sung when another male flies over the male's territory or when a new female appears, whereas the buzzing song is usually sung during boundary disputes between territorial males.[468]

The behavioral repertoire of territorial males includes two flight displays. The male may fly very slowly over his territory on deep wing strokes, with his tail spread and depressed and his body feathers ruffled. This flight display is employed when another male flies across his territory, when a female enters his territory, or when another territorial male is displaying nearby. During boundary disputes, a male may make a short bill-up flight along the boundary of his territory: he flies in a jerky, undulatory pattern with his head and bill pointed upward.[468] During boundary clashes males may also move their heads downward in a nodding or bill-down display or elevate their tails vertically in a tail-lifting display.

Yellow-headed Blackbirds are polygynous; a male typically has a harem of two to five females. Females do not help their mates defend territories, but each female does defend the area around her nest against trespass by all other females. Hostile encounters between females are frequent. Females often assume an asymmetrical song spread and direct their simple song at other females intruding on their nest sites. They typically turn their heads to the left and sing over their left shoulders. Actual fighting may follow. Females direct bill-up flights against other females; they also perform these flights when leaving the nest after carrying material to it or just before copulation. In these flights the female flies on rapid, shallow wingbeats, with legs dangling and bill pointed upward. Tail lifting also occurs in encounters between females.

Courtship interactions between males and females involve symbolic-nesting or nest-site demonstrations and pursuit flights. The symbolic-nesting display of the male is directed at one of his mates or at a newly arrived female. The "male raises his wings over his back (Elevated-wings), flies slowly and awkwardly for some distance (Flapping-flight), then suddenly Drops down into the vegetation where, still holding his wings upright, he Bows and then awkwardly Crawls through the vegetation with tail down and spread, often stopping to Bow and Peck."[477] The female usually follows the male as he crawls. Sometimes this display leads into precopulatory behavior.

Female Yellow-headed Blackbirds build their nests in emergent vegetation growing out of water several feet deep. The bulky nests are constructed of wet aquatic vegetation, which is wrapped and woven around the stems of cattails and tules, about one to three feet above the surface of the water. As the wet building material dries it tightens, forming a strong, compact nest. Often a female will build several nests before getting one that is constructed and positioned to her satisfaction.

Yellow-headed Blackbirds exclude not only Red-winged Blackbirds from their nesting colonies but also Marsh Wrens. They chase adult wrens, and there is some evidence that they crush the nests and kill the young of wrens found nesting nearby. It is not known for sure why blackbirds actively exclude wrens from their nesting colonies. But wrens do have a nasty habit of puncturing the eggs of other birds, and they possibly compete with blackbirds for insect food such as damselflies.

Rusty Blackbird
Euphagus carolinus

The Rusty Blackbird has a breeding range far to the north in Alaska and Canada, and it overwinters in eastern United States. It is of occasional occurrence in the Great Basin. There are several sight records from Utah, as well as a specimen collected in Tooele County.[27] We acquired a sight record for Reno on an Audubon Christmas Bird Count, and I observed an individual in a flock of Brewer's Blackbirds on the University of Nevada campus in Reno during the fall of 1983.

Brewer's Blackbird
Euphagus cyanocephalus

The breeding range of the Brewer's Blackbird extends across the United States and southern Canada, from the West Coast to northern Indiana and Ontario in the Great Lakes region. There is southward and eastward migration away from much of the northern part of the breeding range—some blackbirds overwinter as far east as southeastern United States and as far south as Mexico. The Brewer's Blackbird is a common and abundant permanent resident in the Great Basin. Blackbirds nesting in montane meadows migrate down into the valleys to overwinter. The Brewer's has been expanding its breeding range during the twentieth century in the wake of settlement and the spread of forest removal, cultivation, irrigation, and cattle raising. Its eastward expansion has carried it about seven hundred miles from western Minnesota to Sudbury, Ontario.[478]

The Brewer's Blackbird is an inhabitant of open areas. It frequents shrublands, grasslands, meadows, city and town parks, residential districts, ranches, and the margins of lakes, ponds, and streams. It nests in small colonies. The nest may be located on the ground or up in sedges, shrubs, or trees. In the Great Basin this bird is often found nesting in sagebrush or pinyon pines or willows.

Outside of the breeding season, Brewer's Blackbirds remain together in flocks as they forage, loaf, or roost. Each flock member functions as an individual, even males and females who were mated to each other during the previous breeding season. Studies by Williams revealed that pair bonding gradually gets under way in the flocks in late winter and early spring. Often a blackbird will enter several trial pairing relationships with potential mates before actually pairing with one. Pairs from previous years may reunite. Once a pair bond is formed, the male will guard the female against close approach by other males. He will follow her about, walking within three to six feet of her and perching about eighteen inches from her.[479]

Pair formation involves some mutual displays. When perched the male can initiate mutual displays by performing a ruff-out display. He ruffs out his body feathers, especially those of the breast, neck, head, and upper tail coverts; his wings are partly open and pointing downward; and his tail is spread open and depressed. During this display he utters his *squeee* or *schl-r-r-r-up* call. The female may respond with a much less pronounced ruff-out display. At times, when perched, the male may dart or hop toward the female, precipitating a pursuit flight. Other males may join the chase.

One of the common displays is the head-up display. In this display the tail is pointed downward, the neck and head upward. The bill is in an almost vertical position. Body feathers are compressed, giving the bird a trim appearance.[479] A male will defend his female against the advances of another male by directing this display toward the intruder, who may respond with a head-up display of his own. The head-up posture can be maintained even when the birds are walking along. Females use the head-up display when defending nest sites against intrusion by other females or strange males.

Once a pairing bond is formed, the male guards his female constantly, even accompanying her when she searches for nesting material. Nesting colonies are located in a variety of sites, some wet, some dry. The female builds the nest and incubates the eggs by herself. Sometimes when the female is incubating eggs, the male may acquire another mate or mates. A male may practice polygyny one year and be monogamous the next; however, he may have a primary female with whom he remates each year. The male helps feed the nestlings and fledglings. If a polygynous male has nestlings present in two nests at the same time, he may divide his attention between the two. The nests of a polygynous male are sometimes not all located in a single circumscribed territory but are separated from each other by the nests of females of other males.

Brewer's Blackbirds forage on open ground for their food staples of insects and seeds. They may often be seen foraging among cattle in pastures, and during the winter they forage on lawns in Great Basin cities and towns. On the ground they walk or run, each walking step being accompanied by a forward nod of the head. A foraging flock may move over the ground in a rolling fashion reminiscent of Pinyon Jays or European Starlings—blackbirds at the rear are continually lifting off and flying over the rest of the birds to alight ahead of the moving flock.

Great-tailed Grackle
Quiscalus mexicanus

In recent years this grackle has experienced changes in name and in range. In 1973 the Boat-tailed Grackle, *Cassidix mexicanus,* was divided into two species. The subspecies inhabiting the Atlantic and Gulf Coast regions of the United States were recognized as constituting the Boat-tailed Grackle, *C. major.* The subspecies ranging from southwestern United States into South America were recognized as the Great-tailed Grackle, *C. mexicanus.* Then in 1976 the genus *Cassidix* was merged in the genus *Quiscalus.*

During recent years the Great-tailed Grackle has been extending its range westward and northward from Arizona and southern Texas. In 1964 grackles reached southern California, and by 1969 they were nesting in California.[291] The Great-tail entered southern Nevada in the 1970s and was nesting there by the mid 1970s. Its breeding range has been steadily advancing northward, across Clark County and into Lincoln and Nye counties, toward the southern boundary of the Great Basin in Nevada. By 1978 this grackle was detected in the Great Basin, and it was recorded at the Ruby Lake National Wildlife Refuge in May.[480] In 1980 several grackles were seen at Washington, just south of the Basin in Utah.[401] By 1981 the Great-tail had been seen in Ruby Valley and at Sunnyside in the Nevada part of the Basin and at Bicknell, just to the east of the Basin in central Utah.[481] There is a 1980 record from Malheur National Wildlife Refuge in the northwestern corner of the Basin.[456] Watch for Great-tailed Grackles when you are afield. If the northward spread of this species continues, it may become established as a breeding member of the Great Basin avifauna.

Common Grackle
Quiscalus quiscula

The Common Grackle is an eastern species—its breeding range lies east of the Rocky Mountains in the United States and Canada. This grackle is of rare occurrence in the Great Basin, with most of the records being from the northeastern end in Utah.[27] There are several Nevada records: the Common Grackle has been seen at Ruby Lake and at Lund, and a dead specimen was found near Fallon. It has also been recorded at Malheur National Wildlife Refuge.

Brown-headed Cowbird
Molothrus ater

The breeding range of the Brown-headed Cowbird extends across southern Canada and southward throughout the United States into Mexico. This species is a common summer resident and breeding bird in the valleys and at lower elevations in the mountains of the Great Basin. A few overwinter here.

Back in the 1860s Ridgway found the Brown-headed Cowbird to be a very rare bird in the Great Basin; he wrote, "We found this species to be so rare in the country traversed by the expedition that the list of specimens given below comprises every individual seen during the whole time."[25] The list shows that Ridgway saw only one adult and two juvenile cowbirds and found only two cowbird eggs during the years he spent in the Great Basin. One of the cowbird eggs was found in a Fox Sparrow's nest, the other in a Common Yellowthroat's nest.

Ornithologists agree that the Brown-headed Cowbird was originally a rare species in North America which greatly prospered and increased its range and numbers with the settlement of the continent. In describing the spread of the Brownhead, Bent wrote:

> It deserves the common name cowbird and its former name, buffalo-bird, for its well-known attachment to these domestic and wild cattle. The species is supposed to have been derived from South American ancestors, to have entered North America through Mexico, to have spread through the Central Prairies and Plains with the roving herds of wild cattle, and to have gradually extended its range eastward and westward to the coasts as the forest disappeared, the open lands became cultivated, and domestic cattle were introduced on suitable grazing lands.[466]

Although its English name merely alludes to the ungainly company with which this bird commonly associates, its genus name casts aspersions on the cowbird's character by labeling it a parasite or tramp or vagabond. *Molothrus* "is probably a misspelling of the Greek *molobros.*"[267] The Brown-headed Cowbird is the most notorious North American brood parasite. Its eggs have been found in the nests of well over two hundred other species of birds. When cowbirds occur in high density, they can seriously depress the reproductive success of susceptible host species. The evils of cowbird parasitism are probably most dramatically illustrated by the plight of the Kirtland's Warbler, *Dendroica kirtlandii*, an endangered species of wood-warbler nesting in Michigan. Commenting on the impact the cowbird has recently had on this endangered species, Mayfield wrote:

> The cowbird causes damage at every stage of the nesting process. First, it removes from the nest about as many of the host's eggs as it lays of its own. It usually accomplishes this unnoticed because the action is synchronized with the host's egg-laying, a time when nests are not ordinarily attended. Next, since the cowbird is usually larger than the host, its egg gets more than its share of the heat from the incubating host, thereby reducing the hatching success of the other eggs. Finally, cowbirds hatch two or three days ahead of the host young, and by virtue of their larger size and maturity, they trample nestlings of the host species. At the time of Kirtland's Warbler hatching, cowbird nestlings already in the nest weigh about five times as much as the warbler nestlings and are much stronger and more active.[482]

To lessen the impact of cowbirds on Kirtland's Warblers, they are removed from the warbler's principal nesting areas by trapping. According to Mayfield, the "success of cowbird control has been phenomenal. Parasitism of nests has been reduced to negligible levels, and the production of young has been higher than that reported for any other North American warbler."[482]

Although well over two hundred species are known to have been parasitized by the Brown-headed Cowbird, certain families or groups of birds are more heavily parasitized than others. Various species of warblers, vireos, finches, flycatchers, and thrushes are favorite targets. Raptorial species such as shrikes, pugnacious species such as kingbirds, species who produce precocial chicks such as Killdeers, hole nesters, and species whose nestlings or fledglings have either an unusual diet or an unusual way of feeding are seldom if ever successfully parasitized.[466]

Ornithologists have been intrigued by the question of how female cowbirds find and select host nests to parasitize. There is evidence that female cowbirds frequently locate nest sites by watching for the nest-building activities of potential hosts. It is possible that the sight of nest building may function to stimulate a female cowbird in such a way that her egg laying will be fairly well synchronized with that of her hosts. Once nest building has been detected, the cowbird may make inspection trips to the nest before it is completed and egg laying has commenced. Female cowbirds evidently do not rely entirely on finding all their host nests while in the building stage; they will actively search for completed nests as well.

Usually the cowbird lays her egg in a nest after the host female has commenced egg laying. Egg laying typically occurs early in the morning. The cowbird often removes one or several host eggs from the nest, either the day before or on the day she adds her own egg to the nest. Rarely, she removes the egg a day later. The cowbird usually lays only one egg in a given host nest, unless there is a high density of cowbirds or a scarcity of host nests. Since a wide variety of species laying eggs of various sizes and colors are parasitized, the eggs of the cowbird often stand out in sharp contrast to the rest of the eggs in the host nests.

The question of whether host species recognize the female Brown-headed Cowbird on sight as their natural enemy during the nesting season is an interesting one. There are only a few reports in the ornithological literature of potential host species attacking cowbirds. Red-winged Blackbirds have been seen doing this, as have Song Sparrows, American Robins, and a few others. In a Canadian study the reactions of fifteen potential host species to models of Brown-headed Cowbirds and three kinds of sparrows placed near their nests were observed. All the potential host species tested showed greater aggressiveness toward the cowbird than toward the sparrow models. Further, the intensity of a potential host species' aggressiveness toward cowbird models was generally greater in those species most heavily parasitized by cowbirds in nature.[483]

Back in 1961 Selander and La Rue reported a type of behavior in which a cowbird would lower its head and ruff its neck feathers in the presence of a bird of another species. This was considered to be an interspecific preening invitation, since the other bird would often then preen the displaying cowbird. Thus, the interspecific preening invitation was serving as an appeasement display or, at least, as a way to rechannel the other bird's aggressiveness into a peaceful outlet.[484] Subsequently, Lowther and Rothstein observed that cowbirds directed the interspecific preening invitation or head-down display toward other cowbirds as well as toward birds of other species. Since the display often occurred after an adult cowbird approached or even chased the recipient, it was judged to be more of a threat than an appeasement display.[485] However, the Canadian study using cowbird models showed that potential host species, in general, were "less aggressive towards cowbirds in the bowed posture than towards those in a normal posture."[483] Of course, it is impossible to know exactly both how the recipient bird interprets the head-down or interspecific preening invitation display and whether the recipient's aggressiveness is rechanneled because of fear (from a threat) or appeasement. The head-down display evidently develops early in life—two cowbirds of an estimated age of less than one month have been observed doing it.[485]

It has long been apparent that some species are more tolerant of cowbird parasitism than others. Some birds do not passively accept a cowbird egg in their nest: they eject the egg, or break it, or bury it under nesting material, or abandon the nest. Rothstein has referred to the less tolerant species as rejecters and the more tolerant species as accepters. He carried out experiments in which either artificial or real cowbird eggs were placed in the nests of thirty-one other species. Nine of the host species, with rejection rates of 88 to 100 percent, were classified as rejecters. The remaining host species, with rejection rates of 0 to 42 percent, were classified as accepters. Among the nine rejecters were the Gray Catbird, American Robin, Cedar Waxwing, Western Kingbird, and Northern Oriole. Rothstein discovered that egg ejection by host birds was by far the most prevalent kind of rejection. In ejecting the cowbird eggs, birds usually carried them away from the nest before dumping them, instead of just dumping them over the rim.[486]

Brown-headed Cowbirds are usually encountered in flocks even during the egg-laying season, but individuals, pairs, and courting parties will be seen away from the flocks. Roberts described the antics of several males courting a female in the branches of a tree. Courtship involved "opening and arching the wings, spreading and dipping the tail, stretching the neck upward with the bill pointing skyward, all the time sidling about and finally ending with a profound bow, as though falling from the perch. All these movements are accompanied by a curious medley of clinking, gurgling, squeaking sounds which represent the nuptial song."[487] Sometimes less elaborate courtship takes place on the ground. Many people have studied cowbird behavior, yet there are alternative ways of interpreting their observations. A number of questions about cowbirds still lack unequivocal answers: are they truly territorial and, if so, to what degree; are pair bonds formed or is promiscuity practiced; and, if pair bonds are formed, are the females monogamous or polyandrous?

Following the egg-laying season, larger flocks of cowbirds assemble. Juvenile cowbirds leave their foster parents and seek the company of their own kind soon after they can fly. Cowbird flocks may join flocks of blackbirds and starlings to forage. Their foraging station is on the ground, and their food staple is weed seeds. Fruit and some insects and spiders are also taken. Cowbirds frequently feed in fields in close company with cattle. The movements of cattle may serve to flush grasshoppers and other insects before the bills of the cowbirds.

Northern Oriole
Icterus galbula

The breeding range of the Northern Oriole extends across the southern tip of Canada and southward through the United States into Mexico. The Northern Oriole was created in 1973 by lumping a western species, the Bullock's Oriole, *I. bullockii*, and an eastern species, the Baltimore Oriole, *I. galbula*. These two former species were judged to be conspecific since they interbreed freely in areas where their ranges overlap. The Bullock's form is a common summer resident and breeding bird in the Great Basin. The Baltimore form is of accidental occurrence here; it has been recorded from the eastern part of the Basin.[27]

The Northern Oriole inhabits the valleys and foothills of the Great Basin. It frequents areas where the deciduous trees grow tall along streams, around ranches, and in city parks and residential districts. Evidence of its nesting presence is often not detected until autumn. Then, the leaf fall exposes its deep, hanging, pouch-shaped nests to view—long after the orioles have departed to the south. Sometimes Northern Orioles will nest in the same trees as Western Kingbirds. They may be gaining some protection from nest predators from the nearby pugnacity of the kingbirds.

Among Northern Orioles, the nests of the western Bullock's Oriole are not as pensile as those of the eastern Baltimore Oriole. The pendent nests of the Baltimore Oriole are usually suspended with but their rims attached to the tip of a limb, whereas Bullock's Oriole nests are frequently attached to the twigs by their sides as well. There is evidence that Bullock's Orioles have greater resistance to heat stress than Baltimore Orioles. The insulative capacities of Northern Oriole nests may vary between climatic zones. Studies have shown that in the hottest part of their western range, in the Great Plains region, Baltimore Orioles and hybrid Baltimore-Bullock's Orioles build better-insulated nests than more eastern Baltimore Orioles or more western Bullock's Orioles.[488]

Female as well as male Northerns sing. In reference to this in Bullock's Orioles, Miller wrote:

> The utterances of female Bullock's Orioles while in defense of territory and in association with males in every way are comparable to the songs of males and may be considered as true territorial songs. The song of the female is similar to that of the male in rhythm, pitch, and quality except as regards the concluding notes of the song which in the female are slightly harsher in quality, range over lesser intervals of pitch and show important modifications of the rhythm as compared with those of

the male. Before or during nest building the songs of females on occasion may be even more abundant than the songs of the males.[489]

During the postjuvenal molt, young male Bullock's Orioles acquire "black lores and a narrowly black throat." There is considerable variation in the first prenuptial molt: in young males "new black feathers with olive edges appear on the crown; sometimes only half a dozen such are to be found; in others the crown is entirely covered."[466] Furthermore, "some of these yearling males apparently breed and others don't."[490] In recent years Hooded Orioles, *I. cucullatus*, have been reported as breeding birds in montane aspen stands in the northwestern edge of the Great Basin. If this is true, their breeding range and habits are changing. There is also the possibility that female and yearling male Bullock's Orioles are being misidentified.

Scott's Oriole
Icterus parisorum

The breeding range of the Scott's Oriole lies to the south of the Great Basin. However, there are spring and summer occurrence records for this oriole in the Great Basin from northern Utah[27] and western Nevada.[91] Nesting has even been reported from the Topaz Mountain area in Utah[56] and from near Unionville in Nevada.[491] Although there are more recent than early records of occurrence in the Basin, it is impossible to say whether this reflects an actual increase in numbers of orioles here or just better field coverage by ornithologists.

FAMILY FRINGILLIDAE

As we have already seen, the Family Fringillidae loses two subfamilies to the newly recognized Family Emberizidae in the sixth edition of the AOU checklist. Fringillidae now consists of three subfamilies. The Subfamily Drepanidinae, containing the Hawaiian honeycreepers, is not represented in North America. The Subfamily Fringillinae, containing the Old World fringilline finches, is represented by two species of accidental occurrence in North America—one of which, the Brambling, has been recorded from Sutcliffe at Pyramid Lake. All of our native North American fringillids are cardueline finches, belonging to the Subfamily Carduelinae. The carduelines present in the Great Basin include finches, goldfinches, redpolls, crossbills, and siskins.

The Subfamily Carduelinae evidently arose in Eurasia and subsequently spread over much of the world. Next to Eurasia its greatest centers of species diversity are in Africa and North America. Carduelines are small to medium-size birds. Some are streaked or mottled, others are solidly colored. The males' nuptial plumages are often red or yellow. Carduelines have stout, conical, seed-eating bills. Two species have crossed bills. Although they often visit weed beds in search of seeds, these finches are arboreal birds, mainly frequenting wooded and forested country. They are highly gregarious and outside of the breeding season live in tight flocks; some species even nest in loose colonies. Their flight is undulatory, and they often call while in flight.

Rosy Finch
Leucosticte arctoa

In the fifth edition of the AOU checklist, three North American species of rosy finches were recognized. Two of these three, the Gray-crowned Rosy Finch, *L. tephrocotis*, and the Black Rosy Finch, *L. atrata*, occur in the Great Basin. Since North American rosy finches hybridize in areas of sympatry, they are no longer recognized as separate species. Further, they are now considered to be conspecific with the Rosy Finch, *L. arctoa*, of northeastern Asia. Therefore, in the sixth edition of the checklist they are classified as subspecies of the Rosy Finch.

The Rosy Finch has a circumpolar breeding range in Asia and North America. In North America it extends far southward into mountainous western United States. The Gray-crowned Rosy Finch is the form that nests in the western edge and rim of the Great Basin—in the Carson Range, White Mountains, Sweetwater Mountains, and Sierra Nevada.[492] The Black Rosy Finch is the breeding form in the eastern part of the Basin—in the Ruby Mountains and Snake Range in Nevada[492] and in the Raft River and Wasatch ranges in Utah.[27] Interestingly, the Black Rosy Finch is the breeding form on Steens Mountain, Oregon, in the northwest corner of the Basin.[492] During the winter both Gray-crowned and Black rosy finches can be found at valley and foothill sites throughout the Great Basin.

Rosy Finches breed on top of high mountains, above the upper tree line in the alpine zone or, as it is sometimes called, the arctic-alpine life zone. The behavior of this finch is highly unusual in that males do not defend static real estate. Instead a male accompanies his mate wherever she goes and defends her person against close approach by other males. Thus, the "territory" or defended area moves with the female. This close defense of pair-bonded females by their mates is associated with strongly imbalanced sex ratios in favor of males in both breeding and wintering populations of Rosy Finches. Young finches as well as adults show sex ratio imbalances.

Ratios as high as six to fourteen males to one female have been reported in winter flocks.[493]

On the breeding grounds, agonistic interactions between males are frequent and vigorous—probably accentuated by the imbalanced sex ratios. When females are building nests, males move along the nesting cliff, chirping loudly and engaging in chases. Pair-bonded males must be ever vigilant: any female, mated or unmated, approaching the nesting cliff is fair game for unattached males. Fighting occurs between males, but only when a female is present.[494] When accompanying his mate, a male escorts her from stop to stop, constantly alert, continually chirping. When at rest near his mate, he may assume a threat posture toward other males—he fluffs out his feathers, lowers his head, and opens his bill.[493] His tight defense continues until his mate begins incubating eggs.

French has observed courtship displays being enacted on the ground and on cliffsides. While displaying the male assumes a position with the forepart of his body close to the ground, head back, bill up, and tail straight up. The wings are spread slightly and rapidly vibrated, while the male engages in chirping. Sometimes grass is held in the bill during the performance.[493]

While on the breeding grounds, Rosy Finches forage for food mainly on and about snowbanks and on ground moistened by melting snow. Cliffside foraging also occurs. Much of the food of these finches consists of seeds, although insects can be a major food item. The seeds of sedges are a favorite food. Early in the season last year's seeds are gleaned from wet ground, but by midsummer the new crop is available. Finches process sedge seeds in their bills, removing and discarding the seed coats, before swallowing them. Early in the season they glean cold-paralyzed insects, which have been transported up the mountainside and dumped on the snow by strong upslope winds, off the snowfields and snowbanks. As the season progresses they glean insects from vegetation and chase flying insects such as mayflies, capturing them while in continuous flight as do swallows.[495]

Rosy Finches nest in crevices and holes in cliffsides where the nest is completely sheltered from the elements. The female builds the nest and incubates the eggs. She is escorted by her mate while collecting nesting material, and the male assists in caring for the nestlings. Several weeks after leaving the nest, young finches become independent of their parents. Then small groups of finches form which eventually coalesce into flocks. Altitudinal and geographic migrations occur. During the winter large flocks, often containing both Gray-crowned and Black rosy finches, appear at foothill and valley sites in the Great Basin. By day, flocks or groups from flocks move about, foraging for seeds. Agonistic encounters and chases, involving both sexes, occur between foraging birds. A flock of Rosy Finches does not forage at one spot for very long—the flock is continually lifting off and flying on to another foraging site. Rosy Finch flocks are much like Horned Lark flocks in this respect.

During the winter Rosy Finch flocks seek snug shelter from the weather and may employ the works of others in finding it. If a Cliff Swallow nesting colony is present on the wintering grounds, the finches may roost in the gourd-shaped mud nests of the swallows—this is of no consequence to the swallows, since they have already migrated southward for the winter. Rosy Finches also seek shelter in and around human-made structures. Behle found winter flocks containing both Black and Gray-crowned rosy finches in Salt Lake Valley at Bacchus. Some finches roosted at night on wires, but most sought shelter in sheds, on windowsills and doorsills, and on rafters.[496] Mine tunnels and shafts are used by roosting Rosy Finches. These are apparently their favorite roosting places in the western end of the Great Basin. When I am in the field during the winter months and pass a vertical mine shaft, I toss several rocks down it. As the falling rocks ricochet back and forth off the sides of the shaft, any finches within are flushed out. I have discovered that even during the day, when the flock is out foraging, some individuals may be present at the roost.

Pine Grosbeak
Pinicola enucleator

The breeding range of the Pine Grosbeak lies in the circumboreal coniferous forests of Eurasia and North America. In North America it extends far southward into mountainous western United States. The Pine Grosbeak is a rare to uncommon permanent resident in some of the higher mountain ranges in the Great Basin and in its mountainous western and eastern rims. In Utah records are "numerous from all the major mountain ranges and high plateaus."[56] In Nevada there are records from the Carson Range, West Humboldt Mountains, and Ruby Mountains and from Eureka. Fieldwork is needed to more clearly elucidate the distribution of this species.

Pine Grosbeaks frequent open montane coniferous forests—usually around streams or lakes or meadows. Their foraging station is mainly an arboreal one. Their foraging movements are very deliberate, and a flock may remain in the same trees for some time. Although Pine Grosbeaks do not perform regular migrations, some winter movements occur. Flights of grosbeaks originating farther north appear in northern United States; some years the flights are large enough to constitute minor invasions. In the Great Basin altitudinal movements occur, and small flocks drift down into the edges of valleys. During the winter and early spring Pine Grosbeaks have been seen along the Franktown Road at the eastern base of the Carson Range. In 1983, following extremely heavy snowfalls in the Carson Range, a flock of ten birds was still there in early May. Pine Grosbeaks are a birdwatcher's delight. Not only are they beautiful and out of the ordinary, but they are so very tame that they permit a close approach.

Cassin's Finch
Carpodacus cassinii

The breeding range of the Cassin's Finch extends from the southwestern tip of Canada, through mountainous western United States, into northern Baja California. The Cassin's is a common and widespread summer resident in the montane forests of the Great Basin. Altitudinal movements occur, and finches frequently appear at valley sites during the winter. However, some finches may remain well up in the mountains during the winter, as in the Carson Range and Sierra Nevada. Latitudinal migrations may occur and are apparently pronounced in the eastern end of the Basin. Behle and Perry report the Cassin's Finch as being a "common summer resident showing altitudinal migration; uncommon in winter" in Utah.[27]

The nesting habitat of the Cassin's is open, often dry coniferous forests, including their aspen stands. Breeding finches reach their peak numbers in the cool Canadian life zone, where trees such as lodgepole pine, red fir, mountain hemlock, subalpine fir, Engelmann spruce, limber pine, Douglas fir, and quaking aspen are dominant species. Population densities may vary considerably from year to year in a given area, and the locations of nesting colonies may vary. There is some thought that, like crossbills, these finches shift their nesting localities within a region to exploit local food abundance.

Cassin's Finches are cold-hardy birds. Snow usually still covers much of the area when the birds arrive on their montane breeding grounds, and they are frequently buffeted by June snowstorms. Much of our knowledge of Cassin's Finches has resulted from studies carried out by Samson at Cache Valley and in and around the Beaver Mountains in the northeastern corner of the Great Basin. The sex ratios in Cassin's Finches are strongly imbalanced in favor of males. Young males do not acquire the reddish plumage of an adult male until they are fourteen months old. Hence, on the breeding grounds adult males can be readily distinguished from the grayish brown yearling males and from females of all ages. Yearling males, when in hand, can be distinguished from females by their greater wing length and their lack of an incubation patch during the breeding season.[497]

Samson's observations indicated that yearling males are usually unable to obtain mates, although they are physiologically and behaviorally ready to breed. When pair-bonded adult males disappeared or were experimentally removed, their places were often taken by yearling males. But with a shortage of females the aggressiveness of the adult males practically excludes yearling males from breeding. During the breeding season when not singing, the yearling males form flocks which wander over the breeding grounds, with individual birds leaving their flocks now and then to sing or forage.[497]

Male Cassin's Finches, like male Rosy Finches, do not defend static real estate—instead of establishing territories, males defend their mates from close approach by other males. This defense begins when the female starts selecting a nest site and ends about the time that incubation begins. A male accompanies his female while she selects a nest site, collects nesting material, builds a nest, and forages. That the male has no territorial attachment even to the nest site is shown when his mate is experimentally removed or disappears: he deserts the nest site; he does not defend the site or attempt to attract a new female to it. Once nest construction is under way, a pair-bonded male stops singing. After the female has built her nest, usually on a limb high up in a pine tree, she lays her eggs and does all the incubating. The male feeds the female at the nest and nestlings when present.[497]

At the end of the breeding season, small flocks consisting of one or several family groups are formed. These flocks may forage in the general vicinity of the nesting colonies until early fall. The family flocks may depart for the winter in advance of the yearling male flocks.[497] Samson studied various aspects of winter social behavior in Cache Valley. Although outnumbered by males, females are socially dominant over both young and old males in the winter flocks. This dominance allows them to displace males at choice feeding and roosting sites, and Samson's banding-recapture data suggest that fewer females than males die in wintertime. During agonistic interactions a head-forward display, executed at various intensities, constitutes a threat posture—the displaying finch directs its bill toward the threatened bird. A submissive display consists of a stiff-legged leaning away from the aggressor. Females are tolerant of each other. Fighting is rare and usually involves yearling males.[498]

House Finch
Carpodacus mexicanus

The breeding range of the House Finch lies in western North America, extending from the southwestern tip of Canada, through most of western United States, into southern Mexico. Through introduction the House

Finch has become established as a breeding bird over much of eastern United States and on the major islands of Hawaii. It is a common and widespread permanent resident in the Great Basin. To the Paiutes this bird was We'-to-wich.[25]

The introduction of the House Finch into eastern United States resulted from illegal traffic in cage birds. In 1940 wild finches were unlawfully trapped in southern California and sent to dealers in New York to be sold as "Hollywood finches." When the U.S. Fish and Wildlife Service alertly closed in on the dealers, they turned the House Finches loose to avoid prosecution. The finches, successful in breeding, have now spread over most of eastern United States.[425]

House Finches frequent a variety of habitats in the valleys and foothills of the Great Basin. They are particularly abundant along streams and around buildings in cities and towns. Finches stay within easy flying distance of water and drink frequently during hot weather. Migration does not occur, but foraging flocks do move around during the winter.

House Finches live in flocks outside of the breeding season. As the breeding season approaches, male finches become progressively more intolerant of each other. They move onto the breeding grounds and begin to sing. Chases occur, but males do not establish static territories with well-defined boundaries. Females now appear and begin to associate with the males. Male House Finches are much like male Rosy and Cassin's finches in that they become pair-bonded without the benefit of territories; once pair-bonded, they accompany their mates wherever they go and defend the area immediately around them. Unlike male Rosy and Cassin's finches, however, male House Finches "sporadically and weakly" defend the area immediately around the nest site, once that site is selected. During a study at Berkeley, Thompson calculated the defended area around the nest site to have an average radius of fourteen feet. Since the male accompanies his female as she moves about, foraging or gathering nesting material, his first priority is obviously to defend her, not the nest site. Once incubation is under way, the male's defense of his mate and the nest site wanes. Then a second pair of House Finches may be allowed to build a nest a few feet away.[499]

Thompson recognized four progressive stages in pair formation and courtship in captive House Finches. The first stage involves billing; one member of the pair leans toward the other and gently pecks its closed bill. During the second stage both birds lean toward one another; as they open their bills slightly, the male inserts his bill into the female's bill. During the third stage the male accompanies his mock feeding of the female with regurgitation movements of the throat, although no food is actually regurgitated. At this stage the female begs for food during the mock feeding. In the final stage, as nest building commences, the male actually engages in courtship feeding and feeds the female regurgitated food.[499]

The male House Finch is the most frequently heard songster in the valleys of the Great Basin. Outside of the breeding season, irregular singing occurs over most of the year, even on warm winter days. During the postnuptial molt, a major period of silence occurs. Thompson has described the song of the House Finch as "a rambling warble, ending, when given with full intensity during the breeding season, with a final *tzeep*. This end syllable is given with a rising inflection." Except when he is singing vigorously during the breeding season, the male's song may be weak and incomplete—lacking the terminal *tzeep*. Females occasionally sing the incomplete song. Males sing from perches or while in flight; lacking territorial song perches, sometimes several males sing from the same tree. Thompson described a type of song flight usually involving an unpaired male. The male sings while in butterfly flight—in which the wings are held rigidly and the downstroke is shallow. Still singing vigorously, the male glides the last twenty feet or so to a perch.[499]

House Finches eat very little animal food, even during the warmer part of the year when insects are readily available. Even young nestlings receive little in the way of animal food. This is quite unusual, since most seed-eating birds start their nestlings on insect food. House Finches consume all types of vegetable food, with a preference for seeds. In spring they feed on the leaf and flower buds of shrubs and trees. During the summer they have a fondness for fruit, including many types of cultivated fruit, and they may cause crop damage on occasions. Much of their foraging is on the open ground, where they obtain the seeds of low-growing plants, especially composites.

Red Crossbill
Loxia curvirostra

The Red Crossbill has a circumboreal range in the coniferous forests of Eurasia and North America. In the Old World it extends as far south as northern Africa. In North America it extends from southern Alaska, through Canada, into the mountains of eastern and western United States, and as far south as Baja California. The Red Crossbill is an uncommon permanent resident in the mountainous western and eastern rims of the Great Basin. It is also known from some of the Great Basin mountains and ranges, including the Carson, Virginia, Sweetwater, White, Pine Forest, Quinn Canyon, East Humboldt, Spruce, Schell Creek, Snake, and Deep

Creek.[25, 289, 500] The Red Crossbill is a vagrant and wanders far beyond its normal range, especially during the fall and winter. Sometimes it remains beyond its normal range to breed.

Conifer seeds are the crossbill's food staple. Pine seeds are favored, but fir, hemlock, and spruce seeds are also eaten. The unique bill from which this bird derived its name is well designed for prying ajar the scales on conifer cones; this exposes the seeds, which are on the inner surface of the scales. The tips of the bill cross—the tip of the lower mandible is displaced to one side and curves upward, and the tip of the upper mandible curves downward. Sinistral crossing, where the lower mandible passes to the left of the upper mandible, occurs about as often as dextral crossing, where the lower mandible passes to the right of the upper mandible. The jaw musculature is asymmetrically developed, being more massive on the side on which the lower mandible crosses the upper. Not only do crossbills have either a "left-handed" or a "right-handed" lower mandible and jaw musculature, but they pry cone scales aside in a corresponding fashion. The bird inserts its partially opened bill behind a cone scale. The tip of the lower mandible is then moved away from the stationary tip of the upper mandible and pressed against the central axis of the cone. The pressure of the lower mandible against the immovable central axis of the cone forces the tip of the upper mandible outward, against the inner surface of the cone scale. The tip of the cone scale swings outward, away from the central axis of the cone, and the seeds are removed with the tongue.[501]

The bill design of this species does make certain operations potentially awkward. If a bird wants to drink, it must position its open bill sideways so that its tongue may be employed in lapping up water. Crossbills have a fondness for sodium chloride and salty substances in general. When a bird eats salt-impregnated earth or salt off an icy highway, its open bill must once again be held sideways so that salt can be lapped up with the tongue. One fringe benefit which a crossbill derives from its bill is that it can be used in climbing through tree branches, much like parrots use their bills in climbing.[425]

The main foraging station of the Red Crossbill is on the cone-bearing branches of conifers. When extracting seeds from mature cones, crossbills may hang head-downward. On occasions tender green cones are eaten. Sometimes the buds and seeds of deciduous trees such as alders, willows, elms, poplars, and maples are harvested. Fallen seeds, salt, and water are sought on the ground. A limited amount of insect food is taken. Observations of successful nesting by crossbills in captivity, along with abundant evidence that crossbills often nest successfully during the winter in the wild, indicate that even young nestlings do not have to be fed insect food.[502]

Sinistral (above) and dextral (below) bill crossing in Red Crossbills

In the field the presence of a foraging flock of crossbills is often revealed by their chattering calls or by the restlessness of the flock as, for no apparent reason, it abandons one feeding spot to fly on to another.

Red Crossbills are not rigidly tied to calendar nesting. They may nest at any season of the year, even in the winter, although they nest most frequently in late winter and early spring. When afield on snowshoes I have found them nesting near Tahoe Meadows in the Carson Range when the ground was still covered with five to six feet of snow. Crossbills frequently abandon former nesting haunts and establish new ones. These temporal-spatial irregularities in nesting are the result of irregularities in the occurrence of the crossbills' food staple. In a given locality the conifer seed crop varies greatly in abundance from year to year. Crossbills wander in search of conifer seeds, and, finding an abundant supply, they may remain to nest.

In temperate zone birds the gonads are only seasonally functional, and nesting is thereby restricted to a circumscribed breeding season. In most north temperate zone species which have been studied, the principal *Zeitgeber* (time giver) which activates the gonads is the increasing photoperiod or day length of winter and spring. By subjecting captive birds to long photoperiods under artificial lights, researchers can evoke gonadal recrudescence by midwinter—far in advance of what is happening in nature under shorter photoperiods. Experimental studies by Tordoff and Dawson on captive male crossbills showed that exposure to various light-dark schedules resulted in some enlargements in testicular size. However, none of the males experienced complete recrudescence in gonadal size and function. Tordoff and Dawson concluded that any naturally occurring photoperiod schedule can only partially activate the testes, and they speculated that the *Zeitgeber* which presumably completely activates the testes is the availability of suitable food.[502] This speculation has long been commonly accepted because of the crossbills' temporal irregularities in breeding.

Pine Siskin
Carduelis pinus

The breeding range of the Pine Siskin embraces the coniferous forests of North America, extending from southern Alaska, across Canada, on into northern United States. In the West it extends all the way down into Mexico. The siskin is an uncommon to common breeding species in the montane conifer forests of the Great Basin. Prior to 1976 it had the more attractive scientific name of *Spinus pinus*.

Pine Siskins are gregarious throughout the year. They may nest in loose colonies, and breeding siskins may forage in small groups. After the breeding season ends, siskins move about in search of food; some may associate in foraging flocks with other birds, such as crossbills and goldfinches. Siskins move out of the mountains for the winter—in Utah large flocks appear in the foothills and valleys to feed on weed seeds.[56] In western Nevada siskin flocks are seldom seen during the winter, but in the spring large flocks can be found along the base of the mountains. Winter flocks often wander far beyond their regular range of occurrence in search of food. Their wanderings are especially pronounced in eastern United States.

The food staple of siskins is seeds—particularly conifer seeds. Weed seeds along with the seeds of deciduous trees such as alders and birches are also utilized. Buds and insects are taken as well. Pine Siskins have a bilevel foraging station—on the ground and in trees. Rodgers has described two different techniques of tree foraging. Most commonly a flock enters the crown of a tree and works its way downward to the lower branches. Then the birds fly off on a circular course, soon alighting in another tree crown to repeat the process. A second technique involves a flock foraging on a horizontal course through a sequence of trees.[503] Leapfrogging may occur, as the rear section of a flock flies over the leaders to a new foraging stop. Pine Siskins are quiet when foraging but may utter call notes as they take off or alight.

Lesser Goldfinch
Carduelis psaltria

The breeding range of the Lesser Goldfinch extends from southwestern Washington, through the southwestern quarter of the United States, as far south as northern South America. The Lesser is a rare to somewhat common permanent resident in the Great Basin, but in some regions it is less common in the winter than in the summer. It is apparently not a resident of southeastern Oregon. Although this goldfinch is fairly widespread in the Great Basin, it is certainly not abundant here. Prior to 1976, when the genus *Spinus* was merged in the genus *Carduelis*, its scientific name was *Spinus psaltria*.

The Lesser Goldfinch is mainly a valley bird. Occasional nesting occurs in mountains in the transition life zone at altitudes of up to 6,500 feet or so.[26] Foraging goldfinches wander to still higher altitudes, especially in the fall. Goldfinches nest on or close to open terrain, where weed patches and drinking water are readily available. Lesser Goldfinches are noted for displaying a somewhat polydipsic behavior. Their noticeable drinking habit is forced on them by a combination of factors. They live in hot, arid climates in the summer, are very small in size, and feed mainly on seeds. Weighing in at ten grams or so, goldfinches experience very high rates of evaporative water losses, and they obtain only limited amounts of water of succulence from their food. Lesser Goldfinches use water frequently not only for drinking in the summer but also for bathing. In addition to promoting cleanliness, bathing allows the birds to dissipate some excess body heat through the evaporation of free water in place of body water.

Much of our knowledge of the Lesser Goldfinch is due to studies made by Jean Linsdale at the Hastings Natural History Reservation in California. Linsdale is a name of importance in Great Basin ornithology, since he carried out many field studies here. He compiled the first complete annotated checklist of the birds of Nevada,[74, 91] studied magpies,[338] studied geographic variation in some Nevada birds,[504] studied modifications in the structure and behavior of bird life in Nevada,[505] studied environmental responses of vertebrates in the Great Basin,[506] and worked on a number of other topics.

Linsdale observed four types of male courtship behavior during the establishment of pairing bonds: courtship song, courtship flight, song flight, and a canarylike song. Courtship feeding of the female by the male was of importance in maintaining the pairing bond. Chases between males occurred—males pursued males with spread tails and rapidly beating wings; in shorter flights the males flew slowly, but with the wings moving more rapidly and widely than in normal flights. Males and females also engaged in pursuit flights. Circle flights by males were common: a male would fly out on a high circular course near a perched female, with his wings and tail spread, conspicuously flaunting his light-and-dark markings.[507]

The female does most of the nest building. Her mate may display some interest in the nest site and in the early stages of nest building; he often remains nearby singing. The female does all of the incubating and is fed while on the nest by her mate. After an incubation period of about twelve days, she remains at the nest for several more days to brood her nestlings. Then she too forages for food. In southern California autumnal nesting occurs regularly.

Male Lesser Goldfinches sing in flight as well as from song perches. Linsdale described their song as "a long disorganized series of faintly melodious notes rising and falling many times, but most often rising. Frequently interspersed throughout the long song were sharply rising slurred notes that gave it a quality characteristic of many finch songs."[425] Hoffmann wrote that when disturbed Lessers fly off uttering "a little shivering note like the jarring of a cracked piece of glass." Their other notes include "a plaintive *tee-yee*, both notes on the same pitch, a *tee ee*, the second note higher, and a single plaintive *tee*."[84]

Lesser Goldfinches live in flocks outside of the breeding season. Some birds remain together as pairs within the winter flocks. Sometimes Lesser Goldfinches associate in large mixed flocks with House Finches. Weed seeds form the bulk of their diet, and they are especially fond of the seeds of thistles and dandelions. Their foraging station is consequently positioned close to the ground.

Male Lesser Goldfinches come in two different colors. On the eastern part of their range, males are black-backed. On the western part of their range, including the Great Basin, males are green-backed.

American Goldfinch
Carduelis tristis

The breeding range of the American Goldfinch extends from southern Canada deep into the United States. In the Far West it extends into northern Baja California. The American Goldfinch is an uncommon to common permanent resident in the Great Basin—it is apparently more common in the eastern part than in the western part, where it is quite local in occurrence. Some migration occurs in the Basin. In northern Utah this goldfinch is less common in the winter than in the summer.[27] At some localities it is primarily a migrant or transient. In northeastern California it is more widespread in the winter than in the summer.[93] Prior to 1976 the name of this species was *Spinus tristis*.

The American Goldfinch is primarily an inhabitant of valleys and the lower ends of low-lying canyons. Nesting habitat is found along streams and around the presence of humans, where deciduous trees and shrubs afford nesting sites. Open areas must be situated nearby, where herbaceous plants, especially members of the composite family, flourish to provide a bountiful seed crop. Of all plants thistles are most highly coveted, not only for seeds but also for down to be used in lining nests.

Over most of its breeding range the American Goldfinch is usually the last bird to commence nesting. Nesting typically gets under way in July and August, after the prolonged prenuptial body molt is at an end and when composite seeds and thistle down are plentiful. The nest is built by the female with little or no help from her mate. It is such a tightly woven structure that it will temporarily hold water—unbrooded nestlings have drowned in the nest during rainstorms. The female incubates the eggs and broods the nestlings during the early part of their nest life. She is fed on the nest by her mate.

American Goldfinches live in flocks outside of the breeding season. Migration occurs from the northern part of their range. Where overwintering occurs, flocks wander in search of food. In Utah large winter flocks congregate where sunflower seeds are plentiful.[56] Much of the foraging occurs close to the ground. Goldfinches alight directly on seed heads or on the stems below to feed. Often they flutter their wings and spread or lower their tails to maintain balance when feeding.[508] By late winter and spring, foraging in trees where buds and seeds are available becomes more noticeable. Flying goldfinches are readily identified by their undulating flight and their flight calls of *per-chic-o-ree*. Singing begins in the spring, and flocks may sing in chorus from trees.[425] Flocking continues until nesting commences. Because of its bright yellow body when in nuptial plumage and its canarylike song, the male American Goldfinch is sometimes referred to as the Wild Canary.

Evening Grosbeak
Coccothraustes vespertinus

The breeding range of the Evening Grosbeak lies in the coniferous forests of southern Canada, coastal and mountainous western United States, and northeastern United States. In the West it extends all the way into Mexico. The Evening Grosbeak is a rare to common summer resident on some of the higher mountain ranges in the Great Basin, such as the Warner Mountains, Carson Range, White Mountains, Granite Mountains, Ruby Mountains, and Snake Range.[91, 289] Winter records are available for the West Humboldt Mountains. After the breeding season, grosbeaks wander widely in search of food and often leave their mountain homes and appear in the foothills and valleys of the Great Basin. In Utah this species is reported to be "a common but erratic winter resident in lower valleys"; it has been reported as "nesting in small numbers in conifer and deciduous trees in higher mountains."[56] Not only in Utah but throughout the Basin, Evening Grosbeaks are of erratic occurrence during the winter and spring at foothill and valley sites, and some years major invasions occur. Prior to 1982 the name of this bird was *Hesperiphona vespertina*. *Hesperiphona* is now considered to be congeneric with the Old World genus *Coccothraustes*, to which the Hawfinch, *C. coccothraustes*, and some grosbeaks belong.

The Evening Grosbeak extended its winter and breeding ranges into parts of eastern United States and eastern Canada during historic times. According to Forbush the first record from east of the Great Lakes was at Toronto in 1854. During the final quarter of the nineteenth century, Evening Grosbeaks began increasing in numbers during the winter in the northernmost midwestern states. But until the winter of 1889–90 they were virtually unknown even as far east as Ohio. During that winter a great eastward invasion occurred which extended almost as far as the Atlantic Coast of Massachusetts. Off and on during the winters which followed, Evening Grosbeaks would appear in the East. In 1910–11 another great invasion occurred.[410] Then, in the years that followed, Evening Grosbeaks became established as residents in southeastern Canada and northeastern United States.

Walter Faxon is credited by Forbush with suggesting that the box elder may have facilitated the eastward spread of the Evening Grosbeak, since the buds and seeds of this tree are favored above all others as food by the grosbeaks.[410] By altering the distribution of the box elder, we may have indirectly helped alter the distribution of the grosbeak. Even after the eastward spread of the Evening Grosbeak was accomplished, Peattie observed that "in the states east of the Missouri, Box Elder is not considered a tree of any importance in the wild; it is simply one of the common trees of the riverbanks." But, if the box elder was not widespread on its own in the wild, we made it so in eastern Canada and, to some extent, in eastern United States. Even though the grosbeaks may have welcomed the plantings and may possibly have been tempted eastward by them, the plantings were eventually to be criticized because the box elder is short-lived, has weak wood, and lacks ornamental beauty. Looking back on the scene in the Midwest, Peattie was moved to characterize the plantings by observing that "at one time Box Elder was extensively planted, indeed much over-planted, by persons seeking a quick and cheap effect for mushrooming Middle Western communities."[509]

The preferred nesting habitat of the Evening Grosbeak is in dense, mature coniferous forests where fir or spruce predominates. Outside of the breeding season, grosbeak flocks wander widely in search of food. Although some insect food is taken when available, the bulk of the diet consists of seeds, buds, and fruit from a variety of conifers and from other wild and cultivated trees and shrubs, including juniper berries and pinyon nuts. Evening Grosbeaks frequently drink and bathe. During the winter snow is eaten. Birds continue to bathe even during the winter, seeking ice-free stretches of swiftly flowing streams.

The word grosbeak was derived from the French *grosbec*, meaning thick-billed. The genus name of *Coccothraustes* is Greek, meaning kernel breaker.[267] With its huge conical bill a grosbeak can even crush tough wild cherry pits to extract the kernels within. A grosbeak may swallow some of the crushed pits as well as the kernels—the pieces of pit probably supplement the grit or gravel that the bird swallows to act as millstones in its gizzard. Evening Grosbeaks have a fondness for salt. They will eat salt-impregnated dirt. In regions of the country where salt is spread on roads to melt ice, they will alight to eat the salt. Some are killed by passing cars.

The name Evening Grosbeak is a misnomer, referring to an earlier belief that this grosbeak is active and sings mainly in the evening. This is not so. However, whether by day or by night, the male is not much of a songster. Hoffmann wrote, "The male's attempt at song consists of the ordinary sharp *tseé-a*, followed by a lower harsh *grrree*."[84]

Other Fringillids

The Brambling, *Fringilla montifringilla*, a Eurasian species, was recorded in the fall of 1978 at Sutcliffe, on the shores of Pyramid Lake, Nevada. Dave and Karen Galat obtained photographic documentation, as the Brambling fed on pyracantha and on the ground with House Sparrows.[510]

The breeding range of the Purple Finch, *Carpodacus purpureus*, lies to the north of the Great Basin in the boreal forests of Canada and to the west on the western slopes of the Cascades and Sierra Nevada. Purple Finches have been reported several times from Utah in the northeastern part of the Basin, but no specimens have been collected as documentation.[27] The finch is on the accidental list from Malheur National Wildlife Refuge[61] and on the unconfirmed sighting list from Honey Lake Wildlife Area.[455] Purple Finches have been reported on several Audubon Christmas Bird Counts from Reno. However, I strongly believe that these records have all been based on misidentified House or Cassin's finches. As yet, all the reported sightings in Reno that I have checked out have been misidentifications. As far as I know, documentation does not yet exist for the presence of this finch in western Nevada.

The breeding range of the White-winged Crossbill, *Loxia leucoptera*, lies in the circumboreal conifer forests of Eurasia and North America. In North America it extends across Alaska and Canada, barely penetrating the northern edge of the United States. All our records of the White-winged Crossbill come from the eastern part of the Great Basin. This species has been recorded a number of times in Utah from the Pavant Mountains and in the Salt Lake City area[27] and also from Logan.[418] It has been reported in extreme southeastern Idaho.[456] In August of 1868, Ridgway spotted what he believed to be a male of this species on "the eastern slope of the Ruby Mountains." In September of the same year, he found the White-winged Crossbill to be "common in the lower cañons on the eastern slope of the East Humboldt Mountains, where it inhabited the same localities as the more rare" Red Crossbill.[25] The White-winged Crossbill has a slighter bill than the Red Crossbill and depends more on spruce seeds than pine seeds. The lack of records for the Whitewing in the western part of the Great Basin may be due to the lack of any native spruce trees here. Like the Red

Crossbill, this species also wanders widely in search of seeds and may breed at any season of the year.

The Common Redpoll, *Carduelis flammea*, has a circumpolar breeding range in Eurasia and North America. During the winter in North America, some redpolls come southward irregularly into northern United States. The Common Redpoll is a rare to uncommon winter visitant in the Great Basin, with most of the sightings being from Utah. Prior to 1976 the name of this species was *Acanthis flammea*; then the genus *Acanthis* was merged with *Carduelis*.

There are no documented records for the Hoary Redpoll, *Carduelis hornemanni*, in the Great Basin. But, during late winter on a university field trip, my ornithology class and I saw a flock feeding on the ground in a park along the Truckee River at Verdi, Nevada. We were able to closely approach the flock in our cars for careful study.

FAMILY PASSERIDAE

The Family Passeridae contains the Old World sparrows. It is a former subfamily elevated to family rank. Prior to 1982 it was a subfamily, Passerinae, in the Family Ploceidae, the weavers.

House Sparrow
Passer domesticus

The House Sparrow is not a native North American: it is a Eurasian species which has been spread over much of the world by humans. In 1850 eight pairs of House Sparrows were brought to Brooklyn, New York; they were released in the spring of 1851. In the following years other introductions were made at various sites, some successful, others unsuccessful. These introductions, followed by transplants and the natural spread of colonizing birds, have carried the House Sparrow over much of North America.

The House Sparrow is a common resident in the Great Basin. House Sparrows were introduced into Utah before 1870, and by 1871 they were present around Ogden and in Salt Lake Valley.[56] By 1888 this bird was probably present over most of the Basin, except possibly for the west central and southwestern edges.[511] To the west of the Basin, House Sparrows were introduced into the San Francisco Bay area around 1871 or 1872; by 1915 they had spread over most of California.[26]

Sometimes called the English Sparrow, the House Sparrow was introduced into North America to help control insect pests, but it flourished to such an explosive extent as to become a pest to humans and a menace to native birds. The House Sparrow is thoroughly despised by many bird lovers. It has been written that "the English Sparrow among birds, like the rat among mammals, is cunning, destructive, and filthy"—hardly high praise.[395] The nineteenth-century ornithologist Elliott Coues was, no doubt, the most outspoken critic of the House Sparrow. Coues' greatest ornithological work was his monumental monograph on North American birds, *Key to North American Birds*. In the fifth edition of the key, which he finished writing shortly before his death, Coues described the House Sparrow as follows:

> Repeatedly imported since 1858, and especially in the sixties, during a craze which even affected some ornithologists, making people fancy that a granivorous conirostral species would rid us of insect-pests, this sturdy and invincible little bird has overrun the whole country, and proved a nuisance without a redeeming quality. The original offender in the case is said to have been one Deblois, of Portland, Me., in 1858; but the pernicious activity of Dr. T. M. Brewer affected the city fathers of Boston in 1868–69, and even the Smithsonian Institution at Washington, about the same years. New York had the sparrow-fever in 1860–64, and Philadelphia was not as slow as usual in catching the contagion, in 1869. There is no need to follow the sad record further. Well-informed persons denounced the bird without avail during the years when it might have been abated, but protest has long been futile, for the sparrows have had it all their own way, and can afford to laugh at legislatures, like rats, mice, cockroaches and other parasites of the human race which we must endure. This species, of all birds, naturally attaches itself most closely to man, and easily modifies its habits to suit such artificial surroundings; this ready yielding to conditions of environment, and profiting by them, makes it one of the creatures best fitted to survive in the struggle for existence under whatever conditions man may afford or enforce; hence it wins in every competition with native birds, and in this country has as yet developed no counteractive influences to restore a disturbed balance of forces, nor any check whatever upon its limitless increase. Its habits need not be noted, as they are already better known to every one than those of any native bird whatever, but few realize how many million dollars the bird has already cost us. Nests anywhere about buildings, also in trees, bushes, and vines, built of any rubbish, usually lined with feathers, and making a bulky, unsightly object amidst dirty surroundings; eggs indefinitely numer-

ous, usually 5 to 7, about 0.90 × 0.60, dull whitish thickly marked with dark brown and neutral tints; several broods a year are raised, as the birds breed in and out of season.[512]

The House Sparrow most commonly resides in close proximity to human habitations in either rural or urban areas; only occasionally will it live apart from us. Its common name and its species name of *domesticus*, from the Latin *domus* meaning house, are in reference to its habit of nesting close to domiciles.

The principal vocalization of the House Sparrow is the frequently heard *chirrup*. This call is variable and is used in many situations. Sometimes it is a disyllabic *chirrup*, *chirrip*, *chee-up*, *chillip*, or *chirrip*; sometimes it is abbreviated into a monosyllabic *cheep*, *chirp*, *cheerp*, or *chweep*. When using the *chirrup* as an advertising song to attract a mate to his territory, the male speeds up his "singing" rate above one call per second and uses a higher-pitched *chirrip* or *cheep*. He may also become more rhythmical in his delivery, with refrains such as *cheep chirrip chirrup* or *chirrip cheep chirrup*.[513]

House Sparrows breed the year following their birth. A young male establishes a territory with a nesting site on it, from which he advertises for a mate with his *chirrup* call. When a female is attracted to an advertising male, the male continues to call and to shiver his wings. He will strut in front of the female with his head held high, his breast thrust out, his wings slightly away from his body, and his tail elevated and spread. He may bow up and down. If the female lingers, the male may fly in and out of his nest hole—displaying the presence of a nesting site to her. If she leaves his territory, he may follow her, calling and displaying. Once a male and female have pair-bonded and nested together, they remain together and use the same nesting site or an alternate one, year after year. Following the death of a member of a pair, the remaining member retains the nesting site and, by advertising, soon acquires a new mate.[513]

House Sparrows are cavity nesters by preference. They nest under the eaves of houses, in tile roofs, or in any other available cavity or crevice in buildings. They usurp birdhouses and nest behind shutters or in cavities in stone or brick walls. Occasionally they will nest away from human habitations. Sometimes they usurp the nesting burrows of Bank Swallows. On the Bear River National Wildlife Refuge and at other Great Basin localities, we have found them nesting in the gourd-shaped mud nests of Cliff Swallows in active Cliff Swallow colonies. We have found them nesting in the sides of the stick nests of Golden Eagles. Rarely, Great Basin House Sparrows build domed tree nests of grasses and weeds.

Social birds, House Sparrows often nest in close proximity to each other and are often seen foraging, loafing, or bathing together. They not only bathe in water but frequently engage in communal dust bathing. Often their favorite surfaces of bare, friable, light soil will be thickly pockmarked with the shallow craters they form when dusting. House Sparrows become quite tolerant of human traffic; their favorite long-standing dusting spa on the University of Nevada campus at Reno is located beside a busy sidewalk next to Manzanita Lake. While dusting a sparrow hunkers down or lies down and works dust into its fluffed-up plumage. It uses bill and wing-flicking movements as well as body movements. After thoroughly dusting, the sparrow arises and shakes the dust out of its plumage; it then usually preens itself by drawing feathers through its bill. The function of dusting is not clearly known, but it may serve to dislodge ectoparasites.

House Sparrows feed mainly on vegetable matter. In an extensive study of the stomach contents of 4,848 adult House Sparrows, Kalmbach found that over 96 percent of the contents was vegetable matter.[466] House Sparrows feed on weed seeds but prefer grain when they can obtain it. Before the ascendancy of the automobile, these sparrows obtained much of their grain from horse droppings. House Sparrows are highly injurious to vegetable and fruit crops of all kinds. They also eat the blossoms of apples, peas, beans, and other plants. The damage they do to crops is nowhere counterbalanced by the occasional weed seeds or detrimental insects they eat.

Appendix: Birding in the Great Basin

SOME OF THE prime birding sites in the Great Basin can be reached only on foot. This is especially true of montane sites. A Sierra Club Totebook entitled *Hiking the Great Basin*, by John Hart, has excellent descriptions and maps of the hiking trails in the mountains. This paperback publication will help you plan birding hikes. For those of you who do not wish to attempt any strenuous hiking, many prime birding areas are accessible by car.

East-West Routes across the Great Basin

Two major highways span the Great Basin in an east-west direction: visitors making their first crossing will gain a greater knowledge of the Basin by taking one route when going and the other when returning. The two routes are Interstate 80 between Salt Lake City and Reno and U.S. 50 between Delta and Carson City.

Crossing on Interstate 80

While in Salt Lake City, purchase the paperback publication *Utah Birds*, by William H. Behle and Michael L. Perry, from one of the local bookstores or from the Utah Museum of Natural History at the University of Utah. This publication contains a checklist and seasonal and ecological occurrence charts for Utah birds. It has maps and descriptions of twelve field trips which can be made in the Great Salt Lake region in quest of good birding.

The eastern end of I-80 between Salt Lake City and the Nevada border crosses the Great Salt Lake Desert. Not many birds are usually seen by travelers here. If you have never been in the Great Salt Lake Desert, you may want to cross it. However, if you want better birding and do not mind taking a longer route, I recommend looping around the north end of the desert.

To make the northern loop, take Interstate 15 north out of Salt Lake City. Exit from I-15 at Brigham City to visit the Bear River National Wildlife Refuge. To continue the loop, proceed northward on I-15 from Brigham City until you reach I-84, which will take you to Snowville. State Highway 30 leads west and south from Snowville around the edge of the Great Salt Lake Desert. It will become State Highway 233 when it crosses the Nevada border; it joins I-80 at Oasis, Nevada. The country beyond Snowville is rich in raptors, especially during migration. Migrating raptors are funneled around the edges of the Great Salt Lake Desert. Golden Eagles may be seen, close at hand, perched on utility poles. As a trip bonus, you may see antelope along State Highway 30. Rick Stetter and I have seen them there in pastures feeding with cattle in the fall.

At the Oasis junction, proceed west on I-80 to Wells, Nevada. Here good birding is available at nearby Angel Lake—magnificently positioned in a glacial cirque at 8,378 feet in the East Humboldt Range. A good road leads up to the lake, and a campground is present there. Raptors and montane birds will be seen around the lake. In the late summer the country will be alive with migrating hummingbirds. Angel Lake is also a good site for Clark's Nutcrackers. A hiking trail provides access to other mountain localities. Another good side trip from Wells is a visit to Ruby Lake National Wildlife Refuge to look for Trumpeter Swans before proceeding down I-80 to Elko.

From Elko a birding trip into the nearby Ruby Mountains is a must. The roads from Elko into Lamoille Canyon are good. In my opinion, Lamoille is the most complete and scenic mountain canyon, and the Ruby Mountains are one of the noblest ranges, in all of the Great Basin. A number of hiking trails flare out from Lamoille Canyon. In the past I have found American Dippers nesting at small waterfalls along Thomas Creek. In the latter part of summer the flowering meadows of Lamoille are alive with migrating hummingbirds.

When proceeding westward from Elko on I-80, watch for Snowy Egrets and Golden Eagles—there is a large breeding rookery of egrets just west of town near the Humboldt River, and eagles may be nesting on cliffs along the north side of I-80. Before reaching Carlin, I-80 passes through a tunnel in the east wall of the Humboldt River Canyon. Upon emerging from the tunnel I-80 crosses the river. At this point a section of the old, pre-I-80 highway is still intact along the Humboldt River. Stop and walk up and down the old highway. Both Golden Eagles and Prairie Falcons nest on the canyon walls in the vicinity of the west end of the tunnel. Another approach to these canyon walls would be to exit from I-80 at Carlin. Then drive eastward out of Carlin on the old highway back to the vicinity of the tunnel.

Continue on I-80 to Winnemucca. While in the Winnemucca area, consider a side trip to Paradise Valley and into the unique Santa Rosa Range—where the boreal life zones are missing. Go north from Winnemucca on U.S. 95 and exit on State Highway 290. On Highway 290 you can drive up to Hinkey Summit at 7,867 feet in the Santa Rosa Range. From Winnemucca, Lovelock can be reached on I-80 or by looping east of I-80 through Grass Valley on back roads. If you go by I-80, you can follow in the footsteps of Robert Ridgway and bird at Unionville in the West Humboldt Range; leave I-80 at Mill City and drive along the east side of the range to Unionville. There are hiking trails above Unionville in Buena Vista Canyon which lead deeper into the mountains. After leaving Unionville, you can continue looping south to rejoin I-80. The main attraction along I-80 between Lovelock and Reno will be found at the sloughs and puddles of water to the east of Fernley. Here, by the side of I-80, avocets, stilts, and phalaropes may be seen during the spring and early summer.

Crossing on U.S. 50

One of the main attractions around Delta, Utah, has to be the raptors. This is good Golden Eagle and hawk country. Both the Krider's and Harlan's versions of the Red-tailed Hawk have been sighted here; I have seen a Krider's near Mud Springs. While in the Delta area, make the short drive over to Oak City, and bird up the road leading eastward out of Oak City to a forest camp in Fishlake National Forest. From Delta the shortest route to the Nevada border is via U.S. 50, but this route runs through rather desolate, Common Raven–Horned Lark country. If you are driving a back road vehicle, the birding can be diversified by looping northward from I-80 through Marjum Pass when west of Sevier Lake. A longer, much more interesting birding route loops southward from Delta on State Highway 257 to Milford, then progresses northwestward on State Highway 21 to join U.S. 50 just across the Nevada border. Highway 21 passes close to Old Frisco Mining Town and traverses some good pinyon-juniper woodlands.

Once you cross the Utah-Nevada border, a birding trip into the magnificent Snake Range should have top priority. Baker, Nevada, the gateway to Lehman Caves National Monument and to 13,063-foot Wheeler Peak, is only five miles south of U.S. 50. A paved road from Baker runs up the mountainside to Lehman Caves and to several side roads and campgrounds, then continues all the way up to Wheeler Peak Campground at an altitude of close to 10,000 feet. From the campgrounds, hiking trails lead up to a small mountain glacier, to an ancient bristlecone pine grove, to Stella and Teresa lakes, and above the upper tree line to the tops of Wheeler Peak and Bald Mountain. All the montane life zones, including the arctic-alpine life zone, are readily accessible for birding. During the early morning hours, Golden Eagles can usually be seen perched on utility poles around Baker and along the road leading up to Lehman Caves.

U.S. 50 crosses the Snake Range at Sacramento Pass (7,154 feet). On the west side of the Snake Range, a dirt road runs north from the highway. From this road it is possible to drive and hike in the country around Mount Moriah (12,050 feet) in the north end of the Snake Range. West of the Snake Range U.S. 50 crosses Spring Valley, an area unique in possessing a low-lying patch of Rocky Mountain junipers growing on wet, saline soil in a greasewood–rabbit brush shrubland. West of Spring Valley U.S. 50 crosses the Schell Creek Range at Connors Pass (7,723 feet). While you cross the Snake and Schell Creek ranges, roadside woodlands are available for birding. West of the Schell Creek Range, U.S. 50 runs through Steptoe Valley to enter the town of Ely in the Egan Range. In Steptoe Valley water birds will be seen at Comins Lake along the highway. Using Ely as a base, trips can be made into some of the major canyons in the Egan and Schell Creek ranges in search of montane birds.

Between Ely and Eureka, U.S. 50 passes through excellent raptor country. Woodlands and shrublands are accessible for birding along the roadside. The trip from Eureka to Austin can be a productive one, especially where the highway enters the Toiyabe Range and runs through fine pinyon-juniper woodlands. Watch for Pinyon Jays here. Using Austin as a base, side trips can readily be made to various canyons and creeks in the Toiyabes and to nearby Big Smoky Valley and the Toquima Range. The single most complete birding trip would involve entering the east side of the Toiyabe Range via Kingston Canyon and working upward into the mountains. In a back road vehicle you can drive all the way across the range and drop down into the Reese River Valley. The road below the pass and down the west side is rough and steep and may require some switchbacking to round a curve or two. But this is pristine high country, well worth an all-out effort to reach. Between Austin and Carson City on U.S. 50, side trips in search of water birds can be made to the Stillwater Wildlife Management Area and Lahontan Reservoir. The birding around Fallon is often excellent. While near the Lahontan Reservoir, bird at historic Fort Churchill along the Carson River. From Fort Churchill follow the dirt road (2B) which runs west along the Carson River and rejoins U.S. 50 east of Dayton. Within a few miles west of Fort Churchill, White-throated Swifts nest on high cliffs along the north side of 2B.

North-South Routes through the Great Basin

There are several highly productive north-south birding routes through the Great Basin. U.S. 395, running along the western edge from Burns, Oregon, to Mono Lake, California, is an outstanding birding route. Malheur National Wildlife Refuge, located close to Burns, is a famous water bird and marshbird site. If you drive south from Burns in good weather, I recommend looping southeast from U.S. 395 at Hogback Summit. The Hogback Road runs through Plush and Adel, Oregon. Between Plush and Adel the loop runs along the west side of Hart, Crump, and Pelican lakes. This is good Sandhill Crane nesting country. The American White Pelican, Great Blue Heron, and Great Egret nest on two islands near the west shore of Pelican Lake within view of the road. When in northeastern California, you may wish to bird in the Warner Mountains or search for nesting Snowy Plovers at the three Alkali lakes (Upper, Middle, and Lower). A trip to Eagle Lake to bird and view nesting Ospreys would be worthwhile. The Honey Lake area is tremendous for raptors and good for longspurs in the winter. From U.S. 395 in west central Nevada, the Carson Range, Lake Tahoe, and Pyramid Lake are within easy reach. Along U.S. 395 between Reno and Mono Lake, you will encounter many inviting sites. Be sure to visit the phalaropes, grebes, and gulls at Mono Lake.

The eastern edge of the Great Basin offers a productive north-south array of sites. Birding in the country northeast and east of Great Salt Lake can be stimulating, particularly on the Bear River National Wildlife Refuge. While in the Salt Lake City area, use *Utah Birds* by Behle and Perry in planning field trips. Be sure to drive out over the causeway and bird on Antelope Island in Great Salt Lake. Several rewarding birding routes lead southward. State Highway 68 runs along the west side of Utah Lake. If you go down the east side of Utah Lake, you can search in Provo Canyon for Black Swifts—Al Knorr found them nesting there at five different sites. I prefer a more inland route through Grantsville and Saint John to State Highway 36, then down U.S. 6. Divert into the Little Sahara Recreation Area at White Sand Dunes. I saw a large flock of Pinyon Jays along the entrance road into the area. Continue on to Delta, then to Milford, and then to Cedar City. You will encounter many satisfying birding sites along the way.

An interesting trip through the interior of the Great Basin can be made on U.S. 93 between Jackpot on the Nevada-Idaho border to Caliente in southeastern Nevada. Be sure to bird along Salmon Falls Creek while at Jackpot—it's great raptor country. At Wells visit Angel Lake for montane birds. South of Wells, detour into Ruby Lake National Wildlife Refuge to search for Trumpeter Swans, and look for passerines along Cave Creek. In the Ely area, birding in the Egan and Schell Creek ranges is excellent. In the Pioche-Caliente area visit Cathedral Gorge State Park and Kershaw-Ryan State Recreation Area.

Refuges and Wildlife Management Areas

Prime birding is available the year around on some of the refuges and management areas in the Great Basin. The display is most spectacular during the spring and fall migrations, when throngs of grebes, waterfowl, and shorebirds are on the move. Unique refuges and management areas include Bear River and Fish Springs National Wildlife Refuges in Utah; Ruby Lake National Wildlife Refuge and Stillwater Wildlife Management Area in Nevada, and Malheur National Wildlife Refuge in Oregon. All these sites are administered by the U.S. Fish and Wildlife Service of the Department of the Interior; each has its own bird checklist. Write the refuge managers and request copies of their checklists, maps, and information bulletins before planning any birding trips. In the fall you may have to contend with hunters on or around the refuges and management areas. Be sure and carry insect repellent. Mosquitoes are present at all the wet places and are a bloodthirsty lot. Biting flies may also be encountered.

The Quest for Good Birding

Good birding goes beyond learning to identify birds, compiling life lists, and searching for rare species. There are happenings of quiet beauty, high drama, and even pageantry to be savored, and some birding experiences can be of a spiritual nature. The following list suggests adventures to be sought which possess the essence of what good birding in the Great Basin is all about.

1. Visit Sage Grouse on their strutting grounds in early spring. Request information on the location of accessible strutting grounds from one of the state Fish and Game Departments, the Bureau of Land Management, or local Audubon Society chapters.

2. Witness raptor migration in the late summer and early fall. Go to a known site such as Carson Valley, the Goshute Mountains, or Wellsville. Or search for a new site along the base of some major north-south-trending mountain range or desert edge.

3. Observe concentrations of diurnal raptors from December through February. Overwintering raptors are extremely abundant in the Great Basin, and their hunting behavior is fascinating to watch. Heavy concentrations build up in Carson Valley, Washoe Valley, Truckee Meadows, Honey Lake Basin, Sierra Valley, and valleys throughout eastern Nevada and central Utah.

4. Look for huge migratory concentrations of Eared Grebes along the causeway to Antelope Island in Great Salt Lake or at Mono Lake. An estimated 800,000 grebes may use Mono Lake as a migratory stop.

5. Make a pilgrimage to Mono Lake while it is still alive. Contemplate the Wilson's Phalaropes and California Gulls, as well as the grebes. Up to 100,000 or so phalaropes use this lake as a staging area for their fall migration.

6. Seasonally sample the birding at the great wildlife refuges and wildlife management areas in the Basin. Visit the more remote ones such as Fish Springs and Sheldon as well as the readily accessible ones. Observe the huge fall concentrations of Tundra Swans at Bear River or the lesser concentrations at Malheur, Honey Lake, or Stillwater. Visit Ruby Lake to search for Trumpeter Swans. Visit Malheur to view Sandhill Cranes and to witness Eared and Western Grebe breeding behavior. You will reap a rich harvest of water birds and marshbirds at these sites, as well as some passerine birds. At Bear River and Ruby Lake you may get to view, close at hand, White-faced Ibises.

7. Search for avocets and stilts at pools of brackish or alkali water. These birds are widespread and easy to find. Their courtship, nesting, and feeding behavior makes for fascinating bird-watching.

8. The corvids are the most interesting passerines in the Great Basin—behaviorally speaking. They are common and widespread and easy to locate. Become a student of the Black-billed Magpie or Common Raven or Pinyon Jay or Clark's Nutcracker if you crave intellectually exciting bird-watching. Study the continuing urbanization of the American Crow in the Great Basin. In the early part of 1983, there was an explosive buildup of overwintering and nesting crows in Reno.

9. During the winter, comb the valleys in search of altitudinal and latitudinal migrants and visitants. Keep an eye out for rarities among sparrows, raptors, and waterfowl. Examine Horned Lark flocks for longspurs. Do Burrowing Owls ever overwinter here? Take note of any invasion by Snowy Owls or Bohemian Waxwings. Note how birds forage and utilize shelter. Note their flocking behavior.

10. During the summer determine the composition of the breeding avifaunas on Great Basin mountains. Photograph nests with eggs or nestlings

and juvenile birds as documentation. Work over the entire altitudinal extent of a mountain. On high mountains go above the upper tree line and search for nesting birds in the arctic-alpine life zone.

11. Focus some of your summer birding on montane environments in the interior of the Basin, especially in the region between the Pine Nut and Virginia ranges to the west and the Egan and East Humboldt ranges to the east. Hike about, don't just bird from the limited road system. The closed forest life zones are missing or poorly developed in these interior ranges. Unusual aspects of the altitudinal and longitudinal distribution of birds may be elucidated. Try to solve specific questions: how far west does the Broad-tailed Hummingbird nest, how much of a hiatus exists in the distribution of the Steller's Jay in the interior of the Great Basin? This approach to birding will add intellectual stimulation to your adventures.

Literature Cited

1. "Vegetational Zonation in the Great Basin of Western North America." W. D. Billings. Extrait des C. R. du Colloque sur Les bases écologiques de la régénération de la végétation des zones arides. U.I.S.B. 101–122. 1951.

2. "Bird Energetics: Effects of Artificial Radiation." Sheldon Lustick. Science 163: 387–390. 1969.

3. "Animal Coat Color and Radiative Heat Gain: A Re-evaluation." G. E. Walsberg, G. S. Campbell, and J. R. King. Journal of Comparative Physiology Part B: Biochemical, Systematic, and Environmental Physiology 126(3):211–222. 1978.

4. "Absorption of Radiant Energy in Redwinged Blackbirds (*Agelaius phoeniceus*)." Sheldon Lustick, Sharon Talbot, and Edward L. Fox. Condor 72:471–473. 1970.

5. "Ventilation and Acid-Base Status during Thermal Panting in Pigeons (*Columba livia*)." Marvin H. Bernstein and Felipe C. Samaniego. Physiological Zoology 54:308–315. 1981.

6. "Evaporative Cooling in the Poor-will and the Tawny Frogmouth." Robert C. Lasiewski and George A. Bartholomew. Condor 68:253–262. 1966.

7. "Energy Conserving and Heat Dissipating Mechanisms of the Turkey Vulture." Daniel E. Hatch. Auk 87:111–124. 1970.

8. "Thermal and Caloric Relations of Birds." William A. Calder and James R. King. In Avian Biology. Volume 4. Edited by Donald S. Farner and James R. King. Academic Press. 1974.

9. "Responses to Temperature in the Dipper, *Cinclus mexicanus*." David E. Murrish. Comparative Biochemistry and Physiology 34:859–869. 1970.

10. Arctic Life of Birds and Mammals Including Man. Laurence Irving. Springer-Verlag. 1972.

11. "Observations of a Large Roost of Common Ravens." Richard B. Stiehl. Condor 83:78. 1981.

12. "Radiometric Determination of Feather Insulation and Metabolism of Arctic Birds." James H. Veghte and Clyde F. Herreid. Physiological Zoology 38: 267–275. 1965.

13. The Lives of Desert Animals in Joshua Tree National Monument. Alden H. Miller and Robert C. Stebbins. University of California Press. 1964.

14. "Evaporative Losses of Water by Birds." William R. Dawson. Comparative Biochemistry and Physiology 71A:495–509. 1982.

15. "Physiology and Energetics of Flight." M. Berger and J. S. Hart. In Avian Biology. Volume 4. Edited by Donald S. Farner and James R. King. Academic Press. 1974.

16. "Respiratory Water Loss in Some Birds of Southwestern United States." George A. Bartholomew, Jr., and William R. Dawson. Physiological Zoology 26:162–166. 1953.

17. "Observations on the Temperature Regulation and Water Economy of the Galah (*Cacatua roseicapilla*)." William R. Dawson and Charles D. Fisher. Comparative Biochemistry and Physiology 72A:1–10. 1982.

18. "Renal Function." Eldon J. Braun. Comparative Biochemistry and Physiology 71A:511–517. 1982.

19. "Responses of Brewer's and Chipping Sparrows to Water Restriction." William R. Dawson, Cynthia Carey, Curtis S. Adkisson, and Robert D. Ohmart. Physiological Zoology 52:529–541. 1979.

20. "The Water Economy of the Sage Sparrow, *Amphispiza belli nevadensis*." Ralph R. Moldenhauer and John A. Wiens. Condor 72:265–275. 1970.

21. A New Dictionary of Birds. Edited by Sir A. Landsborough Thomson. McGraw-Hill. 1964.

22. "A Study of the Nesting of Mourning Doves." Margaret Morse Nice. Auk 39: 457–474. 1922.

23. The Avian Egg. Alexis L. Romanoff and Anastasia J. Romanoff. John Wiley and Sons. 1949.

24. Check-list of North American Birds. Sixth edition. Prepared by the Committee on Classification and Nomenclature of the American Ornithologists' Union. Allen Press. 1983.

25. United States Geological Exploration of the Fortieth Parallel. Clarence King, geologist-in-charge. Part 3: Ornithology. Robert Ridgway. Government Printing Office. 1877.

26. The Distribution of the Birds of California. Joseph Grinnell and Alden H. Miller. Pacific Coast Avifauna 27. Cooper Ornithological Club. 1944.

27. Utah Birds: Check-list, Seasonal and Ecological Occurrence Charts and Guides to Bird Finding. William H. Behle and Michael L. Perry. Utah Museum of Natural History. 1975.

28. "Observations on the Habits of Birds at Lake Burford, New Mexico." Alexander Wetmore. Auk 37:221–247. 1920.

29. "Courtship, Hostile Behavior, Nest-Establishment and Egg Laying in the Eared Grebe (*Podiceps caspicus*)." Nancy M. McAllister. Auk 75:290–311. 1958.

30. "A Shower of Grebes." Clarence Cottam. Condor 31:80–81. 1929.

31. "Western Grebe Colony." Robert W. Nero. Natural History 68(5):291–294. 1959.

32. "The Diving and Feeding Activity of the Western Grebe on the Breeding Grounds." George E. Lawrence. Condor 52:3–16. 1950.

33. "A Peculiar Feeding Habit of Grebes." Alexander Wetmore. Condor 22:18–20. 1920.

34. Camps and Cruises of an Ornithologist. Frank M. Chapman. D. Appleton and Company. 1908.

35. "The Grebes of Southern Oregon." William L. Finley. Condor 9:97–101. 1907.

36. "Pesticides and the Reproduction of Birds." David B. Peakill. Scientific American 222(4):72–78. 1970.

37. "Recent Observations on the White Pelican on Gunnison Island, Great Salt Lake, Utah." Jessop B. Low, Lee Kay, and D. I. Rasmussen. Auk 67:345–356. 1950.

38. "The History and Present Status of the Biota of Anaho Island, Pyramid Lake, Nevada." W. Verne Woodbury. M.S. thesis. University of Nevada at Reno. 1966.

39. "Foraging Sites of White Pelicans Nesting at Pyramid Lake, Nevada." Fritz L. Knopf and Joseph L. Kennedy. Western Birds 11:175–180. 1980.

40. "A History of the Bird Colonies of Great Salt Lake." William H. Behle. Condor 37:24–35. 1935.

41. "An Ancient Nesting Site of the White Pelican in Nevada." E. Raymond Hall. Condor 42:87–88. 1940.

42. "Spatial and Temporal Aspects of Colonial Nesting of White Pelicans." Fritz L. Knopf. Condor 81:353–363. 1979.

43. "Pelicans Killed by Lightning." Frederick C. Lincoln. Auk 58:91. 1941.

44. "White Pelicans Killed by Lightning." John W. Sugden. Auk 47:72–73. 1930.

45. The Bird Life of Great Salt Lake. William H. Behle. University of Utah Press. 1958.

46. "White Pelicans Nesting at Honey Lake, California." Ian C. Tait, Fritz L. Knopf, and Joseph L. Kennedy. Western Birds 9:38–40. 1978.

47. "The Current Status of the Double-crested Cormorant in Utah: A Plea for Protection." Ronald M. Mitchell. American Birds 29:927–930. 1975.

48. The Home-Life and Economic Status of the Double-crested Cormorant. Howard L. Mendall. University of Maine Studies, second series, 38. University Press. 1936.

49. "Notes on Water Birds Nesting at Pyramid Lake, Nevada." E. Raymond Hall. Condor 28:87–91. 1926.

50. "A Sojourn among the Wild Fowl at Pyramid Lake." O. J. Gromme. Yearbook of the Public Museum of Milwaukee 10:268–303. 1930.

51. "Birds of Anaho Island, Pyramid Lake, Nevada." Richard M. Bond. Condor 42:246–250. 1940.

52. "Cormorants Killed by Lightning." Alexander Sprunt, Jr. Auk 58:568. 1941.

53. "Voices of a New England Marsh." William Brewster. Bird-Lore 4:43–56. 1902.

54. "Concerning the Nuptial Plumes Worn by Certain Bitterns and the Manner in Which They Are Displayed." William Brewster. Auk 28:90–100. 1911.

55. "Feeding Behavior of North American Herons." James A. Kushlan. Auk 93: 86–94. 1976.

56. Birds of Utah. C. Lynn Hayward, Clarence Cottam, Angus M. Woodbury, and Herbert H. Frost. Great Basin Naturalist Memoirs 1. Brigham Young University Press. 1976.

57. Handbook of North American Birds. Volume 1: Loons through Flamingos. Edited by Ralph S. Palmer. Yale University Press. 1962.

58. "Breeding Biology of the Least Bittern." Milton W. Weller. Wilson Bulletin 73:11–35. 1961.

59. The Birds of Minnesota. Volume 1. Thomas S. Roberts. University of Minnesota Press. 1936.

60. American Birds 35:675–677, 696. 1981.

61. Birds: Malheur National Wildlife Refuge, Oregon. U.S. Fish and Wildlife Service. Government Printing Office. 1981.

62. "Great Blue Heron: Behavior at the Nest." W. Powell Cottrille and Betty Darling Cottrille. Miscellaneous Publications of the Museum of Zoology, University of Michigan 102. 1958.

63. "Breeding Biology of Great Blue Herons and Common Egrets in Central California." Helen M. Pratt. Condor 72:407–416. 1970.

64. "Courtship and Pair Formation in the Great Egret." Jochen H. Wiese. Auk 93:709–724. 1976.

65. Life Histories of North American Marsh Birds. Arthur Cleveland Bent. United States National Museum Bulletin 135. Smithsonian Institution. 1926.

66. "Diversity Typifies Heron Feeding." Andrew J. Meyerriecks. Natural History 71(6):48–59. 1962.

67. American Birds 32:1190. 1978.

68. American Birds 34:915. 1980.

69. "Success Story of a Pioneering Bird." Andrew J. Meyerriecks. Natural History 69(7):46–57. 1960.

70. "An Early Record of the Cattle Egret in Colombia." Alexander Wetmore. Auk 80:547. 1963.

71. "An Egret Observed on St. Paul's Rocks, Equatorial Atlantic Ocean." Vaughan T. Bowen and Geoffrey D. Nicholls. Auk 85:130–131. 1968.

72. The Herons of the World. James Hancock and Hugh Elliott. Harper and Row. 1978.

73. "Adaptiveness of Foraging in the Cattle Egret." Thomas C. Grubb, Jr. Wilson Bulletin 88:145–148. 1976.

74. "A List of the Birds of Nevada." Jean M. Linsdale. Condor 53:228–249. 1951.

75. "The Birds of Carson River Basin." Edited by Elizabeth Ellis and Joanna Nelson. Record-Courier Press. 1952.

76. "Further Analysis of the Social Behavior of the Black-crowned Night Heron." G. K. Noble and M. Wurm. Auk 59:205–224. 1942.

77. American Birds 34:185. 1980.

78. "A Louisiana Heron in Northeastern California." Tim Manolis. California Birds 3:19–21. 1972.

79. American Birds 31:1167. 1977.

80. American Birds 32:1190–1191. 1978.

81. American Birds 33:884. 1979.

82. American Birds 35:964. 1981.

83. American Birds 36:1001. 1982.

84. Birds of the Pacific States. Ralph Hoffmann. Houghton Mifflin. 1927.

85. Handbook of the Birds of Europe, the Middle East and North Africa. Volume 1. Stanley Cramp, chief editor. Oxford University Press. 1977.

86. "The Birds of Lahontan Valley, Nevada." J. R. Alcorn. Condor 48:129–138. 1946.

87. Rails of the World. S. Dillon Ripley. David R. Godine, Publisher. 1977.

88. "The Virginia Rail in Michigan." Lawrence H. Walkinshaw. Auk 54:464–475. 1937.

89. "Territorial Behavior of the American Coot." Gordon W. Gullion. Condor 55:169–186. 1953.

90. The Game Birds of California. Joseph Grinnell, Harold Child Bryant, and Tracy Irwin Storer. University of California Press. 1918.

91. The Birds of Nevada. Jean M. Linsdale. Pacific Coast Avifauna 23. Cooper Ornithological Club. 1936.

92. Birds of Northern California. Guy McCaskie and Paul De Benedictis. Golden Gate Audubon Society. 1966.

93. Birds of Northern California. Second edition. Guy McCaskie, Paul De Benedictis, Richard Erickson, and Joseph Morlan. Golden Gate Audubon Society. 1979.

94. "The Status of the Sandhill Crane in Utah and Southern Idaho." John W. Sugden. Condor 40:18–22. 1938.

95. Population Surveys, Species Distribution, and Key Habitats of Selected Nongame Species. Nevada Department of Wildlife. Job Performance Reports. September 1979 by Gary B. Herron and Paul A. Lucas. September 1980 by Gary B. Herron, San J. Stiver, and Robert Turner. September 1981 by Gary B. Herron, Robert J. Turner, and San J. Stiver.

96. Cranes of the World. Lawrence H. Walkinshaw. Winchester Press. 1973.

97. The Swans. Peter Scott and the Wildfowl Trust. Houghton Mifflin. 1972.

98. "The Whistling Swan in the West with Particular Reference to the Great Salt Lake Valley, Utah." Glen A. Sherwood. Condor 62:370–377. 1960.

99. Ducks, Geese and Swans of North America. Second edition. Second printing, revised. Frank C. Bellrose. Stackpole Books. 1978.

100. Handbook of North American Birds. Volume 2. Edited by Ralph S. Palmer. Yale University Press. 1976.

101. Life Histories of North American Wild Fowl. Arthur Cleveland Bent. United States National Museum Bulletin 130. Smithsonian Institution. 1925.

102. The Trumpeter Swan: Its History, Habits, and Population in the United States. Winston E. Banko. North American Fauna 63. U.S. Fish and Wildlife Service. Government Printing Office. 1960.

103. Birds of the Pacific Northwest. (Formerly titled: Birds of Oregon.) Ira N. Gabrielson and Stanley G. Jewett. Dover Publications. 1970.

104. "Some Additional Records of Birds for Northeastern California." Donald D. McLean. Condor 39:228–229. 1937.

105. The Trumpeter Swan in Alaska. Henry A. Hansen, Peter E. K. Shepherd, James G. King, and Willard A. Troyer. Wildlife Monographs 26. Wildlife Society. 1971.

106. American Birds 33:881. 1979.

107. American Birds 35:209. 1981.

108. Handbook of Waterfowl Behavior. Paul A. Johnsgard. Cornell University Press. 1965.

109. The Waterfowl of the World. Volume 1. Jean Delacour. Arco Publishing Company. 1954.

110. "Reestablishing Aleutian Canada Geese." Paul F. Springer, G. Vernon Byrd, and Dennis W. Woolington. In Endangered Birds. Edited by Stanley A. Temple. University of Wisconsin Press. 1978.

111. "Additions to the List of Nevada Birds." J. R. Alcorn. Condor 45:40. 1943.

112. Honkers. C. S. Williams. D. Van Nostrand Company. 1967.

113. Waterfowl of North America. Paul A. Johnsgard. Indiana University Press. 1975.

114. "Can Counts of Group Sizes of Canada Geese Reveal Population Structure?" Dennis G. Raveling. In Canada Goose Management. Edited by Ruth L. Hine and Clay Schoenfeld. Dembar Educational Research Services. 1968.

115. Wild Geese. M. A. Ogilvie. Buteo Books. 1978.

116. "Formation Flight of Birds." P. B. S. Lissaman and Carl A. Shollenberger. Science 168:1003–1005. 1970.

117. "The Site Guide: Honey and Eagle Lakes, California." Tim Manolis. American Birds 29:19–21. 1975.

118. Birds of Bear River National Wildlife Refuge. U.S. Fish and Wildlife Service. Government Printing Office. 1979.

119. Audubon Field Notes 3:217. 1949.

120. "The Displays of the American Green-winged Teal." F. McKinney. Wilson Bulletin 77:112–121. 1965.

121. The Waterfowl of the World. Volume 2. Jean Delacour. Arco Publishing Company. 1956.

122. "The Evolution of Behavior." Konrad Z. Lorenz. Scientific American 199(6):67–78. 1958.

123. "Studies of Waterfowl in British Columbia, Pintail." J. A. Munro. Canadian Journal of Research 22D:60–86. 1944.

124. Prairie Ducks. Lyle K. Sowls. Wildlife Management Institute. 1955.

125. "Celestial Orientation by Wild Mallards." Frank C. Bellrose. Bird Banding 29:75–90. 1958.

126. Bird Navigation. Second edition. G. V. T. Matthews. Cambridge University Press. 1968.

127. Migration of Birds. Frederick C. Lincoln. Circular 16. U.S. Fish and Wildlife Service. Government Printing Office. 1950.

128. "Ecological Aspects of Ducks Nesting in High Densities among Larids." Kees Vermeer. Wilson Bulletin 80:78–83. 1968.

129. "Ducks Nesting in Association with Gulls—An Ecological Trap?" L. W. Dwernychuk and D. A. Boag. Canadian Journal of Zoology 50:559–563. 1972.

130. "Waterfowl Protection in the Vicinity of Gull Colonies." William Anderson. California Fish and Game 51:5–15. 1965.

131. To Ride the Wind. H. Albert Hochbaum. Harlequin Enterprises. 1973.

132. "The Summer Birds of Washoe Lake, Nevada." Forrest S. Hanford. Condor 5:50–52. 1903.

133. "Notes on Birds Observed in Portions of Utah, Nevada, California." E. W. Nelson. Proceedings of the Boston Society of Natural History 17:338–365. 1875.

134. "The Cinnamon Teal (*Anas cyanoptera* Vieillot): Its Life History, Ecology, and Management." H. E. Spencer, Jr. M.S. thesis. Utah State Agricultural College. 1953.

135. "Nesting Studies of Ducks and Coots in Honey Lake Valley." E. G. Hunt and A. E. Naylor. California Fish and Game 41:295–314. 1955.

136. Ducks, Geese, and Swans of the World. Paul A. Johnsgard. University of Nebraska Press. 1978.

137. "Site Attachment in the Northern Shoveler." Norman R. Seymour. Auk 91: 423–427. 1974.

138. "Breeding Biology of the Gadwall in Northern Utah." John M. Gates. Wilson Bulletin 74:43–67. 1962.

139. The Canvasback on a Prairie Marsh. H. Albert Hochbaum. American Wildlife Institute. 1944. Reprinted in 1981 by the University of Nebraska Press.

140. "Parasitic Egg Laying in the Redhead (*Aythya americana*) and Other North American Anatidae." Milton W. Weller. Ecological Monographs 29:333–365. 1959.

141. "Duck Nesting Studies, Bear River Migratory Bird Refuge, Utah, 1937." Cecil S. Williams and Wm. H. Marshall. Journal of Wildlife Management 2:29–48. 1938.

142. "Pseudo-Sleeping Attitude in Lesser Scaup and Ring-necked Ducks." Daniel W. Anderson. Condor 72:370–371. 1970.

143. "The Breeding Populations of Piscivorous Birds of Eagle Lake." Roger J. Lederer. American Birds 30:771–772. 1976.

144. "The Display Flight of the North American Ruddy Duck." Michael R. Miller, Robert M. McLandress, and Betty Jean Gray. Auk 94:140–142. 1977.

145. "Dry-Land Nesting by Redheads and Ruddy Ducks." Donald E. McKnight. Journal of Wildlife Management 38:112–119. 1974.

146. "The Breeding Status of the Snowy Plover in California." Edited by Gary W. Page and Lynne E. Stenzel. Western Birds 12:1–40. 1981.

147. "Adaptations of the Snowy Plover on the Great Salt Plains, Oklahoma." James R. Purdue. Southwestern Naturalist 21:347–357. 1976.

148. "Salt Water Tolerance and Water Turnover in the Snowy Plover." James R. Purdue and Howard Haines. Auk 94:248–255. 1977.

149. "Thermal Environment of the Nest and Related Parental Behavior in Snowy Plovers, *Charadrius alexandrinus*." James R. Purdue. Condor 78:180–185. 1976.

150. The Shorebirds of North America. Edited by Gardner D. Stout. Viking Press. 1967.

151. Life Histories of North American Shore Birds. Arthur Cleveland Bent. United States National Museum Bulletin 146. Smithsonian Institution. 1929.

152. "Sexual and Agonistic Behaviour in the Killdeer (*Charadrius vociferans*)." R. E. Phillips. Animal Behaviour 20:1–9. 1972.

153. "Belly-Soaking as a Thermoregulatory Mechanism in Nesting Killdeers." Bette J. Schardien and Jerome A. Jackson. Auk 96:604–606. 1979.

154. "The Broken-Wing Behavior of the Killdeer." C. Douglas Deane. Auk 61: 243–247. 1944.

155. "New and Additional Nevada Bird Records." J. R. Alcorn. Condor 43: 118–119. 1941.

156. Comparative Behavior of the American Avocet and the Black-necked Stilt (Recurvirostridae). Robert Bruce Hamilton. Ornithological Monograph 17. American Ornithologists' Union. 1975.

157. The Wind Birds. Peter Matthiessen. Viking Press. 1973.

158. "The Willets of Georgia and South Carolina." Ivan R. Tomkins. Wilson Bulletin 77:151–167. 1965.

159. A Gathering of Shore Birds. Henry Marion Hall. Bramhall House. 1960.

160. The Ecology and Behavior of the Long-billed Curlew in Southeastern Washington. Julia N. Allen. Wildlife Monographs 73. Wildlife Society. 1980.

161. "The Distribution, Migration and Breeding of Shorebirds in Western Nevada." James L. Hainline. M.S. thesis. University of Nevada at Reno. 1974.

162. Life History of North American Shore Birds. Arthur Cleveland Bent. United States National Museum Bulletin 142. Smithsonian Institution. 1927.

163. North American Game Birds of Upland and Shoreline. Paul A. Johnsgard. University of Nebraska Press. 1975.

164. "Social Interactions in Flocks of Courting Wilson's Phalaropes (*Phalaropus tricolor*)." Marshall A. Howe. Condor 77:24–33. 1975.

165. "Behavioral Aspects of the Pair Bond in Wilson's Phalarope." Marshall A. Howe. Wilson Bulletin 87:248–270. 1975.

166. "The Phalarope." E. Otto Höhn. Scientific American 220(6):104–111. 1969.

167. "Notes on Winter Feeding Behavior and Molt in Wilson's Phalaropes." Joanna Burger and Marshall Howe. Auk 92:442–451. 1975.

168. "Observations on the Breeding Biology of Wilson's Phalarope (*Steganopus tricolor*) in Central Alberta." E. Otto Höhn. Auk 84:220–244. 1967.

169. "Wilson's Phalaropes Forming Feeding Association with Shovelers." W. Roy Siegfried and Bruce D. J. Batt. Auk 89:667–668. 1972.

170. "Mono Lake: A Vital Way Station for the Wilson's Phalarope." Joseph R. Jehl, Jr. National Geographic 160:520–525. 1981.

171. "A Nesting Colony of Ring-billed Gulls in California." James Moffitt. Condor 44:105–107. 1942.

172. "A Chapter on the Natural History of the Great Basin, 1800 to 1855 (1)." Vasco M. Tanner. Great Basin Naturalist 1:33–61. 1940.

173. "The Troubled Waters of Mono Lake." Gordon Young. National Geographic 160:504–519. 1981.

174. "The Breeding Status and Migration of the Caspian Tern in Utah." C. Lynn Hayward. Condor 37:140–144. 1935.

175. Life Histories of North American Gulls and Terns. Arthur Cleveland Bent. United States National Museum Bulletin 113. Smithsonian Institution. 1921.

176. American Birds 35:847. 1981.

177. Water Birds of California. Howard L. Cogswell. University of California Press. 1977.

178. "Heermann Gull in Nevada." Michael Wotton and David B. Marshall. Condor 67:83–84. 1965.

179. American Birds 35:848. 1981.

180. American Birds 35:965. 1981.

181. "The Prehistoric Avifauna of Smith Creek Cave, Nevada, with a Description of a New Gigantic Raptor." Hildegarde Howard. Bulletin of the Southern California Academy of Sciences 51, part 2:50–54. 1952.

182. The Role of Olfaction in Food Location by the Turkey Vulture (*Cathartes aura*). Kenneth E. Stager. Los Angeles County Museum Contributions in Science 81. 1964.

183. The Legend of Grizzly Adams: California's Greatest Mountain Man. Richard Dillon. Coward-McCann, Inc. 1966.

184. Raptors of Utah. Second edition. Larry Eyre and Don Paul. Publication 73-7. Utah Division of Wildlife Resources. 1973.

185. Vertebrate Natural History of a Section of Northern California through the Lassen Peak Region. Joseph Grinnell, Joseph S. Dixon, and Jean M. Linsdale. University of California Publications in Zoology 35. University of California Press. 1930.

186. "Osprey Egg and Nestling Transfers: Their Value as Ecological Experiments and as Management Procedures." Paul R. Spitzer. In Endangered Birds. Edited by Stanley A. Temple. University of Wisconsin Press. 1978.

187. Field Studies of the Falconiformes of British Columbia. Frank L. Beebe. Occasional Paper Series 17. British Columbia Provincial Museum. 1974.

188. Life Histories of North American Birds of Prey. Arthur Cleveland Bent. United States National Museum Bulletin 167. Smithsonian Institution. 1937.

189. American Birds 26:487. 1972.

190. "New Bird Records from Malheur National Wildlife Refuge, Oregon." Carroll D. Littlefield. Western Birds 11:181–185. 1980.

191. American Birds 33:793. 1979.

192. American Birds 35:322. 1981.

193. "Bald Eagles Wintering in a Utah Desert." Joseph B. Platt. American Birds 30:783–788. 1976.

194. Population Ecology of Raptors. Ian Newton. Buteo Books. 1979.

195. Eagles, Hawks and Falcons of the World. 2 volumes. Leslie Brown and Dean Amadon. McGraw-Hill. 1968.

196. "Swainson's Hawks on the Laramie Plains, Wyoming." Sidney W. Dunkle. Auk 94:65–71. 1977.

197. "The Attack and Strike of Some North American Raptors." George E. Goslow, Jr. Auk 88:815–827. 1971.

198. Rattlesnakes: Their Habits, Life Histories, and Influence on Mankind. Volume 2. Second edition. Laurence M. Klauber. University of California Press. 1972.

199. "Relationships between Jackrabbit Abundance and Ferruginous Hawk Reproduction." Dwight G. Smith, Joseph R. Murphy, and Neil D. Woffinden. Condor 83:52–56. 1981.

200. "Population Dynamics of the Ferruginous Hawk during a Prey Decline." Neil D. Woffinden and Joseph R. Murphy. Great Basin Naturalist 37:411–425. 1977.

201. "A Survey of Nesting Hawks, Eagles, Falcons and Owls in Curlew Valley, Utah." Joseph B. Platt. Great Basin Naturalist 31:51–65. 1971.

202. "American Rough-legged Hawk in Florida." Alexander Sprunt, Jr. Auk 57: 564–565. 1940.

203. "Where Engineer and Ornithologist Meet: Transmission Line Troubles Caused by Birds." Harold Michener. Condor 30:169–175. 1928.

204. "On the Selective Advantage of Fratricide in Raptors." Christopher H. Stinson. Evolution 33:1219–1225. 1979.

205. "Breeding Biology of the Golden Eagle in Southwestern Idaho." John J. Beecham and M. N. Kochert. Wilson Bulletin 87:506–513. 1975.

206. The Golden Eagle and Its Economic Status. Lee W. Arnold. Circular 27. U.S. Fish and Wildlife Service. Government Printing Office. 1954.

207. "Eagle-Livestock Relationships: Livestock Carcass Census and Wound Characteristics." Robert W. Wiley and Eric G. Bolen. Southwestern Naturalist 16: 151–169. 1971.

208. "Organochlorine Pollutants, Nest-Defense Behavior and Reproductive Success in Merlins." Glen A. Fox and Tom Donald. Condor 82:81–84. 1980.

209. "The Status of the Peregrine Falcon in the Northwest." Morlan W. Nelson. In

Peregrine Falcon Populations: Their Biology and Decline. Edited by Joseph J. Hickey. University of Wisconsin Press. 1969.

210. "The Duck Hawk Breeding in Nevada." Captain L. R. Wolfe. Condor 39: 225. 1937.

211. "Additional Bird Records for Nevada." Fred G. Evenden, Jr. Condor 54: 174. 1952.

212. "The Peregrine Falcon in Utah, Emphasizing Ecology and Competition with the Prairie Falcon." Richard D. Porter and Clayton M. White in collaboration with Robert J. Erwin. Brigham Young University Science Bulletin. Biological Series 18:1–74. 1973.

213. Threatened Wildlife of the United States. Compiled by the Office of Endangered Species and International Activities. Bureau of Sport Fisheries and Wildlife. Government Printing Office. 1973.

214. "Population Trends of the Peregrine Falcon in Great Britain." Derek A. Ratcliffe. In Peregrine Falcon Populations: Their Biology and Decline. Edited by Joseph J. Hickey. University of Wisconsin Press. 1969.

215. "Artificial Increase in Reproduction of Wild Peregrine Falcons." William A. Burnham, Jerry Craig, James H. Enderson, and William R. Heinrich. Journal of Wildlife Management 42:625–628. 1978.

216. "Reintroducing Birds of Prey to the Wild." Stanley A. Temple. In Endangered Birds. Edited by Stanley A. Temple. University of Wisconsin Press. 1978.

217. Newsletter 8. Peregrine Fund. 1980.

218. "A Study of the Prairie Falcon in the Central Rocky Mountain Region." James H. Enderson. Auk 81:332–352. 1964.

219. "Eating Habits of Falcons with Special Reference to Pellet Analysis." Richard M. Bond. Condor 38:72–76. 1936.

220. "A Survey of the Prairie Falcon in Colorado." Harold Webster, Jr. Auk 61: 609–616. 1944.

221. "Observations on the Habits of the Prairie Falcon." John G. Tyler. Condor 25:90–97. 1923.

222. "Some Observations on the Food of the Prairie Falcon." Richard M. Bond. Condor 38:169–170. 1936.

223. "Nesting Density and Success of Prairie Falcons in Southwestern Idaho." Verland T. Ogden and Maurice G. Hornocker. Journal of Wildlife Management 41:1–11. 1977.

224. "Prairie Falcon 'Playing.'" David A. Munro. Auk 71:333–334. 1954.

225. "A Note on the Prairie Falcon." Louis Agassiz Fuertes. Condor 7:34–36. 1905.

226. "Acoustic Location of Prey by Barn Owls (*Tyto alba*)." Roger S. Payne. Journal of Experimental Biology 54:535–573. 1971.

227. "The Hearing of the Barn Owl." Eric I. Knudsen. Scientific American 245(6): 112–125. 1981.

228. "Distributional Data on Certain Owls in the Western Great Basin." Ned K. Johnson and Ward C. Russell. Condor 64:513–514. 1962.

229. "The Supposed Migratory Status of the Flammulated Owl." Ned K. Johnson. Wilson Bulletin 75:174–178. 1963.

230. Birds of the Snake Range. Compiled by Ray Alcorn. Humboldt National Forest. Intermountain Region. Forest Service. Ogden, Utah. No date.

231. "Size Dimorphism and Food Habits of North American Owls." Caroline M. Earhart and Ned K. Johnson. Condor 72:251–264. 1970.

232. "Behavior and Ecology of Burrowing Owls on the Oakland Municipal Airport." Lise Thomsen. Condor 73:177–192. 1971.

233. "Behavior and Population Ecology of the Burrowing Owl, *Speotyto cunicularia*, in the Imperial Valley of California." Harry N. Coulombe. Condor 73: 162–176. 1971.

234. "Food Habits of Burrowing Owls in Southeastern Idaho." R. L. Gleason and T. H. Craig. Great Basin Naturalist 39:274–276. 1979.

235. Birds: Ruby Lake National Wildlife Refuge, Nevada. RF14570-2. U.S. Fish and Wildlife Service. U.S. Department of the Interior. 1981.

236. "Burrowing Owls Wintering in the Oklahoma Panhandle." Kenneth O. Butts. Auk 93:510–516. 1976.

237. "Selected Aspects of Burrowing Owl Ecology and Behavior." Dennis J. Martin. Condor 75:446–456. 1973.

238. New Mexico Birds and Where to Find Them. J. Stokley Ligon. University of New Mexico Press. 1961.

239. "A Spectrographic Analysis of Burrowing Owl Vocalizations." Dennis J. Martin. Auk 90:564–578. 1973.

240. "A Nuptial Song-Flight of the Short-eared Owl." A. Dawes Dubois. Auk 41:260–263. 1924.

241. A Field Study of the Short-eared Owl, *Asio flammeus* (Pontoppidan), in North America. Richard J. Clark. Wildlife Monographs 47. Wildlife Society. 1975.

242. "A Review of the Distribution of Gallinaceous Game Birds in Nevada." Gordon W. Gullion and Glen C. Christensen. Condor 59:128–138. 1957.

243. Grouse and Quails of North America. Paul A. Johnsgard. University of Nebraska Press. 1973.

244. Distribution of American Gallinaceous Game Birds. John W. Aldrich and Allen J. Duvall. Circular 34. U.S. Fish and Wildlife Service. Government Printing Office. 1955.

245. "Drumming Flight in the Blue Grouse and Courtship Characters of the Tetraonidae." Leonard Wing. Condor 48:154–157. 1946.

246. "Territoriality and Non-Random Mating in Sage Grouse, *Centrocercus urophasianus*." R. Haven Wiley. Animal Behaviour Monographs 6(2):87–169. 1973.

247. "Status and Problems of North American Grouse." Conservation Committee of the Wilson Ornithological Society. Wilson Bulletin 73:284–294. 1961.

248. The California Quail. A. Starker Leopold. University of California Press. 1977.

249. "Comparative Ecology of the Mountain and California Quail in the Carmel Valley, California." Ralph J. Gutiérrez. Living Bird 18:71–93. 1980.

250. "Upland Game of California." Donald D. McLean. State of California Department of Fish and Game. California State Printing Office. No date.

251. The Chukar Partridge. Glen C. Christensen. Biological Bulletin 4. Nevada Department of Fish and Game. State Printing Office. 1970.

252. "The Minimum Water Requirements of Mourning Doves." Richard E. MacMillan. Condor 64:165–166. 1962.

253. Pigeons and Doves of the World. Derek Goodwin. Cornell University Press. 1977.

254. "Perch-Cooing and Other Aspects of Breeding Behavior of Mourning Doves." Gary L. Jackson and Thomas S. Baskett. Journal of Wildlife Management 28:293–307. 1964.

255. Birds: Stillwater Wildlife Management Area, Nevada. RF14590-2. U.S. Fish and Wildlife Service. Government Printing Office. 1982.

256. "The Occurrence and Significance of Anomalous Reproductive Activities in Two North American Non-Parasitic Cuckoos *Coccyzus spp*." Val Nolan, Jr., and Charles F. Thompson. Ibis 117:496–503. 1975.

257. "Feeding and Nesting Behavior of the Yellow-billed Cuckoo in the Sacramento Valley." Stephen A. Laymon. Wildlife Management Branch Administrative Report 80-2. State of California Department of Fish and Game. 1980.

258. "Nesting Habits of the Yellow-billed Cuckoo." Norman A. Preble. American Midland Naturalist 57:474–482. 1957.

259. "Physiological Responses to Temperature in the Common Nighthawk." Robert C. Lasiewski and William R. Dawson. Condor 66:477–490. 1964.

260. Life Histories of North American Cuckoos, Goatsuckers, Hummingbirds and

Their Allies. Arthur Cleveland Bent. United States National Museum Bulletin 176. Smithsonian Institution. 1940.

261. "Cock Roosts of Nighthawks." Robert K. Selander and Sherman J. Preece, Jr. Condor 53:302–303. 1951.

262. "Further Observations on the Hibernation of the Poor-will." Edmund C. Jaeger. Condor 51:105–109. 1949.

263. "Torpidity in the White-throated Swift, Anna Hummingbird, and Poor-will." George A. Bartholomew, Thomas R. Howell, and Tom J. Cade. Condor 59:145–155. 1957.

264. "Still More Responses of the Poor-will to Low Temperatures." J. David Ligon. Condor 72:496–498. 1970.

265. "Additional Responses of the Poor-will to Low Temperatures." George T. Austin and W. Glen Bradley. Auk 86:717–725. 1969.

266. "Physiological Responses to Heat Stress in the Poorwill." Robert C. Lasiewski. American Journal of Physiology 217:1504–1509. 1969.

267. Words for Birds. Edward S. Gruson. Quadrangle Books. 1972.

268. "Reactions of Poor-wills to Light and Temperature." Joseph Brauner. Condor 54:152–159. 1952.

269. "Nest Site Movements of a Poor-will." Raymond N. Evans. Wilson Bulletin 79:453. 1967.

270. "Black Swift Breeds in Utah." Owen A. Knorr. Condor 64:79. 1962.

271. "The Geographical and Ecological Distribution of the Black Swift in Colorado." Owen A. Knorr. Wilson Bulletin 73:155–170. 1961.

272. Life Histories of North American Birds: Parrots to Grackles. Charles Bendire. United States National Museum Special Bulletin 3. Smithsonian Institution. 1895.

273. "Further Notes on the White-throated Swifts of Slover Mountains." Wilson C. Hanna. Condor 19:3–8. 1917.

274. Hummingbirds. Crawford H. Greenewalt. Doubleday. 1960.

275. "Taste Preferences, Color Preferences, and Flower Choice in Hummingbirds." F. Gary Stiles. Condor 78:10–26. 1976.

276. Hummingbirds and Their Flowers. Karen A. Grant and Verne Grant. Columbia University Press. 1968.

277. Hummingbirds. Walter Scheithauer. Thomas Y. Crowell Company. 1967.

278. "Regulation of Metabolism during Torpor in 'Temperate' Zone Hummingbirds." F. Reed Hainsworth and Larry L. Wolf. Auk 95:197–199. 1978.

279. "Temperature Relationships and Nesting of the Calliope Hummingbird." William A. Calder. Condor 73:314–321. 1971.

280. "Notes on the Nesting of the Yosemite Fox Sparrow, Calliope Hummingbird and Western Wood Pewee at Lake Tahoe, California." John W. Mailliard. Condor 23:73–77. 1921.

281. "The Timing of Maternal Behavior of the Broad-tailed Hummingbird Preceding Nest Failure." William A. Calder. Wilson Bulletin 85:283–290. 1973.

282. "Factors in the Energy Budget of Mountain Hummingbirds." William A. Calder III. In Perspectives of Biophysical Ecology. Edited by David M. Gates and Rudolf B. Schmerl. Springer-Verlag. 1975.

283. "Feeding Ecology of Hummingbirds in the Highlands of the Chisos Mountains, Texas." Joseph F. Kuban and Robert L. Neill. Condor 82:180–185. 1980.

284. The Ecology and Behavior of the Lewis Woodpecker (*Asyndesmus lewis*). Carl E. Bock. University of California Publications in Zoology 92. University of California Press. 1970.

285. Woodpeckers in Relation to Trees and Wood Products. W. L. McAtee. Biological Survey Bulletin 39. Government Printing Office. 1911.

286. "Methods and Annual Sequence of Foraging by the Sapsucker." James Tate, Jr. Auk 90:840–856. 1973.

287. "Natural History and Differentiation in the Yellow-bellied Sapsucker." Thomas R. Howell. Condor 54:237–282. 1952.

288. A Comparative Life-History Study of Four Species of Woodpeckers. Louise De Kiriline Lawrence. Ornithological Monographs 5. American Ornithologists' Union. 1967.

289. "Controls of Number of Bird Species on Montane Islands in the Great Basin." Ned K. Johnson. Evolution 29:545–567. 1975.

290. "Notes on the Birds of Fort Klamath, Oregon." Dr. J. C. Merrill, U.S.A. Auk 5:351–362. 1888.

291. The Birds of California. Arnold Small. Winchester Press. 1974.

292. "A Quantitative Study of the Foraging Ecology of Downy Woodpeckers." Jerome A. Jackson. Ecology 51:318–323. 1970.

293. "Reproductive Behavior of Downy Woodpeckers." Lawrence Kilham. Condor 64:126–133. 1962.

294. "Differences in Feeding Behavior of Male and Female Hairy Woodpeckers." Lawrence Kilham. Wilson Bulletin 77:134–145. 1965.

295. "Reproductive Behavior of Hairy Woodpeckers. I: Pair Formation and Courtship." Lawrence Kilham. Wilson Bulletin 78:251–265. 1966.

296. "Foraging Behavior of the White-headed Woodpecker in Idaho." J. David Ligon. Auk 90:862–869. 1973.

297. "Effect of Vertebrate Animals on Seed Crop of Sugar Pine." Lloyd Tevis, Jr. Journal of Wildlife Management 17:128–131. 1953.

298. "On the Geographical Ecology and Evolution of the Three-toed Woodpeckers, *Picoides tridactylus* and *P. arcticus*." Carl E. Bock and Jane H. Bock. American Midland Naturalist 92:397–405. 1974.

299. "Climbing and Pecking Adaptations in Some North American Woodpeckers." Lowell W. Spring. Condor 67:457–488. 1965.

300. "A Collecting Trip by Wagon to Eagle Lake, Sierra Nevada Mountains." Harry H. Sheldon. Condor 9:185–191. 1907.

301. "Pendulum Display by Olive-sided Flycatcher." Genevieve M. Tvrdik. Auk 88:174. 1971.

302. Biosystematics of Sibling Species of Flycatchers in the Empidonax Hammondii-Oberholseri-Wrightii Complex. Ned K. Johnson. University of California Publications in Zoology 66(2):79–238. University of California Press. 1963.

303. "Ecological Overlap and the Problem of Competition and Sympatry in the Western and Hammond's Flycatchers." Donald L. Beaver and Paul H. Baldwin. Condor 77:1–13. 1975.

304. Birds of California. Irene Grosvenor Wheelock. A. C. McClurg and Company. 1904.

305. A Field Guide to the Birds East of the Rockies. Fourth edition. Roger Tory Peterson. Houghton Mifflin. 1980.

306. "Nest Site Selection in Eastern and Western Kingbirds: A Multivariate Approach." David I. MacKenzie and Spencer G. Sealy. Condor 83:310–321. 1981.

307. Birds of Colorado. Volume 2. Alfred M. Bailey and Robert J. Niedrach. Denver Museum of Natural History. 1965.

308. "Prairie Horned Lark." Gayle Pickwell. In Life Histories of North American Flycatchers, Larks, Swallows, and Their Allies. Arthur Cleveland Bent. United States National Museum Bulletin 179. Smithsonian Institution. 1942.

309. "Adaptations of Horned Larks (*Eremophila alpestris*) to Hot Environments." Charles H. Trost. Auk 89:506–527. 1972.

310. Birds of the Colorado Valley. Elliott Coues. Government Printing Office. 1878.

311. Studies on the Bank Swallow *Riparia riparia riparia* (Linnaeus) in the Oneida

Lake Region. Dayton Stoner. Roosevelt Wild Life Annals. Volume 4, number 2. Roosevelt Wild Life Forest Experiment Station. 1936.

312. "Predator-Prey Interactions of Adult and Prefledgling Bank Swallows and American Kestrels." Donald Windsor and Stephen T. Emlen. Condor 77: 359–361. 1975.

313. "Territory, Nest Building, and Pair Formation in the Cliff Swallow." John T. Emlen, Jr. Auk 71:16–35. 1954.

314. "Effects of Nasal Tufts and Nasal Respiration on Thermoregulation and Evaporative Water Loss in the Common Crow." Bruce A. Wunder and Joseph J. Trebella. Condor 78:564–567. 1976.

315. "Some Notes on Birds of Elko County, Nevada." George K. Tsukamoto. Condor 68:103–104. 1966.

316. Birds and Mammals of the Sierra Nevada. Lowell Sumner and Joseph S. Dixon. University of California Press. 1953.

317. The Integration of Agonistic Behavior in the Steller's Jay *Cyanocitta stelleri* (Gmelin). Jerram L. Brown. University of California Publications in Zoology 60:223–328. University of California Press. 1964.

318. Animal Life in the Yosemite. Joseph Grinnell and Tracy Irwin Storer. University of California Press. 1924.

319. The Piñon Pine: A Natural and Cultural History. Ronald M. Lanner. University of Nevada Press. 1981.

320. "Coadaptations of the Clark's Nutcracker and the Piñon Pine for Efficient Seed Harvest and Dispersal." Stephen B. Vander Wall and Russell P. Balda. Ecological Monographs 47:89–111. 1977.

321. "Reproductive Interdependence of Piñon Jays and Piñon Pines." J. David Ligon. Ecological Monographs 48:111–126. 1978.

322. "Flocking and Annual Cycle of the Piñon Jay, *Gymnorhinus cyanocephalus*." Russell P. Balda and Gary C. Bateman. Condor 73:287–302. 1971.

323. "Growth, Development, and Food Habits of Young Piñon Jays." Gary C. Bateman and Russell P. Balda. Auk 90:39–61. 1973.

324. Life Histories of North American Jays, Crows, and Titmice. Arthur Cleveland Bent. United States National Museum Bulletin 191. Smithsonian Institution. 1946.

325. "Starling–Piñon Jay Associations in Southern Colorado." Richard G. Beidleman and James H. Enderson. Condor 66:437. 1964.

326. "Flocking Associates of the Piñon Jay." Russell P. Balda, Gary C. Bateman, and Gene F. Foster. Wilson Bulletin 84:60–76. 1972.

327. "Stomach Contents of Clark's Nutcrackers Collected in Western Montana." Mervin Giuntoli and L. Richard Mewaldt. Auk 95:595–598. 1978.

328. "Limber Pine Seed Harvest by Clark's Nutcracker in the Sierra Nevada: Timing and Foraging Behavior." Diana F. Tomback and Kathryn A. Kramer. Condor 82:467–468. 1980.

329. "How Nutcrackers Find Their Seed Stores." Diana F. Tomback. Condor 82: 10–19. 1980.

330. "Emigration Behavior of Clark's Nutcracker." Stephen B. Vander Wall, Stephen W. Hoffman, and Wayne K. Potts. Condor 83:162–170. 1981.

331. "Nesting Behavior of the Clark Nutcracker." L. Richard Mewaldt. Condor 58:3–23. 1956.

332. "Nesting of the Clark Nutcracker in California." James B. Dixon. Condor 36:229–234. 1934.

333. The History of the Lewis and Clark Expedition. Volume 1. Edited by Elliott Coues. Dover Publications. No date.

334. The Journals of Zebulon Montgomery Pike with Letters and Related Documents. Volume 1. Edited and annotated by Donald Jackson. University of Oklahoma Press. 1966.

335. "Seasonal Cycle of Reproductive Physiology in the Black-billed Magpie." Michael J. Erpino. Condor 71:267–279. 1969.

336. "Nest-Related Activities of Black-billed Magpies." Michael J. Erpino. Condor 70:154–165. 1968.

337. "The Use of Magpies' Nests by Other Birds." Robert B. Rockwell. Condor 11:90–92. 1909.

338. The Natural History of Magpies. Jean M. Linsdale. Pacific Coast Avifauna 25. Cooper Ornithological Club. 1937.

339. "Swainson's Hawk in Washington State." J. Hooper Bowles and F. R. Decker. Auk 51:446–450. 1934.

340. "Three Magpies Rob a Golden Eagle." Joseph S. Dixon. Condor 35:161. 1933.

341. Ravens, Crows, Magpies, and Jays. Tony Angell. University of Washington Press. 1978.

342. The Crow and Its Relation to Man. E. R. Kalmbach. U.S. Department of Agriculture Bulletin 621. Government Printing Office. 1918.

343. "The Crow as a 'Villain.'" Ellsworth D. Lumley. In Man's Friend: The Crow. Publication 65. Emergency Conservation Committee. 1937.

344. "Common Raven and Starling Reliance on Sentinel Common Crows." Rich-

ard N. Conner, Irvine D. Prather, and Curtis S. Adkisson. Condor 77:517. 1975.

345. The Common Raven. Richard L. Knight and Mayo W. Call. Technical Note 344. U.S. Department of the Interior, Bureau of Land Management. Government Printing Office. 1980.

346. "Ravens Attracted to Wolf Howling." Fred H. Harrington. Condor 80:236–237. 1978.

347. American Birds 32:1038. 1978.

348. The Audubon Society Field Guide to North American Birds: Western Region. Miklos D. F. Udvardy. Alfred A. Knopf. 1977.

349. "Altitudinal Migration in the Mountain Chickadee." Keith L. Dixon and John D. Gilbert. Condor 66:61–64. 1964.

350. "Dominance-Subordination Relationships in Mountain Chickadees." Keith L. Dixon. Condor 67:291–299. 1965.

351. "The Vertebrate Structure of a Piñon-Juniper Woodland Community in Northwestern Nevada." Howard Ronald Panik. Ph.D. thesis. University of Nevada at Reno. 1976.

352. "Foraging Relationships of Mountain Chickadees and Pygmy Nuthatches." Tim Manolis. Western Birds 8:13–20. 1977.

353. "Behavior of the Plain Titmouse." Keith L. Dixon. Condor 51:110–136. 1949.

354. "Call Notes of the Bush-tit." Joseph Grinnell. Condor 5:85–87. 1903.

355. "Behavior of the Bush-tit in the Breeding Season." Alice Baldwin Addicott. Condor 40:49–63. 1938.

356. "The Energetic Significance of Huddling Behavior in Common Bushtits (*Psaltriparus minimus*)." Susan B. Chaplin. Auk 99:424–430. 1982.

357. "Roosting Aggregations of Bushtits in Response to Cold Temperatures." Susan M. Smith. Condor 74:478–479. 1972.

358. "Reproductive Behavior of White-breasted Nuthatches. II: Courtship." Lawrence Kilham. Auk 89:115–129. 1972.

359. "Use of Blister Beetle in Bill-Sweeping by White-breasted Nuthatch." Lawrence Kilham. Auk 88:175–176. 1971.

360. "Reproductive Behavior of White-breasted Nuthatches. I: Distraction Display, Bill-Sweeping, and Nest Hole Defense." Lawrence Kilham. Auk 85:477–492. 1968.

361. "Communal Roosting of the Pygmy Nuthatch." Owen A. Knorr. Condor 59:398. 1957.

362. "Down-Tree Progress of *Sitta pygmaea*." J. Eugene Law. Condor 31:45–51. 1929.

363. "Insect Food of the Rock Wren." George F. Knowlton and F. C. Harmston. Great Basin Naturalist 3:22. 1942.

364. "Behavior of the Bewick Wren." Edwin V. Miller. Condor 43:81–99. 1941.

365. "Territorial Interactions in Sympatric Song Sparrow and Bewick's Wren Populations." Robert E. Gorton, Jr. Auk 94:701–708. 1977.

366. "Interspecific Territoriality between Bewick's and House Wrens." Richard B. Root. Auk 86:125–127. 1969.

367. Field Book of Wild Birds and Their Music. F. Schuyler Mathews. G. P. Putnam's Sons. 1948.

368. "Bird Nomenclature of the Chippewa Indians." W. W. Cooke. Auk 1:242–250. 1884.

369. "Breeding Biology of the Long-billed Marsh Wren." Jared Verner. Condor 67:6–30. 1965.

370. "Territories, Multiple Nest Building, and Polygyny in the Long-billed Marsh Wren." Jared Verner and Gay H. Engelsen. Auk 87:557–567. 1970.

371. "Destruction of Eggs by the Long-billed Marsh Wren (*Telmatodytes palustris palustris*)." Jaroslav Picman. Canadian Journal of Zoology 55:1914–1920. 1977.

372. "Intraspecific Nest Destruction in the Long-billed Marsh Wren, *Telmatodytes palustris palustris*." Jaroslav Picman. Canadian Journal of Zoology 55:1997–2003. 1977.

373. "Locomotion and Other Behavior of the Dipper." William R. Goodge. Condor 61:4–17. 1959.

374. "Responses to Diving in the Dipper, *Cinclus mexicanus*." David E. Murrish. Comparative Biochemistry and Physiology 34:853–858. 1970.

375. Animal Physiology: Adaptation and Environment. Second edition. Knut Schmidt-Nielsen. Cambridge University Press. 1979.

376. The Mountains of California. John Muir. Doubleday. 1961.

377. "The Niche Exploitation Pattern of the Blue-gray Gnatcatcher." Richard B. Root. Ecological Monographs 37:317–350. 1967.

378. Life Histories of North American Thrushes, Kinglets, and Their Allies. Arthur Cleveland Bent. United States National Museum Bulletin 196. Smithsonian Institution. 1949.

379. "Winter Territoriality and Foraging Behavior of the Townsend's Solitaire." Roger J. Lederer. American Midland Naturalist 97:101–109. 1977.

380. "Winter Territoriality of Townsend's Solitaires (*Myadestes townsendi*) in a Piñon–Juniper–Ponderosa Pine Ecotone." Michael G. Salomonson and Russell P. Balda. Condor 79:148–161. 1977.

381. The Writings of John Burroughs. I: Wake-Robin. John Burroughs. Houghton Mifflin. 1913.

382. "Sensory Mechanisms and Environmental Clues Used by the American Robin in Locating Earthworms." Frank Heppner. Condor 67:247–256. 1965.

383. Bird Song. Aretas A. Saunders. New York State Museum Handbook 7. University of the State of New York. 1929.

384. A Natural History of American Birds of Eastern and Central North America. Edward Howe Forbush and John Bichard May. Bramhall House. 1955.

385. "A Field Study of the Mockingbird's Wing-Flashing Behavior and Its Association with Foraging." Jack P. Hailman. Wilson Bulletin 72:346–357. 1960.

386. "An Ontogeny of Wing-Flashing in the Mockingbird with Reference to Other Behaviors." Robert H. Horwich. Wilson Bulletin 77:264–281. 1965.

387. "On the Functions of Wing-Flashing in Mockingbirds." Robert K. Selander and D. K. Hunter. Wilson Bulletin 72:341–345. 1960.

388. "Mockingbirds, Their Territories and Individualities." Harold Michener and Josephine R. Michener. Condor 37:97–140. 1935.

389. "Bilateral Wing Display in the Sage Thrasher." Terrell D. G. Rich. Wilson Bulletin 92:512–513. 1980.

390. "Nesting of the Sage Thrasher, Sage Sparrow, and Brewer's Sparrow in Southeastern Idaho." Timothy D. Reynolds. Condor 83:61–64. 1981.

391. The Birds of the Deep Creek Mountains of Central Western Utah. William H. Behle. University of Utah Biological Series. Volume 11, number 4. University of Utah. 1955.

392. The Birds of Northeastern Utah. William H. Behle. Occasional Publication 2. Utah Museum of Natural History. 1981.

393. "The Indian Tree Pipit (*Anthus hodgsoni*) Recorded for the First Time in North America." Thomas D. Burleigh. Auk 85:323. 1968.

394. Systematic Revision and Natural History of the American Shrikes (*Lanius*). Alden H. Miller. University of California Publications in Zoology 38(2):11–242. University of California Press. 1931.

395. Birds of America. T. Gilbert Pearson, editor-in-chief. Garden City Publishing Company. 1936.

396. "Starlings Become Established at Fairbanks, Alaska." Brina Kessel. Condor 81:437–438. 1979.

397. "Breeding Range Expansion of the Starling in Utah." Dwight G. Smith. Great Basin Naturalist 35:419–424. 1975.

398. "The Current Status of the Starling in Nevada." Gordon W. Gullion. Condor 58:446. 1956.

399. "A Study of the Breeding Biology of the European Starling (*Sturnus vulgaris* L.) in North America." Brina Kessel. American Midland Naturalist 58:257–331. 1957.

400. Life Histories of North American Wagtails, Shrikes, Vireos, and Their Allies. Arthur Cleveland Bent. United States National Museum Bulletin 197. Smithsonian Institution. 1950.

401. American Birds 34:802. 1980.

402. American Birds 29:96–97. 1975.

403. The Warblers of America. Ludlow Griscom, Alexander Sprunt, Jr., and other ornithologists of note. Devin-Adair Company. 1957.

404. "Breeding Distribution of Nashville and Virginia's Warblers." Ned K. Johnson. Auk 93:219–230. 1976.

405. "Comparative Ethology of the Chestnut-sided Warbler, Yellow Warbler, and American Redstart." Millicent S. Ficken and Robert W. Ficken. Wilson Bulletin 77:363–375. 1965.

406. "The Context of Songs in the Yellow Warbler." Douglass H. Morse. Wilson Bulletin 78:444–455. 1966.

407. "Song Characteristics of the Yellow Warbler." Kenneth G. Bankwitz and William L. Thompson. Wilson Bulletin 91:533–550. 1979.

408. "Feeding Ecology of a Population of Nesting Yellow Warblers." Daniel G. Busby and Spencer G. Sealy. Canadian Journal of Zoology 57:1670–1681. 1979.

409. Life Histories of North American Wood Warblers. Arthur Cleveland Bent. United States National Museum Bulletin 203. Smithsonian Institution. 1953.

410. Birds of Massachusetts and Other New England States. Part 3: Land Birds from Sparrows to Thrushes. Edward Howe Forbush. Massachusetts Department of Agriculture. 1929.

411. "The Ecological Distribution of the Summer Birds of the Carson Range (Eastern Sierra) in the Area of the Lake Tahoe Nevada State Park." James V. Kelleher. M.S. thesis. University of Nevada at Reno. 1970.

412. American Birds 34:187. 1980.

413. American Birds 30:871. 1976.

414. American Birds 32:1193. 1978.

415. Birds of Idaho. Thomas D. Burleigh. Caxton Printers. 1972.

416. American Birds 31:1030. 1977.

417. American Birds 34:917. 1980.

418. American Birds 36:317. 1982.

419. "Prothonotary Warbler and Yellow-shafted Flicker in Nevada." Fred A. Ryser. Condor 65:334. 1963.

420. Birds of the World. Oliver L. Austin, Jr. Golden Press. 1961.

421. American Birds 31:1170. 1977.

422. Audubon Field Notes 22:633. 1968.

423. "Breeding Behavior of the Black-headed Grosbeak." Henry G. Weston, Jr. Condor 49:54–73. 1947.

424. Birds: Salt Lake BLM District. Compiled with assistance from the Utah Audubon Society, University of Utah Museum of Natural History, and the Utah Division of Wildlife Resources. Bureau of Land Management. 1976.

425. Life Histories of North American Cardinals, Grosbeaks, Buntings, Towhees, Finches, Sparrows, and Allies. Arthur Cleveland Bent and collaborators. Compiled and edited by Oliver L. Austin, Jr. United States National Museum Bulletin 237, part 1. Smithsonian Institution. 1968.

426. "Taxonomic Distribution, Origin, and Evolution of Bilateral Scratching in Ground-Feeding Birds." Jon S. Greenlaw. Condor 79:426–439. 1977.

427. "The Double-Scratch as a Taxonomic Character in the Holarctic Emberizinae." C. J. O. Harrison. Wilson Bulletin 79:22–27. 1967.

428. "Songs of the Rufous-sided Towhee." Donald J. Borror. Condor 77:183–195. 1975.

429. Life Histories of North American Cardinals, Grosbeaks, Buntings, Towhees, Finches, Sparrows, and Allies. Arthur Cleveland Bent and collaborators. Compiled and edited by Oliver L. Austin, Jr. United States National Museum Bulletin 237, part 2. Smithsonian Institution. 1968.

430. "The Water Economy of the Black-throated Sparrow and the Rock Wren." Michael Smyth and George A. Bartholomew. Condor 68:447–458. 1966.

431. "Song in a Population of Black-throated Sparrows." Donald B. Heckenlively. Condor 72:24–36. 1970.

432. "Cowbird Parasitism of Sage and Brewer's Sparrows." Terrell D. G. Rich. Condor 80:348. 1978.

433. "Territorial Behavior of the Sage Sparrow: Spatial and Random Aspects." Terrell Rich. Wilson Bulletin 92:425–438. 1980.

434. "Song Pattern Variation in the Sage Sparrow (*Amphispiza belli*): Dialects or Epiphenomena?" John A. Wiens. Auk 99:208–229. 1982.

435. "Territorial Behavior in Savannah Sparrows in Southeastern Michigan." Peter E. Potter. Wilson Bulletin 84:48–59. 1972.

436. American Birds 35:966. 1981.

437. American Birds 34:803. 1980.

438. "Some Ecological Notes on the Grasshopper Sparrow." Robert Leo Smith. Wilson Bulletin 75:159–165. 1963.

439. "The Songs of the Grasshopper Sparrow." Robert Leo Smith. Wilson Bulletin 71:141–152. 1959.

440. Studies in the Life History of the Song Sparrow. 2 volumes. Margaret Morse Nice. Dover Publications. 1964.

441. "Relations between the Sexes in Song Sparrows." Margaret Morse Nice. Wilson Bulletin 45:51–59. 1933.

442. "Sexual Selection and Red-winged Blackbirds." William A. Searcy and Ken Yasukawa. American Scientist 71:166–174. 1983.

443. Life Histories of North American Cardinals, Grosbeaks, Buntings, Towhees, Finches, Sparrows, and Allies. Arthur Cleveland Bent and collaborators. Compiled and edited by Oliver L. Austin, Jr. United States National Museum Bulletin 237, part 3. Smithsonian Institution. 1968.

444. "Anting in Wild Birds, Its Frequency and Probable Purpose." Eloise F. Potter. Auk 87:692–713. 1970.

445. "Mountain White-crowned Sparrows in California." Barbara B. DeWolfe and Robert H. DeWolfe. Condor 64:378–389. 1962.

446. "Snow Conditions and the Onset of Breeding in the Mountain White-crowned Sparrow." Martin L. Morton. Condor 80:285–289. 1978.

447. "Reproductive Cycle and Nesting Success of the Mountain White-crowned Sparrow (*Zonotrichia leucophrys oriantha*) in the Central Sierra Nevada." Martin L. Morton, Judith L. Horstmann, and Janet M. Osborn. Condor 74:152–163. 1972.

448. "Dynamics of Incubation in Mountain White-crowned Sparrows." Eileen Zerba and Martin L. Morton. Condor 85:1–11. 1983.

449. "The Heat Budget of Incubating Mountain White-crowned Sparrows (*Zonotrichia leucophrys oriantha*) in Oregon." Glenn E. Walsberg and James R. King. Physiological Zoology 51:92–103. 1978.

450. "The White-crowned Sparrows (*Zonotrichia leucophrys*) of the Pacific Seaboard: Environment and Annual Cycle." Barbara D. Blanchard. University of California Publications in Zoology 46:1–178. University of California Press. 1941.

451. "Song Dialects in Several Populations of Mountain White-crowned Sparrows (*Zonotrichia leucophrys oriantha*) in the Sierra Nevada." Jorge E. Orejuela and Martin L. Morton. Condor 77:145–153. 1975.

452. "Geographical Variation in Song and Song Dialects of Montane White-crowned Sparrows." Luis F. Baptista and James R. King. Condor 82:267–284. 1980.

453. "Song 'Dialects' in Three Populations of White-crowned Sparrows." P. Marler and M. Tamura. Condor 64:368–377. 1962.

454. "Song Dialects as Barriers to Dispersal in White-crowned Sparrows, *Zonotrichia leucophrys nuttalli*." Myron Charles Baker and L. Richard Mewaldt. Evolution 32:712–722. 1978.

455. "Bird Check List of Honey Lake Wildlife Area." Compiled by F. Hanzlik and J. C. Revill. Revised in 1977 by A. Lapp and C. Holmes, Jr. Resources Agency. State of California Department of Fish and Game.

456. Birds of the Pacific Northwest: Washington, Oregon, Idaho, and British Columbia. Earl J. Larrison. Assisted by Michael D. Johnson and Stanlee Miller. University Press of Idaho. 1981.

457. Systematics and the Origin of Species. Ernst Mayr. Columbia University Press. 1942.

458. "Song Development in Hand-Raised Oregon Juncos." Peter Marler, Marcia Kreith, and Miwako Tamura. Auk 79:12–30. 1962.

459. A Study of Bird Song. Second enlarged edition. Edward A. Armstrong. Dover Publications. 1973.

460. "Integrating Mechanisms of Winter Flocks of Juncos." Winifred S. Sabine. Condor 58:338–341. 1956.

461. "The Winter Society of the Oregon Junco: Intolerance, Dominance, and the Pecking Order." Winifred S. Sabine. Condor 61:110–135. 1959.

462. "Flock Stability in Relation to Social Dominance and Agonistic Behavior in Wintering Dark-eyed Juncos." Martha Hatch Balph. Auk 96:714–722. 1979.

463. "Winter Social Behaviour of Dark-eyed Juncos: Communication, Social Organization, and Ecological Implications." Martha Hatch Balph. Animal Behaviour 25:859–884. 1977.

464. American Birds 30:104. 1976.

465. The Bird List from Sheldon National Wildlife Refuge. U.S. Fish and Wildlife

Service. A list of "over 175 species that have been recorded since the refuge was established." No date.

466. Life Histories of North American Blackbirds, Orioles, Tanagers, and Allies. Arthur Cleveland Bent. United States National Museum Bulletin 211. Smithsonian Institution. 1958.

467. "A Behavior Study of the Red-winged Blackbird. I: Mating and Nesting Activities." Robert W. Nero. Wilson Bulletin 68:5–37. 1956.

468. A Comparative Study of the Behavior of Red-winged, Tricolored, and Yellow-headed Blackbirds. Gordon H. Orians and Gene M. Christman. University of California Publications in Zoology 84. University of California Press. 1968.

469. "A Behavior Study of the Red-winged Blackbird. II: Territoriality." Robert W. Nero. Wilson Bulletin 68:129–150. 1956.

470. "An Experimental Study of Territorial Behavior in Breeding Red-winged Blackbirds." Robert W. Nero and John T. Emlen, Jr. Condor 53:105–116. 1951.

471. "Impact of Marsh Wrens on Reproductive Strategy of Red-winged Blackbirds." Jaroslav Picman. Canadian Journal of Zoology 58:337–350. 1980.

472. "Response of Red-winged Blackbirds to Nests of Long-billed Marsh Wrens." Jaroslav Picman. Canadian Journal of Zoology 58:1821–1827. 1980.

473. The Comparative Biology of the Meadowlarks (*Sturnella*) in Wisconsin. Wesley E. Lanyon. Publications of the Nuttall Ornithological Club 1. Nuttall Ornithological Club. 1957.

474. "Sequence of Songs in Repertoires of Western Meadowlarks (*Sturnella neglecta*)." J. Bruce Falls and John R. Krebs. Canadian Journal of Zoology 53: 1165–1178. 1975.

475. "The Mormon Cricket as Food for Birds." Ira LaRivers. Condor 43:65–69. 1941.

476. "An Ecological Study of the Yellow-headed Blackbird." Thomas A. Burns. M.S. thesis. University of Nevada at Reno. 1964.

477. "Comparative Behavior of the Yellow-headed Blackbird, Red-winged Blackbird, and Other Icterids." Robert W. Nero. Wilson Bulletin 75:376–413. 1963.

478. "Analysis of the Eastward Breeding Expansion of Brewer's Blackbird Plus General Aspects of Avian Expansions." P. H. R. Stepney and Dennis M. Power. Wilson Bulletin 85:452–464. 1973.

479. "Breeding Behavior of the Brewer Blackbird." Laidlaw Williams. Condor 54:3–47. 1952.

480. American Birds 32:1039. 1978.

481. American Birds 35:849. 1981.

482. "Brood Parasitism: Reducing Interactions between Kirtland's Warblers and Brown-headed Cowbirds." Harold F. Mayfield. In Endangered Birds. Edited by Stanley A. Temple. University of Wisconsin Press. 1978.

483. "Behavioral Defenses to Brood Parasitism by Potential Hosts of the Brown-headed Cowbird." Raleigh J. Robertson and Richard F. Norman. Condor 78:166–173. 1976.

484. "Interspecific Preening Invitation Display of Parasitic Cowbirds." Robert K. Selander and Charles J. La Rue, Jr. Auk 78:473–504. 1961.

485. "Head-Down or 'Preening Invitation' Displays Involving Juvenile Brown-headed Cowbirds." Peter E. Lowther and Stephen I. Rothstein. Condor 82: 459–460. 1980.

486. "An Experimental and Teleonomic Investigation of Avian Brood Parasitism." Stephen I. Rothstein. Condor 77:250–271. 1975.

487. The Birds of Minnesota. Volume 2. Thomas S. Roberts. University of Minnesota Press. 1936.

488. "Geographic Variation in the Insulative Qualities of Nests of the Northern Oriole." V. H. Schaefer. Wilson Bulletin 92:466–474. 1980.

489. "Notes on the Song and Territorial Habits of Bullock's Oriole." Alden H. Miller. Wilson Bulletin 43:102–108. 1931.

490. "Prebasic Molt of the Northern Oriole." Spencer G. Sealy. Canadian Journal of Zoology 57:1473–1478. 1979.

491. American Birds 30:985. 1976.

492. "New Breeding Localities for *Leucosticte* in the Contiguous Western United States." Richard E. Johnson. Auk 92:586–589. 1975.

493. "Life History of the Black Rosy Finch." Norman R. French. Auk 76:159–180. 1959.

494. "The Significance of Combat in Male Rosy Finches." Howard Twining. Condor 40:246–247. 1938.

495. "Foraging Behavior and Survival in the Sierra Nevada Rosy Finch." Howard Twining. Condor 42:64–72. 1940.

496. "Notes on *Leucostictes* Wintering in Salt Lake Valley, Utah." William H. Behle. Condor 46:207–208. 1944.

497. "Territory, Breeding Density, and Fall Departure in Cassin's Finch." Fred B. Samson. Auk 93:477–497. 1976.

498. "Social Dominance in Winter Flocks of Cassin's Finch." Fred B. Samson. Wilson Bulletin 89:57–66. 1977.

499. "Agonistic Behavior in the House Finch. Part 1: Annual Cycle and Display Patterns." William L. Thompson. Condor 62:245–271. 1960.

500. "Notes on the Red Crossbill in Nevada." Ned K. Johnson. Condor 60:136–137. 1958.

501. "Social Organization and Behavior in a Flock of Captive, Nonbreeding Red Crossbills." Harrison B. Tordoff. Condor 56:346–358. 1954.

502. "The Influence of Daylength on Reproductive Timing in the Red Crossbill." Harrison B. Tordoff and William R. Dawson. Condor 67:416–422. 1965.

503. "Behavior of the Pine Siskin." Thomas L. Rodgers. Condor 39:143–149. 1937.

504. "Geographic Variation in Some Birds in Nevada." Jean M. Linsdale. Condor 40:36–38. 1938.

505. "Bird Life in Nevada with Reference to Modifications in Structure and Behavior." Jean M. Linsdale. Condor 40:173–180. 1938.

506. "Environmental Responses of Vertebrates in the Great Basin." Jean M. Linsdale. American Midland Naturalist 19:1–206. 1938.

507. "Goldfinches on the Hastings Natural History Reservation." Jean M. Linsdale. American Midland Naturalist 57:1–119. 1957.

508. "Maintenance Behavior of the American Goldfinch." Ellen L. Coutlee. Wilson Bulletin 75:342–357. 1963.

509. A Natural History of Western Trees. Donald Culross Peattie. Houghton Mifflin. 1950.

510. American Birds 33:201. 1979.

511. "Spread of the Starling and English Sparrow." Leonard Wing. Auk 60:74–87. 1943.

512. Key to North American Birds. Fifth edition. Elliott Coues. Dana Estes and Company. 1903.

513. The House Sparrow. J. D. Summers-Smith. Collins. 1963.

Index

Birds of the Great Basin was designed by John Stetter and Robert Blesse, edited by Holly Carver, and proofread by Mary Hill. The text type is Linotron Bembo, set by G&S Typesetters, Austin, Texas, with other display type hand-set in foundry Centaur by Robert Blesse at the Sequoia Press, Truckee, California.